2889

Free Radicals, Oxidative Stress, and Antioxidants

Pathological and Physiological Significance

NATO ASI Series

Advanced Science Institutes Series

A series presenting the results of activities sponsored by the NATO Science Committee, which aims at the dissemination of advanced scientific and technological knowledge, with a view to strengthening links between scientific communities.

The series is published by an international board of publishers in conjunction with the NATO Scientific Affairs Division

A	**Life Sciences**	Plenum Publishing Corporation
B	**Physics**	New York and London
C	**Mathematical and Physical Sciences**	Kluwer Academic Publishers
		Dordrecht, Boston, and London
D	**Behavioral and Social Sciences**	
E	**Applied Sciences**	
F	**Computer and Systems Sciences**	Springer-Verlag
G	**Ecological Sciences**	Berlin, Heidelberg, New York, London,
H	**Cell Biology**	Paris, Tokyo, Hong Kong, and Barcelona
I	**Global Environmental Change**	

PARTNERSHIP SUB-SERIES

1. **Disarmament Technologies**	Kluwer Academic Publishers
2. **Environment**	Springer-Verlag
3. **High Technology**	Kluwer Academic Publishers
4. **Science and Technology Policy**	Kluwer Academic Publishers
5. **Computer Networking**	Kluwer Academic Publishers

The Partnership Sub-Series incorporates activities undertaken in collaboration with NATO's Cooperation Partners, the countries of the CIS and Central and Eastern Europe, in Priority Areas of concern to those countries.

Recent Volumes in this Series:

Volume 293 — Vaccine Design: The Role of Cytokine Networks
edited by Gregory Gregoriadis, Brenda McCormack, and Anthony C. Allison

Volume 294 — Vascular Endothelium: Pharmacologic and Genetic Manipulations
edited by John D. Catravas, Allan D. Callow, and Una S. Ryan

Volume 295 — Prions and Brain Diseases in Animals and Humans
edited by Douglas R. O. Morrison

Volume 296 — Free Radicals, Oxidative Stress, and Antioxidants: Pathological and Physiological Significance
edited by Tomris Özben

Series A: Life Sciences

Free Radicals, Oxidative Stress, and Antioxidants

Pathological and Physiological Significance

Edited by

Tomris Özben

Akdeniz University
Antalya, Turkey

Plenum Press
New York and London
Published in cooperation with NATO Scientific Affairs Division

Proceedings of a NATO Advanced Study Institute on
Free Radicals, Oxidative Stress, and Antioxidants: Pathological and
Physiological Significance,
held May 24–June 4, 1997,
in Antalya, Turkey

NATO-PCO-DATA BASE

The electronic index to the NATO ASI Series provides full bibliographical references (with keywords and/or abstracts) to about 50,000 contributions from international scientists published in all sections of the NATO ASI Series. Access to the NATO-PCO-DATA BASE is possible via a CD-ROM "NATO Science and Technology Disk" with user-friendly retrieval software in English, French, and German (©WTV GmbH and DATAWARE Technologies, Inc. 1989). The CD-ROM contains the AGARD Aerospace Database.

The CD-ROM can be ordered through any member of the Board of Publishers or through NATO-PCO, Overijse, Belgium.

Library of Congress Cataloging-in-Publication Data

On file

ISBN 0-306-45813-6

© 1998 Plenum Press, New York
A Division of Plenum Publishing Corporation
233 Spring Street, New York, N.Y. 10013

http://www.plenum.com

10 9 8 7 6 5 4 3 2 1

Printed in the United States of America

PREFACE

There has been an explosion of research related to free radicals and antioxidants in recent years, and hundreds of laboratories worldwide are actively involved in many aspects of free radicals, oxidative stress, and antioxidants. The literature on these topics increases exponentially every year. Over the last few years, we have been fortunate to witness a widespread recognition of the important role of free radicals in a wide variety of pathological conditions including diseases such as atherosclerosis, cardiovascular and neurological diseases, ischemia, emphysema, diabetes, radiation injury, cancer, etc. In addition, many laboratories are studying the role of free radicals in the inexorable process of aging. Increased evidence involves free radicals with the etiology of various diseases, thereby suggesting the use of antioxidants as a viable therapeutic approach for the treatment of free radical mediated pathologies.

Despite these impressive developments, many important aspects of free radical and antioxidant research are open for investigation. It is important to understand the overall mechanisms involved in free radical mediated physiological and pathological conditions. This knowledge will undoubtedly lead to the development of new therapeutic approaches to prevent or control free radical related diseases.

This book contains the proceedings of the NATO Advanced Study Institute (ASI) on "Free Radicals, Oxidative Stress, and Antioxidants: Pathological and Physiological Significance," which was held in Antalya, Turkey from May 24–June 4, 1997. The scientific program of the ASI was multidisciplinary and included a wide range of issues such as basic concepts, methods, and techniques used in this area and the role of free radicals and antioxidants in various disease states. The present state of knowledge and future trends in free radicals and antioxidant research were presented in depth by lecturers of international standing. The presentation and discussions of the meeting provided the opportunity for investigators from many different areas of basic science and medicine to meet, exchange ideas, information, and techniques, and evaluate the present state and future directions in the field of free radicals, oxidative stress, and antioxidants.

I would like to thank all those who made it possible to organize this meeting. I would like to thank the organizing committee members who had provided invaluable help in the organization of this meeting for which I am grateful. I would like to thank all the speakers who responded to our invitations without hesitation. I thank also all the participants for their enthusiastic participation and their complimentary comments on the success of the Institute. Special thanks are due to the Scientific Affairs Division of NATO for providing a major portion of the grants for the organization of the meeting and also for publication of this book. The contributions of FEBS and UNESCO that were used to support the participation of many young scientists are gratefully acknowledged.

This book presents the contents of the lectures and a selection of the most relevant oral presentations. These proceedings offer a comprehensive account of the most important topics discussed at the Institute. This book is intended to make the proceedings accessible to a large audience.

CONTENTS

MECHANISMS INVOLVED IN FREE RADICAL REACTIONS

FREE RADICALS IN HEALTH AND DISEASE

ANTIOXIDANTS

METHODOLOGY

IRON IN FREE RADICAL REACTIONS AND ANTIOXIDANT PROTECTION

John M. C. Gutteridge

Oxygen Chemistry Laboratory, Unit of Critical Care
Department of Anaesthesia and Adult Intensive Care
Royal Brompton Hospital
Sydney Street, London, SW3 6NP, United Kingdom

1. IRON AND REACTIVE IRON SPECIES (RIS)

Iron is the fourth most abundant element in the Earth's crust and the second most abundant metal (after aluminium). It is the metallic iron at the Earth's centre which accounts for its magnetic field as well as for its overall mass density. This metallic iron was exploited many centuries ago for navigation purposes with the pioneering development of the magnetic compass.

1.1. Body Iron

Aerobic life forms had to use iron, but it was only available to them as insoluble ferric complexes because O_2 oxidised Fe^{2+} to $Fe(III)$ oxides. To overcome the problems of poor bioavailability of iron, micro-organisms evolved to produce siderophores which allowed the capture and assimilation of iron which could then be used as a catalyst for oxygen utilization. Aerobes eventually came to dominate the planet, although it appears that earlier primitive anaerobic life forms may also have used iron-sulphur redox chemistry for energy capture because soluble Fe^{2+} was easily available to them in the anaerobic world to donate electrons. The move of life forms to land from the seas and tidal pools occurred when a protective screen of oxygen and ozone was able to filter out most of the damaging solar UVC radiation. Part of the marine environment is still reflected in blood and body fluids with Na^+, K^+, Ca^{++}, Mg^{++} and Cl^- ions predominating. These ions are present in sea water at concentrations some 10^5–10^6 times greater than those of trace metals such as Fe, Cu, Zn, Mn, Mo, Sn and V. Inspite of aluminium and iron being the most abundant metals in the earth's crust, they do not appear as major ions in surface waters, reflecting their poor solubility at neutral pH values. Iron is present in sea water mainly as colloidal particles of hydrated ferric oxide (finely dispersed rust) representing a true solution concentration of iron ions of around 10^{-9} M. Ever increasing industrialisation is lead-

Free Radicals, Oxidative Stress, and Antioxidants, edited by Özben.
Plenum Press, New York, 1998.

ing to acidification of surface waters, with a consequential rise in the solubility of both iron and aluminium (Martin, 1994), allowing both to enter biological eco-chains with unknown long-term consequences.

An average adult human male contains some 4.5 g of iron, absorbs about 1 mg of iron per day, and when in iron-balance excretes the same amount. Only slight disturbances to the delicate balance between iron intake and iron loss can push the body into conditions of iron overload or iron deficiency. It has been estimated that in the world today some 500 million people are iron deficient and several millions are iron overloaded. No specific mechanisms exist for iron excretion; loss occurring by the turnover of intestinal epithelial cells, in sweat, faeces, urine and by menstrual bleeing in woman. Most of the body's iron (some two thirds) is found in the oxygen-carrying protein haemoglobin, with smaller amounts present in myoglobin, various enzymes, the iron transport proteins transferrin and lactoferrin, and the iron storage proteins ferritin and hemosiderin. Most of our dietary intake of iron is in the form of non-haem ferric ion. This requires reduction and solubilization before absorption from the intestine can occur. Gastric HCI and ascorbate (vitamin C) facilitate absorption in this way. Dietary haem-iron is mostly derived from red-meat products, and can be directly absorbed by the intestine. Cells requiring iron for essential aerobic processes have transferrin receptors which allow uptake of iron into the cell. Once inside the cell, reductive release of iron from transferrin occurs which allows the apotransferrin molecule (devoid of iron) to return to the circulation to continue iron transport functions.

1.2. Reactive Iron Species in the Body

Within cells there normally exists a pool of low molecular mass redox active iron which is essential for the synthesis of iron-requiring enzymes and proteins, and for the synthesis of DNA (Breuer et al., 1996). This pool of iron is the target of iron chelators and is also a form of iron sensed by iron regulatory proteins. The amount, and nature of the ligands attached to this iron remain unknown. However, a recently introduced fluorescence assay based on calcein may enhance our knowledge of intracellular iron pools (Cabantchik et al., 1996). In contrast to the intracellular environment, extracellular compartments do not require, or normally contain, a low molecular mass iron pool. Iron-binding proteins such as transferrin and lactoferrin do not even remotely approach iron saturation in healthy subjects indeed they retain a considerable iron-binding capacity, and are able to remove mononuclear forms of iron that enter extracellular fluids. The differences between intracellular and extracellular, compartments, and their requirements for low molecular mass iron deserves special comment, since it is iron in this form that is the most likely catalyst of biological free radical reactions.

Inside the cell low molecular mass iron need not pose a serious threat as a free radical catalyst, provided that the cell has specific defenses to safely and speedily remove all the $O_2^{\cdot-}$ and H_2O_2 and organic peroxides (such as lipid peroxides) that could react with such iron. This is achieved by intracellular enzymes such as the superoxide dismutases, catalase and glutathione peroxidase and possibly also by thioredoxin-dependent H_2O_2 removal systems (Netto et al., 1996). In the extracellular space, however, we see a different pattern of protection against free radical chemistry. Here, proteins bind, conserve, transport and recycle iron, and whilst doing so keep it in non- or poorly-reactive forms that do not react with H_2O_2 or organic peroxides. Proteins such as transferrin and lactoferrin bind mononuclear iron, whereas haptoglobins bind haemoglobin, and haemopexin binds haem. In addition, plasma contains a ferrous ion oxidising protein (ferroxidase) called caeruloplasmin. By keeping iron in a poorly-reactive state, molecules such as $O_2^{\cdot-}$, H_2O_2, $NO^{\cdot}$, and lipid peroxides can survive long enough to perform important and useful functions as signal, trigger and intercellular messenger molecules.

During situations of iron-overload plasma transferrin can become fully loaded with iron (100% iron saturation) and allow low molecular mass iron to accumulate in the plasma. Such iron, when present in micromolar concentrations, can bind to various added chelating agents such as EDTA, desferrioxamine, and bleomycin that cannot abstract iron from transferrin. This non-transferrin bound iron can be associated with several ligands including citrate (Grootveld et al., 1989) and other organic acids and possibly albumin. Low molecular mass ligands for iron inside the cell are also a subject of considerable debate. ATP (Gurgueira and Meneghini, 1996), ADP, GTP, pyrophosphates, inositol phosphates, aminoacids, and polypeptides have all been proposed.

1.3. Measurement of Reactive Iron Species

1.3.1. Iron-Binding Assay (Bleomycin). In 1981 the author and colleagues introduced the "Bleomycin Assay" as a first attempt to detect and measure chelatable redox active iron that could participate in free radical reactions (Gutteridge et al., 1981). The assay procedure is based on the ability of the metal-ion binding glycopeptide antitumour antibiotic bleomycin to degrade DNA in the presence of an iron salt, oxygen, and a suitable iron reducing agent. The ternary bleomycin-iron-oxygen complex binds tightly to DNA, and during the redox cycling of iron, releases base-propenals from the DNA molecule. These are unstable and rapidly degrade to release malondialdehyde (MDA), which is derived from the deoxyribose sugar (Figure 1). Malondialdehyde can be accurately measured by reacting it with thiobarbituric acid. Binding of the ternary complex to DNA makes the reaction site-directed and prevents most biological antioxidants from interfering. Caeruloplasmin is an exception since it appears to be able to catalyse the oxidation of certain ferrous complexes as well as ferrous ions. To prevent interference from caeruloplasmin the concentration of ascorbate added to the reaction is sufficient to inhibit its ferroxidase activity (Gutteridge, 1991a). The bleomycin assay can be directly applied to most biological fluids, and if it is positive it is reasonable to assume that bleomycin has been able to chelate a low molecular mass form of iron from the sample and that such iron can be redox cycled. Meticulous removal of adventitious iron (known to be present in all laboratory glassware, reagents and chemicals) is essential in order to achieve a sensitivity of detection in the low micromolar range.

1.   Bleomycin (BLM)

$$DNA-BLM + Fe^{2+} + O_2 \longrightarrow DNA_{ox} - BLM - Fe^{(III)} + MDA$$

$$MDA + 2\ TBA \longrightarrow MDA\ (TBA)_2 \quad \text{Pink chromogen A532 nm}$$

2.   Aconitase (ACN)

$$ACN\ (3Fe - 4S)^{+} + Fe^{2+} + e^{-} \longrightarrow ACN\ (4Fe - 4S)^{2+}$$
$$\text{(inactive)} \qquad\qquad\qquad\qquad \text{(active)}$$

$$ACN\ (4Fe - 4S)^{2+} + citrate \longrightarrow isocitrate$$

$$Isocitrate + ICD + NADP^{+} \longrightarrow NADPH\ (A340\ nm) + CO_2 + \text{2-oxoglutarate}$$

Abbreviations: DNA_{ox} = Oxidised DNA;   MDA = malondialdehyde;
TBA = thiobarbituric acid;   ICD = isocitrate dehydrogenase

Figure 1. Assays for reactive iron species.

Blood serum, or plasma, from normal healthy individuals do not contain detectable levels of iron in the bleomycin assay. Indeed they usually give assay values less than the reagent controls, since their iron-binding capacity allows them to remove traces of iron contaminating the reagents used. Such iron is still present despite cleaning-up all reagents by treatments with Chelex resin or by dialysis against the iron-binding protein ovotransferrin (Gutteridge, 1987a). The most likely explanation as to why plasma transferrin can remove iron from the reagents but ovotransferrin dialysis can not, is that iron is present in the reagents in a polynuclear form and not readily chelatable until ascorbate is added to the reaction. Ascorbate releases mononuclear iron which can then bind to ovotransferrin. Bleomycin, with a binding constant of 10^{15} for ferric ions, is not a strong enough chelator to remove iron correctly loaded onto transferrin, or lactoferrin, or into ferritin or haem proteins. Using the bleomycin assay, low molecular mass iron has been detected in a variety of biological fluids as well as in tissue homogenates. By not including ascorbate in the reagents, ferrous salts plus the effect of any endogenous reducing agents can be speciated using the bleomycin assay (Gutteridge, 1991b). Other approaches to the measurement of low Mr iron in biological fluids have been based on alternative ways of measuring chelatable iron, for example the binding of iron by desferrioxamine followed by separation of desferrioxamine and ferrioxamine (Green et al., 1989).

1.3.2. Iron-Activating Assay (Aconitase). Aconitase from pig heart contains an iron-sulphur cluster at its active centre which can change from an inactive form [3Fe–4S] to the active [4Fe–4S] form: i.e.,

$$[3Fe–4S]^+ + Fe^{2+} + e^- \rightarrow [4Fe–4S]^{2+}$$

Iron present in neonatal plasma, not tightly associated with transferrin, is converted to the ferrous state by adding ascorbate (at the same concentration as present in the bleomycin assay), and the resulting ferrous species activates the 3Fe–4S cluster of aconitase. Activation of inactive iron-requiring aconitase was carried out as follows:

Plasma was incubated with PIPES buffer pH 7.4, 200 mM/l, 7.5 mM ascorbate, and 50 mM sodium citrate for 5 mins at 37°C. Thereafter, 100 mM tricarballic acid, 300 mM $MgSO_4$, 20 mM NADP, isocitrate delydrogenase 50 units/ml, and aconitase 126 units/ml were added and incubated at 37°C. Change in absorbance at 340 nm was recorded from zero time for 45 mins. At pH 7.4 RIS in the presence of ascorbate are converted to the ferrous state which can activate the inactive form of aconitase. Active aconitase converts citrate to isocitrate, and isocitrate dehydrogenase converts isocitrate to 2-oxoglutarate and CO_2 with the reduction of $NADP^+$ to NADPH, which is measured at 340 nm (Gutteridge et al., 1996) (Figure 1).

2. ANTIOXIDANT REGULATION

Antioxidants can act at many different stages in an oxidative sequence such as:

- Removing oxygen or decreasing local O_2 concentrations.
- Removing catalytic metal ions.
- Removing key ROS/RNS such as $O_2^{\cdot-}$, H_2O_2, HOCl, singlet O_2 or $ONOO^-$
- Scavenging initiating on such as $^{\cdot}OH$, $RO^{\cdot}$, $RO_2^{\cdot}$
- Breaking the chain of an initiated sequence.

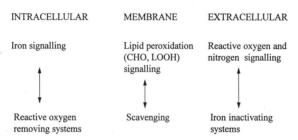

INTRACELLULAR   MEMBRANE   EXTRACELLULAR

Iron signalling   Lipid peroxidation (CHO, LOOH) signalling   Reactive oxygen and nitrogen signalling

Reactive oxygen removing systems   Scavenging   Iron inactivating systems

Figure 2. Antioxidants and signalling.

Many antioxidants have more than one mechanism of action, and many will have complex pharmacological properties as well. Cells have formidable defenses against oxidative damage, many of which may at first sight not seem to be antioxidants. Antioxidant protection can operate at different levels within cells, for example:

- Preventing radical formation.
- Intercepting formed radicals.
- Repairing oxidative damage.
- Increasing elimination of damaged molecules.
- Promoting the death of cells with excessively damaged DNA in order to prevent transformed cells arising.

The different antioxidant strategies used within cells, inside membranes, and in extracellular fluids prompted us to proposed in 1986 that such marked differences were important for humoral signalling (Halliwell and Gutteridge, 1986) see Figure 2.

2.1. General Antioxidant Defenses

The deleterious effects of ROS, RNS, RIS and reactive copper species (RCS) are controlled by antioxidant defences. Some antioxidants are synthesized in the human body: they include enzymes, certain other proteins and low-molecular-mass scavengers. Examples are superoxide dismutases, catalases, peroxidases, thiol-specific antioxidants, metallothioneins, other metal ion-binding and storage proteins, urate, GSH and ubiquinol (Fridovich, 1989; Sies, 1991; Frei, 1994; Gutteridge and Halliwell, 1994; Chubatsu and Meneghini, 1993; Netto et al., 1996; Ernster and Dallner, 1995). These defences protect aerobes against most of the toxic effects of ambient O_2 (21%), but all aerobes (including humans) are injured if exposed to excess O_2 (Balentine, 1982).

It follows that, in general, aerobes have not evolved an excess of antioxidant defences, although often defences are inducible by elevated O_2 if sufficient time for adaptation is allowed (Balentine, 1982). Indeed, antioxidants do not even prevent damage by RNS/ROS at ambient O_2. Animals thus rely on a second line of defence in the form of repair systems, of which the most important maybe those that remove mutagenic lesions in DNA induced by ROS/RNS (Demple and Harrison, 1994). Superimposed on such defenses are inducible proteins such as haem oxygenase-1. Heme oxygenases remove the pro-oxidant haem and in the process produces the antioxidant bilirubin (Stocker, 1990). However, it should be noted that a pro-oxidant form of RIS, is formed which is potentially more redox-active than the haem iron that is removed. The constant assault by ROS/RNS on DNA throughout the long human lifespan may contribute to the age-related development of cancer (Demple and Harrison, 1994). Indeed, ageing itself may involve the cumulative effects of oxidative damage over a lifespan (Ames et al., 1993; Orr and Sohal, 1994), through the constant triggering of oxidative reactions that evolved to ensure survival in the short term (e.g. by killing foreign invaders).

Additional protection is provided by dietary antioxidants. The physiological role of some of these is well-established (e.g. vitamin E, ascorbate) whereas the role of others (e.g. flavonoids, carotenoids) is currently uncertain. Dietary antioxidants appear to be important in delaying/preventing certain human diseases, especially cardiovascular disease and some types of cancer (reviewed in Gutteridge and Halliwell, 1994; Gey, 1995; Block et al., 1992).

2.2. Signalling through Antioxidants

It has been suggested that certain ROS, (and perhaps, by extension RIS and RNS) might be used as "signal, messenger, and trigger molecules" (Halliwell and Gutteridge, 1986; Schreck et al., 1992; Sarafian and Bredesen, 1994; Burdon 1994; Krieger-Brauer and Kather, 1996; Barja, 1993; Saran and Bors, 1989), e.g. mediating the effects of PDGF on smooth muscle cells (Sundaresan et al., 1995), activating adenylate cyclase (Tan et al., 1995), and iron regulatory proteins (Rouault & Klausner, 1996). We are seeing increasing examples of redox regulation of gene expression; not only oxyR and NFkB but also the role of thioredoxin and of AP-1.

2.2.1. Intracellular Signalling. Oxygen metabolism occurs within cells, and it is here that we expect to find antioxidants evolved to deal speedily and specifically with ROS/RNS. Elimination of toxic reduction intermediates of oxygen inside the cell allows a small pool of RIS (the low molecular mass intracellular iron pool) to safely exist to provide iron for the manufacture of iron-containing proteins. Iron is stored in cells within two major proteins; ferritin and haemosiderin. Ferritin is a soluble protein located in the cytosol, whereas haemosiderin is insoluble and found mainly in lysosomes. There is probably some ferritin in every human cell but the largest amounts are present in the liver parenchyma. Iron enters the protein as Fe^{2+}, but is stored as Fe(III) in the central core in structures similar to the mineral ferrihydrite. In order for iron release to occur a reductive conversion back to Fe^{2+} is required. Ferritin is a relatively safe storage form of iron in the body, since little of its iron can be mobilised by $O_2^{\cdot-}$, H_2O_2 and lipid hydroperoxides (LOOH) to participate in radical chemistry (Bolann and Ulvik, 1987). The small amount of iron released under stress by ROS does not appear to be that loaded into the central core (Bolann and Ulvik, 1990). Experiments which show linear relationships between the iron loading of ferritin and its ability to drive ˙OH formation and lipid peroxidation may result from the pool of iron at the surface of the core, or the use of degraded commercial preparations. Haemosiderin is considerably less inclined than ferritin to release chelatable iron (Gutteridge and Hou, 1986) to stimulate free radical reactions essentially because it is insoluble. Thus conversion of ferritin to haemosiderin during conditions of iron-overload may be protective, by limiting the availability of iron for free radical reactions (O'Connell et al., 1986).

2.2.2. Intracellular Iron Signalling. Cells normally accumulate iron via the binding of transferrin to high affinity surface receptors (TfR) followed by endocytosis. There is also a, transferrin-independent pathway of cellular iron uptake that is said to involve a ferri-reductase and an Fe^{2+} transmembrane transport system (reviewed in De Silva et al., 1996). The reductase is proposed to provide iron in a soluble form to the membrane transporter. When non-transferrin bound iron appears in plasma, due to iron-overload or lack of transferrin(apotransferrinaemia), it is rapidly cleared by the membrane-bound transport system constitutively present on parenchymal cells of organs, particularly those of liver, heart, pancreas and the adrenals. This latter system does not require endocytosis of a protein for iron delivery.

The rate of synthesis of TfR and ferritin is regulated at the post-transcriptional level by cellular iron, and co-ordinated by the iron-dependent binding of cytosolic proteins called "the iron responsive element binding proteins "(IRE-BP)" (now known as iron regulatory proteins, IRP) which binds to specific sequences on their mRNAs (Klausner et al., 1993). It appears that low molecular mass iron is capable of acting as a signal to regulate ferritin and TfR synthesis in this way. In the absence of iron, IRE-BP binds to the iron-responsive element (IRE) (the 3^1-untranslated region of the TfR message contains a set of stem-loop structures termed IRE) stabilizing the transcript. When iron is present, the protein dissociates from the IRE and degradation of the mRNA occurs. Recent work has shown that one of the IRE-BPs is identical with to the cytosolic enzyme aconitase (reviewed in Kennedy et al., 1992; Haile et al., 1992a,b). The protein functions as an active aconitase when it has an Fe–S cluster present or as an RNA-binding protein when iron is absent. Switching between these two forms depends on cellular iron-status such that when iron is replete it is an active aconitase, whereas when deprived of iron it has only RNA-binding activity.

2.3. Membrane Signalling

Within the hydrophobic interior of membranes, lipophilic radicals are formed which are usually different from those seen in the intracellular aqueous space. Lipophilic radicals require hydrophobic antioxidants for their removal. α-Tocopherol, a fat-soluble vitamin, is a poor antioxidant outside a membrane but is extremely effective when incorporated into the membrane bilayer (Gutteridge, 1977). Membrane stability and protection against oxidative insult depend very much on the way in which the membrane is assembled from its lipid components. Structural organisation requires that the "correct" ratios of phospholipids and their fatty acids are attached (reviewed in Gutteridge and Halliwell, 1988).

When a cell is damaged, or dies, it is highly likely that its lipids will undergo peroxidation (Halliwell and Gutteridge, 1984). Tissue damage releases RIS/RCS and activates enzymes which catalyse peroxidation of polyunsaturated fatty acids, particularly linoleic acid, leading to a build-up of lipid peroxides (Herold and Spiteller, 1996). Peroxidation of membrane polyunsaturated fatty acids produces a plethora of reactive primary peroxides and secondary carbonyls and it was suggested many years ago by one of the authors that lipid oxidation products such as these, resulting from cell death could act as triggers for new cell growth (Gutteridge and Stocks, 1976). Through the detailed work of Hermann Esterbauer and colleagues (1988) we now have clearer insights into the biological reactivity of lipid oxidation products. 4-Hydroxy-2-nonenal (HNE), a peroxidation product of (n-6) fatty acids (when RIS or RCS are present), is a potent trigger for chemotaxis, can inactivate thiol-containing molecules, and activate certain enzymes (reviewed in Esterbauer et al., 1991). As a general rule low levels of ROS, and possibly reactive carbonyls, activate cellular processes, whilst higher levels turn them off. The resting cell is normally in a reduced state and is progressively activated as oxidation increases up to a maximum (see Figure 3). Too much oxidation deposes all function (Burden et al., 1994; McConkey et al., 1996) until, eventually apoptosis or necrosis is triggered.

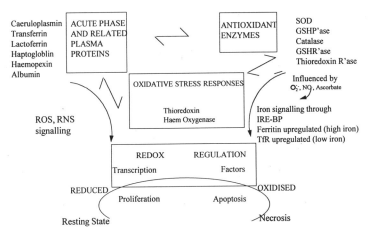

Figure 3. Oxidative stress response.

2.4. Extracellular Signalling

Human body extracellular fluids contain little, or no, catalase activity, and extremely low levels of superoxide dismutase. Glutathione peroxidases, in both selenium-containing and non-selenium-containing forms, are present in plasma but there is little glutathione substrate in plasma (1–2 μM). "Extracellular" superoxide dismutases (EC-SOD) have been identified (Marklund et al., 1982) and shown to contain copper, zinc and attached carbohydrate groups. By allowing the limited survival of $O_2^{\cdot -}$, H_2O_2, lipid peroxides (LOOH) and possibly other ROS/RNS in extracellular fluids the body can utilise these molecules, and others such as nitric oxide (NO), as useful messenger, signal or trigger molecules (Halliwell and Gutteridge, 1986). A key feature of such a proposal is that $O_2^{\cdot -}$, H_2O_2, LOOH, NO˙ and HOCl do not meet with reactive iron or copper, and that extracellular antioxidant protection has evolved to keep iron and copper in poorly or non-reactive forms (Halliwell and Gutteridge, 1986; Gutteridge, 1995).

The major copper-containing protein of human plasma is caeruloplasmin, unique for its intense blue coloration. This protein's ferroxidase activity makes a major contribution to extracellular antioxidant protection by decreasing ferrous ion-driven lipid peroxidation and Fenton chemistry (Gutteridge and Stocks, 1981; Gutteridge et al., 1980).

2.5. Extracellular Control of RIS

2.5.1. Mononuclear Iron. Human apotransferrin (Tf) has two metal-binding lobes the C- and N-terminals which when fully loaded with iron give a ratio of two moles of iron per mole of protein (Fe$_2$ Tf). In order for iron to bind to transferrin an anion, usually bicarbonate (HCO_3^-), is required (Kojiman and Bates, 1981). The bicarbonate co-ordinates with the iron-forming a bridge between the metal and a cationic group on the protein.

Under normal physiological conditions plasma transferrin has a binding constant for ferric ions of around 10^{22}, with the C-lobe having a higher affinity for iron then the N-lobe. In normal healthy individuals transferrin is only up to one-third loaded with iron and retains a considerable iron-binding capacity. This can be calculated in plasma by measuring the total non-haem iron content both before and after saturating the protein with an iron salt.

Lactoferrin is considerably more stable under acid conditions than is transferrin, and is said to hold onto its iron down to pH values as low as 4.0. Iron can be released from transferrin by lowering the pH value; and such iron release is greatly amplified if an iron-reducing molecule and an iron chelator are also present.

Transferrin is mainly synthesised in the liver and secreted into the circulation as the apoprotein, although other tissues can also synthesise transferrin at much lower levels. Transferrin binds ferric ions in the circulation keeping levels of mononuclear iron in the plasma at effectively zero. Iron-loaded transferrin enters cells by binding to high affinity transferrin receptors (TfR) on the surface of cells. The expression of TfRs is determined not only be cellular iron requirements but also by cell growth and differentiation. Changes in TfR gene expression are post-transcriptionally regulated, in response to variations in intracellular iron levels, through the iron regulatory proteins.

Inside the cell, iron is released from transferrin by acidification, through a proton-pumping ATPase, within an endosome (reviewed in Crichton and Charloteaux-Wauters, 1987). The released iron is used for storage, and for the synthesis of DNA and of iron-containing proteins. The remaining apotransferrin is returned to the extracellular environment to continue the cycle of iron-binding and cell delivery.

Enzymes that catalyse the oxidation of ferrous ions, or ferrous complexes, to their ferric state are called ferrous oxidases or ferroxidases. Iron chelators can have ferroxidase-like activities when they displace the equilibrium between ferrous and ferric ions in solution. By binding ferric ions they pull the reaction strongly to the right, i.e.

$$Fe^{2+} \rightarrow Fe\ (III) \rightarrow Ferric\ chelate$$

Many iron chelators show a ferroxidase-like activity, examples being transferrin, bleomycin, EDTA, and desferrioxamine (Harris and Aisen, 1973; Caspary et al., 1979; Goodwin and Whitten, 1965). Caeruloplasmin is the major copper-containing protein of extracellular fluids. It has a relative molecular mass of approximately 132,000 with six or seven copper ions per molecule. Six of these coppers are tightly bound to caeruloplasmin and can only be released at low pH in the presence of a reducing agent whilst the seventh copper is labile and chelatable (reviewed in Gutteridge and Stocks, 1981). It has been suggested that native caeruloplasmin, carrying this seventh copper ion, can under abnormal circumstances interact with cells to become pro-oxidant and contribute to the development of atherosclerosis by causing oxidation of low density lipoprotein (Ehrenwald et al., 1994; Ehrenwald and Fox, 1996).

However, there is no evidence that the protein itself is pro-oxidant to LDL. Caeruloplasmin may be able to supply copper to cells via receptor mediated delivery for incorporation into other copper-containing proteins such as CuZnSOD and cytochrome oxidase. This copper-donor role is sometimes referred to as a copper "transport" function. However, caeruloplasmin does not specifically bind and transport copper in the way that transferrin binds and transports mononuclear iron. Caeruloplasmin catalyses the oxidation of a wide variety of polyamine and polyphenol substrates in vitro. However, with the possible exception of certain bioamines these oxidations appear to have no biological significance in vivo. A role for caeruloplasmin as a ferroxidase enzymes in vivo was first proposed by Frieden and his colleagues (Osaki et al., 1996). The protein can catalyze oxidation of ferrous ions to the ferric state, the electrons being passed onto oxygen to form water. It has been proposed that this ferroxidase activity is essential in vivo for normal iron metabolism, and for the incorporation of ferric ions into transferrin and into ferritin although the latter has recently been suggested as unlikely (Treffry et al., 1995). Several scientists have pointed out that ferrous ions readily autoxidize in neutral aerobic solutions without the requirement for an enzyme to do it (Equations 1–3):

$$4Fe^{2+} + O_2 \rightleftharpoons 4Fe\ (III) + 4O_2^{\cdot-} \tag{1}$$

$$4O_2^{\cdot-} + 4H^+ \rightarrow 2H_2O_2 + 2O_2 \tag{2}$$

$$2H_2O_2 + 2Fe^{2+} \rightarrow 2\,^{\cdot}OH + 2OH^- + 2Fe\ (III) \tag{3}$$

Caeruloplasmin, however, rapidly removes ferrous ions from solutions at neutral pH values, with the complete reduction of oxygen to water, i.e.,

$$4Fe^{2+} + O_2 + 4H^+ \rightarrow 2H_2O + 4Fe\ (III) \tag{4}$$

By transferring 4 electrons at the enzymes active (ferroxidase) centre no reactive forms of oxygen are released into solution (Gutteridge and Stocks, 1981). Compare equations 1–3 with 4.

It has been suggested that a second ferroxidase enzyme exists. This enzyme, ferroxidase (II), is a cupro-lipoprotein complex (Lykins et al., 1977) that is almost completely absent from freshly taken normal human plasma (Gutteridge et al., 1985). Whenever plasma is stored or mishandled, however, a metalloproteinase closely associated with caeruloplasmin rapidly degrades the native protein (132 kDa) into 116 kDa, 50 kDa and 19 kDa fragments. Caeruloplasmin has structural similarities and sequence homologies with factors V and VIII of the coagulation cascade, and with the iron-binding protein lactoferrin. The blood clotting factors V and VIII are normally activated by serine protease cleavage, and the similarity of caeruloplasmin to these clotting factors may in part explain why it is so sensitive to proteolysis. In separated plasma some of these fragments release copper that is chelatable to 1, 10-phenanthroline, and that can cause the oxidation of plasma lipoproteins (Gutteridge et al., 1985). The fact that fresh plasma does not catalyse LDL oxidation argues against the alleged pro-oxidant effect of the intact caeruloplasmin molecule. As fragmentation occurs caeruloplasmin ferroxidase I activity is lost, but a new ferroxidase activity (ferroxidase II) appears which cannot be inhibited by adding azide. Ferroxidase II in human plasma, therefore, appears to be an artefact (Gutteridge et al., 1985).

Caeruloplasmin accounts for most (96% or more) of the total plasma copper, with a normal adult concentration of 0.35 g/l. Caeruloplasmin is a minor acute-phase protein induced in response to tissue damage. Levels will, therefore, change, to some extent, in a wide variety of diseases. The ferroxidase activity of caeruloplasmin was reported by the author and colleagues to be of importance as an antioxidant in vivo (Gutteridge 1978) by rapidly removing Fe^{2+} capable of reacting with H_2O_2 and organic peroxides, to give $^{\cdot}OH$ and alkoxyl radicals ($RO^{\cdot}$) respectively. Caeruloplasmin can also react stoichiometrically with $O_2^{\cdot-}$, H_2O_2 and $^{\cdot}OH$ and non-specifically bind ferric and cupric ions (reviewed in Gutteridge and Quinlan, 1996).

The ferroxidase activity of caeruloplasmin competes effectively for ferrous ions in the Fenton reaction. The second order rate constant for the reaction of caeruloplasmin with ferrous ions has been reported as 27×10^3 $M^{-1}s^{-1}$, which is considerably faster than Fe^{2+} reacts with H_2O_2 (76 $M^{-1}s^{-1}$). It is likely that caeruloplasmin would function in vivo in a similar way to prevent hydroxyl radical formation.

2.5.2. Heme Iron. Haptoglobins are glycoproteins, Mr about 86,000, found in the α,-globulin fraction of human serum. They form stable complexes with hemoglobin, but not with haem, both in vivo and in vitro. The resulting bond is one of the strongest non-covalent protein interactions known, with a stoichiometry of one mole of haptoglobin to one mole of hemoglobin. The amount of haptoglobin present in the total plasma volume has been calculated to be sufficient to bind and conserve 3 g of haemoglobin, ensuring that no pro-oxidant haemoglobin is normally present in the plasma. Hemoglobin can stimulate the process of lipid peroxidation by at least two mechanisms; the heme ring reacts with peroxides to form active oxo-iron species and amino acid radicals (McArthur & Davis, 1993) or when a large excess of peroxide is present it causes fragmentation of the pyrrole rings with the release of RIS (Gutteridge, 1986). Binding of haemoglobin to haptoglobin greatly decreases its pro-oxidant properties (Gutteridge, 1987 b). The concentration of haptoglobins in normal plasma ranges from 0.5–2.0 g/l.

Hemopexin is a plasma β-glycoprotein with a Mr of around 60,000, which tightly binds heme but not hemoglobin. Heme is transported to liver parenchymal cells by a receptor-mediated process involving endocytosis of hemopexin (Smith and Morgan, 1979). Like the iron transport protein transferrin, hemopexin is not degraded when delivering heme to cells, returning to the circulation as an intact protein (Smith and Morgan, 1979). The normal plasma concentration of hemopexin ranges from 0.15–1.3 g/l. Albumin can also tightly bind heme to form a complex known as methemalbumin. The pro-oxidant properties of heme are greatly diminished when it is bound to hemopexin, thereby conserving iron as well as decreasing its reactivity (Gutteridge and Smith, 1988).

The pattern of constitutive antioxidants present inside cells and those present outside cells, and their distinct roles in removing either RIS or ROS leads one to the conclusion that they are operating in concert to facilitate redox signalling. Superimposed on the constitutive antioxidants are inducible antioxidants that respond to tissue damage and oxidative stress (see Figure 3). The lack of clinical success seen with most antioxidant interventions reflects the complex and compensatory responses inherent in such a system.

ACKNOWLEDGMENTS

JMCG thanks the British Lung Foundation, the BOC Group plc, the British Heart Foundation and the Dunhill Medical Trust for their generous financial support.

REFERENCES

Ames, B.N., Shigenaga, M.K. and Hagen. T.M. 1993. Oxidants, antioxidants, and the degenerative disease of aging. *Proc. Natl. Acad. Sci. USA* **90**: 7915–7922.

Balentine, J. 1982. Pathology of Oxygen Toxicity. Academic Press, New York.

Barja, G. 1993. Oxygen radicals, a failure or a success of evolution? *Free Rad. Res Commun.* **18**: 63–70.

Block, G, Patterson, B. and Subar, A. 1992. Fruit, Vegetables and cancer prevention — a review of the epidemiological evidence. *Nutr Cancer* **18**: 1–29.

Bolann, B.J. and Ulvik, R.J. 1987. Release of iron from ferritin by xanthine oxidase. Role of the superoxide radical. *J. Biochem.* **243**: 55–59

Bolann, B.J. and Ulvik, R.J. 1990. On the limited ability of superoxide to release iron from ferritin. *Eur J Biochem.* **193**: 899–904.

Breuer, W. Epsztein, S. and Cabantchik, Z.I. 1996 Dynamics of the cytosolic chelatable iron pool of K562 cells. *FEBS Lett.* **384**: 304–308

Burden, R.H. 1994. Superoxide and H_2O_2 in relation to mammalian cell proliferation. *Free Rad. Biol. Med.* **18**: 775–794.

Cabantchik, Z.I., Glickstein, H., Milgram, P. and Breuer, W. 1996. A Fluorescence assay for assessing chelation of intra cellular iron in a membrane model system and mammalian cells. *Anal. Biochem.* **233**: 221–227.

Caspary, W.J., Lanzo, D.A., Niziak, C., Friedman, R. and Bachur, N.R. 1979. Bleomycin A_2: a ferrous oxidase. *Mol. Pharmacol.* **16**: 256–290.

Chubatsu, L.S. and Meneghini, R. 1993. Methallothionein protects DNA from oxidative damage. *Biochem. J.* **291**: 193–198.

Crichton, R.R. and Charloteaux-Wauters, M. 1987. Iron transport and storage. *Eur. J. Biochem.* **164**: 485–506.

Demple, B. and Harrison, L. 1994. Repairs of oxidative damage to DNA. Enzymology and Biology. *Ann. Rev. Biochem.* **63**: 915–948.

DeSilva, D.M., Askwith, C.C. and Kaplan, J. 1996. Molecular mechanisms of iron uptake in Eukaryotes. *Physiol. Rev.* **76**: 31–47.

Ehrenwald, E., Chisolm, G.M. and Fox, P.L. 1994. Intact human ceruloplasmin oxidatively modifies low density lipoprotein. *J. Clin. Invest.* **93**: 1493–1501.

Ehrenwald, E. and Fox, P.L. 1996. Role of endogenous ceruloplasmin in low density lipoprotein oxdiation by human U937 monocytic cells. *J. Clin. Invest.* **97**: 884–890.

Ernster, E. and Dallner, G. 1995. Biochemical, physiological and medical aspects of ubiquinone function. *Biochem. Biophys. Acta.* **1271**: 195–204.

Esterbauer, H., Schaur, R.J., and Zollner, H. 1991. Chemistry and biochemistry of 4-hydroxynonenal, malonaldehyde and related aldehydes. *Free Rad. Biol. Med.* **11**: 81–128.

Esterbauer, H., Zollner, J. and Schaur, R.J. 1988. Hydroxyl alkenals: cytotoxic products of lipid peroxidation. *ISI Atlas of Sci.* **1**: 311–317.

Frei, B. (ed.) 1994. Natural Antioxidants in Human Health and Disease. Academic Press, San Diego.

Fridovich, I. 1989. Superoxide Dismutase. An adaption to a paramagnetic gas. *J. Biol. Chem.* **264**: 7761–7764.

Gey, F. 1995. Ten-year retrospective on the antioxidant hypothesis of arteriosclerosis-threshold plasma levels of antioxidant micronutrients related to minimum cardiovascular risk. *J. Nutr. Biochem.* **6**: 206–236.

Goodwin, JF. and Whitten, CF. 1965. Chelation of ferrous sulphate solutions by desferrioxamine B. *Nature* **205**: 281–283.

Green, C.J., Gaver, J.D., Healing, G., Cotterill, La., Fuller, B.J., Simpkin, S. 1989. The importance of iron, calcium and free radicals in reperfusion injury: an overview of studies in ischaemic rabbit kidney. *Free Rad. Res. Commun.* **7**: 255–264.

Grootveld, M., Bell, J.D., Halliwell, B., Aruoma, O.I., Bomford, A. and Sadler P.J. 1989. Non-transferrin-bound iron in plasma or serum from patients with idiopathic hemochromatosis. *J. Biol. Chem.* **264**: 4417–4422.

Gurgueira, S.A. and Meneghini, R. 1996. An ATP-dependent iron transport system in isolated rat liver nuclei. *J. Biol. Chem.* **271**: 13616–13620.

Gutteridge, J.M.C. 1977. The membrane effects of vitamin E, cholesterol and their acetates on peroxidative susceptibility. *Res. Commun. Chem. Path. Pharmacol.* **77**: 379–386.

Gutteridge, J.M.C. 1978. Ceruloplasmin: A plasma protein, enzyme and antioxidant. *Annals Clin. Biochem.* **15**: 293–296.

Gutteridge, J.M.C. 1986. Iron promoters of the Fenton reaction and lipid peroxidation can be released from haemoglobin by peroxides. *FEBS Lett.* **201**: 291–925.

Gutteridge, J.M.C. 1987a. A method for removal of trace iron contamination from biological buffers. *FEBS Lett.* **214**: 362–364.

Gutteridge, J.M.C. 1987b. The antioxidant activity of haptoglobin towards haemoglobin stimulated lipid peroxidation. *Biochim. Biophys. Acta* **917**: 219–223.

Gutteridge, J.M.C. 1991a. Plasma ascorbate levels and inhibition of the antioxidant activity of caeruloplasmin. *Clin. Sci.* **81**: 313–317.

Gutteridge, J.M.C. 1991b. Ferrous ions detected in cerebrospinal fluid by using bleomycin and DNA damage. *Clin. Sci.* **82**: 315–320.

Gutteridge, J.M.C. 1995. Signal messenger and trigger molecules from free radical reactions, and their control by antioxidants in: Signalling Mechanisms - from Transcription Factors to Oxidative Stress. (L. Packer, K. Wirtz, eds.) Springer-Verlag, Berlin, pp. 157–164.

Gutteridge, J.M.C. and Halliwell B. 1988. The antioxidant proteins of extracellular fluids. In Cellular Antioxidant Defence Mechanisms (Chow, CK. ed.) CRC Press, Boca Raton. pp. 1–23.

Gutteridge, J.M.C. and Halliwell, B. 1994. Antioxidants in Nutrition, Health and Disease. Oxford University Press: Oxford. pp 1–143.

Gutteridge, JMC. and Hou, YY. 1986. Iron complexes and their reactivity in the bleomycin assay for radical promoting loosely bound iron. *Free Rad. Res.Commun.* **2**: 143–151.

Gutteridge, J.M.C., Mumby, S., Koizumi, M., Taniguchi, N. 1996. "Free" iron in neonatal plasma activates aconi-tase: Evidence for biologically reactive iron. *Biochem. Biophys. Res. Commun.* **229**: 806–809.

Gutteridge, J.M.C. and Quinlan, G.J. 1996. Reactive oxygen species, antioxidant protection and lung injury, in Acute Respiratory Distress Syndrome in Adults. (Evans, TW and Haslett, C. eds). Chapman and Hall, London. pp 167–196.

Gutteridge, J.M.C., Richmond, R. and Halliwell, B. 1980. Oxygen free radicals and lipid peroxidation: Inhibition by the protein caeruloplasmin. *FEBS Lett.* **112**: 269–272.

Gutteridge, J.M.C., Rowley, D.A. and Halliwell, B. 1981. Superoxide-dependent formation of hydroxyl radicals in the presence of iron salts. Detection of "free" iron in biological systems by using bleomycin-dependent degradation of DNA. *Biochem. J.* **199**: 263–265.

Gutteridge, J.M.C. and Smith, A. 1986. Antioxidant protection by hemopexin of haem stimulated lipid peroxida-tion. *Biochem. J.* **256**: 861–865.

Gutteridge, J.M.C. and Stocks, J. 1976. Peroxidation of cell lipids. *J. Med. Lab. Sci.* **53**: 281–285.

Gutteridge, J.M.C. and Stocks, J. 1981. Caeruloplasmin: physiological and pathological perspectives. *CRC Crit. Rev. Clin. Lab. Sci.* **14**: 257–329.

Gutteridge, J.M.C., Winyard, P., Blake, D.R., Lunec, J., Brailsford, S. and Halliwell, B. 1985. The behaviour of caeruloplasmin in stored human extracellular fluids in relation to ferroxidase II activity, lipid peroxidation and phenanthroline-detectable copper. *Biochem. J.* **230**: 517–523.

Haile, D.J., Rouault, T.A., Harford, J.B., Kennedy, M.C., Blondin, G.A., Beinert, G.A. and Klausner, R.D. 1992a. Cellular regulation of the iron-responsive element binding protein: disassembly of the cubane iron-sulfur cluster results in high affinity RNA binding. *Proc. Natl. Acad. Sci. USA* **89**: 11735–11739.

Haile, D.J., Rouault, T.A., Tang, C.K., Chin, J., Harford, J.B, and Klausner, R.D. 1992b. Reciprocal control of RNA-binding and aconitase activity in the regulation of the iron-responsive element binding protein: role of the iron-sulfur cluster. *Proc. Natl. Acad. Sci.* **89**: 7536–7540.

Halliwell, B. and Gutteridge, J.M.C. 1984. Lipid peroxidation, oxygen radicals, cell damage and antioxidant ther-apy. *Lancet.* **1**: 1396–1397.

Halliwell, B. and Gutteridge, J.M.C. 1986. Oxygen free radicals and iron in relation to biology and medicine: some problems and concepts. *Arch. Biochem. Biophys.* **246**: 501–514.

Halliwell, B. and Gutteridge, J.M.C. 1989. Free radicals in Biology and Medicine. Oxford University Press; Oxford.

Harris, D.C. and Aisen, P. 1973. Facilitation of $Fe^{(II)}$ autoxidation by $Fe^{(III)}$ complexing agents. *Biochim. Biophys. Acta.* **329**: 156–158.

Herold, M. and Spiteller, G. 1996. Enzymic production of hydroperoxides of unsaturated fatty acids by injury of mammalian cells. *Chem. Phys. Lipids* **79**: 113–121.

Klausner, R.D., Rouault, T.A. and Harford, J.B. 1993. Regulating the fate of mRNA: The control of cellular iron metabolism. *Cell* **72**: 19–28.

Kennedy, M.N., Mende-Mueller, L., Blondin, G.A. and Beinert, H. 1992. Purification and characterisation of cytosolic aconitase from beef liver and its relationship to the iron-responsive element binding protein (IRE-BP). *Proc. Natl. Acad. Sci. USA* **89**: 11730–11734.

Kojiman, N. and Bates, G.W. 1981. The Formation of Fe^{3+}-Transferrin-$CO_2^{-}/_3$ via the Binding and Oxidation of Fe^{2+}. *J. Biol. Chem.* **256**: 12034–12039.

Kriegor-Brauer, H.I. and Kather, H. 1995. The stimulus-sensitive H_2O_2-generating system present in human fat-cell plasma membranes is multireceptor-linked and under antagonistic control by hormones and cytokines. *Biochem. J.* **307**: 543–548.

Lykins, L.F., Akey, C.W., Christian, E.G., Duval, G.W. and Topham, R.W. 1977. Dissociation and reconstitution of human ferroxidase II. *Biochemistry.* **16**: 693–698.

Marklund, S.L., Holme, E. and Hellner, L. 1982. Superoxide dismutase in extracellular fluids. *Clin. Chim. Acta.* **126**: 41–51.

McArthur, K.M. and Davies, M.J. 1993. Detection and reactions of the globin radical in haemoglobin. *Biochim. Biophys. Acta.* **1202**: 172–181.

McConkey, D.J., Orrenius, S. Jondal, M. 1996. The regulation of apoptosis in thymocytes. *Biochem. Soc. Trans.* **22**: 606–610.

Netto, L.E.S., Chae, H.Z., Kang, S.W., Rhee, S.G. and Stadtman, E.R. 1996. Removal of H_2O_2 by thiol-specific antioxidant enzyme (TSA) is involved with its antioxidant properties. TSA prossesses thiol peroxidase ac-tivity. *J. Biol. Chem.* **271**: 727–731.

O'Connell, M., Halliwell, B., Moorhouse, C.P., Aruoma, O.I., Baum, H. and Peters, T.J. 1986. Formation of hy-droxyl radicals in the presence of ferritin and haemosiderin. *Biochem. J.* **234**: 727–731.

Orr, W.C. and Sohal, R.S. 1994. Extension of life span by over expression of superoxide dismutase and catalase in drosophila melanogaster. *Science* **263**: 1128–1130.

Osaki, S., Johnson, D.A. and Frieden, E. 1996. The possible significance of the ferrous oxidase activity of ceruloplasmin in normal human serum. *J. Biol. Chem.* **241**: 2746–2751.

Rouault, R.A. and Klausner, R.O. 1996. Iron-sulfur clusters as biosensors of oxidants and iron. *TIBS* **21**: 174–177.

Sarafian, T.A. and Bredesen, D.E. 1994. Is apoptosis mediated by reactive oxygen species? *Free Rad. Res.* **21**: 1–8.

Saran, M. and Bors, W. 1989. Oxygen radicals acting as chemical messengers: A hypothesis. *Free Rad. Res. Commun.* **7**: 213–220.

Schreck, R.A., Albermann, K. and Baeuerle, P.A. 1992. Nuclear factor kappa B: an oxidative stress-responsible transcription factor of eukaryotic cells (a review). *Free Rad. Res. Commun.* **17**: 221–237.

Sies, H. ed. 1991. Oxidative Stress, Oxidants and Antioxidants. Academic Press. New York.

Smith, A. and Morgan, W.T. 1979. Haem transport to the liver by haemopexin: Receptor-mediated uptake with recling of the protein. *Biochem. J.* **182**: 47–54.

Stocker, R. 1990. Induction of haemoxygenase as a defense against oxidative stress. *Free Rad. Res. Commun.* **9**: 101–112.

Sundaresan, M., Yu, Z.X., Ferrans, V.J., Irtani, K. and Fintrel, T. 1995. Requirement for generation of H_2O_2 for PDGF signal transduction. *Science* **270**: 296–299.

Tan, C.M., Xenoyannis, Y. and Feldman, R.D. 1995. Oxidant stress enhances adenylyl cyclase activation. *Cire. Res.* **77**: 710–717.

Treffry, A., Gelvan, D., Konijn, A.M. and Harrison, P.M. 1995. Ferritin does not accumulate iron oxidized by caeruloplasmin. *J. Biochem.* **305**: 21–23.

NON-CHELATION DEPENDENT REDOX ACTIONS OF DESFERRIOXAMINE

Ben-Zhan Zhu, Ronit Har-El, Nahum Kitrossky, and Mordechai Chevion*

Department of Cellular Biochemistry
Hebrew University
Hadassah Schools of Medicine and Dental Medicine
Jerusalem 91120, Israel

1. ABSTRACT

Desferrioxamine (DFO) has been widely used as a specific iron chelator. In this study, the non-chelation dependent redox actions of DFO were demonstrated. The protection by DFO against DNA scission caused by tetrachlorohydroquinone arises from a reaction between tetrachlorosemiquinone radical and DFO which yields a DFO nitroxide free radical. DFO was also found to enhance dechlorination of tetrachloro-1,4-benzoquinone. These results suggest two additional mechanisms through which DFO might have effects on biological oxidation reactions.

2. INTRODUCTION

Desferrioxamine (DFO; Desferal®; deferoxamine) is currently the major iron chelator used for the treatment of iron overload (Mclaren et al., 1983). This includes clinical cases of individuals who have ingested toxic oral doses of iron salts or require multiple blood transfusions, such as in the treatment of thalassaemia (Giardina and Grady, 1995; Gabutti and Piga, 1996). DFO is a linear trihydroxamic acid siderophore that forms a kinetically and thermodynamically stable complex with ferric iron, ferrioxamine (Keberle, 1964). Its high binding constant ($\log\beta = 31$) and its redox properties, render the bound iron unreactive for the catalysis of oxygen radicals production (Keberle, 1964; Gutteridge et al., 1979; Graf et al., 1984), as has been implicated in a variety of biological processes (Eaton, 1996; Marx

* Address correspondence to: Prof. Mordechai Chevion, Department of Cellular Biochemistry, Hebrew University-Hadassah Medical School, P.O. Box 12272, Jerusalem 91120, ISRAEL. Tel: 972-2-6758158; 972-2-6758160. Fax: 972-2-6415848; 972-2-6784010.

Free Radicals, Oxidative Stress, and Antioxidants, edited by Özben.
Plenum Press, New York, 1998.

and Van Asbeck, 1996). It has been classically assessed that prevention of damage by DFO was a sufficient proof for the role of loosely-bound iron in the injurious processes. Although DFO has been repeatedly used to probe metal-catalyzed hydroxyl radical formation in biological systems (Marx and Van Asbeck, 1996; Halliwell, 1989; Hershko et al., 1996; Kontoghiorghes, 1995), recent studies demonstrated the ability of this trihydroxamate compound to act as radicals scavenger, in addition to and independent of its iron-binding properties. These include the interaction of DFO with the hydroxyl radical (Hoe et al., 1982), superoxide radical (Halliwell, 1985; Goldstein and Czapski, 1990; Davies et al., 1987), peroxyl radicals (Davies et al., 1987; Darley-Usmar et al., 1989; Shimoni et al., 1994) and ferryl myoglobin radical (Rice-Evans et al., 1989; Green et al., 1993; Cooper et al., 1994) as well as its action as a chain-breaking antioxidant in peroxidizing erythrocyte membranes (Hartley et al., 1990). DFO was also shown to have prooxidant action for alloxan (Grankvist and Marklund, 1983), paraquat (Borg and Schaich, 1986), ferrous iron (Klebanoff et al., 1989), ascorbate (Mordente et al., 1990), bilirubin (De Matteis et al., 1993) and N-hydroperoxide (Dzwigaj and Pezerat, 1995). DFO was also found to be quite susceptible to oxidation by 2,6-dichlorophenolindophenol (Blobstein, et al., 1978), chelated copper (Van Reyk and Dean, 1996), peroxidases in the presence of H_2O_2 (Kanner and Harel, 1987; Morehouse et al., 1987; Soriani et al., 1993) and peroxynitrite (Denicola et al., 1995). DFO-nitroxide radical (DFO$^\bullet$) was detected in most of these reactions.

In this work we propose two new modes of action of DFO, which are based on its protective effect against DNA damage induced by tetrachlorohydroquinone (TCHQ). TCHQ has previously been identified as the main toxic metabolite of both pentachlorophenol (Van Ommen et al., 1986a; Seiler, 1991) and hexachlorobenzene (Van Ommen et al., 1986b; Koss et al., 1987), which, in turn, are highly ubiquitous as environmental pollutants. TCHQ covalently binds to proteins and DNA (Van Ommen et al., 1986a; Witte et al., 1985), induces DNA single strand breaks in isolated DNA (Van Ommen et al., 1986a; Witte et al., 1985), in human fibroblasts (Witte et al., 1985; Carstens et al., 1990), and in V79 cells (Dahlhaus et al., 1995, 1996). TCHQ was also found to induce the formation of 8-hydroxy-2-deoxy-guanosine in V79 cells (Dahlhaus et al., 1995) and in B6C3F1 mice (Dahlhaus et al., 1994).

3. MATERIALS AND METHODS

3.1. Chemicals

Desferrioxamine B (DFO) was obtained as the mesylate salt (Desferal®) from Ciba-Geigy (Basle, Switzerland). Tetrachlorohydroquinone (TCHQ), tetrachloro-1,4-benzoquinone (TCBQ), ethylenediaminetetraacetic acid (EDTA), diethylenetriaminepentaacetic acid (DTPA), 4-(2-hydroxyethyl)-1-piperazinethanesulfonic acid (HEPES), and nitrilotriacetic acid (NTA) were purchased from Sigma. Tetrachorocatechol, tetrachloro-1,2-benzoquinone, and benzohydroxamic acid (BHA) were purchased from Aldrich. TCHQ stock solutions were prepared in methanol.

3.2. DNA Damage

Circular ssDNA (~2 µg) was exposed to different chemical reactants, using a reaction mixture volume of 25 µl and containing 10 mM HEPES buffer (pH 7.4). The incubation period was 1 hr at 37°C. To stop the reaction 1 mM DFO was added and samples were placed on ice. To each sample, loading buffer (4 µl, 0.25% bromophenol blue, 0.25% xylene cyanol and 30% glycerol) was added, and samples were run on 1% agarose gels in

Tris/Borate/EDTA buffer (pH 8.3) at 100 V for 4–5 h. Gels were then stained with ethidium bromide (0.5 µg/ml, 20 min) and photographed.

3.3. Autooxidation of TCHQ

The autooxidation of TCHQ both in the presence and in the absence of DFO was monitored by a UV-VIS spectrophotometer (KONTRON UVIKON 860), in HEPES buffer (10 mM, pH 7.4) at 37°C. The generation of tetrachlorosemiquinone radical (TCSQ$^{\bullet}$), tetrachloro-1,4-benzoquinone(TCBQ) and 2,5-dichloro-3,6-dihydroxy-1,4-benzoquinone (DDBQ) were followed at 455 nm, 292 nm, and 332 nm, respectively.

3.4. Oxygen Consumption and H_2O_2 Generation

Oxygen consumption was monitored with a Clark-type oxygen electrode (Yellow Springs Instrument Co., Yellow Springs, OH) upon mixing TCHQ with or without DFO in HEPES buffer (10 mM, pH 7.4). H_2O_2 generation was indirectly determined by oxygen production upon adding catalase (500 U/ml) to the reaction mixture containing TCHQ and DFO.

3.5. ESR

Samples (0.05–0.1ml) for ESR measurements were drawn by a syringe into a gas-permeable teflon capillary (Zeus Industries, Rarita, NJ) of 0.032 inch inner diameter, 0.0015 inch wall thickness and 8 cm length. The filled capilaries were inserted into a quartz ESR tube (open at both ends) which was then placed vertically in the cavity dewar. During the experiments, gas of desired composition and temperature was blown through the dewar and around the sample. As the teflon capillary is of small diameter and gas permeable, an equilibrium (temperature and gas composition) of the solution was accomplished very quickly. ESR spectra were recorded by using a JES-FE-3XG (JEOL, Tokyo, Japan) spectrometer with 100-KHz field modulation. Spectra were recorded with a microwave power of 4 mW and a modulation amplitude of 0.1 mT or less. Signal intensity (which is directly proportional to the radical concentration for any given species) was determined by measurement of peak to peak line heights on spectra normalized to identical spectrometer settings.

3.6. HPLC Analysis

HPLC analysis of TCBQ and DDBQ was performed with a Varian 5000 Liquid Chromatograph system. TCBQ was analyzed under isocratic conditions using a C_{18} column with a mobile phase (0.7 mL/min) consisting of 60:40 methanol/water (v:v) plus 0.04% trifluoroacetic acid and UV detection at 254 nm. DDBQ was analyzed by C_{18} isocratic HPLC with detection at 320 nm, using a mobile phase of 30:70 methanol/buffered water (12.5 mM potassium phosphate, pH 6.8) containing 5 mM of ion pairing reagent tetrabutylammonium hydride. Both TCBQ and DDBQ were quantified by external standardization against reagent-grade chemicals.

4. RESULTS AND DISCUSSION

DNA single strand breaks were induced by tetrachlorohydroquinone (TCHQ, 0.5 mM). The trihydroxamate iron chelator desferrioxamine (DFO) protected against the dam-

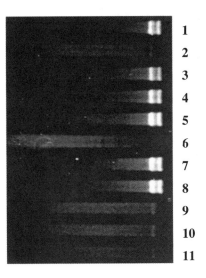

1
2
3
4
5
6
7
8
9
10
11

Figure 1. Protection of TCHQ-induced DNA degradation by DFO and other metal chelators. ~2 μg of single stranded DNA were incubated for 60 min at 37°C in HEPES buffer (10 mM, pH 7.4) with TCHQ (0.5 mM, lane 2) and different concentrations of metal chelators: DFO (lane 3, 0.25 mM; lane 4, 1.0 mM; lane 5, 2.0 mM), benzohydroxamic acid (lane 6, 0.25 mM; lane 7, 1.0 mM; lane 8, 2.0 mM), and DTPA (lane 9, 0.25 mM; lane 10, 1.0 mM; lane 11, 2.0 mM). Untreated ssDNA (lane 1). Degradation reactions were terminated and samples were run on agarose gel, as described in Materials and Methods.

age to DNA while another effective iron chelator diethylenetriaminepentaacetic acid (DTPA) had no effect. The monohydroxamic acid benzohydroxamic acid (BHA) showed a protective effect similar to that of DFO (Fig. 1).

In order to answer the different effects of the two chelators on TCHQ-induced DNA damage, we studied the autooxidation process of TCHQ in the presence of the chelators.

TCHQ has a characteristic UV-absorption spectrum, with an absorbance peak at 326 nm (in HEPES buffer, pH 7.4), which did not change under anaerobic conditions. When followed by ESR, autooxidation of TCHQ generated an ESR signal of tetrachlorosemiquinone radical (TCSQ$^\bullet$, g = 2.0056) which reached maximum intensity at about 10 min (Fig. 2A). The corresponding visible absorption peak of this intermediate is at 455 nm (Naito et al., 1994). Subsequently (for t > 10 min), the absorbance at 455 nm continuously decreased with a concomitant increase at 292 nm, which indicates that autooxidation of TCHQ led to the formation of the oxidation product tetrachloro-1,4-benzoquinone (TCBQ; chloranil), and this was further confirmed by HPLC with pure TCBQ as reference (data not shown). The autooxidation process was also accompanied by the utilization of oxygen, and the generation of hydrogen peroxide, which was substantiated by the catalase-dependent release of oxygen from the reaction mixture (data not shown).

Upon the addition of DFO (50 μM), TCSQ$^\bullet$ was formed and its signal (which was monitored both by ESR and visible spectroscopy) immediately and completely decayed with concurrent formation of TCBQ *and* DFO-nitroxide radical (DFO$^\bullet$), as detected by ESR (Fig. 2B). The DFO$^\bullet$ which contains the structural component $-CH_2-NO^\bullet^-CO-$, gives the 9 line spectrum as a result of splitting of the nitroxide nitrogen coupling (a_N = 0.79 mT) by two protons (a_{2H} = 0.63 mT) from the neighbouring CH_2 group (Morehouse et al., 1987). Interestingly, the absorbance at 292 nm decreased gradually with a corresponding increase in absorbance at 332 nm, indicating that an interaction between TCBQ and DFO might have taken place and that TCBQ was further transformed to an another UV-absorbing substance. The same final product was obtained by reacting pure TCBQ, instead of TCHQ, with DFO. The final product which absorbs at 332 nm was positively identified as 2,5-dichloro-3,6-dihydroxy-1,4-benzoquinone (DDBQ; chloranilic acid) by both UV spectroscopical characteristics (λmax = 332 nm) and its retention time (t_R = 9.05 min) in HPLC, using pure DDBQ as reference.

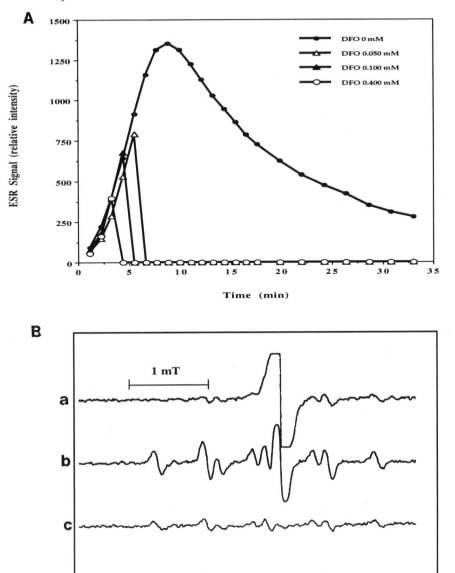

Figure 2. A. Inhibition of TCSQ• formation during TCHQ autooxidation in the presence of DFO: Time course of ESR signal intensity of TCSQ•. Samples contained the indicated concentrations of DFO in HEPES (10 mM, pH 7.4), at 37°C, and the reactions were initiated by the addition of TCHQ (100 μM), followed by a rapid mixing. Each point represents the mean of two separate experiments with the SD less than 5%. **B.** ESR spectra of the TCSQ• and the DFO• (a_N = 0.79 mT, $a_{H(2)}$ = 0.63 mT). The spectra were obtained **a.** 5 min; **b.** 6 min; and **c.** 7 min after the mixing of TCHQ (50 μM) with DFO (200 μM) in HEPES buffer (10 mM, pH 7.4), at 37°C. In **a** and **b**, the nine-line DFO• signal appears superimposed on the large signal of the TCSQ• signal centered at 336.19 mT.

The ESR spectrum of TCSQ• was observed and this semiquinone radical proved relatively stable. Moreover, because the semiquinone radical has no proton in the benzene ring, its ESR spectrum shows a single signal without hyperfine structure (Carsten et al., 1990; Naito et al., 1994). In the presence of DFO, both the maximal concentration *and* the life span of the ESR signal of TCSQ• were markedly reduced (Fig.

2A). For example, when DFO concentration was four times higher than TCHQ (0.1 mM), the maximum intensity of the ESR signal of TCSQ$^{•}$ was less than one third of that of TCHQ without DFO, following complete disapperance of the ESR signal of TCSQ$^{•}$ (Fig. 2A). In contrast, in the absence of DFO, the ESR signal of TCSQ$^{•}$ persisted for more than one hour. These responses were more extensive at higher DFO concentrations, with a greater and faster suppression of the signals corresponding to TCSQ$^{•}$ (Fig. 2A). All these indicate that DFO is a powerful TCSQ$^{•}$ scavenger.

DFO also reduced the consumption of oxygen during TCHQ autooxidation in a concentration-dependent manner. Compared to TCHQ alone, less than one third of oxygen consumption was observed when the molar ratio DFO : TCHQ was 4:1. Under anaerobic conditions, the autooxidation of TCHQ did not take place even in the presence of DFO, suggesting there was no direct reaction between TCHQ and DFO. However, TCBQ, the oxidation product of TCHQ, could react with DFO to produce DFO$^{•}$ with the concomitant disappearance of TCSQ$^{•}$ and formation of DDBQ, even in anaerobic conditions (data not shown). The rates of formation of TCSQ$^{•}$, DFO$^{•}$, TCBQ, DDBQ and the consumption of oxygen, were all dependent on the molar ratios between DFO and TCHQ.

The possibility that the autooxidation of TCHQ was mediated by contaminating transition metals under our experimental conditions was considered and was ruled out. This was based on the following lines of evidence: 1). Special care was taken in 'cleaning' the solutions by Chelex resin and conalbumin A to remove trace amounts of contaminating transition metal ions; 2). The levels of contamination by transition metals in our solutions were determined by Zeeman Atomic Absorption Spectroscopy, and it was found that the contamination of copper is typically below 0.7 nM, and never exceeded 7 nM; likewise, the typical concentration of iron is below 4.7 nM, and never exceeded 47 nM; the higher range of iron contamination comes from DFO, but it is in a non-redox active form; 3). DFO does not stop TCHQ autooxidation, or slows it down, but rather markedly accelerates this reaction. Other transition metal chelators, such as DTPA, EDTA and NTA, or the copper-specific chelator bathocuproinedisulfonic acid, had no effect on the autooxidation of TCHQ; 4). When ferrioxamine (the 1:1 Fe–DFO complex) was added instead of DFO, no acceleration of TCHQ autooxidation was observed. 5). Like other hydroquinones, the autooxidation of TCHQ was a 'self-catalyzed' process, in which the oxidation product, TCBQ, further catalyzes TCHQ autooxidation (data not shown). Thus, we suggest that the iron-free binding sites of DFO, the trihydroxamic acid moieties, are involved in the effects of DFO. This is further substantiated by similar effects observed with other hydroxamic acids, such as rhodotorulic acid (a natural dihydroxamic acid), benzohydroxamic acid and acetohydroxamic acid (both are monohydroxamic acids) (data not shown).

Like TCHQ and TCBQ, tetrachlorocatechol (a minor metabolite of PCP) and its oxidation product tetrachloro-1,2-benzoquinone, showed similar interactions with DFO (data not shown).

In this communication we propose two new modes of action for DFO which are independent of iron-chelation: scavenging of the deleterious semiquinone radical, and stimulation of the hydrolysis (dechlorination) of halogenated substituents on the quinone structure.

The most significant feature of the interaction between DFO and TCHQ is the scavenging of the reactive TCSQ$^{•}$ with the concomitant formation of the much less reactive DFO$^{•}$. The life span of DFO$^{•}$ is rather long (Davies et al., 1987; Morehouse et al., 1987) and no reaction could be observed by ESR between DFO$^{•}$ and H_2O_2 (data not shown). Although there were reports (Davies et al., 1987; Mordent et al., 1990) that DFO$^{•}$ could inactivate some enzymes, the results reported so far provided no evidence to suggest potential deleterious effects on DNA.

Iron does not seem to be involved in TCHQ-induced DNA damage. This was supported by the following points: 1) HEPES buffer was treated carefully prior to the experiment by Chelex resin and conalbumin A to remove trace amounts of metal ions; 2) TCHQ-induced DNA damage was the same in the presence of either DTPA or EDTA as without chelator; and 3) even with the addition of exogenous ferric ions (up to 20 μM), no enhancement of TCHQ-induced DNA damage was observed (Naito et al., 1994).

The DNA damage induced by TCHQ could have been through an organic Fenton reaction which leads to the production of hydroxyl radicals: (TCSQ$^\bullet$ + H$_2$O$_2$ → $^\bullet$OH + TCBQ). Such organic Fenton reactions are thermodynamically feasible and do not require metal ions for catalysis (Koppenol and Butler, 1985; Nohl and Jordan, 1987). It is known that DNA have multiple points of attachment, and that TCHQ could covalently bind to DNA (Van Ommen et al., 1986a; Witte et al., 1985), meeting the requirement of a short distance between the site of $^\bullet$OH generation and the site of its reaction with its target. Oxidative stress might cause an overproduction of TCSQ$^\bullet$. Under these conditions the TCSQ$^\bullet$ species may prevail over reduced transition metal ions in the production of $^\bullet$OH from H$_2$O$_2$. This may well be the case in the present work, where an excess of TCSQ$^\bullet$ is produced and yet iron does not seem to play any major role in the production of $^\bullet$OH radical (unpublished results).

Thus, the protective effects of DFO on TCHQ-induced DNA damage is not due to its binding of iron, but rather to the interaction between DFO and TCHQ, which leads to the scavenging of the reactive TCSQ$^\bullet$, to a lesser oxygen consumption, and thus, to a lesser H$_2$O$_2$ formation, and finally, a lower rate of $^\bullet$OH formation. This is probably the mechanism through which DFO protects against TCHQ-induced DNA damage (Scheme 1). This effect is independent of the iron-chelating property of DFO.

Although there were reports showing that DFO could scavenge different kinds of free radicals, to our knowledge, this is the first report on the ability of DFO to scavenge semiquinone radical. Indeed DFO was found to be a powerful semiquinone radical scavenger in our model system. Semiquinones are intermediates in the oxidation-reduction reactions of quinones and catechols. Quinones are important constituents of the electron-transport chain of both mitochondria and chloroplasts (Nohl et al., 1986), and catechol is an essential part of the backbone structure of L-dopa, dopamine, adrenaline and some other neurochemicals. Nevertheless, it should be kept in mind that DFO could exert toxic effects of DFO in vivo (Freedman et al., 1988; Lee and Wurster, 1995), and caution should also be taken when DFO is used as a "probe" to study the role of iron in the toxicity of other quinones, especially the widely-used antitumor quinones (Powis, 1989). A characteristic feature of the quinone moiety is its ability to undergo reversible redox cycles and form semiquinone radicals and oxygen radicals which were considered as the main cause for their cardiotoxicity and skin toxicity (Powis, 1989). Previous interpretation of the protective effects by DFO are

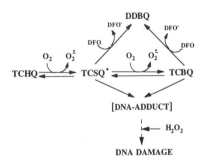

Scheme 1. Proposed mechanism of action of DFO on the protection against DNA damage induced by TCHQ: Non-chelation dependent redox actions of DFO.

often related to the participation or involvement of iron (Hershko et al., 1996; Powis, 1989). This interpretation must be re-examined, particularly in those cases in which DFO was effective while DTPA or other iron chelators were not. We can speculate that DFO exerts its protection against the toxicity of antitumor quinones, at least in part, as a semiquinone radical scavenger, but not as a specific iron chelator.

A recent study showed (Sarr et al., 1995) that TCBQ spontaneously hydrolyzes at physiological pH to DDBQ. TCBQ is a main oxidation product of pentachlorophenol by phenol-oxidizing enzymes in vitro (Samokyszyn et al., 1995; Chung and Aust, 1995) and by fungal metabolism (Ruckdeschel and Renner, 1986; Hammel and Tardone, 1988)). In the past, TCBQ was used as a fungicide and algicide under the trade name Spergon. TCBQ is known to form protein adducts both in vitro and in vivo and has been implicated in the genotoxicity associated with pentachlorophenol (Waidyanatha et al., 1994). While there is much interest in developing technologies to remediate pentachlorophenol-contaminated soil and groundwater, the formation of toxic byproducts such as TCBQ must be considered, if possible, overcome. The final product of TCBQ hydrolysis, DDBQ, appears to be significantly less toxic than the parent compound (Sarr et al., 1995). We show here that DFO significantly stimulates the formation of DDBQ, and thus plays a new role for DFO which was not reported before. The fact that DFO and other hydroxamic acids remarkably enhance TCBQ hydrolysis to form the much less toxic DDBQ, may have implications for waste treatment or remediation processes in which TCBQ is formed from pentachlorophenol.

ABBREVIATIONS

DFO	desferrioxamine (Desferal®; deferoxamine)
DFO•	desferrioxamine-nitroxide radical
TCHQ	tetrachlorohydroquinone
TCSQ•	tetrachlorosemiquinone radical
TCBQ	tetrachloro-1,4-benzoquinone (chloranil)
DDBQ	2,5-dichloro-3,6-dihydroxy-1,4-benzoquinone (chloranilic acid)
DTPA	diethylenetriaminepentaacetic acid
EDTA	ethylenediaminetetraacetic acid
NTA	nitrilotriacetic acid
BHA	benzohydroxamic acid
HEPES	4-(2-hydroxyethyl)-1-piperazinethanesulfonic acid

REFERENCES

Blobstein, S. H., Grady, R. W., Meshnick, S. R.and Cerami, A., 1978, Hydroxamic acid oxidation - pharmacological considerations, *Biochem. Pharmacol.* **27**: 2939–2945.

Borg, D. C.and Schaich, K. M., 1986, Prooxidant action of desferrioxamine: Fenton-like production of hydroxyl radicals by reduced ferrioxamine, *J. Free Rad. Biol. Med.* **2**: 237–243.

Carstens, C. P., Blum, J. K. and Witte, I., 1990, The role of hydroxyl radicals in tetrachloro-hydroquinone induced DNA strand break formation in PM2 DNA and human fibroblasts, *Chem-Biol. Interact.* **74**: 305–314.

Chung, N.and Aust, S. D , 1995, Veratryl alcohol-mediated indirect oxidation of pentachlorophenol by lignin peroxidase, *Arch. Biochem. Biophys.* **322**: 143–148.

Cooper, C. E., Green, E. S. R., Rice-Evans, C. A., Davies, M. J. and Wrigglesworth J. M., 1994, A hydrogen-donating monohydroxamate scavenges ferryl myoglobin radicals, *Free Rad. Res.* **20**: 219–227.

Dahlhaus, M., Almstadt, E.and Appel, K. E., 1994, The pentachlorophenol metabolite tetrachloro-*p*-hydroquinone induces the formation of 8-hydroxy-2-deoxyguanosine in liver DNA of male B6C3F1 mice, *Toxicol. Lett.* **74**: 265–274.

Dahlhaus, M., Almstadt, E., Henschke, P., Luttgert, S. and Appel, K. E., 1995, Induction of 8-hydroxy-2-deoxyguanosine and single strand breaks in DNA of V79 cells by tetrachloro-*p*-hydroquinone, *Mutation Res.* **329**: 29–36.

Dahlhaus, M., Almstadt, E., Henachke, P., Luttgert, S. and Appel, K. E., 1996, Oxidative DNA lesions in V79 cells mediated by pentachlorophenol metabolites, *Arch. Toxicol.* **70**: 457–460.

Darley-Usmar, V. M., Hersey, A. and Garland, L. G., 1989, A method for comparative assessment of antioxidant as peroxyl radical scavengers, *Biochem. Pharmacol.* **38**: 1645–1649.

Davies, M. J., Donkor, R., Dunster, C. A., Gee, C. A., Jonas, S. and Wilson, R. L., 1987, Desferrioxamine and superoxide radicals. Formation of an enzyme-damaging nitroxide, *Biochem. J.* **246**: 725–729.

De Matteis, F., Dawson, S. J. and Gibbs, A. H., 1993, Two pathways of iron-catalyzed oxidation of bilirubin: effect of desferrioxamine and Trolox, and comparison with microsomal oxidation, *Free Rad. Biol. Med.* **15**: 301–309.

Denicola, A., Souza, J. M., Gatti, R. M., Augusto, O. and Radi, R., 1995, Desferrioxamine inhibition of the hydroxyl radical-like reactivity of peroxynitrite: Role of hydroxamic groups, *Free Rad. Biol. Med.* **19**: 11–19.

Dzwigaj, S. and Pezerat, H., 1995, H. Singlet oxygen-trapping reaction as a method of $^{1}O_2$ detection: role of some reducing agents, *Free Rad. Res.* **23**: 103–115.

Eaton, J. W., 1996, Iron: the essential poison, *Redox Report* **2**: 215.

Freedman, N. H., Boyden, M., Talor, M. and Skarf, B., 1988, Neurotoxicity associated with deferoxamine therapy, *Toxicology* **49**: 283–290.

Gabutti, V. and Piga, A., 1996, Results of long-term iron-chelating therapy, *Acta Haematol.* **95**: 26–36.

Giardina, P. J. and Grady, R. W., 1995, Chelation therapy in β-thalassemia: the benefits and limitations of desferrioxamine, *Seminars Hematol.* **32**: 304–312.

Graf, E., Mahoney, J. R., Bryant, R. G.and Eaton, J. W., 1984, Iron-catalyzed hydroxyl radical formation. Stringent requirement for free iron coordination site, *J. Biol. Chem.* **259**: 3620–3624.

Grankvist, K. and Marklund, S. L., 1983, Opposite effects of two metal-chelators on alloxan-induced diabetes in mice, *Life Sci.* **33**: 2535–2540.

Green, E. S. R., Rice-Evans, H., Rice-Evans, P., Davies, M. J., Salah, N. and Rice-Evans, C. A., 1993, The efficacy of monohydroxamates as free radical scavenging agents compared with di- and tri-hydroxamates, *Biochem. Pharmacol.* **45**: 357–366.

Goldstein, S. and Czapski, G., 1990, A reinvestigation of the reaction of desferrioxamine with superoxide radicals. A pulse radiolysis study, *Free Rad. Res. Comm.* **11**: 231–240.

Gutteridge, J. M. C., Richmond, R. and Halliwell, B., 1979, Inhibition of the iron-catalyzed formation of hydroxyl radicals from superoxide and of lipid peroxidation by desferrioxamine, *Biochem. J.* **184**: 469–472.

Halliwell, B., 1985, Use of desferrioxamine as a "probe" for the iron-dependent formation of hydroxyl radicals. Evidence for a direct reaction between desferal and the superoxide radical, *Biochem. Pharmacol.* **34**: 229–233.

Halliwell, B., 1989, Protection against tissue damage in vivo by desferrioxamine: What is its mechanism of action?, *Free Rad. Biol. Med.* **7**: 645–651.

Hammel, K. E. and Tardone, P. J., 1988, The oxidative 4-dechlorination of polychlorinated phenols is catalyzed by extracelluar fungal lignin peroxidases, *Biochemistry* **27**: 6563–6568.

Hartley, A., Davies, M. J. and Rice-Evans, C. A., 1990, Desferrioxamine as a chain breaking antioxidant in sickle cell membrane, *FEBS Lett.* **264**: 145–148.

Hershko, C., Pinson, A. and Link, G., 1996, Prevention of anthracycline cardiotoxicity by iron chelation, *Acta Haematol.* **95**: 87–92.

Hoe, S., Rowley, D. A. and Halliwell, B., 1982, Reactions of ferrioxamine and desferrioxamine with the hydroyl radical, *Chem-Biol. Interact.* **41**: 75–81.

Kanner, J. and Harel, S., 1987, Desferrioxamine as an eletron donor. Inhibition of membrance lipid peroxidation initiated by H_2O_2-activated myoglobin, *Free Rad. Res. Comm.* **3**: 309–317.

Keberle, H., 1964, The biochemistry of desferrioxamine and its relation to iron metabolism, *Ann. N. Y. Acad. Sci.* **119**: 758–768.

Klebanoff, S. J., Waltersdorph, A. M., Michel, B. R. and Rosen, H., 1989, Oxygen-based free radical generation by ferrous iron and deferoxamine, *J. Biol. Chem.* **264**: 19765–19771.

Kontoghiorghes, G. J., 1995, Comparative efficacy and toxicity of desferrioxamine, deferiprone and other iron and aluminium chelating drugs, *Toxicol. Lett.* **80**: 1–18.

Koppenol, W. H. and Butler, J., 1985, Energetics in interconversion reactions of oxyradicals, *Adv. Free Rad. Biol. Med.* **1**: 91–131.

Koss, G., Losekam, M., Seidel, J., Steinbach, K. and Koransky, W., 1987, Inhibitory effect of tetrachloro-*p*-hydro-quinone and other metabolites of hexachlorobenzene on hepatic uroporphyrinogen decarboxylase activity with reference to the role of glutathione, *Ann. N.Y. Acad. Sci.* **514**: 148–159.

Lee, Y. S. and Wurster, R. D., 1995, Deferoxamine-induced cytotoxicity in human neuronal cell lines: Protection by free radical scavengers, *Toxicol. Lett.* **78**: 67–71.

Marx, J. J. M. and Van Asbeck, B. S., 1996, Use of iron chelators in preventing hydroxyl radical damage: adult respiratory distress syndrome as an experimental model for the pathophysiology and treatment of oxygen-radical-mediated tissue damage, *Acta Haematol.* **95**: 49–62.

Mclaren, G. D., Muir, W. A. and Kellermeyer, R. W., 1983, Iron overload disorders: natural history, pathogenesis, diagnosis and therapy, *CRC Crit. Rev. Clin. Lab. Sci.* **19**: 205–265.

Mordente, A., Meucci, E., Miggiano, G. A. D. and Martorana, G. E., 1990, Prooxidant action of desferrioxamine: enhancement of alkaline phosphatase inactivation by interaction with ascorbate system, *Arch. Biochem. Biophys.* **277**: 234–240.

Morehouse, K. M., Flitter, W. D. and Mason, R. P., 1987, The enzymatic oxidation of desferal to a nitroxide free radical, *FEBS Lett.* **222**: 246–250.

Naito, S., Ono, Y., Somiya, I., Inoue, S., Ito, K., Yamamoto, K. and Kawanishi, S., 1994, Role of active oxygen species in DNA damage by pentachlorophenol metabolites, *Mutation Res.* **310**: 79–88.

Nohl, H., Jordan, W. and Youngman R. J., 1986, Quinones in biology: function in electron transfer and oxygen activation, *Adv. Free Rad. Biol. Med.* **2**: 211–279.

Nohl, H. and Jordan, W., 1987, The involement of biological quinones in the formation of hydroxyl radicals via the Haber-Weiss reaction, *Biorg. Chem.* **15**: 374–382.

Powis, G., 1989, Free radical formation by antitumor quinones, *Free Rad. Biol. Med.* **6**: 63–101.

Rice-Evans, C. A., Okunade, G.and Khan, R., 1989, The suppression of iron release from activated myoglobin by physiological electron donors and desferrioxamine, *Free Rad. Res. Comm.* **7**: 45–54.

Ruckdeschel, G. and Renner, G., 1986, Effects of pentachlorophenol and some of its known and possible metabolites on fungi, *Appl. Environ. Microbiol.* **51**: 1370–1372.

Samokyszyn, V. M., Freeman, J. P., Maddipati, K. R. and Lloyd, R. V., 1995, Peroxidase-catalyzed oxidation of pentachlorophenol, *Chem. Res. Toxicol.* **8**: 349–355.

Sarr, D. H., Kazunga, C., Charles, M. J., Pavlovich, J. G. and Aitken, M. D., 1995, Decomposition of tetrachloro-1,4-benzoquinone (*P*-chloranil) in aqueous solution, *Environ. Sci. Technol.* **29**: 2735–2740.

Seiler, J. P., 1991, Pentachlorophenol, *Mutation Res.* **257**: 27–47.

Shimoni, E., Armon, R. and Neeman, I., 1994, Antioxidant properties of deferoxamine, *J. Am. Oil Chem. Soc.* **71**: 641–644.

Soriani, M., Mazzuca, S., Quaresima, V. and Minetti, M., 1993, Oxidation of desferrioxamine to nitroxide free radical by activated human neutrophils, *Free Rad. Biol. Med.* **14**: 589–599.

Van Ommen, B., Adang, A. E. P., Brader, L., Posthumus, M. A., Muller, F. and Van Bladeren, P. J., 1986, The microsomal metabolism of hexachlorobenzene. *Biochem. Pharmacol.* **35**: 3233–3238.

Van Ommen, B., Adang, A., Muller, F. and Van Bladeren, P. J., 1986, The microsomal metabolism of pentachlorophenol and its covalent binding to protein and DNA, *Chem-Biol. Interact.* **60**: 1–11.

Van Reyk, D. M. and Dean, R. T., 1996, The iron-selective chelator desferal can reduce chelated copper, *Free Rad. Res.* **24**: 55–60.

Waidyanatha, S., McDonald, T. A., Lin, P. H. and Rappaport, S. M., 1994, Measurement of hemoglobin and albumin adducts of tetrachrobenzoquinone, *Chem. Res. Toxicol.* **7**: 463–468.

Witte, I., Juhl, U. and Butte, W., 1985, DNA-damaging properties and cytotoxicity in human fibroblasts of tetrachlorohydroquinone, a pentachlorophenol metabolite, *Mutation Res.* **145**: 71–75.

THE EFFECTS OF OXIDATIVE STRESS ON THE REDOX SYSTEM OF THE HUMAN ERYTHROCYTE

İ. Hamdi Öğüş, Mevhibe Balk, Yasemin Aksoy, Meltem Müftüoğlu, and Nazmi Özer

Department of Biochemistry
Faculty of Medicine
Hacettepe University
06100 Sihhiye, Ankara, Turkey

1. INTRODUCTION

Erythrocytes are a group of cells which do not contain a nucleus, mitochondria and other cytoplasmic organelles. Due to the lack of a nucleus and other organelles they can not synthesize proteins and their energy metabolism depends solely on anaerobic glycolysis. The function of these highly differentiated cells is to exchange respiratory gasses between the lung and the tissues. The gas transporting protein, hemoglobin, constitutes 95% of the erythrocyte proteins. Hemoglobin consists of a nonprotein heme group and the protein, globulin. Heme contains iron. For a functional protein this iron must be in the ferrous (Fe^{2+}) state (Mathews and van Holde, 1996; Stryer, 1988).

Erythrocytes are simple cells and exposed to endogenous and exogenous oxidants. Therefore, they have developed protection systems against these oxidants. Erythrocytes perform several important functions for survival. These metabolic functions are;

1. To produce ATP from anaerobic glycolysis for membrane ion pump,
2. To produce NADH for keeping hemoglobin in the reduced state,
3. To produce reducing power, NADPH and NADH, for the neutralization of the effect of endogenous and exogenous oxidants,
4. Erythrocytes have to remove these oxidants using reducing power by the help of enzymes, such as glutathione reductase (GR), glutathione peroxidase (GSH-Px), or by using other systems which do not require reducing power. These systems are catalase (CAT) and superoxide dismutase (SOD). These two enzymes in the absence of reducing power can convert oxidants to less or nontoxic compounds (Figure 1).

Free Radicals, Oxidative Stress, and Antioxidants, edited by Özben.
Plenum Press, New York, 1998.

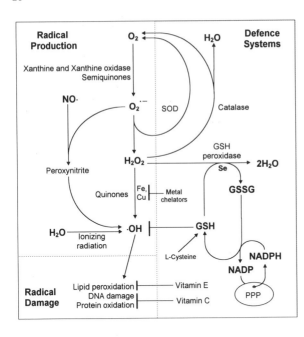

Figure 1. Major reactive oxygen species pathways and antioxidant defence systems. Some of the main pathways in the formation and detoxification of reactive oxygen species. Abbrevations: GSH, reduced glutathione; GSSG, oxidised glutathione; SOD, superoxide dismutase; GR, glutathione reductase. (Modified from M. D. Jacobson).

2. ERYTHROCYTE ENERGY METABOLISM

In erythrocytes ATP and NADH are obtained from anaerobic glycolysis (Embden-Meyerhof metabolic pathway). The ATP is of primary importance for the function of ion pumps in erythrocytes. ATP is also needed for active transport from the plasma membrane. Hence, the concentration of ATP, ADP and AMP is very strictly controlled. The energy status of the cell, the adenylate energy charge or energy charge (E.C.) is calculated using the following formula (Atkinson, 1949).

$$E.C. = ([ATP]+[ADP]/2)/([ATP]+[ADP]+[AMP]) \tag{1}$$

If all adenine nucleotides are in the form of ATP, E.C. is 1.0; whereas if all adenine nucleotides are in the AMP form, E.C. is equal to zero. The energy charge of a cell is buffered between 0.85–0.95.

2,3-Diphosphoglycerate is synthesized and its concentration is regulated by the glycolytic pathway and it is very important in the delivery of O_2 to tissues (Figure 2).

In the glycolytic pathway NAD+ has a very important function as the coenzyme of glyceraldehyde-3-phosphate dehydrogenase. Its concentration is limited so it has to be regenerated for the continuation of glycolysis. This is accomplished by conversion of pyruvate to lactate by lactate dehydrogenase. In erythrocytes NADH has another very important function which is the reduction of oxidized hemoglobin (methemoglobin) to reduced hemoglobin. Everyday about 0.5–3.0% of hemoglobin is converted to methemoglobin. A large fraction of this methemoglobin is reduced by methemoglobin reductase at the expense of NADH, but the remaining methemoglobin is reduced nonenzymatically by ascorbate and glutathione (Wefers, 1988). Some drugs, chemicals or genetic deficiencies may prevent this reduction and cause accumulation of methemoglobin in erythrocytes (Baggot, 1992).

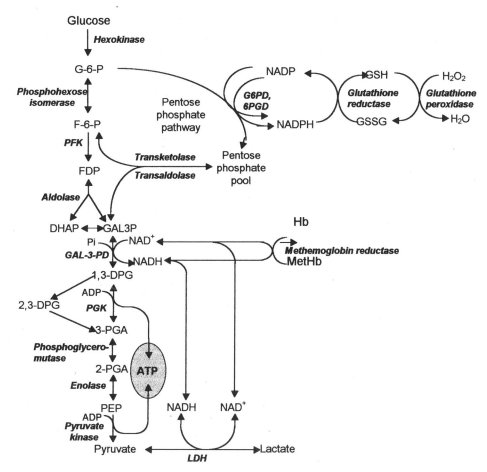

Figure 2. Interaction and summary of the glycolytic and pentose phosphate pathways in the erythrocytes. Abbreviations: PFK, phosphofructokinase; GAL-3-PD, glyceraldehyde 3-phosphate dehydrogenase; G6PD, glucose 6-phosphate dehydrogenase; 6PGD, 6-phosphogluconate dehydrogenase; PGK, phosphoglycerate kinase; LDH, lactate dehydrogenase.

2.1. The Pentose Phosphate Pathway, NADPH Production, and Their Biological Importance

The most important source of reducing power in erythrocytes is the pentose phosphate pathway (PPP). Two enzymes, glucose-6-phosphate dehydrogenase (G6PD) and 6-phosphogluconate dehydrogenase (6PGD) take part in the synthesis of NADPH. G6PD is the control enzyme for PPP. Its deficiency causes a disease called Spherocytic Hemolytic Anemia. In this disease, due to deficiency of G6PD, insufficient NADPH synthesis occurs which makes the erythrocytes more susceptible to oxidants. Due to the oxidation of the membrane lipids the integrity of the plasma membrane is lost and hemolysis occurs (Figure 3) (Orten et al., 1982). The importance of NADPH is not only due to its direct use as reducing power, it is also used for the reduction of oxidized glutathione (GSSG) to reduced glutathione (GSH) by glutathione reductase. In fact, the survival of the erythrocyte is fully dependent on a functional pentose phosphate pathway as a supply for sufficient NADPH and GSH (Yunis and Yasmineh, 1969).

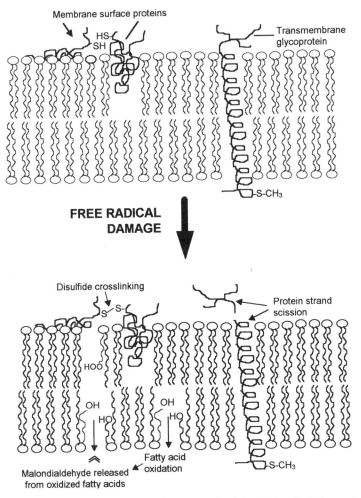

Figure 3. Results of the free radical damage to membranes: protein chain scission, disulfide crosslinking in membrane surface proteins, lipid-lipid crosslinking, lipid peroxidation, and malondialdehyde (MDA) formation.

2.2. Glutathione

Glutathione (γ-glutamylcysteinylglycine, GSH) is a tripeptide. Its oxidized form is a dimer, GSSG. Glutathione metabolism is given in (Figure 4). GSSG is reduced by GR using NADPH. Glutathione has four major functions:

1. Preservation of the thiol groups of proteins in the reduced state,
2. Removal of H_2O_2 catalyzed by GSH-Px,
3. Reduction of some oxidizing compounds, and,
4. Detoxification of xenobiotics by conjugation catalyzed by glutathione S-transferases (GSTs) (Mannervik, 1985).

As noted above one of the major functions of GSH is to preserve the thiol groups of proteins in the reduced state. The protein disulfide reducing function of GSH depends more on the [GSH]/[GSSG] ratio than the size of the glutathione pool in erythrocytes.

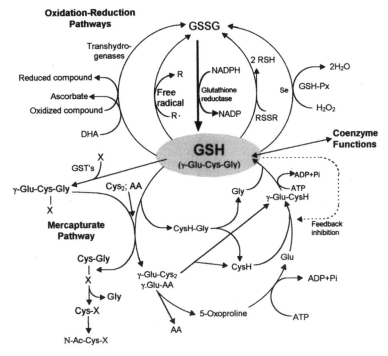

Figure 4. Glutathione: synthesis, and its reactions. Glutathione plays a central role in oxidation-reduction pathways, coenzyme functions and mercapturate pathway. (Adapted from Meister, 1991).

Glutathione, with its 2 mM concentration in erythrocytes, constitutes the largest mobile thiol pool and its reactivity is much higher than the Hb thiol. Due to its relation with GR and GSH-Px, the redox state of glutathione will indirectly reflect the steady state concentrations of oxygen free radicals in erythrocytes (Morell et al., 1964). The erythrocyte redox system extends to membrane proteins and lipids by means of ascorbate and a-tocopherol diacetate respectively.

3. REACTIVE OXYGEN SPECIES

Some of the common reactive oxygen species and some biologically important radicals, their physiological concentrations and half-lives are given in Table 1.

Between molecular oxygen, O_2, and its most stable form, H_2O, there exist three intermediate steps. These intermediate forms, superoxide ($O_2^{\cdot-}$), hydrogen peroxide (H_2O_2) and hydroxyl radical ($^{\cdot}OH$) are obtained by addition of electrons to molecular oxygen one by one (Figure 5). The most reactive form of oxygen radicals is the hydroxyl radical. In addition to these radicals two more reactive oxygen species exist: H_2O_2 and singlet oxygen ($^1O_2$). These are not radicals. Singlet oxygen is spin altered oxygen whereas H_2O_2 can produce oxygen radicals in the presence of metal ions such as iron and copper. These two reactive oxygen species are also potentially toxic.

Radicals can give their unpaired electron to other compounds and may cause chain reactions, polymer breakage and lipid peroxidation. These kinds of reactions in erythrocytes cause the destruction of plasma membrane and hemolysis of the erythrocytes (Figure 3).

Table 1. Some radicals in biological systems and their properties
(Adapted from Sies, 1991, and Pryor, 1986)

Symbol of the radical	Name of the radical	Target molecule[a] and its concentration[b]	Radical concentration[c]	Halftime (s) (at 37°C)
$O_2^{\cdot-}$	Superoxide radical	enzymatic, —	10^{-12}–10^{-11} M	enzymatic
H_2O_2	Hydrogen peroxide	enzymatic, —	10^{-9}–10^{-7}	enzymatic
$^1O_2$	Singlet oxygen	H_2O, —	solvent	10^{-6}
$HO^{\cdot}$	Hydroxyl radical	LH^d, 100 mM	—	10^{-9}
$HOO^{\cdot}$	Hydroperoxide radical (pH 4.8)	—	10^{-15}–10^{-14} M	—
$NO^{\cdot}$	Nitric oxide radical	several biological targets	—	1–10
$ONOO^{\cdot}$	Nitroperoxide radical	—	—	0.05–1
$L^{\cdot d}$	Linolenyl radical	O_2, 2 mM	—	10^{-8}
$RO^{\cdot}$	Alkoxyl radical	LH^d, 100 mM	—	10^{-6}
$ROO^{\cdot-}$	Peroxide radical	LH^d, 1 mM	—	7
$Q^{\cdot-e}$	Quinone radical	enzymatic	—	days

[a]Substrate chosen as representative of typical reactive target molecules for the species in the second column.
[b]Concentration of the substrate that is meant to approximate the sum of all reactive species in the vicinity of the radical and that is chosen to reflect the selectivity of the radical.
[c]Estimated steady state concentration in normal aerobic liver cell.
[d]Linoleate.
[e]The cigarette tar free radical.

4. ERYTHROCYTE ANTIOXIDANT SYSTEMS

The protective systems against cell injury caused by reactive oxygen species or other factors are given on Table 2. These systems, according to their GSH needs, may be classified into two groups.

　　i. GSH independent systems: catalase, superoxide dismutase.
　　ii. GSH dependent systems: NADPH-dependent GR and GSH-Px systems.

It was proposed (Gaetani et al., 1994) that these systems are equally effective in the removal of H_2O_2 in red blood cells, but our results have shown that the erythrocyte catalase activity is so high that its share in the removal of H_2O_2 must be higher: a 10% suspension of erythrocytes destroys 1M H_2O_2 in less than 2 minutes. Both of these systems, catalase and GSH-Px, will easily protect hemoglobin against oxidation.

Glutathione dependent and independent pathways are not completely separate systems because all systems are directly (GSH dependency) or indirectly (through pentose phosphate pathway) coupled. For example, catalase is active if it contains 1 mol NADPH per mole of subunit.

Pyrimidine 5′-nucleotidase, the most sensitive enzyme to oxidation in the erythrocyte, requires a reduced thiol at its active site for activity which is held in the reduced state by the GSH system (Vives et al., 1995).

5. ERYTHROCYTE THIOL REDOX SYSTEM

In adult erythrocytes about 80–85% of the reactive thiols belong to hemoglobin, 10–15% to glutathione, and the remaining 5% to membrane proteins and other thiol compounds (Morell et al., 1964).

No disulfide bond formation is observed between Hb molecules (Mahler and Cordes, 1966) but mixed disulfide formation is common between Hb, glutathione and with

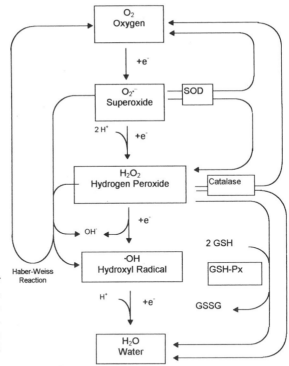

Figure 5. Synthesis and detoxification of the reactive oxygen species. Formation of the reactive oxygen species by consecutive electron addition and their detoxification by glutathione dependent (GSH-Px, glutathione peroxidase) and independent (catalase and SOD) enzymes.

other proteins (Gaetani et al., 1994). So, the thiol groups in the thiol redox system constitute 10–20% of the thiol pool. In erythrocytes, the detailed composition of thiol compounds, their physiological concentrations and their redox potentials are not known. Therefore it is assumed that the GSH/GSSG redox couple forms the basic redox system in erythrocytes.

Due to the high partial oxygen pressure, erythrocytes in circulation face oxidative stress and everyday about 0.5–3.0% of hemoglobin is converted to methemoglobin and autooxidation of hemoglobin will cause the production of superoxide radicals. As a result

Table 2. Cell antioxidant defence systems

Metabolic pathways	Antioxidant defence system
Glycolysis	Reducing equivalent as in form of ATP and NADH
Pentose phosphate pathway	Reducing equivalents as in form of NADPH
Enzymes	
Methemoglobin reductase	$MetHb + NADH^+ + H^+ \rightarrow Hb + NAD^+$
Superoxide dismutase	$2\ O_2^{\cdot-} + 2H^+ \rightarrow H_2O_2 + O_2$
Catalase	$2\ H_2O_2 \rightarrow 2\ H_2O + O_2$
GSH peroxidase (GSH-P)	$H_2O_2 + 2\ GSH \rightarrow GSSG + 2\ H_2O$
GSSG reductase	$GSSG + NADPH \rightarrow 2\ GSH + NADP^+$
GST transferase	$RX + GSH \rightarrow XH + GS\text{-}R$
Small molecular weight compounds	
Ascorbic acid	Reducing agent in water phase
NADH	Reducing agent in water phase
NADPH	Reducing agent in water phase
Vitamin E	Reducing agent in lipid phase

the unsaturated phospholipids found in erythrocytes are oxidized. This will result in the disintegration of the plasma membrane and hemolysis. For protection of erythrocyte against oxidative damage of radicals, several antioxidant systems are developed. To understand the redox status of the erythrocytes it is very important to analyze all the systems which contribute to the thiol metabolism in erythrocytes. In this study, H_2O_2 has been used as the source of oxidative stress, and ascorbate and α-tocopherol diacetate as the externally added common reducing agents. The concentration of GSH, GSSG, protein and total thiol and the activities of the enzymes responsible for the regeneration of reducing power (NADPH, GSH) and the activities of the enzymes responsible for the removal of radicals were measured in the same erythrocyte suspension.

6. MATERIALS AND METHODS

6.1. Preparation of Erythrocytes

Blood samples obtained from healthy volunteers were centrifuged and erythrocytes were washed three times with physiological saline solution. The erythrocytes obtained were incubated in acid-citrate-dextrose solution with or without ascorbate and α-tocopherol acetate and different concentrations of H_2O_2. After incubation for one hour, the erythrocytes were washed twice with physiological saline following centrifugation. The erythrocyte package was hemolyzed and GSH, GSSG, total thiol and protein thiol concentrations; GR, G6PD, CAT, SOD, GSH-Px, GST and pyrimidine 5'-nucleotidase activities were determined in the hemolyzate.

6.2. Determination of Total -SH, Protein-SH, GSH, and GSSG

To prevent the oxidation of GSH erythrocytes were hemolyzed in water containing N-ethylmaleimide (NEM). Excess NEM was removed by ether extraction and the remaining ether in the aqueous phase was evaporated by passing nitrogen gas through the samples (Tietze, 1969). The amount of GSSG in the samples was determined enzymatically using NADPH, DTNB and glutathione reductase (Srivastava and Beutler, 1968). Total -SH was determined by direct titration of samples with DTNB. The concentration of GSH was calculated by subtracting the GSSG value from total -SH. Protein-SH concentration was determined in proteins, using the DTNB reaction, after precipitation with 5% TCA (Ellman, 1959).

6.3. Determination of Enzyme Activities

All enzyme activities were measured, at 37°C, using the same buffer systems at the same concentrations and pH, given in the literature.

The GR assay mixture contained GSSG, NADPH and a suitable aliquot of hemolyzate. The reaction was initiated by the addition of hemolyzate and was monitored by the decrease in absorbance at 340 nm (Tietze,1969).

G6PD activity was measured, using an incubation mixture containing G6P, NADP+, by following the increase in absorbance at 340 nm (Beutler, 1971).

GSH-Px activity was assayed in a coupled system containing GSH, H_2O_2, NADPH and sufficient amount of GR and the absorbance decrease in 340 nm, due to the oxidation of NADPH, was followed (Beutler, 1971).

GST activity was measured by following the increase in absorbance at 340 nm, due to the conjugation of GSH with 1-chloro-2,4-dinitrobenzene (Mannervik, 1985).

Catalase activity was assayed using H_2O_2 as substrate and the activity was followed by measuring the decrease in absorbance at 240 nm (Aebi, 1984).

Superoxide dismutase activity determination was based on nitrobluetetrazolium (NBT) reduction by xanthine oxidase-derived superoxide radicals (Sun et al., 1988).

Pyrimidine 5'-nucleotidase activity was determined using CMP (or UMP) as substrate (Campbell, 1962). Activity was followed by measuring the time-dependent release of phosphate liberated from pyrimidine 5'-phosphate (Ames, 1966) .

7. RESULTS AND DISCUSSION

Erythrocyte levels of GSH and GSSG were found to be 7.87 ± 0.28 and 0.012 ± 0.001 μmole/g hemoglobin, respectively (Table 3). The ratio, [GSH]/[GSSG], was 647.3 ± 46.4. H_2O_2 concentrations lower than 2% did not have any effect on the glutathione pool, whereas 4% H_2O_2, which is equivalent to ca. 1.2 M, caused oxidative stress on erythrocytes. The concentration of GSH and the ratio of [GSH]/[GSSG] decreased to 17.6% and 54.8%, respectively. An increase of 100% in GSSG concentration was observed.

H_2O_2 caused a decrease in GSH concentration (18%), an increase of 100% in GSSG concentration. Only 1% of the decrease in GSH concentration caused by oxidant stress was reflected as a corresponding increase in intracellular GSSG concentration. Probably, the remaining portion of the GSH pool change is converted to non-GSSG disulfides, and because of maintenance of GSH/GSSG ratio by GSSG export, the thiol redox potential of cell is not influenced by oxidant stress (Sies and Akerboom, 1984).

H_2O_2 caused a decrease in total- and protein-SH (11%), and in nonprotein-SH (15%) pools. Both ascorbic acid and α-tocopherol were partially effective in counteracting the effect of H_2O_2 on total- and nonprotein-SH, but were 100% effective on reversing protein thiols (Table 3). As noted above, this might be due to the leakage and/or active export of GSSG from erythrocytes (Sies and Akerboom, 1984). On the other hand, the large molecular weight thiols, proteins, stay in the erythrocytes. Hence, the protein thiol pool was not affected (only 4%) (Table 3).

Under the same conditions, erythrocyte enzyme activities of glutathione reductase, glutathione peroxidase, glucose 6-phosphate dehydrogenase, glutathione S-transferase, catalase, superoxide dismutase, and pyrimidine 5'-nucleotidase activities were decreased 43%, 24%, 24%, 23%, 23%, 13%, and 79%, respectively. Ascorbate and α-tocopherol were equally effective in the protection of pyrimidine 5'-nucleotidase and superoxide dismutase activities against the oxidant stress created by H_2O_2. They retained 100% activity

Table 3. Thiol status of human red blood cells under oxidative stress

Parameter	Control	H_2O_2	ASC + H_2O_2	TOC + H_2O_2
GSH, mmol/g Hb	7.87 ± 0.28	6.49 ± 0.67	7.32 ± 0.35	7.26 ± 0.39
GSSG, mmol/g Hb	0.012 ± 0.001	0.024 ± 0.008	0.019 ± 0.006	0.017 ± 0.005
GSH/GSSG	647.3 ± 46.4	292.5 ± 84.2	424.6 ± 107.8	464.0 ± 113.1
NP-SH, mmol/g Hb	8.67 ± 0.37	7.35 ± 0.54	8.04 ± 0.51	7.89 ± 0.65
PR-SH, mmol/g Hb	25.96 ± 2.05	23.31 ± 2.06	24.93 ± 1.65	25.01 ± 2.17
TOTAL -SH, mmol/mg protein	34.67 ± 2.33	30.71 ± 2.29	32.5 ± 1.60	32.76 ± 2.39

ASC, Ascorbate; TOC, α-tocopherol acetate; NP-SH, non-protein-SH; PR-SH, Protein-SH.

Table 4. Enzyme activity status of human red blood cells under oxidative stress

Enzyme	Control	H_2O_2	ASC + H_2O_2	TOC + H_2O_2
GR, U/g Hb	8.48 ± 0.98	4.88 ± 1.04	7.38 ± 1.36	7.37 ± 1.14
GSH-Px, U/g Hb	42.1 + 10.55	32.19 ± 9.06	38.6 ± 10.25	37.45 ± 9.66
GST, U/g Hb	8.65 ± 1.34	5.81 ± 1.63	7.56 ± 1.42	7.47 ± 1.73
G6PD, U/g Hb	10.66 ± 1.04	8.12 ± 0.48	9.00 ± 0.36	9.05 ± 0.56
Catalase, k/g Hb	292.6 ± 38.00	225.9 ± 38.6	231.9 ± 45.4	268.8 ± 42.9
SOD, U/g Hb	3077 ± 165.6	2683 ± 330.4	2940 ± 162.0	2937 ± 188.4

ASC, Ascorbate; TOC, alpha-tocopherol acetate; GR, glutathione reductase; GSH-PX, glutathione peroxidase, GST, glutathione S-transferase, G6PD, glucose-6-phosphate dehydrogenase; SOD, superoxide dismutase.

in the presence of ascorbate and α-tocopherol. The other enzymes were also protected to varying degrees (85–95%) by these antioxidants (Table 4).

So, the injury created by H_2O_2 treatment could be prevented to an important extent by preincubation of erythrocytes with ascorbic acid and α-tocopherol diacetate.

In order to study the oxidative stress effect of H_2O_2 as such on the activities of the enzymes and on the glutathione pool, catalase has to be inhibited. Therefore we also studied the inhibition of catalase by known inhibitors: Azide, acetate, formate, and aminotriazole (Aebi and Suter, 1969).

Erythrocytes were stable to both H_2O_2 and azide, at milimolar concentrations, when they were added separately. In contrast, erythrocytes were destroyed in the coexistence of both azide and H_2O_2, even at micromolar concentrations. The destructive action of the azide/hydrogen peroxide pair might be due to the formation of ˙N_3 and/or ˙OH radical. Azide probably dissociates iron from heme, and reaction of iron with H_2O_2 results in the formation of ˙OH radical, which in turn causes the oxidative destruction of the plasma membrane.

The binding of azide to catalase was biphasic with K_i values of 0.845 ± 0.055 and 3.17 ± 0.31 µM respectively. The azide inhibition of catalase was reversible and this reversibility was dependent on incubation time. It was also found that hemolyzate contained an endogenous, small molecular weight, inhibitor of catalase. Removal of this inhibitor caused 3–6 fold increase in the activity of human red blood cell catalase.

Aminotriazole (AT) which has been described as a potent inhibitor of catalase was effective at quite high concentrations and the type of the inhibition was reversible. After incubation of red blood cells with 20 mM AT for 70 minutes and after hemolysis, the activity of catalase was measured (after 2000 fold dilution which reduced AT concentration to 10 µM) no difference was observed between the catalase activities of AT treated and untreated RBC's. In hemolysates, 0.2 mM AT caused 22% inhibition of catalase activity. Increasing inhibitor concentration ten times caused an additional increase in inhibition of only 14% (total 36%). The other reported inhibitors acetate and formate caused 5 to 10% inhibition at 10 mM concentration.

As noted above, the protective effect of ascorbate against oxidative stress is very important. The erythrocyte thiol pool by means of ascorbate and cysteamine is in contact with the thiol compounds found in plasma. In peroxidative reactions using transition state elements, ascorbic acid funtions as prooxidant (Haase and Dunkley, 1969). In these reactions ascorbic acid initiates the oxidative reactions by reducing the metal ions. The autooxidation of ascorbate and production of H_2O_2 is increased especially by copper and iron. If the metal ion is part of a biological system then site-specific Fenton reaction may occur (Samuni et al., 1983). It was shown that other antioxidant compounds such as bilirubin, α-tocopherol (Stocker and Frei, 1991), and GSH (Liebler et al., 1986) may also function as prooxidants in the presence of transition state elements. In vitro lipid peroxidation studies have shown that the prooxidant or oxidant behavior of α-tocopherol is determined by its status in the lipid membrane (Wefers and Sies, 1988; Stocker and Frei, 1991; Liebler et al., 1986). The concentration of lipid peroxides in healthy persons is less than 1–5 nM. It was shown that, in the absence of the detectable transition state elements, ascorbic acid will function only as antioxidant (Stocker and Frei, 1991).

As evident from Figure 6, the regulation of the thiol status of erythrocytes is a very complex process and cellular glutathione plays a central role in the battle against oxidants. GSH is used for the reduction of hydrogen peroxide, ascorbate, membrane and cytoplasmic proteins (Hb), cystamine, and dehydroascorbate.

Hydroxyl radical, formed from hydrogen peroxide and superoxide by the catalytic action of iron, will oxidize and form radicals of plasma membrane lipids and tocopherol. These radicals and hydroxyl radical will be removed by ascorbate. Ascorbate is taken into erythrocytes as dehydroascorbate where it is converted into ascorbate by GSH. Ascorbate has very important role as reducing agent for erythrocytes (Stocker and Frei, 1991).

Iron is released either directly from oxidized hemoglobin, methemoglobin and/or the methemoglobin product, hemichromes (Mahler and Cordes, 1966). Hemichromes may also directly act on the plasma membrane of erythrocytes. Ascorbate will also keep toco-

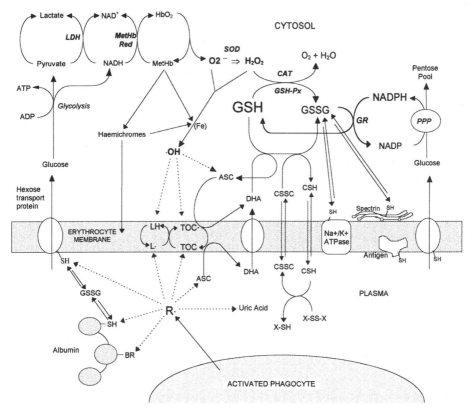

Figure 6. Central role of the glutathione in intracellular and extracellular redox status of the erythrocytes. Abbreviations: DHA, dehydroascorbate; ASC, ascorbate; CSH, cysteamine; CSSC, cystamine; L', lipid radical; LH, lipid; TOC', α-tocopherol radical; TOC, α-tocopherol; PPP, pentose posphate pathway; BR, bilirubin; X-SH, any free or membrane bound thiol; X-SS-X, oxidized form of any free or membrane bound thiol. (Adapted from Hunt and Stocker, 1990).

pherol in the reduced state. This is very important for the removal of lipid peroxyradicals formed in the plasma membrane by the action of intra- or extracellular radicals.

Methemoglobin will be reduced to hemoglobin by methemoglobin reductase at the expense of NADH obtained from anaerobic glycolysis. On the other hand, oxidized glutathione, GSSG, will be reduced by glutathione reductase at the expense of NADPH obtained from the pentose phosphate pathway. GSSG may oxidize intracellular proteins, such as hemoglobin or membrane internal surface (e.g. spectrin, Na^+/K^+ ATPase), external surface (e.g. glucose transport protein) and plasma (e.g. albumin) proteins. About 60% of the thiol groups found on the outside of the erythrocyte membrane belong to the glucose transport system. All, internal, cytoplasmic and membrane, proteins are reduced by GSH. In contrast, the proteins found on the external surface of the plasma membrane and in plasma, require another reducing agent, cysteamine, for reduction, since GSH cannot pass the erythrocyte membrane. The reducing power of GSH is mainly transfered to the extracellular matrix by the transmembrane cystamine/cysteamine shuttle (Figure 6). This transferred reducing power is used to reduce oxidized proteins and other thiols into corresponding reduced forms. Oxidized cysteamine, cystamine, is taken up by the erythrocytes and reduced by the GSH/GSSG redox pair (Reglinski et al., 1988). It was also shown that decreased concentration of intracellular GSH results in extracellular disulfide formation in proteins (Kosower et al., 1982).

Our knowledge about the dynamics of the intracellular glutathione pool; reduced glutathione, oxidized glutathione, other small molecular weight thiols, glutathione mixed disulfide with proteins and their interconversions are limited. Moreover, cells, under oxidative stress, will export oxidized glutathione. The reducing power of GSH is transferred to other thiols and compounds. It is almost impossible to obtain the actual thiol profile of the cell. One can only make some suggestions using the experimentally measured values.

Due to the interaction of intracellular compartments with plasma and ultimately with other cells, the actual thiol pool must involve almost the whole body. This might well be an advantage and a chance for the survival of the organism.

REFERENCES

Aebi, H., Suter, H. 1969, Catalase, in: *Biochemical Methods in Red Blood Cells* (Yunis, J.J., ed.), p.258., Academic Press, New York.

Aebi, H. 1984, Catalase in vitro, in : *Meth. in Enzymol.* (Colowick, S. and Kaplan, C., eds), Volume **105**, pp. 121–126, Academic Press Inc.

Ames, B.N. 1966, Assay of inorganic total phosphate and phosphatases, in: *Methods in Enzymol.* (Colowick, S., Kaplan, N., eds.), Volume: **VIII**, p.115, Academic Press, New York.

Atkinson, D.E. 1949, The energy charge of adenylate pool as a regulatory parameter: Interactions with feedback modifiers., *Biochemistry*, 7:4030–4034.

Baggot, J. 1992, Gas transport and pH regulation, in: *Textbook of Biochemistry with Clinical Correlations.* (Devlin, T.M., ed.), 3rd Edition, p.1031, Wiley-Liss, New York.

Beutler, E. 1971, *Red Cell Metabolism, a Manual of Biochemical Methods*, pp. 42–68, Grune & Stratton, New York.

Campbell, D.M. 1962, Determination of 5′-nucleotidase in blood serum, *Biochem. J.*, **84**:34P.

Ellman, G.L. 1959, Tissue Sulfhydryl Groups, *Arch. Biochem. Biophys.*, **82**:70–77.

Gaetani, G.F., Kirkman, H.N., Mangerini, R., Ferraris, A.M. 1994, Importance of catalase in the disposal of hydrogen peroxide within human erythrocytes, *Blood*, **84(1)**:325–330.

Haase, G., Dunkley, W.L. 1969, Ascorbic acid and copper in linoleate oxidation. II. Ascorbic acid and copper as oxidation catalysts, *J. Lipid Res.*, **10**:561–567.

Hunt, N.H., Stocker, R. 1990, Oxidative stress and the redox status of malaria-infected erythrocytes, *Blood Cells*, **16**:499–526.

Jacobson, M.D 1996, Reactive oxygen species and programmed cell death, *TIBS*, **21**:83–86.

Kosower, N.S., Zipser, Y., Fatlin, Z. 1982, Membrane thiol-disulfide status in glucose-6-phosphate dehydrogenase deficient red cells. Relationship to cellular glutathione, *Biochim. Biophys. Acta*, **691**:345–352.

Liebler, D.C., Kling, D.S., Reed, D.J. 1986, Antioxidant protection of phospholipid bilayers by alpha-tocoperol control of alpha-tocopherol status and lipid peroxidation by ascorbic acid and glutathione, *J. Biol. Chem.*, **261**:12114–12119.

Mahler, H.R., Cordes, E.H. 1966, *Biological Chemistry*, 2nd Ed., pp. 117, 665–666, Harper & Row, Publishers, Inc., New York.

Mannervik, B., Guthenberg, C. 1981, Glutathione transferase: Human Placenta, in: *Meth. Enzymol.* (Colowick, S. and Kaplan, C., Eds),Volume **77**, pp. 231–233.

Mannervik, B. 1985, The isoenzymes of glutathione transferase, in: *Advances in Enzymology and Related Areas of Molecular Biology* (Meister, A., Ed.), Volume 57, pp. 357–417.

Mathews, C.K., van Holde, K.E. 1996, *Biochemistry*, 2nd ed., p.216, The Benjamin/Cummings Publish. Co., Inc., New York.

Meister, A. 1991, Glutathione deficiency produced by inhibition of its synthesis, and its reversal; Applications in research and therapy., *Pharm. Ther.*, **51**:155–194.

Morell, S.A., Ayers, V.E., Greenwalt, T.J., Hoffman, P. 1964, Thiols of the erythrocyte:1. Reaction of N-ethylmaleinide with intact erythrocytes, *J. Biol. Chem.*, **239(8)**:2696–2705.

Orten, J.M., Neuhaus, O.W. 1982, *Human Biochemistry*, 10th ed., p. 457, The C.V. Mosby Co., St. Louis.

Pryor, W.A. 1986, Oxy-radicals and related species: Their formation, life-times, and reactions, *Ann. Rev. Physiol.*, **48**:657–667.

Reglinski, J., Hoey, S., Smith, W.E., Sturrock, R.D. 1988, Cellular response to oxidative stress at sulfhydryl group receptor sites on the erythrocyte membrane, *J. Biol. Chem.*, **263**:12360–12366.

Samuni, A., Aronovitch, J., Godinger, D., Chevion, M., Czapski, G. 1983, On the cytotoxicity of vitamin C and metal ions. A site-specific Fenton mechanism, *Eur. J. Biochem.*, **137**:119–124.

Sies, H. 1991, Oxidative stress: Introduction, in: *Oxidative Stress: Oxidants and Antioxidants.* (Sies, H., ed.), p. xvii, Academic Press Ltd., London.

Sies, H., Akerboom, T.P.M. 1984, Glutathione disulfide (GSSG) efflux from cells and tissues, in: *Methods Enzymol.* (Colowick, S., Kaplan, N., eds.),Volume 105, pp. 445–451, Academic Press, New York.

Srivastava, S.K., Beutler, E. 1968, Accurate measurement of oxidized glutathione content of human, rabbit, and rat red blood cells and tissues, *Anal.Biochem.*, **25**:70–76.

Stocker, R., Frei, B. 1991, Endogenous antioxidant defences in human blood plasma, in: *Oxidative Stress: Oxidants and Antioxidants* (Sies, H., Ed.), p. 235, Academic Press Ltd., London.

Stryer, L. 1988, *Biochemistry*, 3rd ed., p.170, W.H.Freeman &Co., New York.

Sun, Y., Larry, W., Oberley, L., Li, Y. 1988, A simple method for clinical assay of superoxide dismutase, *Clin. Chem.*, **34(3)**:497–500. .

Tietze, F. 1969, Enzymic method for quantitative determination of nanogram amounts of total and oxidized glutathione: Applications to mammalian blood and other tissues., *Anal. Biochem.*, 27:502–522.

Vives, C.J.L., Miguel, G.A., Pujades, M.A., Miguel, S.A., Cambiasso, S., Linares, M., Dibarrart, M.T., Calvo, M.A. 1995, Increased susceptibility of microcytic red blood cells to in vitro oxidative stress., *Eur. J. Haematol.*, **55(5)**:327–331.

Wefers, H., Sies, H. 1988, The protection by ascorbate and glutathione against microsomal lipid peroxidation is dependent on vitamin E, *Eur. J. Biochem.*, **174(2)**:353–357.

Yunis, J.J., Yasmineh, W. 1969, Glucose metabolism in human erythrocytes, in: *Biochemical Methods in Red Cell Genetics* (Yunis, J.J., ed.), p. 8., Academic Press, New York, New York 10003.

4

FREE RADICALS AS REAGENTS FOR ELECTRON TRANSFER PROCESSES IN PROTEINS

I. Pecht[1] and O. Farver[2]

[1]Department of Immunology
The Weizmann Institute of Science
Rehovot 76100, Israel
[2]Department of Chemistry
Royal Danish School of Pharmacy
Copenhagen, Denmark

1. INTRODUCTION

Free radicals produced from amino acids and polypeptide side chains were originally investigated in order to resolve the molecular basis of light or radiation induced damage. However, as discussed below they have also turned out to be effective tools for studying electron transfer processes within polypeptide matrices and yielded important insights into the mechanism of these reactions. Production and study of free radicals in proteins by pulse radiolysis was first introduced in the sixties. A major method developed for the purpose of investigating radiation chemistry, found a broad range of important applications in different fields of chemistry and biochemistry reaching far beyond the subject to which it was first applied (Adams, Fielden and Michael, 1975). The method is based on the excitation and ionization of solvent molecules by short pulses of high energy electrons. Introducing radiation (e.g. 5–10 MeV) into dilute aqueous solutions of a given solute causes primary changes in the solvent. Thus, water molecules undergo conversion into OH radicals, hydrated electrons and to a lesser extent H atoms, H_2 and H_2O_2 are also produced (the yields are usually presented as G values, i.e. number of chemical species produced per 100 eV of absorbed energy: $e_{aq}^- = 2.9$; OH = 2.8; H = 0.55; $H_2O_2 = 0.75$ $H_2 = 0.45$). The hydrated electrons and OH radicals present thermodynamic extremes of reducing and oxidizing potentials, respectively. Hence, they provide the possibility to initiate a wide range of electron transfer processes. However, their extreme reactivity is leading to non selective reactions. Hence, they are usually converted to less reactive ones, by proto-

Free Radicals, Oxidative Stress, and Antioxidants, edited by Özben.
Plenum Press, New York, 1998.

cols devised by radiation chemists. This is illustrated by one useful reactions sequence, converting the e_{aq}^- (with a reduction potential of -2.8 V) to the considerably milder reductant, the CO_2^- radical ($E° = -1.8$ V): First, the e_{aq}^- is converted, in N_2O saturated solutions, into an additional equivalent of OH radicals by the following reaction;

$$e_{aq}^- + N_2O \xrightarrow{\quad H_2O \quad} N_2 + OH + OH^- \tag{1}$$

The two equivalents of OH radicals then react with formate ions, also added to the solution, to produce two equivalents of the CO_2^- radical:

$$HCO_2^- + OH \rightarrow H_2O + CO_2^- \tag{2}$$

By analogy, other reducing and oxidizing agents can be produced and employed (Buxton and Sellers, 1973). A technically important advantage of pulse radiolysis is that the whole range of optical spectrum is usually available for monitoring the reactions of interest. This is the case since the reactive species is derived from the solvent rather than by the excitation of a given solute as is the case for flash-photolysis. Taken together, the wide range of chemical reactivity of the reagents produced combined with a time resolution that usually extends from nanoseconds to minutes and the convenience of spectrophotometrically monitoring of the reactions, made pulse radiolysis a method of choice for investigation a wide range of reactions notably of free radicals.

The potential of pulse radiolysis for studying biological redox processes, particularly of macromolecules, has been recognized rather early. It was initially employed for investigating radiation induced radicals in proteins and nucleic acids and later on, also as an effective tool for resolving electron transfer processes to and within proteins. Cytochrome c, being a well characterized electron mediating protein was the first to be examined by this method for its reactivity with hydrated electrons. Two groups (Land and Swallow, 1971; Faraggi and Pecht, 1971a) initiated these studies and found the reduction of Fe^{3+} site to be a diffusion controlled process (5×10^{10} M^{-1} s^{-1}). Several groups (e.g. Wilting, Braams, Nauta and van Buuren, 1972; Lichtin, Shafferman and Stein, 1973; Pecht and Faraggi, 1972) then extended investigation of this protein. Soon thereafter, other electron mediating proteins with different redox centers, like copper ions or non-heme iron (Faraggi and Pecht, 1971b; Pecht and Faraggi, 1971) were also examined for their reactivity with e_{aq}^- yielding similar, rather high, diffusion limited rate constants. These results illustrated the applicability of the method, yet, at the same time indicated drawbacks of using the hydrated electron as a reductant having excessive driving force and hence limited specificity, leading to additional side reactions. Hence, future studies increasingly employed milder reducing or oxidizing agents than the e_{aq}^- or OH radicals respectively (Klapper and Faraggi, 1979).

Two main interests were guiding investigators who applied pulse radiolysis to redox proteins: The pursuit of the reaction mechanisms of these proteins as well as the more general problem of resolving the parameters which determine rate constants of electron transfer within proteins (Klapper and Faraggi, 1979; Pecht and Goldberg, 1975; Faraggi and Klapper, 1990). Obviously, these two motives do overlap and complement each other as illustrated by cases described below. The fast progress attained in the last two decades in resolving three dimensional structures of an increasing number of redox-active proteins provided the insights essential for a meaningful analysis and interpretation of the kinetic results.

2. PROTEINS' INTRINSIC DISULFIDE RADICALS SERVE AS DONORS IN INTRAMOLECULAR ELECTRON TRANSFER IN PROTEINS; AZURINS AS MODELS

The family of "blue" single copper proteins called azurins mediate electrons in the energy conversion systems of several bacteria (Adman, 1991; Clarke *et al.*, 1991). Though azurins isolated from different bacteria exhibit great homology in their sequences, differences do exist, hence providing variation in reactivity and redox potentials (Adman, 1991). All azurins sequenced to date contain a conserved disulfide bridge (Cys3–Cys26) at one end of their barrel-shaped structure, 2.5 nm from the copper binding site, present at the opposite end of the protein (Fig. l) (Adman, 1991; Clarke *et al.*, 1991). Using pulse radiolytically produced CO_2^- radicals, this disulfide is reduced to the $RSSR^-$ radical. This transient species was found to decay by an intramolecular electron transfer process to the copper(II) center (Farver and Pecht, 1989; Farver and Pecht, 1994). This process has been investigated in detail in both wild type (wt) and in single-site mutated azurins (Farver and Pecht, 1989, 1994, 1992; Farver *et al.*, 1992, 1993). The former were isolated from different bacteria, exhibited a range of differences in their structure and hence in other properties (Farver and Pecht, 1989, 1994). Results of their internal ET were therefore less amenable to a rigorous structure-reactivity correlation than the latter. Thus, most recent studies employed single-site mutants of *Pseudomonas aeruginosa* (Pae) azurin where changes in specific features of the protein were modulated.

As described above, pulse radiolytic reductions are performed in N_2O saturated aqueous solutions containing formate ions. Under these conditions, the primary products of water decomposition by the radiation pulse are practically all converted into CO_2^- radical ions (Klapper and Faraggi, 1979) which react at diffusion controlled rates with the two redox-active sites present in azurin: The Cu(II) ion and the disulphide bridge (Faraggi and Pecht, 1971b; Farver and Pecht, 1989). This is illustrated by absorption changes observed

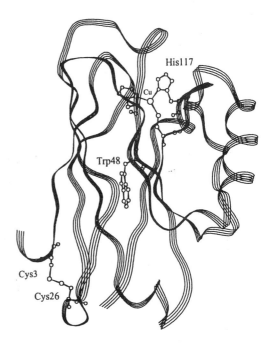

Figure 1. Three-dimensional structure of the polypeptide backbone of *Pseudomonas aeruginosa* azurin with some amino acid residues of particular interest included. Coordinates were obtained from Adman, 1991; Clarke *et al.*, 1991.

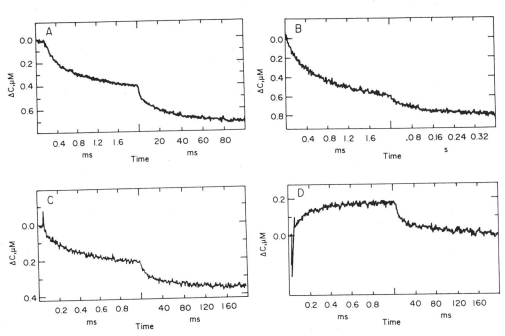

Figure 2. Time resolved absorption changes observed upon reaction with CO_2^- radicals at 625 nm in solutions of *Pseudomonas aeruginosa* (A), *Alcaligenes faecalis* (B) and *Pseudomonas fluorescence* (C) azurins. Changes at 410 nm in *Pseudomonas fluorescence* are shown in (D). T=285 K; pH 7.0; Pulse width 0.5 ms.

in solutions of oxidized, single-site mutant of Pae azurin following the pulse radiolytically produced CO_2^- radicals (Fig. 2): When the reaction is monitored at the main absorption band of the Cu(II) (ε_{625} = 5,700 M^{-1}cm^{-1}), an initial, relatively fast phase of decrease in absorbancy is observed. This was found to be a second order process corresponding to a direct, bimolecular reduction of the Cu(II) center by CO_2^- radicals. A subsequent slower phase of Cu(II) reduction is illustrated by Fig. 2a. At 410 nm (Fig. 2b) a fast increase in absorption is observed which is due to the disulfide reduction and formation of RSSR$^-$ radical ion, having an intense absorption band at this wavelength ($\varepsilon_{410} \cong 10,000$ M^{-1}cm^{-1}) (Adams *et al.*, 1975; Buxton and Sellers, 1973; Klapper and Faraggi, 1979). The formation process of this radical is also a bimolecular process with a rate constant similar to that of the fast Cu(II) reduction phase, both these are due to the parallel bimolecular reduction by CO_2^- radicals of the Cu(II) site, and of the partially exposed single disulphide bridge (Farver and Pecht, 1989, 1994). The 410 nm absorption was found to decay in a first order process with a rate constant identical to that of the slow reduction of the 625 nm band. Both slower absorption changes are therefore due to one process; the intramolecular electron transfer from the disulfide radical anion to the Cu(II) center.

$$\begin{array}{c} RSSR^--Az-Cu(II) \\ \nearrow \qquad \downarrow \qquad\qquad + CO_2 \\ RSSR-Az-Cu(II) + CO_2^- \qquad\qquad\qquad\qquad \\ \searrow \qquad \\ RSSR-Az-Cu(I) \end{array} \qquad (3)$$

Kinetic and thermodynamic parameters were determined for this internal ET process in different wt and single-site mutated Pae. azurins and the results are summarized in

Table 1. Kinetic and thermodynamic data for the intramolecular reduction of Cu(II) by RSSR$^-$ in azurins

Azurin	k_{298} (s^{-1})	E' (mV)	$-\Delta G°$ (kJ mol^{-1})	$DH^{\neq}$ (kJ mol^{-1})	$DS^{\neq}$ (J K^{-1}mol^{-1})
Wild type					
Ps. aer.	44 ± 7	304	68.9	47.5 ± 2.2	−56.5 ± 3.5
Ps. fluor.	22 ± 3	347	73.0	36.3 ± 1.2	−97.7 ± 5.0
Alc. spp.	28 ± 1.5	260	64.6	16.7 ± 1.5	−17.1 ± 1.8
Alc. faec.	11 ± 2	266	65.2	54.5 ± 1.4	−43.9 ± 9.5
Mutant					
D23A	15 ± 3	311	69.6	47.8 ± 1.4	−61.4 ± 6.3
F110S	38 ± 10	314	69.9	55.5 ± 5.0	−28.7 ± 4.5
F114A	72 ± 14	358	74.1	52.1 ± 1.3	−36.1 ± 8.2
H35Q	53 ± 11	268	65.4	37.3 ± 1.3	−86.5 ± 5.8
I7S	42 ± 8	301	68.6	56.6 ± 4.1	−21.5 ± 4.2
M44K	134 ± 12	370	75.3	47.2 ± 0.7	−46.4 ± 4.4
M64E	55 ± 8	278	66.4	46.3 ± 6.2	−56.2 ± 7.2
M121L	38 ± 7	412	79.3	45.2 ± 1.3	−61.5 ± 7.2
V31W	285 ± 18	301	68.6	47.2 ± 2.4	−39.7 ± 2.5
W48A	35 ± 7	301	68.6	46.3 ± 5.9	−58.3 ± 6.0
W48F	80 ± 5	304	68.9	43.7 ± 6.7	−61.9 ± 9.7
W48S	50 ± 5	314	69.9	49.8 ± 4.9	−44.0 ± 3.5
W48Y	85 ± 5	323	70.7	52.6 ± 6.9	−30.2 ± 3.6
W48L	40 ± 4	323	70.7	48.3 ± 0.9	−51.5 ± 5.7
W48M	33 ± 5	312	69.7	48.4 ± 1.3	−50.9 ± 7.4

Table 1. For example mutants which have redox potentials close to that of the wild type (wt) (E° = 304 mV) but differ in residues proximal to the copper site were investigated. In one of these, the Met64 residue of the wt protein is substituted by Glu ([M64E]-azurin) with a Cu(II)/Cu(I) redox potential of 278 mV (Farver *et al.*, 1992). In the two other mutants, Ser has been introduced instead of either Ile7 ([I7S]azurin) or Phe110 ([F110S]azurin) with redox potentials at pH 7.0 of 301 and 314 mV, respectively (Farver and Pecht, 1992; Farver *et al.*, 1993). Results showed little, if any, variation in rate constants (cf. Table 1). In order to try and examine potential structural features involved in the ET the algorithm developed for identifying long-range ET (LRET) pathways (Beratan *et al.*, 1991; Onouchic *et al.*, 1992) has been applied to the different azurins studied. For those azurins for which no three dimensional structure is available, putative structures were calculated by energy minimization. In line with results of direct structure determination of single-site mutants, these calculations suggested that the mutations had a rather limited influence on the protein's overall structure, except for the immediate loci of the substitution (Baker, 1988; Nar *et al.*, 1991a; van Pouderoyen *et al.*, 1994; Gilardi *et al.*, 1994). Results of the pathways analysis suggest that similar hypothetic LRET routes may be operative in all: Two main pathways were identified and are illustrated in Fig. 3. One pathway (Fig. 3, left side) proceeds through the polypeptide backbone from Cys3 to Asn10 then, via the hydrogen bond from the carbonyl of Asn10 to amide of the His46, the imidazole of which is a copper ligand. Another pathway (Fig. 3, right side) proceeds directly from Cys3 via a hydrogen bond to Thr30 and then through the polypeptide further from Val31 to Trp48, by a 0.40 nm through-space jump. Then again via the polypeptide to Val49 and by another H-bond to Phe111, followed by a backbone linkage to the copper ligand Cys112.

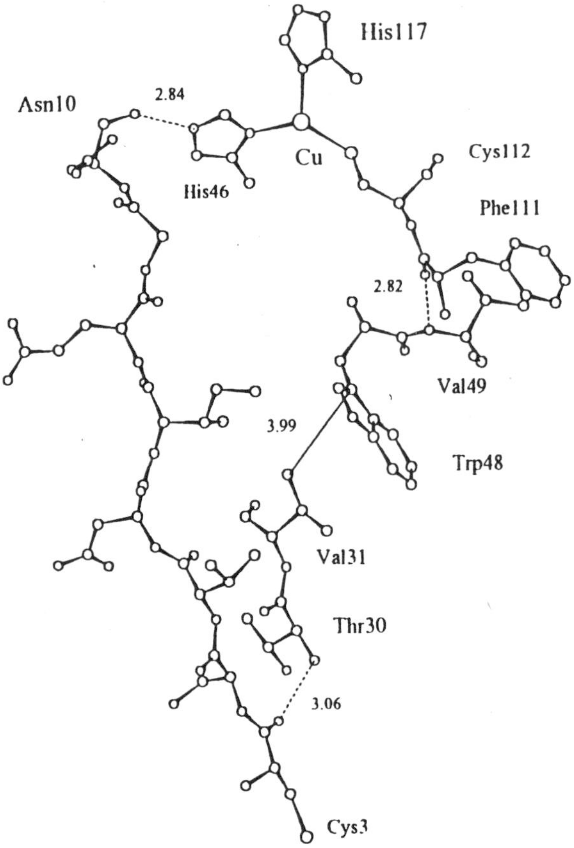

Figure 3. Calculated ET pathways from the disulfide to the copper center in *Pae.* azurin, using the methodology of Beratan and Onuchic (Beratan *et al.*, 1991). Hydrogen bonds are shown by broken lines and the through space jump by a thin line. Some distances (in Å) are also indicated in the figure.

Two features of the intramolecular LRET in azurins deserve attention: The first is that rather small structural changes take place in the copper coordination site during its shuttling between divalent and monovalent oxidation states (Baker, 1988), this should yield a rather low Frank–Condon barrier for electron transfer. The second being that the Cys3–Cys26 disulfide bridge has only a structural role and is most probably not serving a physiological redox function. Hence, to the medium, separating the electron donor (RSSR⁻) and acceptor (Cu(II)), did not undergo evolutionary selection for optimal performance of this ET process. Azurins provide, therefore, an interesting model system where structure-activity relationship for LRET within a protein matrix can be examined in considerable detail (Farver and Pecht, 1994; Marcus and Sutin, 1985). The distance between the RSSR⁻ and the Cu(II) is expected to be the same for all azurins studied so far. In contrast, the driving force for the intramolecular LRET and the properties of the medium separating the reaction partners can be varied.

The semi-classical Marcus equation can be applied to electron transfer between spatially fixed and oriented redox centers (Marcus and Sutin, 1985). Values obtained by employing this theoretical framework are in good agreement for both wild type and single-site mutants of Pae. azurin (Farver and Pecht, 1989) thus supporting the applicability of this analysis of the LRET process. The mechanism of intramolecular electron transfer through matrices of biological macromolecules, mainly proteins but also nucleic acids has attracted considerable interest (Onouchic *et al.*, 1992). Moreover, the question whether electron transfer proceeds through-space or through-bond pathways is still an issue of major discussion (Marcus and Sutin, 1985; Jortner and Bixon, 1993; Moser *et al.*, 1992; Farid *et al.*, 1993). From the refined crystallographic structure of the wt azurins (Baker, 1988; Nar *et al.*, 1991a) we calculate for the shortest edge-to-edge distance, $(r-r_0) = 2.46$ nm between the copper ligating thiolate of Cys112 and the sulfur of Cys26. Using this value leads to $\beta = 10 \pm 0.5$ nm^{-1}. Moser et al. calculated from their analysis of a large number of biological electron transfer systems based on a through space model a β value of 14 nm^{-1} (Moser et al., 1992). This difference between the two values seems too large to be accounted for by experimental uncertainties. Moreover, results of our studies of LRET in different wt azurins (Farver and Pecht, 1989, 1992), yield a calculated maximal (i.e. for $\Delta G° = -\lambda$) LRET rate constant of 300 s^{-1} while using the correlation line of Moser et al. (1992) with a β of 14 nm^{-1} this rate constant should be two orders of magnitude smaller (Farver and Pecht 1992; Moser *et al.*, 1992; Farid *et al.*, 1993). This discrepancy strongly suggests that it is the through-bond LRET model which applies to the examined process in azurins where the donor and acceptor are probably coupled by superexchange via the bridging orbitals (Farver and Pecht, 1994, 1992; Farver *et al.*, 1992, 1993).

Considerable experimental evidence has been accumulated for electron transfer over long distances through saturated bonds where electronic interactions decrease exponentially with the distance. In the highly interconnected β-sheet structure of azurin the redox centers will probably couple strongly with the protein. The LRET pathway calculations for azurin indicate that the two most likely routes are those discussed above. A more detailed examination of the two pathways is instructive: A comparison between covalent tunneling lengths of the above two pathways (or, equivalently, their two electronic coupling factors) in the different azurins suggests that the "Trp48 pathway" is some ten-fold less effective. However, electronic interactions between the Cu(II) ion and its ligands were not taken into account in this pathway analysis. Theoretical studies have recently proposed that a high degree of anisotropic covalency exists in the blue single-copper protein, plastocyanin which would promote ET through the thiolate ligand (Christensen *et al.*, 1990; Lowery *et al.*, 1993). By similar arguments, it can be estimated from the Y_{HOMO} ligand coefficients obtained by Larsson *et al.* (1995) for azurin that ET through the thiolate would be enhanced ~150 fold more than that via one of the His ligands. This means that the "Trp48" and "His 46" pathways would have about equal contributions to the process. It is also noteworthy that the pathway analysis treats the van der Waals contact between Val31 and Trp48 only as a single point interaction between $C_{\gamma 1}$ of valine and $C_{\delta 2}$ of tryptophan. The refined structure coordinates of azurin (Nar *et al.*, 1991b) show however that at least six of the indole ring atoms are in close contact (≤ 0.43 nm) with the valine side chain. Finally the "Trp pathway" involves an aromatic residue raising the possible involvement of a conjugated π-orbital system in the electronic coupling. So far, the pathway calculation, however, only takes σ-bonds into account.

The possible involvement of aromatic residues in ET was examined in single-site mutated azurins where Trp48 has been substituted by other amino acids. The results are shown in Table 1 together with the standard free energies of reaction ($\Delta G°$) and activation parameters, and demonstrate that substitution of Trp48 by other amino acids has only a negligible effect on the kinetic parameters when corrected for changes in driving force. More recently, however, an additional mutant was prepared where Val31 was substituted by Trp. This "double-Trp" azurin (V31W) (Farver *et al.*, 1997) has the two indole residues placed in neighbouring positions. 2D-NMR measurements of this mutant demonstrated that the region located between Trp 48 and the copper center in this protein maintains the same structure as its wild-type protein equivalent. Moreover, energy minimization calculations have also been performed and show a close (van der Waals contact) of the two indole rings, consistent with observed NOEs between protons of the two indoles (Farver *et al.*, 1997). The LRET between the disulfide and the Cu(II) ion was found to take place, like in all other azurins studied so far, also in the V31W azurin mutant and was found to proceed with a rate constant of 285 s^{-1} at 298 K and pH 7.0, i.e. several fold faster than any other azurin studied so far. Thus, in addition to suggesting a role for aromatic residues in the ET process, it is also in line with being the "Trp48" pathway the dominant ET route, since the alternative one, through His46, would not be affected by this mutation.

Analysis of the LRET process in a range of different azurins are thus in very good agreement with a through bond tunneling model. Furthermore, the above results illustrate the usefulness of employing free radicals to the study of electron transfer processes in single site mutated proteins. They enabled resolving the role of driving force and reorganization energies in determining the ET rates. Moreover, these results clearly demonstrate the applicability of the general Marcus theory (Marcus and Sutin, 1985) for assessing these parameters and that the electron transfer process can be rationalized in terms of a through-bond model.

3. INTRAMOLECULAR ELECTRON TRANSFER AS PART OF ENZYMATIC REACTION CYCLE

Studies of intramolecular ET in redox enzymes may provide insights into both, these enzymes' respective reaction mechanisms and of electron transfer within protein polypeptide matrices that were most probably selected by evolution. Several such processes were studied employing pulse radiolysis and free radicals as reagents (Pecht and Faraggi, 1971; Klapper and Faraggi, 1979; Faraggi and Pecht, 1973; Zaitzeva *et al.*, 1996; Anderson *et al.*, 1986). One of the very early attempts to resolve the electron uptake mechanism by the blue copper oxidase, ceruloplasmin, showed that the diffusion controlled decay process of the e_{aq}^- in this protein's solutions, is paralleled by the formation of transient optical absorptions due to electron adducts of protein residues, primarily of cystine disulfide bonds (Faraggi and Pecht, 1973). The monomolecular decay of the RSSR$^-$ absorption was then found to proceed with the same rate constant as that at which this enzyme's type 1 Cu(II) absorption band was reduced. These results were assigned to the high reactivity of the e_{aq}^-, the exposure of the disulfide bonds and to the relatively inaccessible type 1 Cu(II) site leading to an indirect, intramolecular ET pathway from exposed surface residues (Faraggi and Pecht, 1973). The three dimensional structure of human ceruloplasmin has recently been determined by X-ray crystallography and the relatively exposed disulfide bridges were clearly resolved confirming the above interpretation (Zaitzeva *et al.*, 1996).

Another effective application of pulse radiolysis to the study of the mechanism of action of a multicentered redox enzyme is that of xanthine oxidase (Anderson *et al.*, 1986; Hille and Anderson, 1991). This enzyme has been extensively investigated by different kinetic methods, including stopped flow and flash photolysis. Use of pulse radiolytically produced radicals (produced by reduction of N-methylnicotinamide or 5-deazalumiflavin), enabled monitoring directly several phases of the internal electron flow, starting from the enzyme's molybdenum center, which is the preferred site of electron uptake, to the iron-sulfur center and finally ending at the flavin site which is where dioxygen is reduced (Anderson *et al.*, 1986). Intramolecular ET rates of up to 8.5×10^3 s^{-1} were observed and examination of these rates in terms of the three dimensional structure of this enzyme would be of considerable interest.

The three dimensional structure of two blue copper enzymes, ascorbate oxidase (AO) (Kroneck *et al.*, 1982; Messerschmidt *et al.*, 1989, 1992, 1993) and nitrite reductase (Godden *et al.*, 1991; Kukimoto *et al.*, 1994; Dodd *et al.*, 1997) resolved marked similarity in the covalent linkage between the electrons uptake site the type 1 Cu(II) to that their substrate reduction site. This prompted several labs to examine the internal electron flow using free radicals as reductants.

Blue copper oxidases catalyze the specific one-electron oxidation of substrates (e.g., L-ascorbate) by O_2 which is reduced to water (Kroneck *et al.*, 1982). Their minimal catalytic unit consists of four copper ions bound to distinct sites in the protein, designated type 1, 2 and 3 respectively (T1,2,3). Early pulse radiolysis studies of the blue oxidases, laccase and ceruloplasmin have mainly focused on the kinetics of T1[Cu(II)] reduction and contributed to the notion that this site acts as the electron uptake port from substrates (Faraggi and Pecht, 1971b; Pecht and Faraggi, 1971; Faraggi and Pecht, 1973).

AO was shown to exist as a dimer of identical 70 kDa subunits, each containing a catalytic unit of one T1, one T2 and one T3 copper sites. The latter two were found to be proximal forming a trinuclear center where dioxygen is reduced to water. The reduction potentials of the T1 and T3 Cu(II)/Cu(I) couples are practically identical, 350 mV at pH 7.0 (Kroneck *et al.*, 1982). The catalytic cycle of this enzyme is suggested to proceed by a sequential mechanism where single electrons are transferred from the reducing substrates to the T1[Cu(II)]. ET from the latter then takes place to the trinuclear copper center, which serves as the dioxygen binding and reducing site. Intramolecular ET from T1[Cu(I)] to T3[Cu(II)] therefore seems to be the essential step required for the four electrons (and four protons) necessary for O_2 reduction to two water molecules.

Three groups have independently studied the intramolecular ET from T1[Cu(I)] to T3[Cu(II)] in AO by photochemical and pulse-radiolysis methods and have reported similar rate constants for this process (Meyer *et al.*, 1991; Farver and Pecht 1992b; Kyritsis *et al.*, 1993). Photochemically produced lumiflavin semiquinone was shown to reduce T1[Cu(II)] in a fast second order process that was followed by a partial reoxidation of the T1[Cu(I)] site in a first order reaction with a rate constant of 160 s^{-1} (Meyer *et al.*, 1991). The flavin absorption in the near UV region prevented monitoring changes at 330 nm where the T3[Cu(II)] absorbs. Still, the authors interpreted the latter process as being due to intramolecular ET between the T1 and T3 sites. Pulse radiolysis study of AO enabled the independent direct monitoring of the process as observed at both 610 nm (T1) and 330 nm (T3) bands (Farver and Pecht, 1992b). The CO_2^- radicals were used as primary electron donors and reduced the type 1 Cu(II) site in a bimolecular, diffusion controlled process. The ensuing processes of T1[Cu(I)] reoxidation and T3[Cu(II)] reduction occurred with identical rate constants which were concentration independent, confirming their assignment as intramolecular ET from the T1 to T3 sites. However, two to three distinct phases were observed in this intramolecular ET process having rate constants of 200 s^{-1} for the fastest phase and 2 s^{-1} for the slowest. Similar ET rates were later on determined in another pulse radiolysis study using different organic radicals as reductants, yet monitoring the 610 nm chromophore only (Kyritsis *et al.*, 1993).

One of the T1-copper ligands in the three-dimensional model of AO is the Cys507 thiolate while imidazoles of the two neighboring His506 and His508 coordinate to the T3 copper ions. This led to the proposal that the shortest ET pathway from T1[Cu(I)] to T3[Cu(II)] takes place via Cys507 and His506 or His508 (Messerschmidt *et al.*, 1989, 1992, 1993). Both pathways consist of nine covalent bonds yielding a total distance of 1.34 nm. Performing the pathway calculations as developed by Beratan and Onuchic

(Beratan *et al.*, 1991) supports this notion and suggests a further path via a H-bond between the carbonyl oxygen of Cys507 and His506.

The dissimilatory, copper containing nitrite reductases (NiRs) are trimers of 109 kDa subunits, each containing two copper binding sites. One is a type 1 and the other a type 2 site. Three dimensional structures have been determined for Cu NiRs isolated from three different bacteria and found to be very similar, showing a ~ 0.65 Å r.m.s. deviation of their Cα positions (Godden *et al.*, 1991; Kukimoto *et al.*, 1994; Dodd *et al.*, 1997). The T1 and T2 copper sites were found to be 12.5 Å apart and connected by a covalent linkage which is rather similar to that linking the T1 and T3 sites in AO (Messerschmidt *et al.*, 1989, 1992, 1993). This similarity of covalent bonding between the ET centers and the simpler nature of the acceptor site, i.e. T2 in NiR called for investigating the internal ET between them. Pulse radiolytically produced $NMNA^-$ or CO_2^- radicals were found to reduce the T1 Cu(II) sites only. However this was followed by its reoxidation in a first order process assigned to an internal ET to the T2 Cu(II) site. The specific rates for this process were found to be in the range of 450 to 1900 s^{-1} (Suzuki *et al.*, 1997; Farver *et al.*, submitted) for the different NiRs examined. More recently we have also determined the activation parameters of the internal ET in the NiR isolated from *Alcaligenes xylosoxidans* and found them very similar to those of the faster phase of ET in AO (Farver and Pecht, 1992b; Farver *et al.*, submitted).

The above results of the internal ET in the different blue copper proteins illustrate the broad range of applications of pulse radiolitically produced free radicals for such studies. Future applications to engineered proteins and to other model systems would certainly extend our understanding of the parameters controlling ET in protein matrices as well as of mechanistic aspects of redox processes in biological macromolecules.

ACKNOWLEDGMENTS

Support of the work done by O.F. and I.P. by a grant from the German–Israeli Foundation (I-0320-211.05/93) is gratefully acknowledged. Part of the work was supported by the Danish Natural Science Research Foundation (The Bioinorganic Program). The co-ordinates of the azurin structures were kindly provided by Drs. Robert Huber and Herbert Nar.

REFERENCES

Adams, G.E., Fielden, E.M., and Michael, B.D., eds.1975, *Fast proceses in Radiation Chemistry and Biology*; J., Wiley, N.Y.

Adman, E.T.,1991, *Adv. Protein Chem.*, **42**:195–197.

Anderson, R.F., Hille, R., and Massey, V.,1986, *J. Biol. Chem.* **261**:15870–15876.

Baker, E.N., 1988, *J. Mol. Biol.* **203**:1071–1095.

Beratan, D.N., Betts, J.N., and Onuchic, J.N., 1991, *Science*, **252**:1285–1288.

Buxton, G.V., and Sellers, R.M., 1973, *J. Chem. Soc. Farraday Trans.* **69**:555–559.

Christensen, H.E.M., Conrad, L.S., Mikkelsen, K.V., Nielsen, M.K., and Ulstrup, J. 1990, *Inorg. Chem.* **29**:2808–2816

Clarke, M.J. et al.1991, Eds. *Struct. Bonding* **75**:1.

Dodd, F.R., Hasnain, S.S., Abraham, Z.H.L., Eady, R.E., and Smith, B.E.,1997, *Acta Cryst.* (in press).

Faraggi, M., and Klapper, M.H., 1990, in: *Excess Electrons in Dielectric Media*; (Ferradini, C., and Jay-Gerin, J.-P., eds.), pp. 397–423, CRC Press.

Faraggi, M., and Pecht, I., 1971b, *Biochem. Biophys. Res. Comm.*, **45**:842–848.

Faraggi, M., and Pecht, I., 1973, *J, Biol. Chem.* **248**:3146–3149.

Farid, R.S., Moser, C.C., and Dutton, P.L., *Curr. Opinion Struc. Biol.*, 1993, **3**:225.

Farver, O. and Pecht, I., *Biophys. Chem.*, 1994, **50**:203–216.

Farver, O., and Pecht, I., *J. Am. Chem. Soc.* 1992, **114**:5764–5767.

Farver, O., and Pecht, I., *Proc. Natl. Acad. Sci. U.S.A.,* 1992, **89**:8283–8287.

Farver, O., and Pecht, I., *Proc. Natl. Acad. Sci. USA*, 1989, **86**:6968–6972.

Farver, O., Skov, L.K., Young, S., Bonander, N., Karlsson, B.G., Vänngård, T., and Pecht, I., 1997, *J. Am. Chem. Soc.* (in press).

Farver, O., Eady, R.R., Abraham, Z.L.H. and Pecht, I., 1997, Submitted.

Farver, O., Skov, L.K., Pascher, T., Karlsson, B.G., Nordling, M., Lundberg, L.G., Vaangard, T., and Pecht, I., *Biochemistry* 1993, **32**:7317–7322.

Farver, O., Skov, L.K., Van de Kamp, M., Canters, G.W., and Pecht, I., 1992, *Eur. J. Biochem.*, **210**:399–403.

Gilardi, G., Mei, G., Rosato, N., Canters, G.W., and Finazzi-Agro, A., 1994, *Biochemistry* **33**:1425–1431.

Godden, J.W., Turley, S., Teller, D.C., Adman, E.T., Liu, M.Y.,; Payne, W.J., and LeGall, J.,1991, *Science* **253**:438–442.

Hille, R., and Anderson, R.F.,1991, *J. Biol. Chem.*, **266**:5608–5615.

Jortner, J., and Bixon, M., 1993, *Mol. Cryst. Liquid. Cryst.* **234**:29–41.

Klapper, M.H., and Faraggi, M., 1979, *Quart. Rev. Biophys.* **12**:465–519.

Kroneck, P.M.H., Arnstrong, F.A., Merkle, H., and Marchesini, A.,1982, *Adv. Chem. Ser.* **220**:223–48.

Kukimoto, M., Nishiyama, M., Murphy, M.E.P., Turley, S., Adman, E.T., Horinouchi, S., and Beppu, T.,1994, *Biochemistry* **33**:5246–5252.

Kyritsis, P., Messerschmidt, A., Huber, R., Salmon, G.A., and Sykes, A.G., 1993, *J. Chem. Soc. Dalton Trans.* 731–735.

Land, E.J., and Swallow, A.J., 1971, *Arch. Biochem. Biophys.*, **145**:365–372.

Larsson, S., Broo, A., and Sjölin, L., 1995, *J. Phys. Chem.* **99**:4860–4865

Lichtin, N.N., Shafferman, A., and Stein, G., 1973, *Science* **179**:680–683.

Lowery, M.D., Guckert, J.A., Gebhard, M.S., and Solomon, E.I., 1993, *J. Am. Chem. Soc.* **115**:3012–3013

Marcus, R.A., and Sutin, N., 1985, *Biochim. Biophys. Acta* 811:265–322.

Messerschmidt, A., Ladenstein, R., Huber, R., Bolognesi, M., Avigliano, L., Petruzelli, R., Rossi, A., and Finazzi-Agro, A., 1992, *J. Mol. Biol.* **224**:179–205.

Messerschmidt, A., Luecke, H., and Huber, R.,1993, *J. Mol. Biol.* **230**: 997–1014.

Messerschmidt, A., Rossi, A., Ladenstein, R., Huber, R., Bolognesi, M., Gatti, G., Marchesini, A., Petruzelli, R., and Finazzi-Agro, A., 1989, *J. Mol. Biol.* **206**:513–29.

Meyer, T.E., Marchesini, A., Cusanovich, M.A., and Tollin, G., 1991, *Biochemistry* **30**:4619–23.

Moser, C.C., Keske, J.M., Warncke, K., Farid, R.S., and Dutton, P.L., 1992, *Nature* **355**:796–802.

Nar, H., Messerschmidt, A., Huber, R., van de Kamp, M., and Canters, G.W., 1991b *J. Mol. Biol.* **221**:765–772.

Nar, H., Messerschmidt, A., Huber, R., van de Kamp, M., and Canters, G.W., 1991a, *J. Mol. Biol.* **218**:427–447.

Onuchic, J.N., Beratan, D.N., Winkler, J.R., and Gray, H.B., 1992, *Annu. Rev. Biophys. Biomol. Struct.* **21**:349–377.

Pecht, I., and Faraggi, M.,1971, *Nature* **233**:116–118.

Pecht, I., and Faraggi, M., 1971a, *FEBS Lett.* **13**, 221-

Pecht, I., and Faraggi, M., 1972, *Proc. Natl. Acad. Sci. USA* **69**:902–906.

Pecht, I., and Goldberg, M., 1975, in: *Fast Processes in Radiation Chemistry and Biology* (Michael, B.D., ed.), pp. 274–284, J. Wiley, N.Y.

Suzuki, S., Deligeer, Yamaguchi, K., Kataoka, K., Kobayashi, K., Tagawa, S., Kohzuma, T., Shidara, S., and Iwasaki, H., 1997, *J. Biol. Inorg. Chem.* **2**:265–274.

Van Pouderoyen, G., Mazumdar, D.M., Hunt, N.I., Hill, H.A.O., and Canters, G.W., 1994, *Eur. J. Biochem.* **222**:583–588.

Wilting, J., Braams, R., Nauta, H., and van Buuren, K.J.H.,1972, *Biochim. Biophys. Acta* **283**:543–547.

Zaitzeva, I., Zaitzev, V., Card, C., Moshkov, K., Bax, B., Ralph, A., and Lindley, P., 1996, *J. Biol. Inorg. Chem.* **1**:15–23.

5

FREE RADICAL MEDIATED OXIDATION OF PROTEINS

Earl R. Stadtman*

Laboratory of Biochemistry
National Heart, Lung, and Blood Institute
National Institutes of Health
Building 3, Room 222
Bethesda, Maryland 20892

1. ABSTRACT

Living organisms are constantly bombarded with a battery of oxygen free radicals and other forms of reactive oxygen, leading to the modification of proteins. These modifications include: (a) fragmentation of the polypeptide chain, (b) formation of intra- and inter-molecular cross-linkages, direct oxidation of amino acid residue side chains, (c) derivatization of amino groups of lysine by reducing sugars or their oxidation products (glycation), (d) derivatization of lysine, histidine, or cysteine residues by lipid oxidation products (malondialdehyde, 2,3 unsaturated aldehydes), and (e) nitration of tyrosine residues. The generation of carbonyl derivatives (aldehydes, ketones) by some of these reactions may serve as markers of oxidative protein damage in aging and disease. The nitration of tyrosine residues can seriously compromise major mechanisms of enzyme regulation and signal transduction. And the formation of protein-protein cross linkages can lead to the accumulation of protease resistant protein polymers and inhibitors of proteases that degrade the oxidized forms of oxidatively modified proteins.

2. INTRODUCTION

It is now well established that the modification of proteins by various reactive oxygen species (ROS) is implicated in aging and a number of pathological conditions (Stadtman and Oliver, 1991; Stadtman, 1992, 1995; Sohal, Agarwal, Dubey, and Orr, 1993; Ames, Shigenaga, and Hagen, 1993; Halliwell and Gutteridge, 1990). The ROS involved in these processes may be produced as by-products of normal electron transport processes,

---

* Telephone: 301-496-4096. Fax: 301-496-0599.

Free Radicals, Oxidative Stress, and Antioxidants, edited by Özben.
Plenum Press, New York, 1998.

by metal-catalyzed oxidation (MCO) systems, by activation of neutrophils or macro-phages, by exposure to hyperoxia, by exposure to x-irradiation, UV-irradiation, or gamma-radiation, by forced exercise, ischemia-reperfusion, rapid correction of hyponatremia, exposure to cigarette smoke, chronic alcohol ingestion, estrogen administration, magnesium deficiency, exposure to atmospheric pollutants (ozone, nitrogen dioxide, etc.), or as a consequence of arginine catabolism (for review, see Stadtman, this volume, chapter 13). Collectively, these diverse processes lead to formation of a battery of free radical species, including hydroxyl radical ($H\dot{O}$), superoxide anion radical ($O_2^{\cdot-}$), nitric oxide radical ($N\dot{O}$), thiyl radical ($R\dot{S}$), alkyl radical ($R^{\cdot}$), alkylperoxy radical ($RO\dot{O}$), alkoxyl radical ($RC\dot{O}$), metal ions (Fe^{2+}, Cu^{2+}), as well as a number of non-radical species, including hydrogen peroxide (H_2O_2), alkylperoxides (ROOH), ozone (O_3), singlet oxygen (^1O), per-oxynitrite ($ONOO^-$), nitronium ion (NO_2^+), hypochlorous acid (HOCl), and a number of reactive carbonyl compounds derived from the oxidation of lipids (malondialdehyde, 4-hydroxy-2-nonenal), and the reaction of sugars or their oxidation products with protein amino acid side chains (glycation, glycoxidation reactions). There seems little doubt that among these, $H\dot{O}$ is by far the most damaging form of ROS. Hydroxyl radicals can be formed: (a) by the homolytic cleavage of water by ionizing radiation (x-rays, gamma-rays) according to the overall reaction 1, in which aqueous electrons (e_{aq}^-) and an excited state of water (H_2O^+) are intermediates (Swallow, 1960): (b) by reactions of Fe^{2+} or Cu^+ with hy-drogen peroxide, reaction 2 (Halliwell and Gutteridge, 1989); (c) by homolytic cleavage of peroxynitrite, reaction 3 (Beckman, Beckman, Chen, Marshall, and Freeman, 1990; van der Vliet, Eiserich, O'Neil, Halliwell, and Cross, 1995); (d) by reaction of ozone with phe-nols (PH), reaction 4 (Grimes, Perkins, and Boss, 1983; Pryor, 1994).

$$H_2O \longrightarrow H^{\cdot} + HO^{\cdot} \tag{1}$$

$$H_2O_2 + Fe^{2+} \text{ or } Cu^+ \longrightarrow HO^{\cdot} + OH^- + Fe^{3+} \text{ or } Cu^{2+} \tag{2}$$

$$HONOO \longrightarrow \dot{O}H + N\dot{O}_2 \tag{3}$$

$$PH + O_3 \longrightarrow H\dot{O} + P + O_2 \tag{4}$$

Basic mechanisms involved in the oxidation of proteins by $H\dot{O}$ were established by the pioneering studies of Swallow (1960), Garrison, Jayko, and Bennett (1962), Garrison (1987), and Schuessler and Schilling (1984), who subjected proteins in aqueous solution to ionizing radiation (x-rays, gamma-rays) under conditions where only $H\dot{O}$, or $O_2^{\cdot-}$, or a mixture of both were formed. Results of these studies demonstrated that exposure of pro-teins to ROS formed under these conditions can lead to oxidation of amino acid side chains, to formation of covalently cross-linked protein aggregates, and to fragmentation of the polypeptide chain. In the meantime, it has become obvious that many of the basic prin-ciples established in these early studies with ionizing radiation apply also to alterations of proteins by metal-catalyzed oxidation systems.

3. OXIDATION OF PROTEIN BACKBONE

As illustrated in Fig. 1, oxidation of a protein main chain, in the presence of oxygen, is initiated by the $H\dot{O}$-dependent abstraction of the alpha-hydrogen atom from any one of the

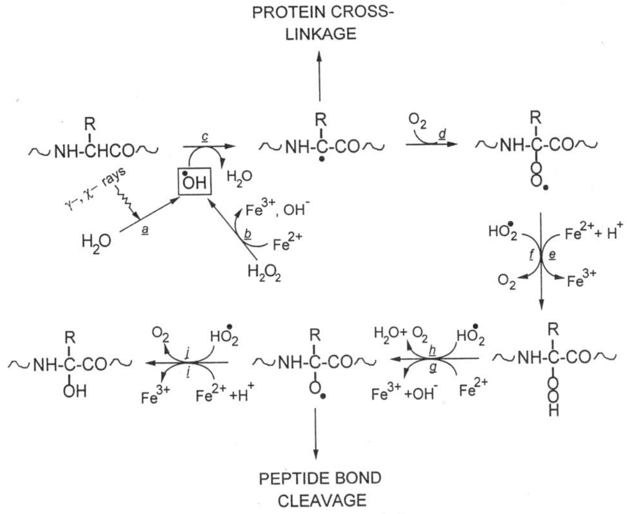

Figure 1. Free radical mediated oxidation of protein carbonyls.

amino acid residues to form an alkyl radical derivative (reaction c, Fig. 1). This is followed by addition of O_2 to the alkyl radical to form an alkylperoxyl radical (reaction d), which can either react with the protonated form of superoxide anion ($HO_2^\bullet$) or with Fe^{2+} (reactions e and f), to form the alkylperoxide derivative. The alkylperoxide may react further with either $HO_2^\bullet$ or Fe^{2+} to form an alkoxy radical (reactions g and h), which may undergo peptide bond cleavage (see below), or undergo further reduction by $HO_2^\bullet$ or Fe^{2+} to the corresponding hydroxy derivative (reactions i and j). Although not shown in Fig. 1, the protein alkyl-, peroxyl-, and alkoxyl-radical intermediates may also abstract a hydrogen atom from amino acid residues in the same or a different protein molecule to generate another alkyl radical, capable of undergoing an analogous series of reactions. Moreover, in the absence of oxygen, two different alkyl protein radicals may react with one another to generate inter- or intra-protein cross-linked derivatives (Garrison, 1987). It is noteworthy that this reaction scheme is based on the results of studies in which proteins were exposed to ionizing radiations under conditions where $HO^\bullet$, $O_2^{\bullet-}$, and $HO_2^\bullet$ are the reactive oxygen species formed. However, many of the steps mediated by these ROS can also be catalyzed by Fe^{2+} as illustrated in Fig. 1 (reactions e, g, j) or Cu^+ (not shown) (for review, see Borg and Schaich, 1988).

4. PEPTIDE BOND CLEAVAGE

Oxygen radical-mediated cleavage of the polypeptide chain can occur by at least four well-established mechanisms: (a) cleavage of alkoxyl peptide derivatives by the alpha-amidation pathway (Garrison, 1987); (b) cleavage of an alkoxyl peptide derivative by the diamide pathway (Garrison, 1987); (c) oxidation of glutamyl or aspartyl side chains (Garrison, 1987); (d) oxidation of proline side chains (Uchida, Kato, and Kawakishi, 1990). As illustrated in Fig. 2, cleavage of a protein alkoxyl derivative by the diamide pathway (Fig. 2, route a) leads to two peptides. The N-terminal amino acid residue of the peptide fragment obtained from the C-terminal portion of the protein exists as an isocyanate derivative, whereas the C-terminal amino acid residue of the peptide derived from the N-terminal portion of the protein exists as a diamide derivative. However, when cleavage occurs by the

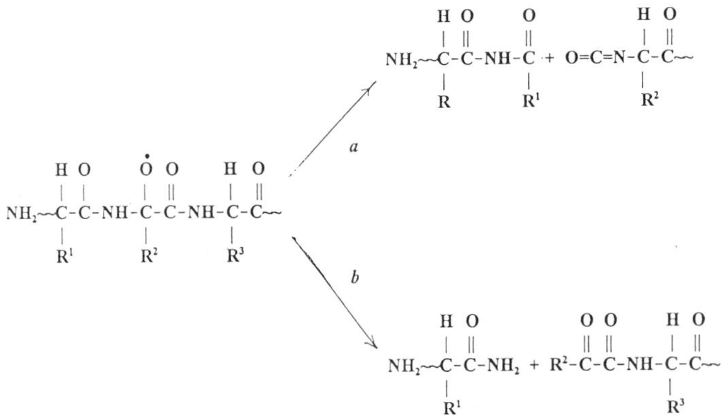

Figure 2. Oxidative cleavage of polypeptide chains by (a) the diamide pathway and (b) the α-amidation pathway.

alpha-amide pathway (Fig. 2, route b) the N-terminal amino acid residue of the peptide derived from the C-terminal portion of the protein exists as an alpha- ketoacyl derivative, and the C-terminal amino acid residue of the other peptide exists as an amide derivative.

Peptide bond cleavage by the glutamate oxidation pathway is initiated by the HO·- dependent abstraction of a hydrogen atom from the gamma-carbon of the glutamyl side chain, followed by a series of reactions analogous to those described in Fig. 1 (reactions d, f, and h). This leads eventually to elimination of oxalic acid and peptide bond cleavage by a mechanism in which the N-terminal amino acid residue of the peptide fragment derived from the C-terminal portion of the protein is blocked by a pyruvyl moiety, according to the overall reaction 5 (Garrison, 1987).

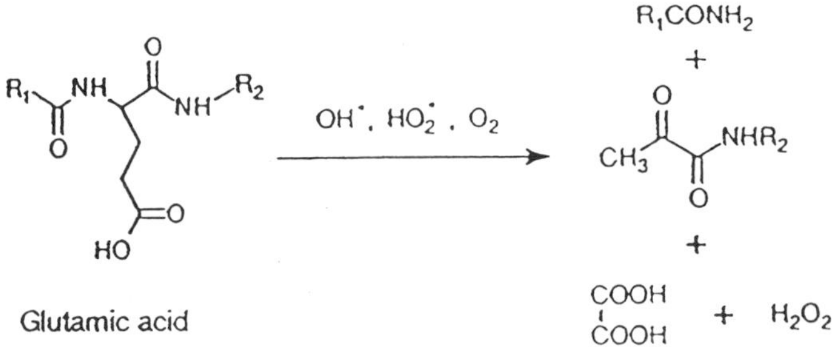

$$\tag{5}$$

Schuessler and Schilling (1984) observed that the number of peptides formed during the exposure of proteins to ionizing radiation is approximately equal to the number of prolyl residues present. They therefore proposed that the oxidation of proline residues leads to peptide bond cleavage. Subsequently, Uchida et al. (1990) demonstrated that the oxidation of proline residues leads to the formation of the 2-pyrrolidone derivative and is associated with peptide bond cleavage according to the overall reaction 6. They showed further that

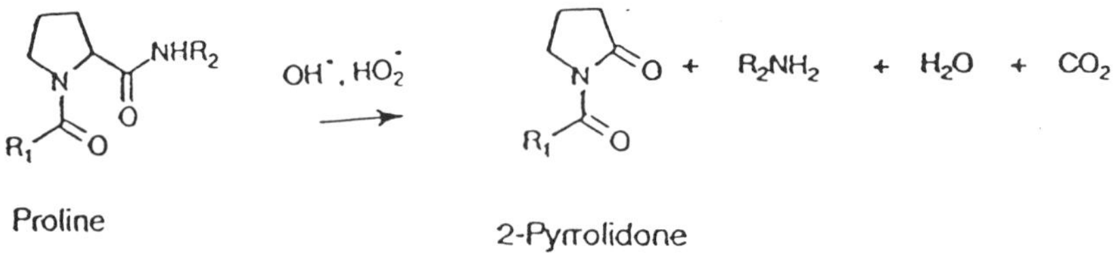

$$\tag{6}$$

upon acid hydrolysis, 2-pyrrolidone is converted to 4-aminobutyric acid. The presence of 4-aminobutyric acid in acid hydrolysates of proteins is therefore presumptive evidence for peptide cleavage by the 2-pyrrolidone pathway.

5. OXIDATION OF AMINO ACID RESIDUE SIDE CHAINS

All amino acid residues of proteins are susceptible to oxidation by $H\dot{O}$ generated by ionizing radiation (Swallow, 1960; Garrison, 1987; Garrison et al., 1962; Schuessler and Schilling, 1984) or by high concentrations of the Cu^{2+}/H_2O_2 or Fe^{2+}/H_2O_2 MCO systems (Huggins, Wells-Knecht, Detorie, Baynes and Thorpe, 1993; Nuezil, Gebiki, and Stocker, 1993). Nevertheless, unambiguous characterization of the products formed has been established in only a few cases, as are summarized in Tables 1, 2, and 3.

The aromatic amino acid residues are particularly susceptible to oxidation by various forms of ROS. As shown in Table 1, phenylalanine is converted to both mono- and dihydroxy derivatives. Tyrosine is converted to the 3,4-dihydroxy derivative (dopa); this is of special interest because such derivatives can undergo redox cycling and the production of ROS (Simpson, Gieseg, and Dean, 1993; Waite, 1995). Oxidation of tyrosine residues

Table 1. Oxidation of aromatic amino acid residues

Residue	Products	References
Phenylalanine	2,3 Dihydroxy-phenlanine, 2-, 3- and 4-Hydroxyphenylalanine	Solar, 1985; Garrison, 1987; Maskos, Rush and Koppenol, 1992, 1992a; Huggins et al., 1993; Kaur and Halliwell, 1994
Tyrosine	3,4-Dihydroxyphenylalanine	Fletcher and Okada, 1961; Garrison, 1987; Maskos et al., 1992a; Huggins et al., 1993; Gieseg, Simpson, Charlton, Duncan, and Dean, 1993; Simpson, Gieseg, and Dean, 1993; Wells-Knecht, Huggins, Dyer, Thorpe and Baynes; 1993
Tyrosine	Dityrosine (2,2'-biphenyl-derivatives)	Verweij, Christianse, and Van Steveninck, 1982; Garrison, 1987; Gieseg, Simpson, Charlton, Duncan, Li, Francis, and Goldstein, 1993; Huggens et al., 1993; Wells-Knecht et al., 1993; van der Vliet, Eiserich, O'Neil, Halliwell, and Cross, 1995; Ischiropoulos and Al-Mehdi, 1995
Tyrosine	3-Nitrotyrosine	Beckman, Ischiropoulos, Zhu, van der Woerd, Smith, Chen, Harrison, Martin, and Tsai, 1992; Heinecke, Li, Daehnke III, and Goldstein, 1993; van der Vliet et al., 1995; Pryor and Squadrito, 1995; Eiserich, Cross, Jones, and Halliwell, 1996; Berlett, Friguet, Yim, Chock, and Stadtman, 1996
Tyrosine	Chlorotyrosine	Domigan, Charlton, Duncan, Winterbourne, and Kettle, 1995, Eiserich et al., 1996
Tryptophan	2-,4-,5-,6-,7-Hydroxy-tryptophan	Armstrong and Swallow, 1969; Maskos et al., 1992
Tryptophan	Kynurenine, 3-Hydroxy-kynurenine; Oxindole; Hydropyrroloindole; N-Formylkinurenine; 3-Hydroxy-kynurenine	Winchester and Lynn, 1970; Kuroda, Sakiyama, and Narita, 1975; Knight and Mudd, 1984; Guptasarma, Balasubramanian, Matsago, Saito, 1992; Berlett, Levine, and Stadtman, 1996; Kato, Kawakishi, Aoki, Itakura, and Osawa, 1997
Histidine	2-Oxohistidine, 4-OH-glutamate, aspartate, asparagine	Farber and Levine, 1986; Uchida and Kawakishi, 1993; Uchida and Kawakishi, 1994; Berlett et al., 1996

Table 2. Oxidative conversion of amino acid residues to carbonyl derivatives

Residue	Carbonyl product	References
Lysine	2-Amino-adipic-semialdehyde	Berlett, Miller, Szweda, and Stadtman[a]
Arginine	Glutamic-semialdehyde	Amici, Tsai, Levine, and Stadtman, 1989; Climent, Tsai, and Levine, 1989
Proline	Glutamic-semialdehyde	Amici et al., 1989
Threonine	2-amino-3-keto-butyric acid	Taborsky, 1973
Glutamic acid	Pyruvic acid	Garrison, 1987
Aspartic acid	Pyruvic acid	Garrison, 1987

[a]B.S. Berlett, D.G. Miller, L. Szweda, and E.R. Stadtman, unpublished results.

can also lead to tyrosyl radicals that can react with one another to form intra- and/or inter-protein-protein cross-linkages (dityrosine derivatives). It has been proposed that the presence of such 2,2′-biphenyl derivatives may serve as a marker of ROS-mediated oxidative damage (Giulivi and Davies, 1993; Huggins, Wells-Knecht, Detorie, Baynes; and Thorpe, 1993; Heinecke, Li, Francis, and Goldstein, 1993). Moreover, because myeloperoxidase catalyzes the generation of dityrosine cross-linkages, it has been suggested that the presence of dityrosine may be used to pinpoint targets where phagocytes inflict oxidative damage *in vivo* (Heinecke, Li, Daehnke III, and Goldstein, 1993). Tryptophan is highly sensitive to oxidation by gamma irradiation, which leads to various hydroxy derivatives, formylkynurenine, and 3-hydroxy kynurenine (Table 1). But in the presence of ozone, ultraviolet irradiation, high levels of Fe^{2+} and H_2O_2, or peroxynitrite, tryptophan is converted mainly to kynurenine and N-formylkynurenine. It is noteworthy that tyrosine and tryptophan residues are not major targets for oxidation by MCO systems at physiological concentrations transition metal ions (Stadtman, 1990), presumably because these residues are not present at metal binding sites; i.e. at sites where metal-catalyzed oxidation takes place. To the contrary, histidine, arginine, and lysine residues are sensitive targets for MCO-catalyzed reactions because they are often located at metal binding sites on proteins. Histidine residues are converted to 2-oxohistidine, asparagine, or aspartic acid (Table 1).

6. GENERATION OF CARBONYL DERIVATIVES

It is of special significance that the oxidation of some amino acid residues leads to carbonyl derivatives. Thus, as shown in Table 2, upon oxidation, the side chains of lysine,

Table 3. Oxidative modification of sulfur-containing amino acid residues

Residue	Products	References
Cysteine	Nitrosothiols	Stamler, 1994; Mohr, Stamler and Brune, 1994; Arnelle and Stamler, 1995
Cysteine	CyS–SCy; CyS–SG	Swallow, 1960; Garrison, 1987; Zhou and Gafni, 1991; Brodie and Reed, 1990; Takahashi and Goto, 1990
Cysteine	Thiol radicals	Garrison, 1987
Methionine	Methionine sulfoxide; Methionine sulfone	Garrison and Jayko, 1962; Vogt, 1995; Armstrong and Swallow, 1969; Brot and Weissbach, 1983; Winterbourne, 1985; Pryor and Squadrito, 1995; Berlett et al., 1996; Levine, Mosoni, Berlett, and Stadtman 1996a; Chao, Ma, and Stadtman, 1997

arginine, proline, and threonine are converted directly to aldehyde or ketone derivatives, and the oxidation of glutamic acid and aspartic acid residues leads to a peptide bond cleavage reaction in which the N-terminal amino acid of one peptide fragment is blocked by a pyruvyl moiety, reaction 5 (Garrison, 1987). In addition, as noted above, oxidative cleavage of the polypeptide chain by the alpha-amidation pathway leads to the formation of a peptide in which the N-terminal amino acid is blocked by a 2-keto-acyl derivative (Fig. 2). Direct oxidation of amino acid residues is not the only mechanism for the introduction of carbonyl groups into proteins. Reactions of proteins with reducing sugars or their oxidation products (glycation, glycoxidation reactions) or with lipid oxidation products (malondialdehyde, 2,3-unsaturated aldehydes, especially 4-hydroxy-2-nonenal) will also lead to protein adducts possessing a carbonyl (aldehydic) function.

6.1. Reaction with Lipid Peroxidation Products

As shown in Fig. 3, cysteine sulfhydryl groups, the free amino group of lysine residues, or the imidazole moiety of histidine residues of proteins may undergo Michael addition-type reactions with the double bond of 2,3-unsaturated aldehydes, such as 4-hydroxy-2-nonenal (HNE), to form adducts that possess a carbonyl function (Schuenstein and Esterbauer, 1993; Uchida and Stadtman, 1993; Friguet, Stadtman, and Szweda, 1994). Antibodies prepared against various Michael addition complexes of HNE have been used to demonstrate that the accumulation of HNE protein derivatives is associated with a number of diseases (review by Stadtman, this volume, chapter 13). Reaction of only one of the two aldehyde moieties of malondialdehyde (MDA) with the free amino group of a lysine residue of a protein provides another route for the introduction of a carbonyl group into proteins (Burcham and Kuhan, 1996).

6.2. Glycation–Glycoxidation

Reaction of the carbonyl group of reducing sugars with the free amino group of lysine residues of proteins leads to the formation of a Schiff-base derivative, which can undergo spontaneous Amadori rearrangement (Cerami, Vlassara, and Brownlee, 1987) to form ketoamine (Fig. 4). The Amadori product is highly sensitive to metal-catalyzed oxidation (glycoxidation) leading eventually to the N-carboxymethyl lysine derivative (Wells-Knecht et al., 1993), or it may react with arginine residues to form pentosidine protein cross-linked adducts (Monnier, 1990; Monnier, Gerhardinger, Marion, and Taneda, 1995). Furthermore, the Amadori product can undergo dehydration and oxidative fragmentation to yield deoxyosones that may be converted to highly fluorescent, pigmented cross-linked derivatives of poorly defined structures (Monnier, 1990). Research in this area has been recently reviewed by Kristal and Yu, 1992 and by Baynes, 1996.

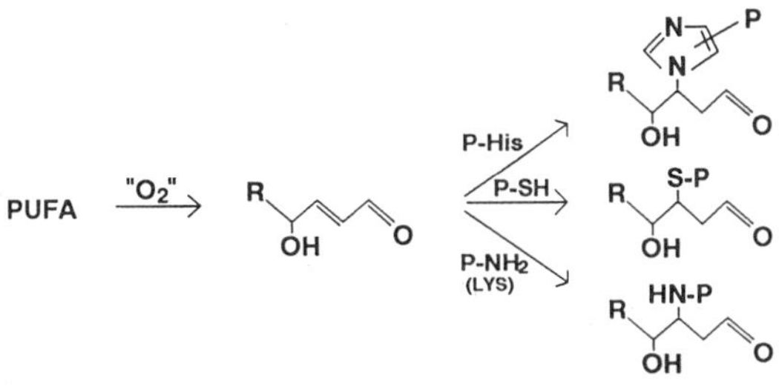

Figure 3. Formation of protein carbonyl derivatives by reaction with 4-hydroxy-2-nonenal. Abbreviations: PUFA, polyunsaturated fatty acids; P–His, histidine reaction of protein; P–NH$_2$-amino group of lysine residues; P–SH, sulfhydryl group of cysteine residues.

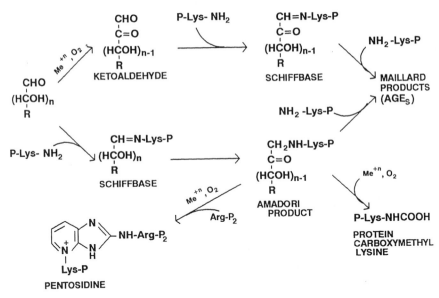

Figure 4. Protein modification by reaction of reducing sugars on their oxidation products with lysine residues of proteins: Glycation-glycoxidation reactions. Abbreviations: P–Lys–NH$_2$, lysine amino groups of proteins; arg–P$_2$, arginine residues of proteins. Based on reports of Monnier, 1990; Wolff and Dean, 1987; Baynes, 1996.

7. OXIDATION OF SULFUR-CONTAINING AMINO ACID RESIDUES

Cysteine and methionine residues of proteins are particularly susceptible to oxidation (Table 3) by almost all forms of reactive oxygen, including ozone (Pryor and Uppu, 1993; Berlett *et al.*, 1996; Mudd, Leavitt, Ongun, and McManus 1969), metal-catalyzed reactions (Takahashi and Goto, 1990; Zhou and Gafni, 1991), hydrogen peroxide (Brot and Weissbach, 1983; Levine, Mosoni, Berlett, and Stadtman, 1996; Vogt, 1995), peroxy radicals (Chao, Ma, and Stadtman, 1997), gamma radiation (Swallow, 1960; Garrison, 1987; Armstrong and Swallow, 1996), peroxynitrite (Pryor and Squadrito, 1995; Berlett *et al.*, 1996), hypochlorous acid (Winterbourne, 1985; Vogt, 1995). Significantly, the oxidation of cysteine and methionine residues are the only oxidative modifications that can be reversed by endogenous enzymes, and in both cases the cyclic oxidation-reduction reactions can form the basis of an antioxidant function. For example, oxidation of cysteine residues by H$_2$O$_2$ (reaction 8) leads to the formation of a disulfide derivative (RSSR^1), which can be reduced by the action of glutathione-disulfide exchange reactions (reactions 9 and 10), leading to the formation of oxidized glutathione (GSSG), from which GSH can be regenerated by the action of glutathione reductase, GR, (reaction 11). The net result of these reactions is described by reaction 12; i.e. an NADPH peroxidase activity.

$$RSH + R^1SH + H_2O_2 \longrightarrow RSSR^1 + 2H_2O \tag{8}$$

$$RSSR^1 + GSH \longrightarrow RSH + GSSR^1 \tag{9}$$

$$GSSR^1 + GSH \longrightarrow R^1SH + GSSG \tag{10}$$

$$GSSG + NADPH + H^+ \xrightarrow{\text{GR}} 2GSH + NADP^+ \qquad (11)$$

$$\text{Sum: } H_2O_2 + NADPH + H^+ \longrightarrow NADP^+ + 2H_2O \qquad (12)$$

In an analogous series of reactions, methionine sulfoxide (MESOX) residues, formed by the reaction with various ROS, such as H_2O_2 (reaction 13), can be reduced back to methionine by a reaction catalyzed by methionine sulfoxide reductase (MSR) in which thioredoxin, $T(SH)_2$, serves as an electron donor (reaction 14). When coupled with reactions catalyzed by thioredoxin reductase (TR) reaction 15, the overall reaction is again reaction 12.

$$H_2O_2 + \text{Methionine} \longrightarrow \text{MeSOX} + H_2O \qquad (13)$$

$$\text{MeSOX} + T(SH)_2 \xrightarrow{\text{MSR}} \text{Methionine} + T(S)_2 \qquad (14)$$

$$T(S)_2 + NADPH + H^+ \xrightarrow{\text{TR}} T(SH)_2 + NADP^+ \qquad (15)$$

$$\text{Sum: } H_2O_2 + NADPH + H^+ \longrightarrow NADH^+ + H_2O \qquad (12)$$

For purposes of illustration, H_2O_2 was used as the oxidant in the above reactions; however, this can be replaced by any form of ROS capable of oxidizing cysteine or methionine residues in proteins. In effect then, these systems could serve as ROS scavenger systems. Based on these considerations, it was proposed that some methionine residues in proteins may undergo cyclic oxidation and reduction when exposed to ROS and thereby serve as "built-in" antioxidant systems for the protection of proteins from more extensive ROS-mediated damage (Levine et al., 1996). This concept is supported by the fact that, for some proteins, oxidation of the methionine residues that are most susceptible to oxidation has little or no effect on their biological activities (Levine et al., 1996). For the same reasons, some cysteine residues in proteins might have an antioxidant function. Finally, it is noteworthy that the oxidation of methionine residues in proteins converts the proteins to a more hydrophobic conformation and renders them more susceptible to proteolytic degradation by the multicatalytic protease (Levine et al., 1996; Chao et al., 1996).

8. PEROXYNITRITE-MEDIATED PROTEIN MODIFICATION

The discovery that nitric oxide ($\overset{\cdot}{\text{NO}}$) is a normal product of arginine catabolism and serves as an important messenger in the regulation of a number of important physiological processes, together with the subsequent discovery that $\overset{\cdot}{\text{NO}}$ reacts at near diffusion-controlled rates with $O_2^{\cdot-}$ to form the highly toxic peroxynitrite (reaction 7), has focused attention on the metabolic consequences of $\overset{\cdot}{\text{NO}}$ and peroxynitrite (for reviews, see: Beckman, Chen, Ischiropoulos, and Crow, 1994; Pryor and Squadrito, 1995; Katayama, 1996).

$$\overset{\cdot}{\text{NO}} + O_2^{\cdot-} \longrightarrow OONO^- \qquad (7)$$

Reaction of peroxynitrite (PN) with proteins leads to the nitration of tyrosine residues (Beckman et al., 1992; van der Vliet et al., 1995; Pryor and Squadrito, 1995; Berlett et al., 1996), formation of tyrosine-tyrosine cross-linkages (van der Vliet et al., 1995), oxidation of tryptophan (Kato, Kawakishi, Aoki, Itakura, and Osawa, 1997), methionine (Pryor and Squadrito, 1995; Berlett *et al.*, 1996), and cysteine residues (Pryor and Squadrito, 1995; Rubbo, Denicola, and Radi, 1994).

8.1. Tyrosine Nitration

The cyclic interconversion of tyrosine residues in regulatory proteins between unmodified and phosphorylated (Hunter, 1995) or nucleoditylated forms (Stadtman, Chock, and Rhee, 1994) plays a central role in cellular regulation of key metabolic processes. The nitration of tyrosine residues in regulatory proteins could therefore seriously compromise one of the most important mechanisms of cellular control. The potential importance of nitration of tyrosine residues in regulatory proteins is highlighted by the results of studies showing: (a) that nitration of *E. coli* glutamine synthetase converts the enzyme to a form which has regulatory properties similar to those obtained *in vivo* by the adenylylation of a single tyrosine residue per subunit (Berlett *et al.*, 1995); and (b) that nitration of the tyrosine residue in a substrate for the lck kinase prevents its phosphorylation by the enzyme (Kong, Yim, Stadtman, and Chock, 1996). Significantly, in contrast to phosphorylation or nucleotidylation of tyrosine residues, which are reversible processes *in vivo*, the nitration of tyrosine residues is irreversible and therefore will freeze regulatory proteins in a fixed configuration. Nitration of tyrosine residues in proteins by peroxynitrite is, however, dependent upon the presence of CO_2 (Lymar and Hurst, 1995; Uppu, Squadrito, and Pryor, 1996; Denicola, Freeman, Trujillo, and Radi, 1996; Lymar, Jiang, and Hurst, 1996; Lymar and Hurst, 1996; Berlett and Stadtman, 1996a). Because peroxynitrite also catalyzes the conversion of methionine residues to methionine sulfoxide, the effect of CO_2 on the peroxynitrite-mediated oxidation of methionine residues in glutamine synthetase was also examined. It was found that the ability of peroxynitrite to nitrate tyrosine residues in glutamine synthetase is almost completely dependent upon the presence of CO_2 and that, at physiological concentrations of CO_2, peroxynitrite does not oxidize methionine residues (Berlett and Stadtman, 1996a). Based on the results of earlier detailed kinetic studies (Pryor, Jin, and Squadrito, 1994), it was concluded that the oxidation of methionine by peroxynitrite (PN) involves the intermediate conversion of peroxynitrite to an activated form (PN*) of unknown structure. It follows, therefore, that the nitration of tyrosine residues and the oxidation of methionine residues in proteins are competitive processes that are dependent upon the concentration of CO_2, as illustrated in Fig. 5.

9. FORMATION OF PROTEIN–PROTEIN CROSS-LINKAGES

In addition to the modifications described in the foregoing sections, the oxidation of proteins can give rise to protein-protein cross-linked derivatives. Cross-linking can occur by

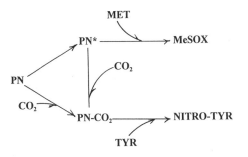

Figure 5. Effect of carbon dioxide on the oxidation of methionine residues and the nitration of tyrosine residues of proteins by peroxynitrite (PN).

at least six different mechanisms: (a) by reaction of two alkyl radical protein derivatives produced in the oxidation of the protein backbone or amino acid side chains (reaction 16):

$$P^1R_2\dot{C} + P^2R_2\dot{C} \longrightarrow P^1R_2CCR_2P^2 \tag{16}$$

(b) by the combination of two tyrosyl radicals (reaction 17)

$$P^1-tyr^{\bullet} + P^2-tyr^{\bullet} \longrightarrow P^1-tyr-tyr-P^2 \tag{17}$$

(c) by the reaction of malondialdehyde with amino groups of lysine residues in two different protein molecules (reaction 18)

$$P^1-NH_2 + P^{2-}-NH_2 + CH_2(CHO)_2 \longrightarrow P^{1-}-N=CCH_2C=N-P^2 + 2H_2O \tag{18}$$

(d) by reaction of the HNE-Michael addition product (HNE-CHO) of one protein with the amino group of a lysine residue in a second protein (reaction 19)

$$P^1-HNE-CHO + P^2-NH_2 \longrightarrow P^1-HNE-C=N-P^2 + H_2O \tag{19}$$

(e) by reaction of the carbonyl group of the glycation product (P^1R_2CHO) of one protein with the amino group of lysine in a second protein (reaction 20)

$$P^1R_2CHO + P^2-NH_2 \longrightarrow P^1R_2C=N-P^2 + H_2O \tag{20}$$

(f) by the oxidation of cysteine residues in two different protein molecules (reaction 21)

$$P^1SH + P^2SH + ROS \longrightarrow P^1SSP^2 + \text{reduced ROS} \tag{21}$$

The physiological significance of these reactions is underscored by the fact that cross-linked proteins are resistant to proteolytic degradation by the multicatalytic protease (Guilivi and Davies, 1993; Grune, Reinheckel, Joshi, and Davies, 1995; Friguet, Szweda, and Stadtman, 1994), and in fact may inhibit the degradation of the oxidized forms of other enzymes by the multicatalytic protease (Friguet *et al.*, 1994). Such reactions may therefore contribute to the accumulation of the oxidized forms of proteins during aging and some diseases.

REFERENCES

Ames, B. N., Shigenaga, M. K., and Hagen, T. M., 1993, Oxidants, antioxidants, and the degenerative diseases of aging, *Proc. Natl. Acad. Sci. USA* **90**, 7915–7922.

Amici, A., Levine, R. L., Tsai, L., and Stadtman, E. R., 1989, Conversion of amino acid residues in proteins and amino acid homopolymers to carbonyl derivatives by metal-catalyzed reactions, *J. Biol. Chem.* **264**, 3341–3346.

Armstrong, R. C. and Swallow, A. J., 1969, Pulse- and gamma-radiolysis of aqueous solutions of tryptophan, *Radiat. Res.* **41**, 563–579.

Arnelle, D. R. and Stamler, J. S., 1995, NO$^+$, NO$^{\bullet}$, and NO$^-$ donation by S-nitrosothiols: Implication for regulation of physiological functions by S-nitrosylation and acceleration of disulfide formation, *Arch. Biochem. Biophys.* **318**, 279–285.

Baynes, J. W., 1996, Perspectives in diabetes. Role of oxidative stress in development of complications in diabetes, *Diabetes* **40**, 405–411.

Beckman, J. S., Beckman, T. W., Chen, J., Marshall, P. A., and Freeman, S., 1990), Apparent hydroxyl radical production by peroxynitrite: Implications for endothelial injury from nitric oxide and superoxide, *Proc. Natl. Acad. Sci. USA* **87**, 1620–1624.

Beckman, J. S., Chen, J., Ischiropoulous, H., and Crow, J. P., 1994, Oxidative chemistry of peroxynitrite, *Methods Enzymol.* **233**, 229–239.

Beckman, J. S., Ischiropoulos, H., Zhu, L., van der Woerd, M., Smith, C., Chen, J., Harrison, J., Martin, J.C., and Tsai, M., 1992, Kinetics of superoxide dismutase- and iron-catalyzed nitration of phenolics by peroxynitrite, *Arch. Biochem. Biophys.* **298**, 438–445.

Berlett, B. S. and Stadtman, E. R., 1996, Carbon dioxide stimulates nitration of tyrosine residues and inhibits oxidation of methionine residues in *Escherichia coli* glutamine synthetase by peroxynitrite, *FASEB J.* **10**, Abstract #585.

Berlett, B. S., Friguet, B., Yim, M. B., Chock, P. B., and Stadtman, E. R., 1996, Peroxynitrite-mediated nitration of tyrosine residues of *Escherichia coli* glutamine synthetase mimic adenylylation: Relevance to signal transduction, *Proc. Natl. Acad. Sci. USA* **93**, 1776–1780.

Berlett, B. S., Levine, R. L., and Stadtman, E. R., 1996a, A comparison of the effects of ozone on the modification of amino acid residues in glutamine synthetase and bovine serum albumin, *J. Biol. Chem.* **271**, 4177–4182.

Borg, D. C. and Schaich, K. M., 1988, Iron and iron-derived radicals, In: *Oxygen radicals and tissue injury* (Halliwell, B., ed.), pp. 20–26, Proceedings of an Upjohn Symposium, Federation of American Societies for Experimental Biology, Bethesda, MD.

Brodie, E. and Reed, D. J., 1990, Cellular recovery of glyceraldehyde-3-phosphate dehydrogenase activity and thiol status after exposure to hydroperoxide, *Arch. Biochem. Biophys.* **277**, 228–233.

Brot, N. and Weissbach, H., 1983, Biochemistry and physiological role of methionine sulfoxide reductase in proteins, *Arch. Biochem. Biophys.* **233**, 271–288.

Burcham, P. C. and Kuhan, Y. T., 1996, Introduction of carbonyl groups into proteins of the lipid peroxidation product, malondialdehyde, *Biochem. Biophys. Res. Commun.* **220**, 996–1001.

Cerami, A., Vlassara, H., and Brownlee, M., 1987, Glucose and aging, *Sci. Am.* **256**, 90–96.

Chao, C.-C., Ma, Y.-S., and Stadtman, E. R., 1997, Modification of protein surface hydrophobicity and methionine oxidation by oxidative stress, *Proc. Natl. Acad. Sci. USA* **94**, 2969–2974.

Climent, I. and Levine, R. L., 1991, Oxidation of the active site of glutamine synthetase: Conversion of arginine-344 to -glutamyl semialdehyde, *Arch. Biochem. Biophys.* **189**, 371–375.

Denicola, A., Freeman, B. A., Trujillo, M., and Radi, R., 1996, Peroxynitrite reaction with carbon dioxide/bicarbonate: Kinetics and influence on peroxynitrite-mediated oxidations, *Arch. Biochem. Biophys.* **333**, 49–58.

Domigan, N. M., Charlton, T. S., Duncan, M. W., Winterbourne, C. C., and Kettle, A. J., 1995, Chlorination of tyrosyl residues in peptides by myeloperoxidase and human neutrophils, *J. Biol. Chem.* **270**, 16542–16548.

Eiserich, J. P., Cross, C. E., Jones, D., and Halliwell, B., 1996, Formation of nitrating and chlorinating species by reaction of nitrite with hypochlorous acid, *J. Biol. Chem.* **271**, 19199–19206.

Farber, J. M. and Levine, R. L., 1986, Sequence of a peptide susceptible to mixed-function oxidation: Probable cation binding site in glutamine synthetase, *J. Biol. Chem.* **261**, 4574–4578.

Fletcher, G. L. and Okada, S., 1961, Radiation-induced formation of dihydroxy phenylalanine from tyrosine and tyrosine-containing peptides in aqueous solution, *Radiat. Res.* **15**, 349–351.

Friguet, B., Szweda, L., and Stadtman, E. R., 1994, Susceptibility of glucose-6-phosphate dehydrogenase modified by 4-hydroxy-2-nonenal and metal-catalyzed oxidation to proteolysis by the multicatalytic protease, *Arch. Biochem. Biophys.* **311**, 168–173.

Garrison, W. M., 1987, Reaction mechanisms in the radiolysis of peptides, polypeptides, and proteins, *Chem. Rev.* **87**, 381–398.

Garrison, W. M., Jayko, M. E., and Bennett, W., 1962, Radiation-induced oxidation of proteins in aqueous solution, *Radiat. Res.* **16**, 487–502.

Gieseg, S. P., Simpson, J. A., Charlton, T. S., Duncan, M. W., and Dean, R. T., 1993, Protein-bound 3,4-dihydroxyphenylalanine is a major product formed during hydroxyl radical damage to proteins, *Biochemistry* **32**, 4780–4786.

Giulivi, C. and Davies, K. J. A., 1993, Dityrosine and tyrosine oxidation products are endogenous markers for the selective proteolysis of oxidatively modified red blood cell hemoglobin by (the 19S) proteosome, *J. Biol. Chem.* **268**, 8752–8759.

Grimes, H. D., Perkins, K. K., and Boss, W. F. , 1983, Ozone degrades into hydroxyl radical under physiological conditions: A spin trapping study, *Plant Physiol.* **72**, 1016–1020.

Grune, T., Reinheckel, T., Joshi, M., and Davies, K. J. A., 1995, Proteolysis in cultured liver epithelial cells during oxidative stress, *J. Biol. Chem.* **270**, 2344–2351.

Guptasarma, P., Balasubramanian, D., Matsugo, S., and Saito, I., 1992, Hydroxyl radical mediated damage to proteins, with special reference to the crystallins, *Biochemistry* **31**, 4296–4302.

Halliwell, B. and Gutteridge, J. M. C., 1990, Role of free radicals and catalytic metal ions in human disease: an overview, *Methods Enzymol.* **186**, 1–85.

Heinecke, J. W., Li, W., Daehnke III, H. L., and Goldstein, J. A., 1993, Dityrosine, a specific marker of oxidation, is synthesized by the myeloperoxidase-hydrogen peroxide system of human neutrophils and macrophages, *J. Biol. Chem.* **268**, 4069–4077.

Huggins, T. G., Wells-Knecht, M. C., Detorie, N. A., Baynes, J. W., and Thorpe, S. R., 1993, Formation of O-tyrosine and dityrosine in proteins during radiolytic and metal-catalyzed oxidation, *J. Biol. Chem.* **268**, 12341–12347.

Hunter, T., 1995, Protein kinases and phosphatases: The Ying and Yang of protein phosphorylation and signaling, *Cell* **80**, 225–236.

Katayama, Y., 1996, Nitric oxide: Mysterious messenger—Its chemistry, biology, and new reagents for research, *Dojindo Newsletter* **3**, Dojindo Laboratories, Bethesda, MD, pp. 3–18.

Kato, Y., Kawakishi, S., Aoki, T., Itakura, K., and Osawa, T., 1997, Oxidative modification of tryptophan exposed to peroxynitrite, *Biochem. Biophys. Res. Commun.* **234**, 82–84.

Kaur, H. and Halliwell, B., 1994, Aromatic hydroxylation of phenylalanine as an assay for hydroxyl radicals, *Anal. Biochem.* **220**, 11–15.

Knight, K. L. and Mudd, J. B., 1984, The reaction of ozone with glyceraldehyde-3-phosphate dehydrogenase, *Arch. Biochem. Biphys.* **229**, 259–269.

Kong, S.-K., Yim, M. B., Stadtman, E. R., and Chock, P. B., 1996, Peroxynitrite disables the tyrosine phosphorylation regulatory mechanism: Lymphocyte-specific tyrosine kinase fails to phosphorylate nitrated cdc2(6–20)NH$_2$ peptide, *Proc. Natl. Acad. Sci. USA* **93**, 3377–3382.

Kristal, B. S. and Yu, B. P., 1992, An emerging hypothesis: Synergistic induction of aging by free radicals and Maillard reactions, *J. Gerontol.* **47**, B104–B107.

Kuroda, M., Sakiyama, F., and Narita, K., 1975, Oxidation of tryptophan in lysozyme by ozone in aqueous solution, *J. Biol. Chem.* **78**, 641–651.

Levine, R. L., Mosoni, L., Berlett, B. S., and Stadtman, E. R., 1996, Methionine residues as endogenous antioxidants in proteins, *Proc. Natl. Acad. Sci. USA* **93**, 15036–15040.

Lymar, S. V. and Hurt, J. K., 1995, Rapid reaction between peroxynitrite ion and carbon dioxide: Implications for biological activity, *J. Am. Chem. Soc.* **117**, 8867–8868.

Lymar, S. V. and Hurst, J. K., 1996, Carbon dioxide: Physiological catalyst for peroxynitrite-mediated cellular damage or cellular protectant? *Chem. Res. Toxicol.* **9**, 845–850.

Lymar, S. V., Jiang, Q., and Hurst, J. K., 1996. Mechanisms of carbon dioxide-catalyzed oxidation of tyrosine by peroxynitrite, *Biochemistry* **35**, 7855–7861.

Maskos, Z., Rush, J. D., and Koppenol, W. H., 1992, The hydroxylation of tryptophan, *Arch. Biochem. Biophys.* **296**, 514–520.

Maskos, Z., Rush, J. D., and Koppenol, W. H., 1992, The hydroxylation of phenylalanine and tyrosine: A comparison with salicylate and tryptophan, *Arch. Biochem. Biophys.* **296**, 521–529.

Mohr, S., Stamler, J. S., and Brune, B., 1994, Mechanism of covalent modification of glyceraldehyde-3-phosphate dehydrogenase at its active site thiol by nitric oxide, peroxynitrite, and related nitrosating agents, *FEBS Lett.* **348**, 223–227.

Monnier, V., 1990, Nonenzymatic glycosylation, the Maillard reaction and the aging process, *J. Gerontol.* **45**, B106–B111.

Monnier, V., Gerhardinger, C., Marion, M. S., and Taneda, S., 1995, Novel approaches toward inhibition of the Maillard reaction *in vivo*: Search, isolation, and characterization of prokaryotic enzymes which degrade glycated substrates. In: *Oxidative Stress and Aging* (Cutler, R. G., Packer, L., Bertram, J., and Mori, A., eds.), pp 141–149, Birkhauser Verlag, Basel, Switzerland.

Neuzil, J., Gebiki, J. M., and Stocker, R., 1993, Radical-induced chain oxidation of proteins and its inhibition by chain-breaking antioxidants, *Biochem. J.* **293**, 601–606.

Pryor, W. A., 1994, Mechanisms of radical formation from reactions of ozone with target molecules in the lung, *Free Radical Biol. Med.* **17**, 451–465.

Pryor, W. A. and Squadrito, G., 1995, The chemistry of peroxynitrite: A product from the reaction of nitric oxide with superoxide, *Am. J. Physiol.*, L699–L722.

Pryor, W. A. and Uppu, R. M., 1993, A kinetic model for the competitive reactions of ozone with amino acid residues in proteins in reverse micelles, *J. Biol. Chem.* **268**, 3120–3126.

Pryor, W. A., Jin, X., and Squadrito, G. L., 1994, One- and two-electron oxidations of methionine by peroxynitrite, *Proc. Natl. Acad. Sci. USA* **91**, 11173–11177.

Rubbo, H., Denicola, A., and Radi, R., 1994, Peroxynitrite inactivates thiol-containing enzymes of *Trypanosoma cruzi* energetic metabolism and inhibits cell respiration, *Arch. Biochem. Biophys.* **308**, 96–112.

Schuenstein, E. and Esterbauer, H., 1979, Formation and properties of reactive aldehydes, In: *Submolecular Biology of Cancer*, pp 225–244, CIBA Foundation Series 67, Excerpta Medica, Elsevier, Amsterdam.

Schuessler, H. and Schilling, K., 1984, Oxygen effect in radiolysis of proteins, Part 2, Bovine serum albumin, *Int. J. Radiat. Biol.* **45**, 267–281.

Simpson, J. A., Giesseg, S. P., and Dean, R. T., 1993, Free radical and enzymatic mechanisms for the generation of protein bound reducing moieties, *Biochem. Biophys. Acta* **1156**, 190–196.

Solar, S., 1985, Reactions of OH with phenylalanine in neutral aqueous solutions, *Radiat. Phys. Chem.* **26**, 103–108.

Stadtman, E. R., 1990, Metal ion-catalyzed oxidation of proteins: Biochemical mechanism and biological consequences, *Free Rad. Biol. Med.* **9**, 315–325.

Stadtman, E. R., 1992, Protein oxidation and aging, *Science* **257**, 1220–1224.

Stadtman, E. R., Chock, P. B., and Rhee, S. G., 1981, Interconvertible enzyme cycles in cellular regulation, *Curr. Top. Cell. Regul.* **18**, 79–83.

Stadtman, E. R. and Oliver, C. N., 1991, Metal catalyzed oxidation of proteins: Physiological consequences, *J. Biol. Chem.* **266**, 2005–2008.

Stamler, J. S., 1994, Redox signaling: Nitrosylation and related target interactions of nitric oxide, *Cell* **78**, 931–936.

Swallow, A. J., 1960, Effect of ionizing radiation on proteins. RCO groups, peptide bond cleavage, inactivation, –SH oxidation, In: *Radiation Chemistry of Organic Compounds* (Swallow, A. J., ed.), pp 211–224, Pergamon Press, New York.

Taborsky, G., 1973, Oxidative modification of proteins in the presence of ferrous iron and air. Effect of ionic constituents of the reaction medium on the nature of the oxidation products, *Biochemistry* **12**, 1341–1348.

Takahashi, R. and Goto, S., 1990, Alteration of aminoacyl-tRNA synthetase with age: Heat labilization of the enzyme by oxidative damage, *Arch. Biochem. Biophys.* **277**, 228–233.

Uchida, K. and Kawakishi, S., 1993, 2-oxohistidine as a novel biological marker for oxidatively modified proteins, *FEBS Lett.* **332**, 208–210.

Uchida, K. and Stadtman, E. R., 1993, Covalent modification of 4-hydroxynonenal to glyceraldehyde-3-phosphate, *J. Biol. Chem.* **268**, 6388–6393.

Uchida, K., Kato, Y., and Kawakishi, S., 1990, A novel mechanism for oxidative damage of prolyl peptides induced by hydroxyl radicals, *Biochem. Biophys. Res. Commun.* **169**, 265–271.

Uppu, R. M., Squadrito, G. L., and Pryor, W. A., 1996, Acceleration of peroxynitrite oxidations by carbon dioxide, *Arch. Biochem. Biophys.* **327**, 335–343.

van der Vliet, A., Eiserich, J. P., O'Neill, C. A., Halliwell, B., and Cross, C. E., 1995, Tyrosine modification by reactive nitrogen species. A closer look, *Arch. Biochem. Biophys.* **319**, 341–349.

Verweij, H., Christianse, K., and van Steveninck, J., 1982, Ozone-induced formation of O,O'-dityrosine cross-links in proteins, *Biochim. Biophys. Acta* **701**, 180–184.

Vogt, W., 1995, Oxidation of methionine residues in proteins: Tools, targets, and reversal, *Free Rad. Biol. Med.* **18**, 93–105.

Waite, H., 1995, Precursors of quinone tanning: DOPA-containing compounds, *Methods Enzymol.* **258**, 1–20.

Wells-Knecht, M. C., Huggins, T. G., Dyer, G., Thorpe, S. R., and Baynes, J. W., 1993, Oxidized amino acids in lens protein with age, *J. Biol. Chem.* **268**, 12348–12352.

Winchester, R. V. and Lynn, K. R., 1970), X- and -radiolysis of some tryptophan dipeptides, *Int. J. Radiat. Biol.* **17**, 541–549.

Winterbourne, C. C., 1985, Comparative reactivities of various biological compounds with myeloperoxidase–hydrogen peroxide–chloride, and similarity of the oxidant to hypochloride, *Biochim. Biophys. Acta* **840**, 204–210.

Zhou, J. Q. and Gafni, A., 1991, Exposure of rat muscle phosphoglycerate kinase to a nonenzymatic MFO system generates the old form of enzyme, *J. Gerontol.* **46**, B217–B221.

6

NITRIC OXIDE REGULATION OF MEMBRANE AND LIPOPROTEIN OXIDATION IN THE VASCULATURE

A Radical Hypothesis

Bruce A. Freeman,[1,2] Jason Eiserich,[1] and Valerie O'Donnell[1]

[1]Departments of Anesthesiology
[2]Biochemistry and Molecular Genetics
University of Alabama at Birmingham
Birmingham, Alabama 35233

1. ABSTRACT

Nitric oxide ($\cdot$NO, nitrogen monoxide) exerts potent actions in the regulation of cell function and tissue viability that extend far beyond the recognized ability of $\cdot$NO to mediate signal transduction via activation of guanylate cyclase. Chemical reactions, cell culture systems, animal models and clinical studies have all revealed an ability of $\cdot$NO to modulate oxygen radical reactions and pathologic processes associated with inflammation. The focus of this article is to reveal the reactions of $\cdot$NO with reactive oxygen species and lipid radicals (LO$\cdot$ and LOO$\cdot$) formed during membrane and lipoprotein oxidation. The products of these reactions, LONO/LNO$_2$ and LOONO/LONO$_2$ adducts, are potentially reactive and can serve as mediators of inflammation and tissue injury. Because of the high reactivity of $\cdot$NO with lipophilic radicals and the interactions that $\cdot$NO can have with lipophilic antioxidants, the reactions of $\cdot$NO with membrane and lipoprotein oxidant defense mechanisms will be explored as well.

2. THE CENTRAL ISSUE

Free radicals and other oxidizing species have the capacity to directly mediate toxic reactions or initiate secondary reactions that become self-sustaining by regenerating

Free Radicals, Oxidative Stress, and Antioxidants, edited by Özben.
Plenum Press, New York, 1998.

65

propagating radicals. The best defenses against these events are to directly scavenge the initiating radical or to terminate the radicals that sustain propagation, via an integrated network of tissue antioxidant mechanisms. We propose that ·NO, produced in high concentrations during tissue pathology, can play a central role in this defensive network by a) serving as a terminator for lipid radicals (LO· and LOO·) formed during membrane and lipoprotein oxidation and b) protecting a-tocopherol from oxidation and rereducing oxidized a-tocopherol.

In diverse forms of vascular injury, free radical species including $O_2^{\cdot-}$, ·NO and lipid oxidation products play important homeostatic roles in mediating inflammatory reactions and maintaining vascular tone. In particular, ·NO displays a potent regulatory role in oxidative stress by both exacerbating and blunting the cytotoxic actions of reactive oxygen species, at the same time leading to production of the potent oxidant $ONOO^-$ following reaction with $O_2^{\cdot-}$. Nitric oxide also contributes to the formation of more stable nitrogen-containing adducts of aromatic amino acids in proteins, carbohydrates, DNA and lipids of tissues undergoing oxidative stress. It is our contention that these recently described nitrogen-containing products, often present in high concentrations in tissues, have unique biological and signal transduction properties and will serve as footprints for ·NO-mediated events in oxidant tissue injury.

3. OXYGEN RADICALS AND TISSUE INJURY

Aerobiosis permits efficient cell energy metabolism and poses the threat of highly reactive and potentially toxic oxygen byproducts. During normal cellular aerobic metabolism, about 98% of molecular oxygen is fully reduced to $2H_2O$ by 4 e^- transfer at mitochondrial cytochrome c oxidase, with no release of partially reduced intermediates. The remaining O_2 consumption (Freeman and Crapo, 1981) includes 1 or 2 e^- reduction of O_2 to $O_2^{\cdot-}$ and hydrogen peroxide (H_2O_2). Diverse cell components are responsible for $O_2^{\cdot-}$ and H_2O_2 production. Membrane-bound e^- transport systems (mitochondrial respiratory chain, endoplasmic reticular cytochrome P450 system) "leak" e^-, reducing O_2 to $O_2^{\cdot-}$ (Freeman and Crapo, 1982). A membrane-bound hemeprotein in both neutrophil and macrophage NADPH oxidase complexes is the primary source of inflammatory cell-derived $O_2^{\cdot-}$. A similar oxidase in vascular cells (Clancy, Leszczynska-Piziak, and Abramson, 1992; Mohazzab, Kaminski, and Wolin, 1995; Pagano, Ito, Tomheim, Gallop, Tauber and Cohen, 1995) is proposed to serve as a key source of reactive oxygen species in the vascular compartment. Other proteins including hemoglobin, xanthine oxidase (XO) and nitric oxide synthases (Freeman and Crapo, 1981; Schmidt, Hofmann, Schindler, Shutenko, Cunningham, and Feelisch, 1996; Xia, Dawson, Dawson, Snyder, and Zweier, 1996) are also critical sources of $O_2^{\cdot-}$ and H_2O_2, with the spontaneous or enzymatically catalyzed dismutation of $O_2^{\cdot-}$ often yielding H_2O_2. We have recently observed XO binding to vascular endothelium and showed that $O_2^{\cdot-}$ derived from endothelial-bound XO (McAdams, Vickers and Freeman,submitted; White, Darley-Usmar, McAdams, Berrington, Gore, Thomson, Parks, Tarpey and Freeman, 1996) attenuates endothelial-dependent, ·NO-mediated vascular relaxation in an animal model of atherosclerosis. Normally, endogenous tissue antioxidant defenses such as the superoxide dismutases (SOD), catalase, the glutathione peroxidase system and soluble or lipophilic scavengers (uric acid, ascorbate, thiols, a- and g-tocopherol, b-carotene) (Freeman and Crapo, 1982) maintain intracellular concentrations of reactive oxygen species in the nM range or less. In spite of this, there are pathological situations that activate oxidant production and overwhelm tissue antioxidant capabilities, thus facilitating target molecule reactions and causing toxicity due to impairment of fundamental metabolic and structural elements of tissues. The singular oxidative characteristics of different cell types and tissues will result in a unique spectrum of oxidizing species being produced at various tissue sites. From this, complex secondary tissue reactions of free radicals are expected.

4. SITE-DIRECTED ANTIOXIDANTS ARE THE BEST FREE RADICAL DEFENSE

Since most tissue oxidant production occurs intracellularly, at cell surfaces (e.g., the site of inflammatory cell margination) or in the interstitial matrix and yields species of short diffusional distances, it is important to target antioxidant interventions at or near the site of production of reactive species. Native CuZn SOD (Turrens, Crapo and Freeman, 1984) has a number of limitations as an exogenously administered scavenger of $O_2^{\cdot-}$, including a short circulating half-life (< 6 min) and a net negative charge at physiological pH, causing electrostatic repulsion from highly anionic cell surfaces and limited intracellular entry of therapeutic concentrations of SOD. The intracellular delivery of SOD and catalase, following conjugation with polyethylene glycol or pH-sensitive liposome entrapment, has lent resistance to oxidant stress in cell systems, animals subjected to cardiovascular or cerebrovascular ischemia-reperfusion injury and animal models of vascular dysfunction (Beckman, Minor, White, Rosen, Repine and Freeman, 1988; Buckley, Tanswell and Freeman, 1987; Freeman, Young and Crapo, 1983; Laursen, Rajagopalan, Tarpey, Freeman and Harrison, 1997; Liu, Beckman, Freeman, Hogan and Hsu, 1989; Munzel, Sayegh, Freeman, Tarpey and Harrison, 1995; Tamura, Chi, Driscoll, Hoff, Freeman, Gallagher and Lucchesi, 1988; Turrens et al., 1984). An extracellular SOD variant termed EC-SOD-C has been shown to a) have an extended circulating half-life, b) bind to cell surface and interstitial matrix glycosaminoglycans (GAGs) and c) exhibit more potent tissue-protective properties during oxidant stress than cytoplasmic CuZn SOD (Marklund, 1992; Abrahamsson, Brandt, Marklund and Sjoqvist, 1992; Inoue, Watanabe, Morino, Tanaka, Amachi and Sasaki, 1990). For example, a GAG-binding CuZn-SOD chimera (Nakazono, Watanabe, Matsuno, Sasaki, Sato and Inoue, 1991) induces a reduction of blood pressure in spontaneously hypertensive rats that was insensitive to native CuZn SOD. It was concluded that GAG-bound SOD intercepts $O_2^{\cdot-}$ that would otherwise "inactivate" ·NO from acting in its capacity as endothelial-derived relaxing factor (EDRF). Finally, a meso-substituted metalloporphyrin, Mn TBAP (5,10,15,20 tetrakis (4-benzoic acid) manganese (III) porphyrin) is proving to be an extremely facile site-directed antioxidant. It is cell-permeable, scavenges not only $O_2^{\cdot-}$, but also H_2O_2 and $ONOO^-$ at a very high rate constant (e.g., has peroxidase activity, unpublished observations), is stable and can be derivatized to bind to cell surfaces (Day and Crapo, 1996; Faulkner, Liochev and Fridovich, 1994). It not only protects cells from oxidant stress (Groves and Marla, 1995), but has been observed to keep Mn SOD knockout mice alive long term that otherwise die in 5–7 days (Irwin Fridovich, personal communication). The anionic vascular cell surface and interstitial matrix is a critical locus for oxidant reactions not only with cellular targets, but also with ·NO, yielding secondary species including $ONOO^-$ and oxidized lipid-·NO adducts. Thus, optimal inhibition of tissue oxidant injury is best accomplished by positioning enzymatic or enzyme mimetic antioxidants (which have higher rate constants than low molecular weight scavengers) at the cell glycocalyx or intracellularly.

5. NITRIC OXIDE—A FREE RADICAL SIGNAL TRANSDUCING AGENT

Nitric oxide production, first associated with vascular endothelium (Furchgott and Zawadski, 1980) and now linked with a multitude of cell types (Moncada and Higgs, 1991), activates cellular cyclic guanosine monophosphate (cGMP) synthesis by inducing a structural change in guanyl cyclase (Ignarro, 1992). This occurs via reaction with enzyme thiol or heme-iron moieties and is catalyzed not only by ·NO, but also by ·OH, thiol-reactive agents, carbon monoxide (CO) and other oxidation states of ·NO (Fukuto, Chiang, Hszieh, Wong and Chaudhur, 1992; Mittal and Murad, 1977; Wu, Brune, von Appen and Ullrich, 1992). The chemical properties of ·NO that result in its role as a mediator of smooth muscle relaxation, platelet inhibition, neurotransmission, and immune function (Lancaster, 1992) include a) the presence of a reactive unpaired electron, b) the ability to react with molecular oxygen and metalloproteins, giving tissues a nonenzymatic method for ·NO removal, c) direct or indirect reactivity with heme, iron-sulfur and thiol-containing proteins, influencing both its signal transduction and toxic properties (Schmidt, 1992), d) charge neutrality and a low Stoke's radius, permitting facile transmembrane diffusion and e) a long biological half-life in spite of its reactivity (Stamler, Singel, and Loscalzo, 1992). Nitric oxide is synthesized by oxidative deamidation of arginine to citrulline by constitutive and inducible forms of ·NO synthase (NOS). This family of enzymes are homologous to cytochrome P450 reductase and require NADPH, heme, flavins, tetrahydrobiopterin (BH_4) and molecular oxygen as cofactors (Bredt, Hwang, Glass, Lowenstein, Reed and Snyder, 1991; White and Marletta, 1992). The constitutive and inducible forms of ·NO synthase from a number of species have been cloned and sequenced including the neuronal, endothelial, smooth muscle cell and macrophage enzymes (Forsterman, Pollock, Schmidt, Heller and Murad, 1991; Lowenstein, Glass, Bredt and Snyder, 1992; Marsden, Schappert, Chen, Flowers, Sundell, Wilcox, Lamas and Michel, 1992).

Interestingly, NOS isoforms will autooxidize to yield $O_2^{·-}$ and consequently, both H_2O_2 and $ONOO^-$, especially under conditions of limiting substrate concentration (Schmidt *et al.*, 1996; Xia *et al.*, 1996). Furthermore, it is now becoming apparent that $ONOO^-$ not only directly induces oxidant-mediated cell injury, but also reversibly nitrosates thiols and other possible target molecules in low yields. This can result in subsequent ·NO-dependent vasorelaxant and tissue-protective reactions, the occurrence of which will critically depend on $ONOO^-$ concentration and the general tissue oxidative milieu (Lefer, Scalia, Campbell, Nossuli, Hayward, Salamon, Grayson and Lefer, 1997; Wu, Pritchard, Kaminski, Fayngersh, Hintze and Wolin, 1994). The biochemical reactions of ·NO during oxidative stress are thus complex and remain controversial. For example, the form in which ·NO is released from cells to serve as a bioactive molecule is disputed, with candidates being ·NO, S-nitrosothiol derivatives, iron-nitrosyl complexes, $ONOO^-$ and nitroxyl (HNO) or nitroxyl anion (NO^-) (Fukuto *et al.*, 1992; Knudsen, Svane and Tottrup, 1992; Murphy and Sies, 1991; Myers, Minor, Guerra, Bates and Harrison, 1990; Palmer, Ferrige and Moncada, 1987; Vanin, 1991; Xia *et al.*, 1996). The reversible nitrosothiol derivatives of ·NO lend more stability and potency to the vasorelaxant effects of ·NO, with the mechanisms of RSNO formation and subsequent ·NO release/action debated (Arnelle and Stamler, 1995; Karoui, Hogg, Frejaville, Tordo and Kalyanaraman, 1996; Stamler, Jaraki, Osborne, Simon, Keaney, Vita, Single, Valeri and Loscalzo, 1992a; Stamler, Simon, Osborne, Mullins, Jaraki, Michel, Singel and Loscalzo, 1992b; Tamir, Lewis, de Rojas, Dean, Wishnok and Tannenbaum, 1993). Also, SOD may catalyze the interconversion of ·NO and NO^- to protect ·NO/EDRF activity by an electron transfer mechanism which is in addition to and separate from scavenging of $O_2^{\cdot-}$ by dismutation (Murphy and Sies, 1991; Xia *et al.*, 1996). Importantly, $\cdot$NO is still a precursor and/or product of the different redox forms of this species, so we generally write "$\cdot$NO" but should be attuned to both nuances of its reactivities and emerging concepts about its chemical properties.

6. TISSUE INJURY MEDIATED BY $\cdot$NO/ONOO

Macrophage-derived $\cdot$NO has been recognized for a number of years to be immuno-modulatory, with the pathogen-killing activity of macrophages having L-arginine depend-ence, NOS/$\cdot$NO inducibility by lipopolysaccharide (LPS) or interferon-g (IFg) and concomitant production of the $\cdot$NO oxidation products NO_2^- (nitrite) and NO_3^- (nitrate) (Munoz-Fernandez, Fernandez and Fresno, 1992; Sherman, Loro, Wong, and Tashkin, 1991; Stuehr and Marletta, 1985). An important (if not principal) role for $\cdot$NO has been es-tablished in macrophage tumoricidal activity and the killing of invading microbes and parasites (Green, Mellouk, Hoffman, Meltzer and Nacy, 1990; Nussler, Drapier, Renia, Pied, Miltgen, Gentilini and Mazier, 1991; Padgett and Pruett, 1992; Stuehr and Nathan, 1989). The major mechanisms attributed to explain $\cdot$NO-mediated cytotoxic actions involve destruction of enzyme 4Fe-4S centers to yield inactive Fe–S–NO derivatives, (e.g., mitochondrial aconitase, NADH:ubiquinone oxidoreductase and succinate:ubiqui-none reductase inactivation), inhibition of DNA synthesis by inhibiting ribonucleotide reductase, inhibition of protein synthesis and nitrosylation of thiols (Curran, Ferrari, Kis-pert, Stadler, Stuehr, Simmons and Billiar, 1991; Hibbs, Taintor, Vavrin and Rachlin, 1988; Kwon, Stuehr and Nathan, 1991; Lancaster and Hibbs, 1990; Lepoivre, Flaman and Henry, 1992; Nathan, 1992; Stadler, Billiar, Curran, Stuehr, Ochoa and Simmons, 1991). Recent data compellingly shows that $\cdot$NO-induced toxicity also invokes a role for $O_2^{\cdot-}$, with previously proposed mechanisms of both $O_2^{\cdot-}$ and $\cdot$NO-mediated tissue injury often modified to now include a contributory if not predominant role for $ONOO^-$. For example, it was simultaneously reported by two groups that inhibition of aconitase, via oxidation of the 4Fe-4S center, was induced by $ONOO^-$ and not by $\cdot$NO as was previously viewed (Castro, Rodriguez and Radi, 1994; Hausladen, and Fridovich, 1994). We first proposed $ONOO^-$ as a biological oxidant, with $ONOO^-$ serving as both a potent one and two elec-tron oxidant, displaying hydroxyl radical-like reactivity and readily oxidizing thiols and lipids (Beckman, Beckman, Chen, Marshall and Freeman, 1990; Radi, Beckman, Bush and Freeman, 1991a; Radi, Beckman, Bush and Freeman, 1991b). Because the reaction of $O_2^{\cdot-}$ with $\cdot$NO has a slightly greater rate constant than that of SODs for scavenging $O_2^{\cdot-}$, $ONOO^-$ will be formed whenever tissue $\cdot$NO concentration approaches SOD concentra-tions. Recently, it has been observed that $\cdot$NO also mediates the nitration of biomolecules during oxidative stress (tyrosine, tryptophan, guanosine, g-tocopherol) (Alvarez, Rubbo, Kirk, Barnes, Freeman and Radi, 1996; Beckman, Ischiropoulos, Zhu, van der Woerd, Smith, Chen, Harrison, Martin and Tsai, 1992; Beckman, Ye, Anderson, Chen, Accavitti, Tarpey and White, 1994; Christen, Woodall, Shigenaga, Southwell-Keely, Duncan and Ames, in press; Leeuwenburgh, Hardy, Hazen, Wagner, Oh-ishi, Steinbrecher and He-inecke, 1997; Shigenaga, Lee, Blount, Christen, Shigeno, Yip and Ames, in press).

During inflammation, nitration reactions can be mediated by $ONOO^-$ via an activated isomerization intermediate, nitronium ion, $\cdot NO_2$ or complex of $ONOO^-$ with CO_2 (Denicola, Freeman, Trujillo and Radi, 1996; Ischiropoulos, Zhu, Chen, Tsai, Martin, Smith and Beck-man, 1992; Radi, Cosgrove, Beckman and Freeman, 1993; Van der Vliet, Eiserich, O'Neill, Halliwell and Cross, 1995; Yermilov, Yoshie, Rubio and Ohshima, 1996). Alternatively, we have recently reported facile aromatic nitrating reactions catalyzed by myeloperoxidase (MPO). Hypochlorous acid, an MPO-derived product released by activated neutrophils and monocytes, reacts with NO_2^- to form nitryl chloride (NO_2Cl), a product capable of both nitrating and chlorinating phenolic compounds including tyrosine (Eiserich, Cross, Jones, Halliwell and Van der Vliet, 1996). In addition, $NO_2^{\cdot-}$ can be directly oxidized by MPO compounds I and II to form nitrating species (presumably $\cdot NO_2$) (van der Vliet, Eiserich, Halliwell and Cross, 1997). Indeed, our recent observations demonstrating that activated human neutrophils can catalyze these unique reaction pathways (Eiserich, Hristova, Cross, Jones, Halliwell and Van der Vliet, submitted) underscores their physiological relevance and contribution to the formation of reactive nitrogen intermediates at sites of inflammation and tissue injury. These reaction pathways are of particular relevance to vascular diseases, since a) monocytes and other phagocytic cells implicated in the pathogenesis of atherosclerosis express MPO and produce HOCl upon stimulation (Lampert and Weiss, 1983), b) large amounts of active MPO are present within human atherosclerotic lesions (Daugherty, Dunn, Rateri and Heinecke, 1994), and c) antibodies specific for both chlorinated proteins (Hazell, Arnold, Flowers, Waeg, Malle and Stocker, 1996) and nitrotyrosine (Beckman *et al.*, 1994; Leewenburgh *et al.*, 1997) recognize epitopes in atherosclerotic lesions. However tissue nitration reaction mechanisms occur, in depth study is warranted because a) we and others are learning this is a quantitatively extensive event during oxidative stress (Gow, Duran, Malcolm and Ischiropoulos, 1996), b) nitrated tyrosines and tryptophans are highly immunogenic, c) nitrated aromatics can redox cycle via nitroaromatic anion radical intermediates and generate further reactive oxygen species, d) nitration of biomolecules will disrupt structure and catalytic functions of proteins, d) nitration of tyrosine impairs tyrosine kinase-mediated signal transduction and e) nitrated biomolecules can be metabolized to other bioactive intermediates (e.g., $\cdot NO$).

7. TISSUE PROTECTION MEDIATED BY $\cdot NO$

It is now becoming apparent that $\cdot NO$ can also display antioxidant qualities. For example, cytoprotection, rather than cytotoxicity, occurs in pulmonary cells and rats exposed to enhanced rates of production of both $O_2^{\cdot-}$ and $\cdot NO$ (Gutierrez, Chumley, Rivera and Freeman, 1996). Since the reaction of $\cdot NO$ with $O_2^{\cdot-}$ yields the potent oxidant $ONOO^-$, from a purely chemical point of view it would follow that a) an even broader array of target molecules would become susceptible to the toxic effects of reactive oxygen species when $\cdot NO$ is present and b) $\cdot NO$ potentiates the toxicity of reactive oxygen species. While this is sometimes the case, $\cdot NO$ also exerts direct or indirect antioxidant actions in biological systems subjected to oxidant stress. Examples of this reactivity are extensive:

7.1. Nitric Oxide Reactions with Metals

Because $\cdot NO$ readily serves as an iron ligand, $\cdot NO$ can modulate the prooxidant effects of iron. The reversible binding of $\cdot NO$ to the iron of reduced heme proteins (e.g., cytochrome P450 isoforms) (Hori, Masuya, Tsubaki, Yoshikawa and Ichikawa, 1992) results in the formation of $\cdot NO$-heme complexes. It has been postulated that $\cdot NO$ can exert a protective role towards metal complex and metalloprotein-catalyzed lipid oxidation via formation of catalytically inactive Fe-NO complexes, thus modulating the pro-oxidant effects of iron and other transition metals (Kanner, Harel and Granit, 1991). Importantly, very high (nonbiological) $\cdot NO$ concentrations are required to observe metal-dependent antioxidant effects of ·NO (Kanner *et al.*, 1991; Rubbo, Parthasarathy, Kalyanaraman, Barnes, Kirk and Freeman, 1995). The rates of ·NO reaction with most metal centers are orders of magnitude slower than for the almost diffusion-limited reaction of ·NO with either $O_2^{·-}$ or LO·/LOO· (Huie and Padmaja, 1993; Padmaja and Huie, 1993). Also, ·NO can mediate prooxidant reactions with transition metals by reducing ferric iron complexes (Brieland, Clarke, Karmil, Phan and Fantone, 1992; Reif and Simmons, 1990). Thus, modulation of radical reactions by ·NO-metal complex formation (aside from binding to guanylate cyclase) only partially explains the many oxidant-protective actions of ·NO.

7.2. Nitric Oxide Redirects Peroxynitrite-Mediated Oxidative Reactions

Peroxynitrite-mediated lipid peroxidation occurs by predominantly iron-independent mechanisms and can be inhibited by ·NO (Laskey and Mathews, 1996; Radi, Beckman, Bush and Freeman, 1991; Rubbo, Radi, Trujillo, Telleri, Kalyanaraman, Barnes, Kirk and Freeman, 1994). Nitric oxide may also inhibit $ONOO^-$-induced oxidant injury (e.g., thiol oxidation) by directly reacting with the trans form of $ONOO^-$ to yield both nitrosating/ nitrating intermediates and finally, less reactive products (Pfeiffer, Gorren, Schmidt, Werner, Hansert, Bohle and Mayer, 1997). Of the several functional groups in the biological milieu which can react directly with $ONOO^-$, thiols are of particular interest. While $ONOO^-$ irreversibly oxidizes thiols to higher oxidation states, ·NO can also mediate S-nitrosylation of thiols to form $ONOO^-$-resistant S-nitrosothiol adducts (Stamler, 1995). Finally, both dihydrorhodamine oxidation and benzoic acid hydroxylation induced by ·NO, $O_2^{·-}$ and $ONOO^-$ is inhibited by ·NO, emphasizing that ·NO can orchestrate switching between oxidation, hydroxylation and nitrosation reactions during oxidative stress (Miles, Bohle, Glassbrenner, Hansert, Wink and Grisham, 1996).

7.3. Nitric Oxide Modulates PMN Function and Inflammation in an Antioxidant-like Manner

A number of model systems for inflammation, oxidant lung injury, vascular disease (atherogenesis, restenosis following angioplasty) and surgical problems (ischemia-reperfusion injury, graft reanastomosis) — that include a pathogenic role for oxidant injury — reveal that endogenous ·NO biosynthesis or exogenously added sources of ·NO often inhibit oxidant-dependent damage at both molecular and tissue structural/functional levels. Many, if not all of these studies, have inflammatory injury as a common denominator. It is sometimes reported that PMN-derived $O_2^{·-}$ is "inactivated" or "scavenged" by ·NO (Miller and Britigan, 1995). These investigations were correct, in that $O_2^{·-}$-mediated cytochrome c reduction by PMNs was inhibited by added ·NO (at stupendous concentrations), but did not consider that $ONOO^-$, an even more potent oxidant, is the product of a "diversionary" reaction of $O_2^{·-}$ with ·NO. More carefully interpreted studies show that ·NO directly inhibits PMN NADPH oxidase (Beckman *et al.*, 1988), possibly via cytochrome b_{558} reaction and inactivation, but again only at high concentrations. Nitric oxide also inhibits leukocyte adhesion to vascular endothelium (Albina and Reichner, 1995), modulates PMN-dependent loss of microvascular barrier function (Flavahan, 1992) and inhibits platelet aggregation (Cooke, Singer, Tsao, Zera, Rowan and Billingham, 1992) all key steps in inflammatory tissue injury.

The protective effects of ·NO towards in vivo models of reperfusion injury, when ·NO is administered as a bolus of an ·NO donating drug, are often ascribed to ·NO inhibition of inflammatory cell margination and function (Kubes, Suzuki and Granger, 1991; Kurose, Wolf, Grisham and Granger, 1994; Lefer, Nakanishi and Vinten-Johansen, 1993; Niu, Ibbotson and Kubes, 1996; Siegfried, Carey, Ma and Lefer, 1992). In some of these models, inhibition of endogenous cell ·NO synthesis enhanced injury as well. Mechanisms underlying this ·NO-mediated phenomenon include both acute events and more delayed processes involving regulation of integrin gene expression. Acutely, ·NO administration to reperfused ischemic tissues results in stimulation of vessel wall, circulating platelet and neutrophil cGMP levels. This results in increased blood flow and oxygen delivery to tissues, as well as alterations in shear forces on the vessel wall, critical for regulating vascular-inflammatory cell interactions and secondary gene expression events (Yaqoob, Edelstein and Schrier, 1996). The translocation of P-selectin to the platelet surface and/or the function of P-selectin becomes inhibited by ·NO as well, resulting in attenuation of platelet and neutrophil aggregation, thus secondary "downstream" inflammatory cell-derived oxidant stress to the vasculature. It has even been recently reported that low concentrations of ONOO⁻ inhibit P-selectin expression on endothelium by secondary generation of ·NO from undescribed nitrosated/nitrated intermediates, an issue to be addressed in our experimental aims. Isolated organ preparations treated with inhibitors of ·NO synthesis showed increased vessel wall neutrophil adhesion and margination that was inhibitable by antibodies to CD18, the b-subunit of the integrin receptor for intercellular cell adhesion molecule-1 (ICAM-1) on neutrophils. Mast cell degranulation is also inhibited by ·NO, which will limit the release of other proinflammatory mediators such as histamine and platelet activating factor (Kurose *et al.*, 1994). In many of these studies, it is stressed that the "balance" between ·NO and O_2^- is critical in tissue-protective versus injurious outcomes, with underlying biochemical mechanisms often being indeterminate (Gergel, Misik, Ondrias and Cederbaum, 1995; Niu *et al.*, 1996; Yaqoob *et al.*, 1996).

7.4. Nitric Oxide Can Indirectly Act as an Antioxidant via Regulation of Redox Sensitive Proinflammatory Gene Expression

Oxidative signals play critical roles in the activation of gene expression of key inflammatory mediators. In particular, the redox-sensitive modulation of the expression and function of vascular genes is being revealed to play a central role in the initiation and propagation of atherosclerosis (Wintztum and Steinberg, 1991). This process involves regulatory events that are mediated by oxidative reactions in at least three key points, all of which can be modulated by ·NO. First, the oxidative modification of LDL by reactive oxygen species (seeding hydroperoxides) and enzymes such as lipoxygenase generates novel epitopes on LDL to yield minimally modified and further oxidized LDL (mmLDL and LDLox, respectively). These oxidized forms of LDL are recognized by macrophage scavenger receptors and become concentrated both intracellularly and indirectly at sites of macrophage accumulation (Wintztum and Steinberg, 1991). Nitric oxide can both stimulate as well as inhibit many of these lipid oxidative events. A second critical oxidative event in atherogenesis is the activation of proinflammatory gene transcription by oxidized lipids (Khan, Harrison, Olbrych, Alexander and Medford, 1996; Marui, Offerman, Swerlick, Kunsch, Ahmad, Alexander and Medford, 1993). Specifically, fatty acid hydroperoxides found in oxidized LDL and other plasma lipoproteins (e.g., 13-hydroperoxy-octadecadienoic acid) activate (at least) NFkB, with this transcription factor then mediating gene expression of an integrin termed vascular cell adhesion molecule-1 (VCAM-1). Similarly, cytokines such as interleukin-1a, interleukin-1b and tumor necrosis factor-a induce gene expression of VCAM-1 via redox-sensitive mechanisms (De Caterina, Libby, Peng, Thannickal, Rajavashisth, Gimbrone, Shin and Liao, 1995). In both oxidized lipid and cytokine-mediated activation of VCAM-1 gene expression, ·NO and other inhibitors of both lipid peroxyl radical formation and reaction inhibit oxidative activation of NFkB and gene expression of VCAM-1 (De Caterina et al., 1995; Khan et al., 1996; Marui et al., 1993). Additionally, ·NO indirectly inhibits NFkB-mediated gene expression by inducing the expression of and stabilizing the inhibitory protein for NFkB, IkBa, as well as inhibiting DNA binding of NFkB (Matthews, Botting, Panico, Morris and Hay, 1996; Peng, Libby and Liao, 1995). The third and final oxidant-protective event that ·NO mediates in vascular injury is a consequence of ·NO-mediated inhibition of integrin gene expression. Numerous studies in diverse model systems have shown that both antibody blocking of integrins and inhibition of integrin gene expression profoundly inhibit inflammatory cell-induced injury (Fujita, Morita and Murota, 1994). In summary, ·NO can often be vasoprotective because of inhibition of a) LDL lipid oxidation, b) the oxidative activation of proinflammatory gene transcriptional factors and c) extents of oxidative injury induced by inflammatory cells.

7.5. *In Vitro* Models of Nitric Oxide-Mediated Cytoprotection

The concept of ·NO-mediated cytoprotection from oxidative stress was affirmed by the seminal observation that rodent lung fibroblasts (V79 cells) and dopaminergic mesencephalic neuronal cells were protected by ·NO from toxicity, induced by either purine plus xanthine oxidase or addition of H_2O_2 to culture medium (Maragos, 1991; Wink, Cook, Pacelli, DeGraff, Gamson, Liebman, Krishna and Mitchell, 1996; Wink, Hanbauer, Krishna and DeGraff, Gamson and Mitchell, 1993). Subsequently, it was observed that V79 cells also became more resistant to the toxicity of organic hydroperoxides (LOOH) added to culture medium if chemical sources of ·NO were present during LOOH exposure (Wink, Cook, Krishna, Hanbauer, DeGraff, Gamson and Mitchell, 1995). There is both support for this concept as well as contrasting data provided in Preliminary Studies that emphasizes the critical role of the "balance" between relative rates of production of ·NO and $O_2^{·-}$ on cell and tissue responses.

This article has presented supporting evidence for the the concept that ·NO can play a critical role in regulating vascular and tissue injury. To summarize key issues:

- The reaction of ·NO with $O_2^{·-}$ redirects of the reactivity of $O_2^{·-}$ to other oxidative pathways which may be more or less cytotoxic. This modifies the physiologic actions of ·NO and often yields nitrogen-containing adducts of target molecules.
- Site-directed enzymatic or enzyme-mimetic antioxidants will potently inhibit the generation and reactions of ·NO-derived secondary reactive species.
- Nitric oxide will modulate the inflammatory and oxidant-producing potential of both tissue and reticuloendothelial cells via modulation of redox-sensitive gene expression of proinflammatory proteins.

It remains important to reveal the molecular mechanisms of ·NO action and the critical mitigating factors which manifest the toxic qualities of ·NO or alternatively, mediate its tissue-protective properties, if ·NO is to be used extensively as a drug.

REFERENCES

Abrahamsson, T., Brandt, U., Marklund, S. L., and Sjoqvist, P. O., 1992, Vascular bound recombinant extracellular superoxide dismutase type C protects against the detrimental effects of superoxide radicals on endothelium-dependent arterial relaxation. *Circ. Res.* **70**:264–271.

Albina, J. E., and Reichner, J. S.,1995, Nitric oxide in inflammation and immunity. *New Horizons* **3**: 46–64.

Alvarez, B., Rubbo, H., Kirk, M., Barnes, S., Freeman, B. A., and Radi, R., 1996, Peroxynitrite-dependent tryptophan nitration. *Chem. Res. Toxicol.* **9**: 390–396.

Arnelle, D. R., and Stamler, J. S., 1995, NO^+, $NO\cdot$ and NO^- donation by S-Nitrosothiols: Implications for regulation of physiological functions by S-nitrosylation and acceleration of disulfide formation. *Arch. Biochem. Biophys.* **318**: 279–285.

Beckman, J. S., Beckman, T. W., Chen, J., Marshall, P. A., and Freeman, B. A., 1990, Apparent hydroxyl radical production by peroxynitrite: Implications for endothelial injury from nitric oxide and superoxide. *Proc. Natl. Acad. Sci.* **87**:1620–1624.

Beckman, J. S., Ischiropoulos, H., Zhu, L., van der Woerd, M., Smith, C. D., Chen, J D., Harrison, J., Martin, J. C., and Tsai, M., 1992, Kinetics of SOD and iron catalyzed nitration of phenolics by peroxynitrite. *Arch. Biochem. Biophys.* **298**: 438–445.

Beckman, J. S., Minor, R. L., White, C. W., Rosen, G. M., Repine, J. R., and Freeman, B. A., 1988, Superoxide dismutase and catalase conjugated to polyethylene glycol increases endothelial enzyme activity and oxidant resistance. *J. Biol. Chem.* **263**: 6884–6892.

Beckman, J. S., Ye, Y., Anderson, P. G., Chen, J., Accavitti, M. A., Tarpey, M. M., and White, R., 1994, Extensive nitration of protein tyrosines in human atherosclerosis detected by immunohistochemistry. *Biol. Chem. Hoppe-Seyler.* **375**: 81–88.

Bredt, D. S., Hwang, P. M., Glass, C. E., Lowenstein, C., Reed, R. R., and Snyder, S. H., 1991, Cloned and expressed nitric oxide synthase structurally resembles cytochrome P-450 reductase. *Nature* **351**:714–718.

Brieland, J. K., Clarke, S. J., Karmil, S., Phan, S. H., and Fantone, J. C., 1992, Transferrin: A potential source of iron for oxygen free radical-mediated endothelial cell injury. *Arch. Biochem. Biophys.* **294**:265–270.

Buckley, B. J., Tanswell, A. K., and Freeman, B. A., 1987, Liposome mediated augmentation of catalase in type II alveolar epithelial cells protects against hydrogen peroxide injury. *J. Appl. Physiol.* **63**: 359–367.

Castro, L., Rodriguez, M., and Radi, R., 1994, Aconitase is readily inactivated by peroxynitrite, but not by its precursor, nitric oxide. *J. Biol. Chem.* **269**: 29409–29415.

Christen, S., Woodall, A. A., Shigenaga, M. K., Southwell-Keely, P. T., Duncan, M. W., and Ames, B. N., g-Tocopherol traps mutagenic electrophiles such as NOx and complements a-tocopherol: physiological implications. *Proc. Natl. Acad. Sci. USA* (In Press).

Clancy, R. M., Leszczynska-Piziak, J., and Abramson, S. B., 1992, Nitric oxide, an endothelial cell relaxation factor, inhibits neutrophil superoxide anion production via a direct action on the NADPH oxidase. *J. Clin. Invest.* **90**:1116–1121.

Cooke, J. P., Singer, A. H., Tsao, P., Zera, P., Rowan, R., and Billingham, M. E., 1992, Antiatherogenic effects of L-arginine in the hypercholesterolemic rabbit. *J. Clin. Invest.* **90**: 1168–1172.

Curran, R. D., Ferrari, F. K., Kispert, P. H., Stadler, J., Stuehr, D. H., Simmons, R. L., and Billiar, T. R., 1991, Nitric oxide and nitric oxide-generating compounds inhibit hepatocyte protein synthesis. *FASEB J.* **5**:2085–2092.

Daugherty, A., Dunn, J. L., Rateri, D. L., and Heinecke, J., 1994, Myeloperoxidase, a catalyst for lipoprotein oxidation, is expressed in human atherosclerotic lesions. *J. Clin. Invest.* **94**: 437–444.

Day, B. J., and Crapo, J. D., 1996, A metalloporphyrin superoxide dismutase mimetic protects against paraquat-induced lung injury in vivo. *Toxicol. Appl. Pharmacol.* **140**: 94–100.

De Caterina, R., Libby, P., Peng, H., Thannickal, V., Rajavashisth, T., Gimbrone, M. A., Shin, W., and Liao, J. K., 1995, Nitric oxide decreases cytokine-induced endothelial activation: Nitric oxide selectively reduces endothelial expression of adhesion molecules and proinflammatory cytokines. *J. Clin. Invest.* **96**: 60–68.

Denicola, A., Freeman, B. A., Trujillo, M., and Radi, R., 1996, Peroxynitrite reaction with carbon dixide/bicarbonate: kinetics and influence on peroxynitrite-mediated oxidations. *Arch. Biochem. Biophys.* **333**: 49–58. Appendix J

Eiserich, J. P., Cross, C. E., Jones, A. D., Halliwell, B., and Van der Vliet, A., 1996, Formation of nitrating and chlorinating species by reaction of nitrite with hypochlorous acid: a novel mechanism for nitric oxide-mediated protein modification. *J. Biol. Chem.* **271**: 19199–19208. Appendix G

Eiserich, J. P., Hristova, M., Cross, C. E., Jones, A. D., Halliwell, B., and van der Vliet, A., Neutrophil-mediated conversion of nitrite into the nitrating and chlorinating inflammatory oxidant NO_2Cl. *Nature* (Submitted, 1997).

Faulkner, K. M., Liochev, S.I., and Fridovich, I., 1994, Stable Mn(III) porphyrins mimic superoxide dismutase in vitro and substitute for it in vivo. *J. Biol. Chem.* **269**: 23471–23476.

Flavahan, N. A., 1992, Atherosclerosis or lipoprotein-induced endothelial dysfunction. Potential mechanisms underlying reduction in EDRF/nitric oxide activity. *Circulation* **85**: 1927–1938.

Forstermann, U., Pollock, J. S., Schmidt, H. H. H. W., Heller, M., and Murad, F., 1991, Calmodulin-dependent endothelium-derived relaxing factor/nitric oxide synthase activity is present in the particulate and cytosolic fractions of bovine aortic endothelial cells. *Proc. Natl. Acad. Sci. USA* **88**: 1788–1792.

Freeman, B. A., and Crapo, J. D., 1981, Hyperoxia increases oxygen radical production in rat lungs and lung mito-chondria. *J. Biol. Chem.* **256**:10986–10992.

Freeman, B. A., and Crapo, J. D., 1982, Biology of disease: Free radicals and tissue injury. *Lab. Invest.* **47**:412–426.

Freeman, B. A., Young, S. L., and Crapo, J. D., 1983, Liposome-mediated augmentation of superoxide dismutase in endothelial cells prevents oxygen injury. *J. Biol. Chem.* **258**: 12534–12542.

Fujita, H., Morita, I., and Murota, S., 1994, A possible mechanism for vascular endothelial cell injury elicited by activated leukocytes: A significant involvement of adhesion molecules, CD11/CD18 and ICAM-1. *Arch. Biochem. Biophys.* **309**: 62–69.

Fukuto, J. M., Chiang, K., Hszieh, R., Wong, P., and Chaudhur, G., 1992, The pharmacological activity of nitroxyl: A potent vasodilator with activity similar to nitric oxide and/or endothelium-derived relaxing factor. *J. Pharmacol. Exp. Ther.* **263**:546–551.

Furchgott, R. F. and Zawadski, J.B., 1980, The obligatory role of endothelial cells in the relaxation of arterial smooth muscle by acetylcholine. *Nature* **288**: 373–376.

Gergel, D., Misik, V., Ondrias, K., and Cederbaum, A., 1995, Increased cytotoxicity of 3-morpholinosydnonimine to HepG2 cells in the presence of superoxide dismutase. *J. Biol. Chem.* **270**: 20922–20929.

Gow, A. J., Duran, D., Malcolm, S., and Ischiropoulos, H., 1996, Effects of peroxynitrite-induced protein modifi-cations on tyrosine phosphorylation and degradation. *FEBS Lett.* **385**: 63–66.

Green, S. J., Mellouk, S., Hoffman, S. L., Meltzer, M. S., and Nacy, C. A., 1990, Cellular mechanisms of nonspe-cific immunity to intracellular infection: cytokine-induced synthesis of toxic nitrogen oxides from L-ar-ginine by macrophages and hepatocytes. *Immunol. Lett.* **25**:15–20.

Groves, J. T., and Marla, S. S., 1995, Peroxynitrite-induced DNA strand scission mediated by a manganese por-phyrin. *J. Am. Chem. Soc.* **117**: 9578–9579.

Gutierrez, H. H., Chumley, P., Rivera, A.,and Freeman, B. A., 1996, Nitric oxide regulation of superoxide-depend-ent lung cell injury: oxidant-protective actions of endogenously produced and exogenously administered nitric oxide. *Free. Rad. Biol. Med.* **21**: 43–52.

Hausladen, A. and Fridovich, I., 1994, Superoxide and peroxynitrite inactivate aconitases, but nitric oxide does not. *J. Biol. Chem.* **269**: 29405–29408.

Hazell, L. J., Arnold, L., Flowers, D., Waeg, G., Malle, E., and Stocker, R., 1996, Presence of hypochlorite-modi-fied proteins in human atherosclerotic lesions. *J. Clin. Invest.* **97**: 1535–1544.

Hibbs, Jr. J. B.,Taintor, R. R., Vavrin, Z., and Rachlin, E. M., 1988, Nitric oxide: A cytotoxic activated macro-phage effector molecule. *Biochem. Biophys. Res. Commun.* **157**:87–94.

Hori, H., Masuya, F., Tsubaki, M., Yoshikawa, S., and Ichikawa, Y., 1992, Electronic and stereochemical charac-terizations of intermediates in the photolysis of ferric cytochrome P450 nitrosyl complexes. *J. Biol. Chem.* **267**:18377–18381.

Huie, R. E., and Padmaja, S., 1993, Reaction of NO with $O_2^{\cdot-}$. *Free Rad. Res. Comm.* **18**:195–199.

Ignarro, L.J., 1992, Haem-dependent activation of cytosolic guanylate cyclase by nitric oxide: a widespread signal transduction mechanism. *Biochem. Soc. Transactions.* **20**: 465–469.

Inoue, M., Watanabe, N., Morino, Y., Tanaka, Y., Amachi, T., and Sasaki, J., 1990, Inhibition of oxygen toxicity by targeting superoxide dismutase to endothelial cell surface. *FEBS Lett.* **269**:89–92.

Ischiropoulos, H., Zhu, L., Chen, J., Tsai, M., Martin, J., Smith, C., and Beckman, J. S., 1992, Peroxynitrite-medi-ated nitration of tyrosine catalyzed by superoxide dismutase. *Arch. Biochem. Biophys.* **298**: 431–437.

Kanner, J., Harel, S., and Granit, R., 1991, Nitric oxide as an antioxidant. *Arch. Biochem. Biophys.* **289**: 130–136.

Karoui, H., Hogg, N., Frejaville, C., Tordo, P., and Kalyanaraman, B., 1996, Characterization of sulfur-centered radical intermediates formed during the oxidation of thiols and sulfite by peroxynitrite. *J. Biol. Chem.* **271**: 6000–6009.

Khan, B. V., Harrison, D. G., Olbrych, M. T., Alexander, R. W., and Medford, R. M., 1996, Nitric oxide regulates vascular cell adhesion molecule 1 gene expression and redox-sensitive transcriptional events in human vas-cular endothelial cells. *Proc. Natl. Acad. Sci. USA* **93**:9114–9119.

Knudsen, M. A., Svane, D., and Tottrup, A., 1992, Action profiles of nitric oxide, S-nitroso-L-cysteine, SNP, and NANC responses in opossum lower esophageal sphincter. *Am. J. Physiol.* **262**: G840–846.

Kubes, P., Suzuki, M.,and Granger, D. N., 1991, Nitric oxide: An endogenous modulator of leukocyte adhesion. *Proc. Natl. Acad. Sci. USA* **88**: 4651–4655.

Kurose, I., Wolf, R., Grisham, M. B., and Granger, D. N., 1994, Modulation of ischemia/reperfusion-induced mi-crovascular dysfunction by nitric oxide. *Circ. Res.* **74**: 376–382.

Kwon, N. S., Stuehr, D. H., and Nathan, C. F., 1991, Inhibition of tumor cell ribonucleotide reductase by macro-phage-derived nitric oxide. *J. Exp. Med.* **174**:761–767.

Lampert, M. B., and Weiss. S. J., 1983, The chlorinating potential of the human monocyte. *Blood* **62**: 645–651.

Lancaster, J. R., 1992, Nitric oxide in cells: This simple molecule plays Janus-faced roles in the body, acting as both messenger and destroyer. *Am. Sci.* **80**:248–259.

Lancaster, J. R., and Hibbs, J. B. Jr., 1990, EPR demonstration of iron-nitrosyl complex formation by cytotoxic activated macrophages. *Proc. Natl. Acad. Sci. USA* **87**:1223–1227.

Laskey, R. E., and Mathews, W. R., 1996, Nitric oxide inhibits peroxynitrite-induced production of hydroxyeicosatetraenoic acids and F2-isoprostanes in phosphatidylcholine liposomes. *Arch. Biochem. Biophys.* **330**:193–198.

Laursen, J. B., Rajagopalan, S.,Tarpey, M., Freeman, B. A., and Harrison, D. G., 1997, A role of superoxide in angiotension II-but not catecholamine-induced hypertension. *Circulation* **95**: 588–593.

Leeuwenburgh, C., Hardy, M. M., Hazen, S. L., Wagner, P., Oh-ishi, S., Steinbrecher, U. P., and Heinecke, J. W., 1997, Reactive nitrogen intermediates promote low density lipoprotein oxidation in human atherosclerotic intima. *J. Biol. Chem.* **272**: 1433–1436.

Lefer, D. J., Nakanishi, K., and Vinten-Johansen, J., 1993, Endothelial and myocardial cell protection by a cysteine-containing nitric oxide donor after myocardial ischemia and reperfusion. *J. Cardiovasc. Pharmacol.* **22** (Suppl. 7): S34–S43.

Lefer, D. J., Scalia, R., Campbell, B., Nossuli, T., Hayward, R., Salamon, M., Grayson, J., and Lefer, A. M., 1997, Peroxynitrite inhibits leukocyte-endothelial cell interactions and protects against ischemia-reperfusion injury in rats. *J. Clin. Invest.* **99**: 684–691.

Lepoivre, M., Flaman, J. M., and Henry, Y., 1992, Early loss of the tyrosyl radical in ribonucleotide reductase of adenocarcinoma cells producing nitric oxide. *J. Biol. Chem.* **267**: 22994–23000.

Liu, T. H., Beckman, J. S., Freeman, B. A., Hogan, E. L., and Hsu, C. Y., 1989, Polyethylene glycol-conjugated superoxide dismutase and catalase reduce ischemic brain injury. *Am. J. Physiol.* **260**: H589–593.

Lowenstein, C. J., Glass, C. S., Bredt, D. S., and Snyder, S H., 1992, Cloned and expressed macrophage nitric oxide synthase contrasts with the brain enzyme. *Proc. Natl. Acad. Sci. USA* **89**: 6711–6715.

Maragos, C. M., 1991, Complexes of nitric oxide with nucleophiles as agents for the controlled biological release of nitric oxide. *J. Med. Chem.* **34**: 3242–3247.

Marklund, S., 1992, Regulation of cytokines of extracellular superoxide dismutase and other superoxide dismutase isoenzymes in fibroblasts. *J. Biol. Chem.* **267**: 6696–6701.

Marsden, P. A., Schappert, K. T., Chen, H. S., Flowers, M., Sundell, C. L., Wilcox, J. N., Lamas, S.,and Michel, T., 1992, Molecular cloning and characterization of human endothelial nitric oxide synthase. *FEBS Lett.* **307**: 287–293.

Marui, N., Offermann, M., Swerlick, R., Kunsch, C., Ahmad, M., Alexander, R., and Medford, R. M., 1993, VCAM-1 gene transcription and expression are regulated through an antioxidant-sensitive mechanism in human vascular endothelial cells. *J. Clin. Invest.* **92**: 1866–1874.

Matthews, J. R., Botting, C. H., Panico, M., Morris, H. R., and Hay, R. T., 1996, Inhibition of NF-kB DNA binding by nitric oxide. *Nucleic Acids Research* **24**:2236–2242.

McAdams, M., Vickers, S., and Freeman, B. A., A specific receptor for xanthine oxidase on vascular endothelium. *FEBS Lett.* (submitted).

Miles, A. M., Bohle, D. S., Glassbrenner, P. A., Hansert, B., Wink, D. A., and Grisham, M. B., 1996, Modulation of superoxide-dependent oxidation and hydroxylation reactions by nitric oxide. *J. Biol. Chem.* **271**:40–47.

Miller, R. A., and Britigan,.B. E., 1995. The formation and biologic significance of phagocyte-derived oxidants. *J. Invest. Med.* **43**: 39–49.

Mittal, C. K., and Murad, F., 1977, Activation of guanylate cyclase by superoxide dismutase and hydroxyl radical: A physiological regulator of guanosine 3',5'-monophosphate formation. *Proc. Natl. Acad. Sci. USA* **74**: 4360–4364.

Moncada, S., and Higgs, E. A., 1991, Endogenous nitric oxide: physiology, pathology and clinical relevance. *Eur. J. Clin. Inv.* **21**: 361–374.

Mohazzab, K. M., Kaminski, P. M., and Wolin,.M. S., 1995, NADH oxidoreductase is a major source of superoxide anion in bovine coronary artery endothelium. *Am. J. Physiol.* **266**, H2568–H2572.

Munoz-Fernandez, M. A., Fernandez, M A.,and Fresno,.M , 1992, Synergism between tumor necrosis factor-a and interferon-g on macrophage activation for the killing of intracellular Trypanosoma cruzi through a nitric oxide-dependent mechanism. *Eur. J. Immunol.* **22**:301–307.

Munzel, T., Sayegh, H., Freeman, B. A., Tarpey, M. M., and Harrison, D. G., 1995, Evidence for enhanced vascular superoxide anion production in nitrate tolerance: a novel mechanism underlying tolerance and cross tolerance. *J. Clin. Invest.* **95**: 187–194.

Murphy, M. E., and Sies, H., 1991, Reversible conversion of nitroxyl anion to nitric oxide by superoxide dismutase. *Proc. Natl. Acad. Sci. USA* **88**: 10860–10864.

Myers, R. R., Minor, R. L., Guerra, R., Bates, J. N., and Harrison, D. G., 1990, Vasorelaxant properties of the endothelial-derived relaxant factor more closely resemble S-nitrosocysteine than nitric oxide. *Nature* **365**: 161–163.

Nakazono, K., Watanabe, N., Matsuno, K., Sasaki, J., Sato, T., and Inoue, M., 1991, Does superoxide underlie the pathogenesis of hypertension? *Proc. Natl. Acad. Sci. USA* **88**: 10045–10048.

Nathan, C., 1992, Nitric oxide as a secretory product of mammalian cells. *FASEB J.* **6**:3051–3064.

Niu, X. F., Ibbotson, G., and Kubes, P., 1996, A balance between nitric oxide and oxidants regulates mast cell-dependent neutrophil-endothelial cell interactions. *Circ. Res.* **79**: 992–999.

Nussler, A., Drapier, J. C., Renia, L., Pied, S., Miltgen, F., Gentilini, M., and Mazier, D., 1991, L-arginine-dependent destruction of intrahepatic malaria parasites in response to tumor necrosis factor and/or interleukin-6 stimulation. *Eur. J. Immunol.* **21**:227–230.

Padgett, E. L., and Pruett, S. B., 1992, Evaluation of nitrite production by human monocyte-derived macrophages. *Biochem. Biophys. Res. Commun.* **186**:775–781.

Padmaja, S., and Huie, R. E., 1993, The reaction of nitric oxide with organic peroxyl radicals. *Biochem. Biophys. Res.Comm.* **195**, 539–544.

Pagano, P., Ito, Y., Tornheim, K., Gallop, P., Tauber, A., and Cohen, R., 1995, An NADPH oxidase superoxide-generating system in the rabbit aorta. *Am. J. Physiol.* **268**, H2274–2280.

Palmer, R. M. J., Ferrige, A. G., and Moncada, S., 1987, Nitric oxide release accounts for the biological activity of endothelium-derived relaxing factor. *Nature* **327**: 524–526.

Peng, H., Libby, P., and Liao, J. K., 1995, Induction and stabilization of IkB by nitric oxide mediates inhibition of NFkB. *J. Biol. Chem.* **270**: 14214–14219.

Pfeiffer, S., Gorren, A. C. F., Schmidt, K., Werner, E. R., Hansert, B., Bohle, D. S., and Mayer, B., 1997, Metabolic fate of peroxynitrite in aqueous solution: reaction with nitric oxide and pH-dependent decomposition to nitrite and oxygen in a 2:1 stoichiometry. *J. Biol. Chem.* **272**: 3465–3470.

Radi, R., Beckman, J. S., Bush, K. M., and Freeman, B. A., 1991a, Peroxynitrite oxidation of sulfhydryls: the cytotoxic potential of endothelial-derived superoxide and nitric oxide. *J. Biol. Chem.* **266**:4244–4250.

Radi, R., Beckman, J. S., Bush, K. M., and Freeman, B. A., 1991b, Peroxynitrite-induced membrane lipid peroxidation: the cytotoxic potential of superoxide and nitric oxide. *Arch. Biochem. Biophys.* **288**:481–487.

Radi, R., Cosgrove, T. P., Beckman, J. S., and Freeman, B. A., 1993, Peroxynitrite-induced luminol chemiluminescence. *Biochem. J.* **290**: 51–57.

Reif, D. W., and Simmons,.R. D , 1990, Nitric oxide mediates iron release from ferritin. *Arch. Biochem. Biophys.* **283**:537–541.

Rubbo, H., Parthasarathy, S., Kalyanaraman, B., Barnes, S., Kirk, M., and Freeman, B. A., 1995, Nitric oxide inhibition of lipoxygenase-dependent liposome and low density lipoprotein oxidation: Termination of radical chain propagation reactions and formation of nitrogen-containing oxidized lipid derivatives. *Arch. Biochem. Biophys.* **324**: 15–25.

Rubbo, H., Radi, R., Trujillo, M., Telleri, R., Kalyanaraman, B., Barnes, S., Kirk M.,and Freeman, B. A., 1994, Nitric oxide regulation of superoxide and peroxynitrite-dependent lipid peroxidation: Formation of novel nitrogen-containing oxidized lipid derivatives. *J. Biol. Chem.* **269**: 26066–26075.

Schmidt, H., 1992, NO·, CO and ·OH, endogenous guanylyl cyclase-activating factors. *FEBS Lett* **397**:102–107.

Schmidt, H. H. H. W., Hofmann, H., Schindler, U., Shutenko, Z. S., Cunningham, D. D., and Feelisch, M., 1996, No ·NO from NO synthase. *Proc. Natl. Acad. Sci. USA* **93**: 14492–14497.

Sherman, M. P., Loro, M. L., Wong, V. Z., and Tashkin, D. P., 1991, Cytokine- and Pneumocystis carinii-induced L-arginine oxidation by murine and human pulmonary alveolar macrophages. *J. Protozool.* **38**:234S–236S.

Shigenaga, M. K., Lee, H., Blount, B., Christen, S., Shigeno, E. T., Yip, H., and Ames, B. N., Inflammation and NO_x-induced nitration: assay for 3-nitrotyrosine by HPLC with electrochemical detection. *Proc. Natl. Acad. Sci. USA* (In Press).

Siegfried, M. R., Carey, C., Ma, X.,and Lefer, A. M., 1992, Beneficial effects of SPM-5185, a cysteine-containing nitric oxide donor in myocardial ischemia-reperfusion. *Am. J. Physiol.* **263**: H771–777.

Stadler, J., Billiar, T. R., Curran, R. D., Stuehr, D. J., Ochoa, J. B., and Simmons, R. L., 1991, Effect of exogenous and endogenous nitric oxide on mitochondrial respiration of rat hepatocytes. *Am. J. Physiol.* **260**:C910–916.

Stamler, J. S., 1995, S-nitrosothiols and bioregulatory actions of nitrogen oxides through reactions with thiol groups. *Cur. Top. Mic. Immunol.* **196**: 19–36.

Stamler, J. S., Jaraki, O., Osborne, J., Simon, D. I., Keaney, J., Vita, J., Single, D., Valeri, C. R., and Loscalzo, J., 1992a, Nitric oxide circulates in mammalian plasma primarily as an S-nitroso adduct of serum albumin. *Proc. Natl. Acad. Sci. USA* **89**: 7674–7677.

Stamler, J. S., Simon, D. I., Osborne, J. A., Mullins, M. E., Jaraki, O., Michel, T., Singel, D. J., and Loscalzo, J., 1992b, S-nitrosylation of proteins with nitric oxide: Synthesis and characterization of biologically active compounds. *Proc. Natl. Acad. Sci. USA* **89**:444–448.

Stamler, J. S., Singel, D. J., Loscalzo, J., 1992c, Biochemistry of nitric oxide and its redox-activated forms. *Science* **258**: 1898–1902.

Stuehr, D. J., and Marletta, M. A., 1985, Mammalian nitrate biosynthesis: Mouse macrophages produce nitrite and nitrate in response to Escherichia coli lipopolysaccharide. *Proc. Natl. Acad. Sci. USA* **82**: 7738–7742.

Stuehr, D. J., and Nathan, C. F., 1989, Nitric oxide - a macrophage product responsible for cytostasis and respiratory inhibition in tumor target cells. *J. Exp. Med.* **169**:1543–1555.

Tamir, S., Lewis, R. S., de Rojas, W. T., Dean, W. M., Wishnok, J. S., and Tannenbaum, S. R., 1993, The influence of delivery rate on the chemistry and biological effects of nitric oxide. *Chem. Res. Toxicol.* **6**: 895–899.

Tamura, Y., Chi, L., Driscoll, E. M. Jr., Hoff, P. T., Freeman, B. A., Gallagher, K. P., and Lucchesi, B. R., 1988, Superoxide dismutase conjugated to polyethylene glycol provides sustained protection against myocardial ischemia/reperfusion injury in the canine heart. *Circ. Res.* **63**: 944–959.

Turrens, J. F., Crapo, J. D., and Freeman, B. A., 1984, Protection against oxygen toxicity by intravenous injection of liposome-entrapped catalase and superoxide dismutase. *J. Clin. Invest.* **73**:87–95.

Van der Vliet, A., Eiserich, J. P., O'Neill, C. A., Halliwell, B., and Cross, C. E., 1995, Tyrosine modification by reactive nitrogen species: a closer look. *Arch. Biochem. Biophys.* **319**: 341–349.

Van der Vliet, A., JP Eiserich, J. P., B Halliwell, B., and CE Cross, C. E., 1997, Formation of reactive nitrogen species during peroxidase-catalyzed oxidation of nitrite: a potential additional mechanism of nitric oxide-dependent toxicity. *J. Biol. Chem.* **272**: 7617–7625.

Vanin, A. F., 1991, Endothelium-derived relaxing factor is a nitrosyl iron complex with thiol ligands. *FEBS Lett.* **289**: 1–3.

White, R., Darley-Usmar, V., McAdams, M., Berrington, W. R., Gore, J., Thomson, J. A., Parks, D. A., Tarpey, M. M., and Freeman, B. A., 1996, Circulating plasma xanthine oxidase contributes to vascular dysfunction in hypercholesterolemic rabbits. *Proc. Natl. Acad. Sci. USA* **93**: 8745–8749.

White, K. A., and Marletta, M. A., 1992, Nitric oxide synthase is a P-450 type hemoprotein. *Biochemistry* **31**:6627–6631.

Wink, D. A., Cook, J., Krishna, M. C., Hanbauer, I., DeGraff, W., Gamson, J., and Mitchell, J. B., 1995, Nitric oxide protects against alkyl peroxide-mediated cytotoxicity: Further insight into the role nitric oxide plays in oxidative stress. *Arch. Biochem. Biophys.* **319**: 402–407.

Wink, D. A, Cook, J. A., Pacelli, R., DeGraff, W., Gamson, J., Liebmann, J., Krishna, M. C., and Mitchell, J. B., 1996, The effect of various nitric oxide-donor agents on hydrogen peroxide-mediated toxicity: a direct correlation between nitric oxide formation and protection. *Arch. Biochem. Biophys.* **331**: 241–248.

Wink, D. A., Hanbauer, I., Krishna, M. C., DeGraff, W., Gamson, J., and Mitchell, J. B., 1993, Nitric oxide protects against cellular damage and cytotoxicity from reactive species. *Proc. Natl. Acad. Sci. USA* **90**: 9813–9817.

Witztum, J. L., and Steinberg, D., 1991, Role of oxidized low density lipoprotein in atherogenesis. *J. Clin. Invest.* **84**: 1086–1095.

Wu, X. B., Brune, B., von Appen, F., and Ullrich, V., 1992, Reversible activation of soluble guanylate cyclase by oxidizing agents. *Arch. Biochem. Biophys.* **294**:75–82.

Wu, M., Pritchard, Jr. K. A.,, Kaminski, P. M., Fayngersh, R. P., Hintze, T. H., and Wolin, M. S., 1995, Involvement of nitric oxide and nitrosothiols in relaxation of pulmonary arteries to peroxynitrite. *Am. J. Physiol.* **266**: H2108–H2113.

Xia, Y., Dawson, V. L., Dawson, T. M., Snyder, S. H., and Zweier, J. L., 1996, Nitric oxide synthase generates superoxide and nitric oxide in arginine-depleted cells leading to peroxynitrite-mediated cellular injury. *Proc. Natl. Acad. Sci. USA* **93**: 6770–6774.

Yaqoob, M., Edelstein, C. L., and Schrier, R. W., 1996, Role of nitric oxide and superoxide balance in hypoxia-reoxygenation proximal tubular injury. *Nephrol. Dial. Transplant.* **11**: 1743–1746.

Yermilov, V., Yoshie, Y., Rubio, J., and Ohshima, H., 1996, Effects of carbon dioxide/bicarbonate on induction of DNA single-strand breaks and formation of 8-nitroguanine, 8-oxoguanine and base-propenal mediated by peroxynitrite. *FEBS Lett.* **399**: 67–70.

7

APOLIPOPROTEIN E OXIDATION THROUGH ENZYMES LOCALIZED INTO THE BRAIN

B. Leininger-Muller, C. Jolivalt, R. Herber, P. Bertrand, and G. Siest

Centre du Médicament
Université Henri Poincaré Nancy I
30 Rue Lionnois, 54000 Nancy, France

1. INTRODUCTION

The cerebral tissue is highly susceptible to oxidative damage as it essentially contains lipids and proteins, which are potential targets for free radicals. Oxidation of proteins results as an increase of carbonyl groups through the formation of aldehydes and ketones (Stadtman and Oliver, 1991) and accumulation of oxidized proteins has been observed in Alzheimer's disease (AD) brains (Smith et al., 1991).

The brain contains numerous enzymes involved in xenobiotic metabolism found primarily in the liver, albeit at relatively low specific activities (Minn et al., 1991). For several years, we have studied the cellular and regional localization in the rat brain, showing in particular their preferential localization in the blood-brain barrier (Ghersi-Egea et al., 1988), other interfaces like leptomeninges and choroid plexus (Leininger-Muller et al., 1994), as well as circumventricular organs, lacking a blood-brain barrier (Ghersi-Egea et al., 1992). We also studied subcellular distribution of cytochrome P450 and measured activities of conjugating enzymes in human brain (Ghersi-Egea et al., 1993). The metabolic processes involved in the biotransformation of xenobiotics can be classified into phase I (functionalization) and phase II (conjugation) reactions. The substrates of phase I reactions are generally lipophilic molecules which undergo oxidative reactions leading to more polar metabolites. These molecules can then serve as substrates for phase II enzymes. However, phase I metabolism can also generate toxic reactive metabolites and/or free radicals. We previously reported that superoxide radical formation is largely mediated by NADPH cytochrome P450 (c) reductase reductive pathway in the rat brain (Ghersi-Egea et al., 1991) and that the free radicals produced lead to a diminution of the cytochrome P450 dependent activity itself.

More recently, we compared drug-metabolizing enzymes activities in AD and control brains. The activity of one phase II enzyme, the glutathione *S*-transferase, was increased in AD samples, while phase I enzymes related activities globally decreased (Leininger-Muller et al., submitted).

Free Radicals, Oxidative Stress, and Antioxidants, edited by Özben.
Plenum Press, New York, 1998.

Among other enzymes catalysing cerebral oxidation reactions, we recently showed the presence of the myeloperoxidase (MPO) enzymatic system in the human brain (Jolivalt et al., 1996). In the presence of H_2O_2 and Cl^-, this heme protein catalyses peroxidation of chloride to hypochlorite which then rapidly interacts with oxidizable moieties of proteins(Drozdz and Naskalski, 1988). The MPO enzymatic system is known to be of importance for the inflammatory processes involved in atherosclerosis. In the brain of AD patients, it could be implicated in the formation of amyloid plaques which also depends from inflammatory processes (Kisilevsky, 1993).

Among the proteins accumulating in cerebral amyloid deposits, we focused our attention on apolipoprotein E (apo E) which is a polymorphic protein corresponding to the product of three alleles ε2, ε3 and ε4, differing from each other by a single mutation (Mahley, 1988). It is the major apolipoprotein synthetized in the brain and its main function is to locally recycle cholesterol and phospholipids (Poirier, 1994). We previously showed *in vitro* that recombinant apo E is another target protein of superoxide radicals produced through the cytochrome P450 metabolic pathway (Leininger-Muller et al., 1995). Our interest in apo E in AD came with the high linkage of ε4 allele with AD (for a review, see Siest et al., 1995) and from the possible role of its oxidation in the apo E- amyloid peptide (Aβ) interactions, as well as by its ability to interfere with *A*β fibrillogenesis (Soto et al., 1996).

Thus, in order to precise how apo E function and interaction with other proteins could be modified by oxidative stress, we studied *in vitro* the oxidation of the three recombinant isoforms (E2, E3, E4) produced in our lab and the consequence of oxidative modifications on Aβ fibrillation.

2. MATERIALS AND METHODS

The 3 isoforms of recombinant human apo E (E2, E3, E4) cDNAs were cloned in modified pARHS-2 and produced in *E. coli* BL21(DE3) as a fusion protein of about 44 kDa. This 44 kDa apo E results from the fusion of apo E and a peptide containing a polyhistidine sequence, which allows the recombinant proteins to be purified by single-step affinity chromatography on nickel gel (Barbier et al., 1997).

In vitro oxidation assays with MPO, isolated from polymorphonuclear neutrophils, were performed as previously described (Jolivalt et al., 1996), with slight modifications of pH adapted to the formation of apo E/PL discoidal complexes. Chloramines were formed by the reaction of MPO in 55 mM phosphate buffer (pH 4.5), 100mM NaCl and 10 mM leucine with different concentrations of H_2O_2 (0; 1.25; 4; 5.75; 7.6 mM) (Figure 1). Then, native apo E was oxidized by the chloramines at physiological pH (Zcliczynski et al., 1971). Proteins were submitted to SDS-PAGE in non-reducing conditions and tranferred to PVDF (Millipore) membranes.

The protein carbonyl content was measured by forming labelled protein hydrazone derivatives using 2,4-dinitrophenyl hydrazine (DNPH) (Keller et al., 1993). The dinitro-

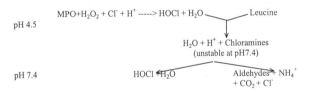

Figure 1. Myeloperoxidase related reactions.

phenyl moities formed on apo E were revealed by enhanced chemiluminescence (ECL, Pierce) after blot incubation with rabbit anti-DNP antibodies (Sigma) followed by incubation with HRP-labelled anti-rabbit antibodies (Sigma). After stripping, western blots were incubated with monoclonal anti-apo E antibodies (kind gift from Dr. Y. Marcel, Canada) followed by incubation with secondary anti-mouse antibodies labelled with HRP (Sigma), and then revealed by ECL.

Aβ (Bachem) aggregation was followed using the thioflavine T (thT) fluorescent assay as previously described (Naiki et al., 1989; Wisniewski et al., 1994). Experimental solutions were incubated for 6 to 24 hours according to a molar ratio apo E/Aβ of 1/100. The incubated samples were added to 50 mM glycine (pH 9)/2 µM thT (Sigma). Emission spectra were obtained using a Cytofluor 2300 (Pharmacia) spectrofluorimeter with $\lambda ex = 435$ and $\lambda em = 485$ nm.

Oxidized and non oxidized apo E was resolved by HPLC. Peaks areas were quantified at 215 nm. Analytical reverse-phase was done on a C-4 column as a gradient of acetonitrile in 0.1% TFA from 40 to 60% during 30 minutes, with a flow rate of 1 ml/min.

3. RESULTS

Upon oxidation, the apo E3-specific band migrated with a higher apparent molecular weight (MW) (Figure 2). These shifts were visualized for the 3 apo E isoforms, and are differential between the apo E isoforms. The difference observed between the main apo E band before and after oxidation is more important for apo E4 than apo E3 than apo E2 (not shown).

We compared HPLC elution profiles of oxidized apo E3 with the non oxidized protein. The peaks seen on the HPLC chromatograph are designated as shown on Figure 3. In presence of MPO without H_2O_2, the major peak of apo E appears with a retention time of 27 min (Figure 3a). When H_2O_2 is added to the incubation medium, the peak corresponding to apo E appears at 20 min. Several peaks which could correspond to peptidic fragments are detected after oxidation at earlier retention times (Figure 3b).

The consequences of apo E oxidation on Aβ fibrillation were studied with the thT assays (Figure 4). When Aβ is incubated alone during 24 hours, an Aβ fibrillation was measured. After addition of 0.9 µM of apo E3, there is a diminution of 60% of the rate of fibrillation after 24 hours. When oxidized apo E was added in the same conditions, there was no significant difference with the native form.

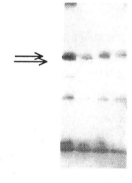

Figure 2. Immunoblotting analysis of apo E3 after oxidation at pH 7.4, using anti-DNP antibodies. Line 1: apo E3; line 2: apo E3 + MPO; line 3: apo E3 + MPO + 0.025 mM H_2O_2; line 4: apo E3 + MPO + 0.5 mM H_2O_2. Arrows indicate apo E band.

1 2 3 4

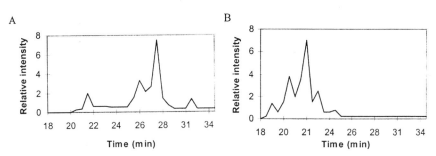

Figure 3. HPLC profiles of apo E3 (panel A and oxidized apo E3 (panel B).

4. DISCUSSION

Apo E4 is associated with AD (Siest et al., 1995). Moreover, oxidative process are increased in AD whose apo E could be a target (Montine et al., 1996). The apo E allele specificity linked to pathological aging points to the differential susceptibility of apo E isoforms to oxidation. Indeed, we showed that apo E4 is more oxidized than apo E3, itself more than apo E2. This differential oxidation is characterized by a differential mobility in SDS-PAGE. As it is known, oxidation may modify amino acid residues, which in turn modify the function and/or the protein structure. The difference between the 3 apo E isoforms could be due to a difference in their structure, leading to a differentially exposing oxidizable amino acid residues. Indeed, the amino acid residues cystein, methionine, tryptophane, proline, lysine, histidine, tyrosine are the more susceptible to oxidation, in a decreasing order and it is known that discrete amino acid residues may mediate isoform-specific conformational changes which alter apo E interaction with lipids and receptors (Weisgraber, 1994, Dong et al., 1996). As shown by Anantharamaiah et al. (1988) through a combination of reverse-phase high performance liquid chromatography (RP-HPLC) and cyanogen bromide cleavage, oxidation of apo AI methionine alters the secondary structure of the protein. By RP-HPLC, we observed the apo E structural modification by the decrease of the retention time of apo E3 after oxidation.

Furthermore, the differential susceptibility of apo E isoforms to oxidation may be related to their primary structure as they differ in their cysteine content: cysteine residues may act as internal antioxidants, and apo E4 is Arg112–Arg158, apo E3 is Arg112–Cys158, and apo E2 is Cys112–Cys158. In addition, methionine residues were recently proposed to also act as internal antioxidants in proteins (Levine et al., 1996). Depending on the exposure of the methionine residues (8 in apo E), we might expect a differential capacity of apo E isoforms to resist to reactive species attack. Cyanogen bromide cleavage experiments are in progress to precise the impact of methionine residues in apo E oxyda-

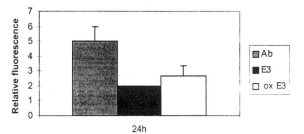

Figure 4. Thioflavine T assays of $A\beta_{1-40}$ fibrillation in the presence of non oxidized apo E3 and oxidized apo E3. The relative fluorescence of native $A\beta$ was 2 (n=3). Ab: amyloid β, E3: non oxidized apo E3, ox E3: oxidized apo E3.

tion. Such properties could correlate with the differential antioxidant properties of apo E isoforms (Miyata and Smith, 1996).

As these oxidative modifications of apo E suggested modification of its structure and its function, we looked at the consequences on Aβ fibrillation. Recombinant apo E3 was able to inhibit Aβ fibrillogenesis. According to the litterature, apo E has been proposed to inhibit or to increase fibrillation. These discrepancies are probably linked to the use of different sources of purified apo E (Webster and Rogers, 1996). Moreover, the molar ratio apo E/Aβ differs from one work to another. It has been suggested that inhibition of Aβ fibrillation results from apo E sequestering of soluble Aβ, thereby preventing its transition to the insoluble form (Goldgaber et al., 1993). Here, oxidation of apo E3 did not affect its inhibitory effect on Aβ fibrillation, which might suggest that the oxidative modifications of the structure of apo E3 do not affect the amino acid residues implicated in apo E/Aβ interactions. Now, it remains to define if the relation with Aβ fibrillogenesis depends on the isoform of apo E and new experiments are now underway.

These observations could have important pathophysiological implications for AD, as they show the implication of the apo E polymorphism in its susceptibility to oxidation by the MPO, an enzymatic system present in the brain. although apo E oxidation does not modify the effect of apo E on Aβ fibrillation, other functional consequences of apo E oxidation have to be addressed. Most notably, we are now analysing the consequences of apo E oxidation on its ability to associate with lipids.

ACKNOWLEDGMENTS

This study was supported by a grant from Ipsen Foundation (Paris), PIR (Pôle Interuniversitaire Régional, Recherche biologique médicale et santé) and the Région Lorraine. We thank Dr. A. Dergunov (Russia) for helpful discussion.

REFERENCES

Anantharamaiah, G.M., Hughes, T.A., Iqbal, M., Gawish, A., Neame, P.J., Medley, M.F. and Segrest, J.P., 1988, Effect of oxidation on the properties of apolipoproteins AI and AII, *J. Lip. Res.* **29**:309–318.

Barbier, A., Visvikis, A., Mathieu, F., Diez, L., Havekes, L., Siest, G., in press, Characterization of three human apolipoprotein E isoforms (E2, E3 and E4) expressed in *Escherichia coli*, Eur. J. Clin. Chem. Clin. Biochem.

Dong, L.M., Parkin, S., Trakhanov, S.D., Rupp, B., Simmons, T., Arnold, K.S., Newhouse, Y.M., Innerarity, T.L. and Weisgraber, K.H., 1996, Novel mechanism for defective receptor binding of apolipoprotein E2 in type III hyperlipoproteinemia, *Nature Struct. Biol.* **3**:718–722.

Drozdz, R. and Naskalski, J.W., 1988, Action of myeloperoxidase-hydrogen peroxide-chloride system on the egg white lysozyme, *Acta Biochim. Pol.* **35**: 277–286.

Ghersi-Egea, J.F., Leininger-Muller, B., Minn, A., and Siest, G., 1992, Drug metabolizing enzymes in the rat pituitary gland, *Prog. Brain Res.* **91**:373–378.

Ghersi-Egea, J. F., M. H. Livertoux., A. Minn, R. Perrin, and G. Siest. 1991. Enzyme-mediated superoxide radical formation initiated by exogenous molecules in rat brain preparations, *Toxicol. Appl Pharmacol* **110**:107–117.

Ghersi-Egea, J.F., Perrin, R., Leininger-Muller, B., Grassiot, M.C., Jeandel, C., Floquet, J., Cuny, G., Siest, G. and Minn, A., 1993, Subcellular localizatin of cytochrome P450, and activities of several enzymes responsible for drug metabolism in the human brain, *Biochem. Pharmacol.* **45**: 647–658.

Ghersi-Egea, J.F., Tayarani, Y., Lefauconnier, J.M. and Minn, A.,1988, Enzymatic protection of the brain: role of 1-naphthol-UDP-glucuronosyltransferase from cerebral tissue and cerebral microvessels, in *Cellular and Molecular aspects of Glucuronidation*, (G. Siest, J. Magdalou, and B. Burchell, eds), pp. 169–175, Colloque INSERM/John Libbey.

Goldgaber, D., Schwarzman, A.I., Bhasin, R., Gregori, L., Schmechel, D., Saunders, A.M., Roses, A.D. and Strittmatter, W.J., 1993, Sequestration of amyloid β-peptide, *Ann. NY Acad. Sci.* **695**:139–143.

Jolivalt, C., Leininger-Muller, B., Drozdz, R., Naskalski, J.W.and Siest, G., 1996, Apolipoprotein E is highly susceptible to oxidation by myeloperoxidase, an enzyme present in the brain, *Neurosci. Lett.* **210**:61–64

Keller, R.J., Halmes, N.C., Hinson, J.A. and Pumford, N.R., 1993, Immunochemical detection of oxidized proteins, *Chem Res Toxicol* **6**:430–33

Kisilevsky, R., 1993, Inflammation-associated amyloidogenesis. Lessons for Alzheimer's amyloidogenesis, *Mol. Neurobiol.* **8**: 65–66.

Leininger-Muller, B., Jolivalt, C., Pillot, T., Lagrange, P., Livertoux, M.H., Grassiot, M.C., Minn, A. and Siest, G., 1995, Apolipoprotein E oxidation and functional consequences. in Apolipoprotein E and Alzheimer's Disease, Research and perspectives in Alzheimer's disease, (Roses et al. eds) Springer-Verlag.

Leininger-Muller, B., Ghersi-Egea, J.F., Siest, G. and Minn, A., 1994, Induction and immunological characterization of the uridine diphosphate-glucuronosyltransferase conjugating 1-naphthol in the rat choroid plexus, *Neurosci. Lett.* **175**:37–40.

Levine, R.L., Mosoni, L., Berlett, B.S. and Stadtman, E.R., 1996, Methionine residues as endogenous antioxidants in proteins, *Proc. Natl. Acad. Sci. (USA)* **93**:15036–15040.

Mahley, R.W., 1988, Apolipoprotein E : cholesterol transport protein with expanding role in cell biology, *Science* **240**:622–630.

Minn, A., Ghersi-Egea, J.-F., Perrin, R., Leininger, B., Siest, G., 1991, Drug metabolizing enzymes in the brain and cerebral microvessels, *Brain Res. Rev.* **16**:65–82.

Miyata, M. and Smith, J.D., 1996, Apolipoprotein E allele-specific antioxidant activity and effects on cytotoxicity by oxidative insults and β-amyloid peptides, *Nature Genet.* **14**:55–61.

Montine, T.J., Huang, D.Y., Valentine, W.M., Amarnath, V., Saunders, A., Weisgraber, K.H., Graham, D.G.and Strittmatter, W.J., 1996, Crosslinking of apolipoprotein E by products of lipid peroxidation, *J. Neuropathol. Expal. Neurol.* **55**:202–210.

Naiki, H., Higuchi, K., Hosokawa, M. and Takeda, T., 1989, Fluorometric determination of amyloid fivrils in vitro using the fluorescent dye, thioflavine T, *Anal. Biochem.* **177**: 244–249.

Poirier, J., 1994, Apolipoprotein E in animal models of brain injury and in Alzheimer's disease, *Trends Neurosci.* **12**:525–530.

Siest, G., Pillot, T., Régis-Bailly, A., Leininger-Muller, B., Steinmetz, J., Galteau, M.M. and Visvikis, S., 1995, Apolipoprotein E: An important gene and protein to follow in laboratory medicine, *Clin. Chem.* **41**:1068–1086.

Smith, C.D., Carney, J.M., Starke-Reed, P.E., Oliver, C.N., Stadtman, E.R., Floyd, R.A. and Markesbery, W.R., 1991, Excess brain protein oxidation and enzyme dysfunction in normal aging and in Alzheimer's disease, *Proc. Natl. Acad. Sci.* USA **88**: 10540–10543.

Soto, C., Golabek, A., Wisniewski, T. and Castano, M., 1996, Alzheimer's β-amyloid peptide is conformationally modified by apolipoprotein E in vitro, *Neuroreport* **7**: 721–725.

Stadtman, E.R. and Oliver, C.N., 1991, Metal-catalyzed oxidation of proteins. Physiological consequences, *J. Biol. Chem.* **266**: 2005–2008.

Webster, S. and Rogers, J., 1996, Relative efficacies of amyloid β (Aβ) binding proteins in Aβ aggregation, *J. Neurosci. Res.* **46**: 58–66.

Weisgraber, K.H., 1994, Apolipoprotein E: structure-function relationships, *Adv. Prot.Chem.* **45**:249–302.

Wisniewski, T., Castano, E.M., Golabek, A., Vogel, T. and Frangione, B., 1994, Acceleration of Alzheimer's fibril formation by apolipoprotein E in vitro, *Am. J. Pathol.* **145**: 1030–1036.

Zcliczynski, J.M., Stelmaszynska, T., Domanski, J., Ostrowski, W., 1971, Chloramines as intermediates of oxidation reaction of amino acids by myeloperoxidase, *Bioch. Biophys. Acta* **235**:419–424.

8

FLAVONOID RADICALS

Wolf Bors,[1] Christa Michel,[1] Werner Heller,[2] and Heinrich Sandermann, Jr.[2]

[1]Institut für Strahlenbiologie,
[2]Institut für Biochemische Pflanzenpathologie
GSF Forschungszentrum für Umwelt und Gesundheit
D-85764 Neuherberg, Germany

1. INTRODUCTION

Flavonoids belong to the recently popular phytochemicals (Huang, Osawa, Ho, and Rosen, 1994; Ho, Osawa, Huang, and Rosen, 1994; Manach, Régérat, Texier, Agullo, Demigné, and Rémésy, 1996), plant products with potential benefit for human health. Since the compounds exist as ubiquitous secondary plant metabolites, they are an important part of human diet (Das and Ramanathan, 1992; Hertog, Feskens, Hollman, and Katan, 1993; Kühnau, 1976; Stavric and Matula, 1992). They are also considered as the active principles in many medicinal plants (Wollenweber, 1988; Xin, Zhao, Li, and Hou, 1990). Due to the pronounced antioxidative potential of flavonoids, there is considerable interest in the structure and reactions of the flavonoid aroxyl radicals as obligatory intermediates during radical-scavenging reactions. Indeed the preferred mechanistic interpretation of the antioxidative effect of flavonoids lies in the radical-scavenging properties of these compounds, both in model systems and under in vitro conditions (see Bors, Michel, and Saran, 1994; Bors, Heller, Michel, and Stettmaier, 1996; Rice-Evans, Miller, and Paganga, 1996).

2. SCAVENGING RATE CONSTANTS

Generation of flavonoid radicals ist optimally achieved by pulse radiolysis (Bors et al., 1994; Bors et al., 1996; Bors, Heller, Michel, and Saran, 1990; Erben-Russ, Bors, and Saran, 1987; Györgyi, Antus, Blazovics, and Földiak, 1992b; Steenken and Neta, 1982), a method which combines selective formation of individual radicals (Bors, Saran, Michel, and Tait, 1984) and observation of fast absorption changes with kinetic spectroscopy in the micro- and milli-second time range due to the generation and decay of intermediate radicals (Saran, Vetter, Erben-Russ, Winter, Kruse, Michel, and Bors, 1987). Rate constants with ten different oxidizing radicals have been determined for kaempferol and almost as many for quercetin, the two most common flavonoid aglycones (Bors et al., 1990; Bors et al., 1994). Almost consistently, kaempferol exhibited higher rate constants than

Free Radicals, Oxidative Stress, and Antioxidants, edited by Özben.
Plenum Press, New York, 1998.

quercetin, yet at pH 11.5 the corresponding aroxyl radical is less stable than that of quercetin by two orders of magnitude (Bors and Saran, 1987).

Despite of the numerous reports of flavonoids acting as specific scavengers for $O_2^{\cdot-}$ (Baumann, Wurm, and von Bruchhausen, 1980; Chen, Zheng, Jia, and Yu, 1990; Hu, Calomme, Lasure, de Bruyne, Pieters, Vlietinck, and vanden Berghe, 1995; Huguet, Manez, and Alcaraz, 1990; Robak and Gryglewski, 1988; Sichel, Corsaro, Scalia, di Bilio, and Bonomo, 1991), only few rate constants with $O_2^{\cdot-}$ have been determined by pulse radiolysis, all being rather low (Bors et al., 1994). Similarly low values were obtained with catechins and $O_2^{\cdot-}$, using the competitive inhibition of chemiluminescence of lucigenin analogs (Suzuki, Goto, Oguni, Mashiko, and Nomoto, 1991). We have to assume that the discrepancies between claims and the actual data arose from unspecific sources of $O_2^{\cdot-}$. If we consider $O_2^{\cdot-}$ as co-existing with hydroxyl and/or peroxyl radicals, the results of the scavenging reactions might be rationalized.

Aside from scavenging superoxide anions (Baumann et al., 1980; Chen et al., 1990; Hu et al., 1995; Huguet et al., 1990; Robak and Gryglewski, 1988; Sichel et al., 1991), flavonoids react with peroxyl radicals both in aqueous (Torel, Cillard, and Cillard, 1986) and organic solutions (Belyakov, Roginsky, and Bors, 1995; Roginsky, Barsukova, Remorova, and Bors, 1996), alkoxyl radicals (Bors et al., 1990), and the NO radical (Krol, Czuba, Threadgill, Cunningham, and Pietsz, 1995; van Acker, Tromp, Haenen, van der Vigh, and Bast, 1995). Adding the studies with other inorganic electrophilic radicals (Bors et al., 1990; Györgyi, Blazovics, Feher, and Földiak, 1990; Salah, Miller, Paganga, Tijburg, Bolwell, and Rice-Evans, 1995), it becomes readily apparent that flavonoids are highly effective scavengers of all types of oxidizing radicals. Efficient scavenging of hydroxyl radicals (Bors et al., 1990; Husain, Cillard, and Cillard, 1987) needs no special mentioning, except with regard to the possible contribution of the metal-chelating properties of flavonoids (Afanas'ev, Doroshko, Brodskii, Kostyuk, and Potapovich, 1989; Puppo, 1992).

The results with the flavans (Suzuki et al., 1991) are an example for the application of the generally available method of competition kinetics as an alternative to pulse radiolysis to determine rate constants (Bors, Michel, and Saran, 1985). Based on the known absolute rate constants of a reference substance with a given radical (Bors, Michel, and Saran, 1992b), usually obtained by pulse radiolysis, relative scavenging rates resulting from competition plots can be transformed into absolute rate constants. This approach was used to determine the reaction rates of flavonoids with photolytically generated *tert*-butoxyl radicals (Bors et al., 1990; Bors, Heller, Michel, and Saran, 1992a). Various other reference substances exist that are suitable for competition studies with ·OH radicals and/or $O_2^{\cdot-}$, the dye 2,2'-azino-*bis*-(3-ethylbenzthiazoline)-6-sulfonate (ABTS) and crocin (Bors et al., 1992b) being the most versatile ones.

3. UV/VIS-SPECTRA OF AROXYL RADICAL INTERMEDIATES

Steenken and Neta (1982) first showed uv/vis-transient spectra of flavonoids and of a number of other polyphenols, which they obtained upon oxidation of the phenolates at pH 13.5 with ethylene glycol radicals. In our own studies we preferred azide radicals (·N₃) to generate the flavonoid aroxyl radicals, as did György and co-workers for silybinin (Györgyi et al., 1992b; György, Antus, and Földiak, 1992a). Originally working at pH 11.5 because of the low water solubility of both the flavonoid aglycones and the linoleic acid and its hydroperoxide (Erben-Russ et al., 1987), we eventually changed the pH to near physiological levels, pH 8.5–9 (Bors, Michel, and Schikora, 1995).

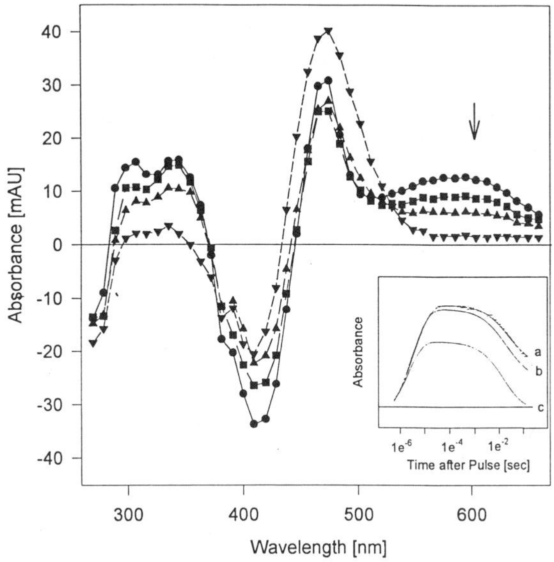

Figure 1. Transient spectra and kinetic traces of the fisetin radical at pH 8.5. The aroxyl radical was generated by the attack of azide ($^{\bullet}N_3$) radicals in N_2O-saturated solutions; the difference spectra are dose-normalized and were taken at various times after the pulse: (●) 44μs, (■) 882μs, (▲) 2ms, (▼) 18ms; sodium azide 10 mmolar, fisetin 55 μmolar, ascorbate 50 μmolar. *Inset*: kinetic traces at 586 nm (arrow in spectrum) for different concentrations of ascorbate: **a**: zero, **b**: 10 μmolar, **c**: 50 μmolar; absorbance (full scale): 60 mAU at a pulse dose of 27.55 Gy.

Pulse radiolysis in combination with kinetic spectroscopy revealed that B-ring-localized o-semiquinones are the major radical species observed after univalent oxidation of flavanols, flavanones and dihydroflavonols. In effect, a saturated 2,3-bond causes an interruption of the π-electron system between the carbonyl group and the B-ring. In contrast, the presence of a 2,3-double bond in flavones and flavonols caused an extensive delocalization of the odd electron in the radical over all three ring systems, evident from the bathochromic (red) shift of the radical absorption as compared to the non-radical parent compound (Bors et al., 1990; Bors et al., 1992a), respectively. Figure 1 depicts the dose-normalized transient spectra of the aroxyl radical of fisetin at different times after the pulse in the presence of a near equimolar amount of ascorbate, generated at pH 8.5 with azide radicals. The inset pertains to the change in the kinetics of the aroxyl radical at 586nm at different concentrations of ascorbate as discussed below.

4. TRANSIENT KINETICS AND REACTIVITIES OF FLAVONOID AROXYL RADICAL INTERMEDIATES

Concerning the reactivities of the flavonoid radicals, at least the prerequisite of sufficient stability for optimal antioxidative capacity seems to be fulfilled in aqueous solutions, with further stabilization if B-ring o-semiquinone structures are present (Bors and Saran, 1987). In early studies performed at pH 11.5, we found second-order decay rate constants ranging over three orders of magnitude and in a few cases first-order decay of flavonoid radicals (Bors and Saran, 1987). We postulated, that it was basically the presence of a B-ring catechol group which stabilized the aroxyl radical (Bors et al., 1990; Bors et al., 1992a). Yet, studies at pH 8.5 showed that all aroxyl radicals investigated so far had similar decay rate constants (Bors et al., 1992a).

Since most flavonoid aglycones are only poorly soluble in water, it was of interest to study their behavior in micellar systems. We observed a change in the kinetic behavior towards less effective antioxidative potential at higher concentrations and proposed this to be due to the formation of secondary radicals in the presence of oxygen which are capable of propagating chain reactions (Belyakov et al., 1995; Roginsky et al., 1996). This change in mechanism may also accounts for the observation, that flavonoids in general do not behave like simple phenolic antioxidants in organic solvents. Both the stoichiometric factor deviating from a value of two and a contrasting influence of the scavenger concentration was optimally resolved if we included such secondary radical formation in our kinetic model.

5. REDOX REACTIONS OF FLAVONOID RADICALS

The interaction of flavonoid and their radicals with other redox-active compounds is of considerable interest, as such reactions may be the normal set of circumstances under *in vivo* conditions. Thus far, only interactions with ascorbate (Bors et al., 1995) and trolox c as a water-soluble model of α-tocopherol (Jovanovic, Steenken, Tosic, Marjanovic, and Simic, 1994; Jovanovic, Hara, Steenken, and Simic, 1995; Jovanovic, Steenken, Hara, and Simic, 1996) have been studied. Many flavonoids, especially aglycones, are lipophilic so that membrane association and interaction with tocopherols are of interest. As shown for the interaction of flavonoids with ascorbate, 2,3-unsaturated compounds acted in a way comparable to the synergistic effect observed for ascorbate and α-tocopherol. In fact, with flavonoids as superior radical scavengers and the ascorbate radical being quite stable and amenable to enzymatic removal, such a combination would also be preferential from a metabolic standpoint.

In the case of ascorbate, the numerous reactions which were considered to take place between the respective ionic and radical species, of which few could be directly observed, led to the application of computer simulation. i.e. a kinetic modeling approach (Bors et al., 1995). As depicted in the inset of Fig. 1, trace **a** is obtained in the absence of ascorbate and shows the digitized experimental data points while the solid line depicts the theoretical curve obtained from the kinetic model. Similarly the solid lines **b** and **c** represent the change in the decay kinetics in the presence of 30 and 50 μmolar amounts of ascorbate, respectively. This approach furthermore provided us with individual rate constants for the equilibrium reactions and eventually the corresponding univalent redox potentials of the FlO$^{\bullet}$/FlO^{-} couples (Bors et al., 1995). As it turned out, only dihydroquercetin was able to reduce the ascorbate radical, whereas the flavone and flavonol radicals investigated all oxidized ascorbate to its radical anion.

Redox interactions of some flavonoids (Jovanovic et al., 1994) and procyanidins (Jovanovic et al., 1995) with trolox c as a water-soluble model of α-tocopherol were studied with the intention to determine redox potentials for these redox couples. These authors also carefully investigated the pH dependence of these parameters, establishing a number of dissociation constants for the various compounds. Dissociation of specific flavonoid hydroxy groups does not only control the redox potential, but also seems to be essential for their reactivity as radical scavengers (Erben-Russ et al., 1987).

6. AUTOXIDATION

The polyphenolic structure of flavonoids obviously makes them prone to autoxidation reactions, primarily those with catechol or pyrogallol groups. This susceptibility, which is particularly evident in alkaline solutions, was the topic of early studies (Brown, Rajananda, Holroyd, and Evans, 1982; Hathway and Seakins, 1957; Pelter, Bradshaw, and Warren, 1971) and was applied in a number of EPR investigations (Cotelle, Bernier,

Catteau, Pommery, Wallet, and Gaydou, 1996; Jensen and Pedersen, 1983; Kuhnle, Windle, and Waiss, 1969). Our studies in slightly alkaline solution (pH < 9) were unlikely to result in further hydroxylation. Such hydroxylation in 3-position has been observed by EPR spectroscopy of the respective secondary radicals for B-ring pyrogallol structures above pH 13 (Cotelle, Bernier, Hénichart, Catteau, Gaydou, and Wallet, 1992; Cotelle et al., 1996) and in 2'-position for quercetin and luteolin (Cotelle et al., 1996). Luteolin, in effect, allows both 3- and 2'-positions to be further hydroxylated.

The pro-oxidative effect of flavonoids, based on the formation of oxygen radicals during autoxidative redox cycling (Canada, Giannella, Nguyen, and Mason, 1990; Hodnick, Milosavljevic, Nelson, and Pardini, 1988; Yoshioka, Sugiura, Kawahara, Fujita, Makino, Kamiya, and Tsuyuma, 1991), has been suggested to be the reason for their mutagenicity—a topic which has received major attention (Brown, 1980; MacGregor, 1986; Nagao, Morita, Yahagi, Shimizu, Kuroyanagi, Fukuoka, Yoshihira, Natori, Fujino, and Sugimura, 1981; Pardini, 1995). Two studies involving flavonoid and chalcone oxides (Rashid, Mullin, and Mumma, 1986; Sweeny, Iacobucci, Brusick, and Jagannath, 1981), and the particularly detailed investigations on the correlation of quercetin mutagenicity with the production of oxygen radicals (Hatcher and Bryan, 1985; Rueff, Laires, Gaspar, Borba, and Rodrigues, 1992; Ueno, Kohno, Haraikawa, and Hirono, 1984) all suggest the formation of reactive oxygen species (ROS) as the likely source of the mutagenic effects. Superoxide anions, formed during autoxidation of quercetin or during futile redox cycling of quinone methides (Sweeny et al., 1981), either induce mutagenicity or are scavenged by other antioxidants (Minnunni, Wolleb, Müller, Pfeifer, and Äschbacher, 1992; Teel and Castonguay, 1992) or quercetin itself (Ueno et al., 1984).

The goal of this presentation was to document the reactivities of flavonoid radicals and the implications for their various functions. After being generated in radical-scavenging reactions, the relative stability of the aroxyl radicals is of utmost importance to their antioxidative potential. Another criterion relevant to this function is their univalent reduction potential, in effect, their readiness to interact with other biologically important redox compounds, such as ascorbate or α-tocopherol. Finally, the formation of reactive oxygen species during the autoxidation of these polyhydroxylated substances, which essentially requires the intermediary formation of flavonoid radicals, may well be the basis for their mutagenic and cytotoxic effects. There is much recent interest in isoflavonoids and lignans as phytoestrogens. The formation and possible role of radicals derived from these phenylpropanoid compounds remain to be elucidated.

ACKNOWLEDGMENTS

We appreciate the stimulating and productive discussions with Manfred Saran and Kurt Stettmaier.

REFERENCES

Afanas'ev, I.B., Dorozhko, A.I., Brodskii, A.V., Kostyuk, V.A., and Potapovitch, A.I., 1989, Chelating and free radical scavenging mechanisms of inhibitory action of rutin and quercetin in lipid peroxidation, *Biochem. Pharmacol.* **38**:1763–1769

Baumann, J., Wurm, G., von Bruchhausen, F., 1980, Hemmung der Prostaglandin-Synthetase durch Flavonoide und Phenolderivate im Vergleich mit deren O_2^- Radikalfängereigenschaften. *Arch. Pharmacol.* **313**:330–337

Belyakov, V.A., Roginsky, V.A., and Bors, W., 1995, Rate constants for the reaction of peroxyl free radical with flavonoids and related compounds as determined by the kinetic chemiluminescence method, *J. Chem. Soc., Perkin II*, **1995**:2319–26

Bors, W., and Saran, M., 1987, Radical scavenging by flavonoid antioxidants, *Free Radical Res. Comm.* **2**:289–294

Bors, W., Saran, M., Michel, C., and Tait, D., 1984, Formation and reactivities of oxygen free radicals. in: *Advances on Oxygen Radicals and Radioprotectors*, (A. Breccia, C.L. Greenstock, and M. Tamba, eds.), pp. 13–27, Ed. Scient. Lo Scarabeo, Bologna

Bors, W., Michel, C., and Saran, M., 1985, Determination of kinetic parameters of oxygen radicals by competition studies. In: *CRC Handbook of Methods for Oxygen Radical Research*. (R.A. Greenwald, ed.), pp. 181–188, CRC Press, Boca Raton, FL

Bors, W., Heller, W., Michel, C., and Saran, M., 1990, Flavonoids as antioxidants: determination of radical scavenging efficiencies, *Meth. Enzymol.* **186**:343–354

Bors, W., Heller, W., Michel, C., and Saran, M., 1992a, Structural principles of flavonoid antioxidants. In: *Free Radicals and the Liver*, (G. Csomos and J. Feher, eds.), pp. 77–95, Springer, Berlin

Bors, W., Michel, C., and Saran, M., 1992b, Determination of rate constants for antioxidant activity and use of the crocin assay. In: *Lipid-Soluble Antioxidants: Biochemistry and Clinical Applications*. (A.S.H. Ong and L. Packer, eds.), pp. 52–64, Birkhäuser, Basel

Bors, W., Michel, C., and Saran, M., 1994, Flavonoid antioxidants: rate constants for reactions with oxygen radicals. *Meth Enzymol* **234**:420–429

Bors, W., Michel, C., and Schikora, S., 1995, Interaction of flavonoids with ascorbate and determination of their univalent redox potentials: a pulse radiolysis study, *Free Radical Biol. Med.* **19**:45–52

Bors, W., Heller, W., Michel, C., and Stettmaier, K., 1996, Flavonoids and Polyphenols: Chemistry and Biology, in: *Handbook of Antioxidants*, (E. Cadenas and L. Packer, eds.), pp. 409–466, Marcel Dekker, New York, NY

Brown, J.P., 1980, A review of the genetic effects of naturally occurring flavonoids, anthraquinones and related compounds. *Mutat. Res.* **75**:243–277

Brown, S.B., Rajananda, V., Holroyd, J.A., and Evans, E.G.V., 1982, A study of the mechanism of quercetin oxygenation by ^{18}O labelling. A comparison of the mechanism with that of haem degradation, *Biochem. J.* **205**:239–244

Canada, A.T., Giannella, E., Nguyen, T.D., and Mason, R.P., 1990, The production of reactive oxygen species by dietary flavonols, *Free Radical Biol. Med.* **9**:441–449

Chen, Y., Zheng, R., Jia, Z., and Ju, Y., 1990, Flavonoids as superoxide scavengers and antioxidants, *Free Radical Biol. Med.* **9**:19–21

Cotelle, N., Bernier, J.L., Hénichart, J.P., Catteau, J.P., Gaydou, E., and Wallet, J.C., 1992, Scavenger and antioxidant properties of ten synthetic flavones, *Free Radical Biol. Med.* **13**:211–219

Cotelle, N., Bernier, J.L., Catteau, J.P., Pommery, J., Wallet, J.C., and Gaydou, E.M., 1996, Antioxidant properties of hydroxy-flavones, *Free Radical Biol. Med.* **20**:35–43

Das, N.P., and Ramanathan, L., 1992, Studies on flavonoids and related compounds as antioxidants in food. In: *Lipid-Soluble Antioxidants: Biochemistry and Clinical Applications*, (A.S.H. Ong and L. Packer, eds.), pp. 295–306, Birkhäuser, Basel

Erben-Russ, M., Bors, W., and Saran, M., 1987, Reactions of linoleic acid peroxyl radicals with phenolic antioxidants: a pulse radiolysis study. *Int. J. Radiat. Biol.* **52**:393–412

György, I., Blazovics, A., Feher, J., and Földiak, G., 1990, Reactions of inorganic free radicals with liver protecting drugs, *Radiat. Phys. Chem.* **36**:165–167

György, I., Antus, S., and Földiak, G., 1992a, Pulse radiolysis of silybin: one-electron oxidation of the flavonoid at neutral pH, *Radiat. Phys. Chem.* **39**:81–84

György, I., Antus, S., Blazovics, A., and Földiak, G., 1992b, Substituent effects in the free radical reactions of silybin: radiation-induced oxidation of the flavonoid at neutral pH, *Int. J. Radiat. Biol.* **61**:603–609

Hatcher, J.F., and Bryan, G.T., 1985, Factors affecting the mutagenic activity of quercetin for *Salmonella typhimurium* TA98: metal ions, antioxidants and pH, *Mutat. Res.* **148**:13–23

Hathway, D.E., and Seakins, J.W.T., 1957, Autoxidation of polyphenols. III. Autoxidation in neutral aqueous solution of flavans related to catechin, *J. Chem. Soc.*, **1957**:1562–1566

Hertog, M.G.L., Feskens, E.J.M., Hollman, P.C.H., and Katan, M.B., 1993, Dietary antioxidant flavonoids and risk of coronary heart disease in the Zutphen Elderly Study, *Lancet* **342**:1007–11

Ho, C.T., Osawa, T., Huang, M.T., and Rosen, R.T. (eds.), 1994, *Food Phytochemicals for Cancer Prevention. II. Teas, Spices, and Herbs*. ACS Sympos. Ser. 547, ACS Press, Washington, DC

Hodnick, W.F., Milosavljevic, E.B., Nelson, J.H., and Pardini, R.S., 1988, Electrochemistry of flavonoids. Relationships between redox potentials, inhibition of mitochondrial respiration, and production of oxygen radicals by flavonoids, *Biochem. Pharmacol.* **37**:2607–11

Hu, J.P., Calomme, M., Lasure, A., de Bruyne, T., Pieters, L., Vlietinck, A., and vanden Berghe, D.A., 1995, Structure-activity relationship of flavonoids with superoxide scavenging activity, *Biol. Trace Elem. Res.* **47:**327–331

Huang, M.T., Osawa, T., Ho, C.T., and Rosen, R.T. (eds.), 1994, *Food Phytochemicals for Cancer Prevention. I. Fruits and Vegetables.* ACS Sympos. Ser. **546,** ACS Press, Washington, DC

Huguet, A.I., Manez, S., and Alcaraz, M.J., 1990, Superoxide scavenging properties of flavonoids in a non-enzymic system, *Z. Naturforsch.* **45c:**19–24

Husain, S.R., Cillard, J., and Cillard, P., 1987, Hydroxyl radical scavenging activity of flavonoids, *Phytochemistry* **26:**2489–91

Jensen, O.N., and Pedersen, J.A., 1983, The oxidative transformations of (+)-catechin and (-)- epicatechin as studied by ESR, *Tetrahedron* **39:**1609–15

Jovanovic, S.V., Steenken, S., Tosic, M., Marjanovic, B., and Simic, M.G., 1994, Flavonoids as antioxidants, *J. Am. Chem. Soc.* **116:**4846–51

Jovanovic, S.V., Hara, Y., Steenken, S., and Simic, M.G., 1995, Antioxidant potential of gallocatechins. A pulse radiolysis and laser photolysis study. *J. Am. Chem. Soc.* **117:**9881–9888

Jovanovic, S.V., Steenken, S., Hara, Y., and Simic, M.G., 1996, Reduction potentials of flavonoid and model phenoxyl radicals. Which ring in flavonoids is responsible for antioxidant activity? *J. Chem. Soc., Perkin Trans. II,* **1996:**2497–2504

Krol, W., Czuba, Z.P., Threadgill, M.D., Cunningham, B.D.M., and Pietsz, G., 1995, Inhibition of nitric oxide (NO) production in murine macrophages by flavones, *Biochem. Pharmacol.* **50:**1031–1035

Kühnau, J., 1976, The flavonoids. A class of semi-essential food components: their role in human nutrition, *World Rev. Nutr. Diet.* **24:**117–191

Kuhnle, J.A., Windle, J.J., and Waiss, A.C., 1969, EPR spectra of flavonoid anion-radicals, *J. Chem. Soc. B,* **1969:**613–616

MacGregor, J.T., 1986, Mutagenic and carcinogenic effects of flavonoids, in: *Plant Flavonoids in Biology and Medicine,* (V. Cody, E. Middleton, and J.B. Harborne, eds.), pp. 4110424, A.R. Liss, New York, NY

Manach, C., Régérat, F., Texier, O., Agullo, G., Demigné, C., and Rémésy, C., 1996, Bioavailability, metabolism and physiological impact of 4-oxo-flavonoids, *Nutrit. Res.* **16:**517–544

Minnunni, M., Wolleb, U., Müller, O., Pfeifer, A., and Äschbacher, H.U., 1992, Natural antioxidants as inhibitors of oxygen species induced mutagenicity, *Mutat. Res.* **269:**193–200

Nagao, M., Morita, N., Yahagi, T., Shimizu, M., Kuroyanagi, M., Fukuoka, M., Yoshihira, K., Natori, S., Fujino, T., and Sugimura, T., 1981, Mutagenicities of 61 flavonoids and 11 related compounds, *Environ. Mutagen.* **3:**401–419

Pardini, R.S., 1995, Toxicity of oxygen from naturally occurring redox-active pro-oxidants, *Arch. Insect Biochem. Physiol.* **29:**101–118

Pelter, A., Bradshaw, J., and Warren, R.F., 1971, Oxidation experiments with flavonoids, *Phytochemistry* **10:**835–850

Puppo, A., 1992, Effect of flavonoids on OH radical formation by Fenton-type reactions: influence of the iron chelator, *Phytochemistry* **31:**85–88

Rashid, K.A., Mullin, C.A., and Mumma, R.O., 1986, Structure-mutagenicity relationships of chalcones and their oxides in the *Salmonella* assay, *Mutat. Res.* **169:**71–79

Rice-Evans, C.A., Miller, N.J., and Paganga, G., 1996, Structure-antioxidant activity relationships of flavonoids and phenolic acids, *Free Radical Biol. Med.* **20:**933–956

Robak, J., and Gryglewski, R.J., 1988, Flavonoids are scavengers of superoxide anions, *Biochem. Pharmacol.* **37:**837–841

Roginsky, V.A., Barsukova, T.K., Remorova, A.A., and Bors, W., 1996, Moderate antioxidative efficiencies of flavonoids during peroxidation of methyl linoleate in homogeneous and micellar solutions, *J. Am. Oil Chem. Soc.,* **73:**777–786

Rueff, J., Laires, A., Gaspar, J., Borba, H., and Rodrigues, A., 1992, Oxygen species and the genotoxicity of quercetin, *Mutat. Res.* **265:**75–81

Salah, N., Miller, N.J., Paganga, G., Tijburg, L., Bolwell, G.P., and Rice-Evans, C., 1995, Polyphenolic flavanols as scavengers of aqueous phase radicals and as chain-breaking antioxidants, *Arch. Biochem. Biophys.* **322:**339–346

Saran, M., Vetter, G., Erben-Russ, M., Winter, R., Kruse, A., Michel, C., and Bors, W., 1987, Pulse radiolysis equipment: a setup for simultaneous multiwavelength kinetic spectroscopy, *Rev. Sci. Instrum.* **58:**363–368

Sichel, G., Corsaro, C., Scalia, M., di Bilio, A.J., and Bonomo, R.P., 1991, In vitro scavenger activity of some flavonoids and melanins against O_2^- *Free Radical Biol. Med.* **11:**1–8

Stavric, B., and Matula, T.I., 1992, Flavonoids in foods: their significance for nutrition and health, in: *Lipid-Soluble Antioxidants: Biochemistry and Clinical Applications,* (A.S.H. Ong and L. Packer, eds.), pp. 274–294, Birkhäuser, Basel

Steenken, S., and Neta, P., 1982, One-electron redox potentials of phenols. Hydroxy- and aminophenols and related compounds of biological interest, *J. Phys. Chem.* **86:**3661–67

Suzuki, N., Goto, A., Oguni, I., Mashiko, S., and Nomoto, T., 1991, Reaction rate constants of tea leaf catechins with superoxide: superoxide-dismutase (SOD)-like activity measured by *Cypridina* luciferin analogue chemiluminescence, *Chem. Express* **6:**655–658

Sweeny, J.G., Iacobucci, G.A., Brusick, D., and Jagannath, D.R., 1981, Structure-activity relationships in the mutagenicity of quinone methides of 7-hydroxy-flavylium salts for *Salmonella typhimurium*, *Mutat. Res.* **82:**275–283

Teel, R.W., and Castonguay, A., 1992, Antimutagenic effects of polyphenolic compounds, *Cancer Lett.* **66:**107–113

Torel, J., Cillard, J., and Cillard, P., 1986, Antioxidant activity of flavonoids and reactivity with peroxy radicals, *Phytochemistry* **25:**383–385

Ueno, I., Kohno, M., Haraikawa, K., and Hirono, I., 1984, Interaction between quercetin and O_2^- radical. Reduction of the quercetin mutagenicity, *J. Pharm. Dyn.* **7:**798–803

van Acker, S.A.B.E., Tromp, M.N.J.L., Haenen, G.R.M.M., van der Vijgh, W.J.F., and Bast, A., 1995, Flavonoids as scavengers of nitric oxide radical, *Biochem. Biophys. Res. Comm.* **214:**755–759

Wollenweber, E., 1988, Occurrence of flavonoid aglycones in medicinal plants. in: *Plant Flavonoids in Biology and Medicine II: Biochemical, Cellular, and Medicinal Properties,* (V. Cody, E. Middleton, J.B. Harborne, and Beretz, A., eds.), pp. 45–55, A.R. Liss, New York, NY

Xin, W.J., Zhao, B.L., Li, X.J., and Hou, J.W., 1990, Scavenging effects of chinese herbs and natural health products on active oxygen radicals, *Res. Chem. Intermed.* **14:**171–183

Yoshioka, H., Sugiura, K., Kawahara, R., Fujita, T., Makino, M., Kamiya, M., and Tsuyumu, S., 1991, Formation of radicals and chemiluminescence during the autoxidation of tea catechins, *Agric. Biol. Chem.* **55:**2717–23

9

THE MOLECULAR MECHANISM OF INTERACTION OF H_2O_2 WITH METMYOGLOBIN

Dimitrios Galaris, Stelianos Kokkoris, Ioannis Toumpoulis, and Panagiotis Korantzopoulos

Laboratory of Biological Chemistry
University of Ioannina Medical School
451 10 Ioannina, Greece

1. INTRODUCTION

Slight changes in the intacellular redox equilibrium is a physiological situation utilized by nature to regulate many important cell functions (Sundaresan et al., 1995). However, when the level of the oxidizing substances produced exceeds a certain threshold it becomes deleterious for the cells (Halliwell and Gutteridge, 1989). The main source of oxidizing agents is the monovalent reduction of oxygen to superoxide anion and the subsequent formation of reactive oxygen intermediates (ROI) (Figure 1).

Our knowledge about the molecular mechanism(s) underlying the toxicity of ROI and their byproducts is limited. Among these agents H_2O_2 which is continuously generated and decomposed in the cells plays a central role. Increased generation or decreased removal of this molecule has been shown to lead to lipid peroxidation, protein oxidation and DNA damage. It has to be stressed, however, that H_2O_2 itself is a rather weak oxidizing agent incapable to induce these effects directly. It is generally believed that free metal ions, mostly Cu and Fe are involved in the transmission of H_2O_2-induced cell toxicity by Fenton-like reactions. The existence of free Cu or Fe ions in the cells, however, is not experimentally proved. On the contrary, organisms are equipped with a panoply of metal chelating proteins (ferritin, transferrin, lactofferin, seruloplasmin and others) able to bind the metal ions in a way that render them incapable of participating in Fenton-like reactions. Consequently, intense research has been directed toward elucidating new mechanisms of H_2O_2-mediated cell toxicity.

In this chapter the molecular mechanism of interaction of metmyoglobin with H_2O_2 will be reviewed. The possible role of this hemoprotein as a mediator of H_2O_2-induced toxicity or as a defense protein will be discussed.

Free Radicals, Oxidative Stress, and Antioxidants, edited by Özben.
Plenum Press, New York, 1998.

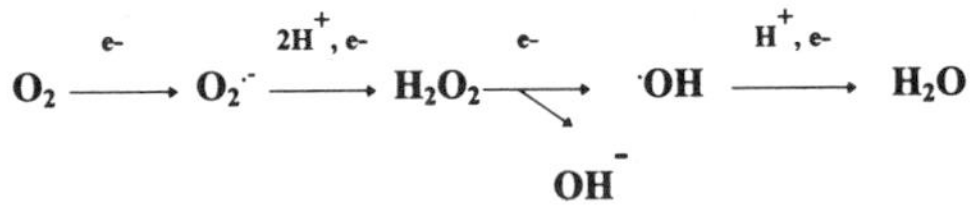

$$O_2 \xrightarrow{\text{e-}} O_2^{\cdot} \xrightarrow{\text{2H}^+, \text{e-}} H_2O_2 \underset{\text{OH}^-}{\overset{\text{e-}}{\searrow}} \cdot OH \xrightarrow{\text{H}^+, \text{e-}} H_2O$$

Figure 1. Monovalent reduction of oxygen to water.

2. THE INTERACTION OF H_2O_2 WITH METMYOGLOBIN

2.1. Is the Iron Bound in the Heme of Myoglobin Capable of Catalyzing Fenton-like Reactions?

The most of the iron in the body is found bound in the heme of heme-containing proteins hemoglobin and myoglobin. The reaction of H_2O_2 with these proteins was studied intensively in early fifties by Chance (1952), George and Irvine (1952) and others. Research in this area later subsided due to the fact that catalase was much more efficient to react and decompose H_2O_2. Moreover, it was realized that the interaction of H_2O_2 with hemoglobin and myoglobin was extremely complicated by intra-molecular electron trafficking and modification of the protein (King and Winfield, 1963; Yonetani and Schleyer, 1967; Rao, Wilks, and Ortiz de Montellano, 1993). Consequently, these interactions were studied only in order to understand the mechanistic details of the reactions of heme-proteins with peroxides (Catalano, Choe, and Ortiz de Montellano, 1989; Kelman, DeGray, and Mason, 1994). Moreover, it was realized that the presence of hemoglobin or myoglobin could mediate the initiation of lipid peroxidation by H_2O_2 and in this way play a role in conditions of oxidative stress in muscle as is the case in reoxygenation after a period of ischemia (Kanner and Harel, 1985; Grisham, 1985). It was also shown that metmyoglobin apart from H_2O_2-induced lipid peroxidation, could mediate the oxidation and inactivation of proteins (Evans and Halliwell, 1995).

The formation of highly reactive species, possibly hydroxyl radicals (·OH), during these interactions should explain the above observations (Sadrzadreh et al., 1984). However, the formation of hydroxyl radicals (·OH) by a Fenton-like mechanism, although intensively examined, was not experimentally proved. These reactive species could be detected only in the case of heme destruction and liberation of free iron from the proteins (Puppo and Halliwell, 1988; Harel and Kanner, 1988). Heterolytic scission of H_2O_2 with two electron oxidation of myoglobin seems to be closer to reality (Allenntoff et al., 1992). However, only ferryl-myoglobin ($Fe^{+4}=O$) was observed while the second oxidizing equivalent was transiently detected as a globin-centered free radical by using EPR spectroscopy (Galaris, Cadenas and Hochstein, 1989b; Giulivi and Cadenas, 1994). This free radical was not localized at a specific position but could move freely from one to another oxidizable amino acid (Kelman et al., 1994; Gunther et al., 1995). It is still unclear today which of the two reactive centers of the molecule i.e. the ferryl-group or the protein-centered radicals are responsible for myoglobin-mediated H_2O_2 toxicity.

2.2. The Molecular Mechanism of the Reaction between Metmyoglobin and H_2O_2 in the Presence or Absence of Ascorbate

The interaction of H_2O_2 with metmyoglobin was easily followed by detecting the changes in the optical spectrum of metmyoglobin after the addition of H_2O_2 (Galaris et al., 1988; Galaris et al., 1989b,c). The final spectrum was stable for a relatively long time.

This spectrum represents the ferryl state, since the existence of the free radical in the polypeptide chain did not affect the absorption properties of the molecule. The time course of this reaction could be easily followed in many different wavelengths but a region around 560 nm was utilized most of the time because the difference in absorption between met- and ferryl-forms of myoglobin in this wavelength was the biggest.

Addition of ascorbic acid to previously formed ferrylmyoglobin led to the reduction of this form back to metmyoglobin (Galaris et al., 1989b). The rereduced form, however, was not exactly identical with the original metmyoglobin. The main difference was a shift of the peak at 630 nm toward lower wavelengths (around 618 nm). This observation might indicate the modification of myoglobin molecule during the course of the reaction. Interestingly, when ascorbate was present in the reaction medium before the addition of H_2O_2 the transient appearance of a spectrum similar to that of ferrylmyoglobin was observed which returned to metmyoglobin without any modification after the exhaustion of the added H_2O_2. It is interesting to note, however, that the stoichiometry of H_2O_2 reduced to ascorbic acid oxidized was close to 1 while that of ferrylmyoglobin reduced to ascorbic acid oxidized was close to 2. Based on these results a cyclic process was proposed in which the reduction of H_2O_2 was connected to the oxidation of ascorbic acid through ferryl- and metmyoglobin interconversions (Figure 2).

2.3. Other Antioxidants

A vast number of antioxidants have been tested according to Figure 2 and shown to be able to reduce ferrylmyoglobin back to the met-form (Giulivi and Cadenas, 1994; Giulivi, Romero, and Cadenas, 1992; Turner et al., 1991). In most of the cases metmyoglobin was first oxidized by excess H_2O_2, the rest of which was decomposed by subsequent addition of catalase. The various antioxidant substances were then added and their capacity to reduce the ferryl form of myoglobin back to metmyoglobin was checked by following the decrease in absorption mainly at 560 nm (Galaris et al., 1989b; Giulivi and Cadenas, 1993; Ostal, Daneshvar, and Skibsted, 1996). For most of these agents a univalent reduction mechanism was proposed and their free radical products were detected using Electron Spin Resonance or other techniques. In many cases the free radicals formed reacted further with molecular O_2 leading to formation of hydroxyl radicals and occasionally to new $O_2^{\cdot-}$ and H_2O_2 (Galaris, Cadenas and Hochstein, 1989a; Romero et al., 1992).

This reaction has been utilized by the RANDOX company in a technique for measuring the total antioxidant status in biological fluids. The principle of this method was based on the assumption that any antioxidant present in these fluids is going to compete

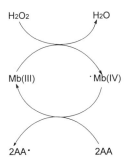

Figure 2. General scheme illustrating the redox cycle of metmyoglobin-catalyzed reduction of H_2O_2 by ascorbate.

with the reducing agent ABTS for the reaction with H_2O_2-activate myoglobin and in this way decrease the absorption change due to the oxidation of this compound.

2.4. The Point of Interaction of Antioxidants during the Oxidation of Metmyoglobin by H_2O_2

The exact point of action of the various antioxidants in the multistep process of H_2O_2-induced oxidation of metmyoglobin gained little attention in most of the above mentioned studies. In most of these reports the reactions of various reducing agents with preoxidized ferrylmyoglobin in the presence or absence of catalase was performed and reaction rate constants were estimated based on the rate of reduction of the heme iron as followed spectrophotometrically (Giulivi and Cadenas, 1994). In the cases that ascorbate or other antioxidants were present from the beginning of the reaction, the absorbance at 560 nm was used as a reliable value representing the ratio [ferryl]/[metmyoglobin] (Galaris et al. 1989b; Larahjinja, Almeida, and Madeira, 1995). However, we recently observed that changes in the absorbance in the presence or absence of ascorbate are highly variable depending on the wavelength selected (Galaris et al., unpublished data). These results strongly support the idea that at least one species, photometrically distinct from ferrylmyoglobin, was formed during the course of the reaction and that the presence of ascorbate changed the equilibrium.

2.5. Formation of Mb(form I)

Careful analysis of the kinetics of myoglobin-catalyzed reduction of H_2O_2 by ascorbic acid revealed a somewhat unusual behavior. It was observed that the ratio [Mb(IV)]/[Mb(III)], as evaluated from the absorption at 560 nm, was proportional to the amount of H_2O_2 added but independent from the concentration of ascorbic acid present (Galaris and Korantzopoulos, 1997). Only after the total oxidation of ascorbic acid the remaining H_2O_2 was able to completely oxidize metmyoglobin. The same pseudofirst order kinetics were observed when the overall rate of the reaction was measured by following the oxidation of ascorbic acid at 265 nm.

The results of these experiments in accordance with previous observations clearly indicated that the process of metmyoglobin oxidation by H_2O_2 was a multistep process which, in the presence of ascorbate, is interrupted leading ultimately to a cyclic set of reactions. In the absence of ascorbic acid, reactive products were formed and intramolecular interactions took place leading to covalent bonds between various amino acids or amino acids and heme (Galaris and Korantzopoulos, 1987; Osava and Korzekwa, 1991; Osava and Williams, 1996; Hanan and Shaklai, 1995).

Based on these observations, we modified the original molecular mechanism of the reaction and proposed the existence of an intermediate which we called Mb(form I) (Figure 3).

The above proposed mechanism was strongly supported also from experiments with manganese-reconstituted myoglobin (Mondal and Mitra, 1996). Mb(form I) represents the counterpart of compound I of classical peroxidases but in contrast to them it is rather unstable leading, in the absence of ascorbic acid and possibly other antioxidants, to electron withdrawal from neighboring oxidizable amino acids. The globin-centered free radicals formed were in their turn unstable themselves abstracting electrons from other amino acids leading to the formation of new free radicals located in the polypeptide chain and covalent bonds between these radicals. In the presence of ascorbic acid protein-centered free radicals were not observed but were replaced by an increased steady state concentra-

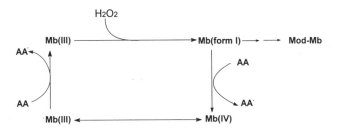

Figure 3. Scheme illustrating the formation and subsequent reactions of Mb(form I).

tion of ascorbic radical indicating the reduction of Mb(form I) in a univalent mode (Giulivi and Cadenas, 1993).

Interestingly, the reduction of Mb(IV) buck to Mb(III) follows also pseudofirst order kinetics (Galaris and Korantzopoulos, 1997) supporting the idea of the electron tunneling mechanism as proposed in Figure 3.

2.6. Is Singlet Oxygen Produced during the Reaction?

The production of $^1O_2$ during the interaction of metmyoglobin with H_2O_2 was initially proposed on the basis of a) product analysis of the cholesterol oxidation when cholesterol was present during the course of the reaction, and b) analysis of the low level chemilluminescence which is detected during the first minute after the initiation of the reaction (Galaris et al., 1988). In a later paper, however, this proposal was disputed on the basis of stereochemical studies using recombinant sperm whale myoglobin with or without point mutations which resulted in replacement of the oxidizable amino acids (Rao et al., 1994).

Recently we initiated an effort in order to better understand the mechanism of H_2O_2/Mb(III)-induced peroxidation of unsaturated fatty acids. By following the time course of the peroxidation of arachidonic acid we observed that it took place in two distinct phases: In the first phase, which lasted only for about one min, a fast lipid peroxidation rate was observed while in the second phase the peroxidation lasted for more than 35 min but was about ten times less effective than that of the first phase (Galaris et al., 1997). While the second phase was linearly dependent on both metmyoglobin and H_2O_2 concentrations, the first phase demonstrated saturation kinetics and did not take place at all at concentrations of H_2O_2 lower than that of metmyoglobin. It was proposed that two different oxidizing agents are formed during the reaction responsible for each of the two phases of peroxidation.

Interestingly, when ascorbate was present, both phases were inhibited completely indicating the inhibition of the formation of both oxidizing agents even in the presence of very low concentrations of ascorbate.

Based on these results we proposed that singlet oxygen is formed at an initial phase of the reaction only in the case in which H_2O_2 was used in excess over metmyoglobin. In this case residual H_2O_2 was able to react with the Mb(form I) resulting ultimately in a catalatic-type reaction (Figure 4).

The above proposed mechanism in figure 4 was also supported by experiments in which the change of O_2 concentration was detected during the interaction of H_2O_2 with metmyoglobin (Galaris et al., 1997). A transient O_2 liberation was apparent directly after the addition of H_2O_2 to a solution containing metmyoglobin. However, a ratio of $[H_2O_2]/[Mb(III)]$ higher than 1.5 was needed in order to detect this evolution. When the

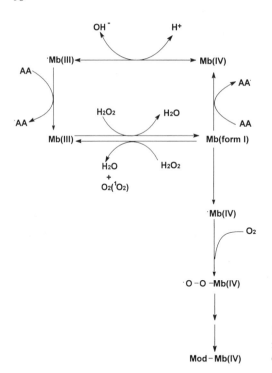

Figure 4. General scheme illustrating the proposed mechanisms for generation of different reactive species during the reaction of metmyoglobin with H_2O_2.

ratio was lower than 1.5 a continuous O_2 consumption was observed, while at a ratio exactly 1.5 neither evolution nor consumption were detected. In the presence of ascorbate both evolution and consumption were completely inhibited.

The evolution of O_2 during the first minute of the interaction may be attributed to a catalatic-type reaction as proposed in Figure 4. If a part of the O_2 liberated in this step was in the form of singlet oxygen it would also explain the induction of the first transient phase of arachidonic acid peroxidation. Moreover, the consumption of O_2 might be attributed to the reaction of this molecule with globin-centered radicals resulting in formation of peroxyl radicals (Fig. 4). The latter have been proposed to be able to initiate lipid peroxidation (Kelman et al., 1994; Newman et al., 1991) and in this way might be regarded responsible for the second slow phase.

2.7. Are These Reactions of Physiological Significance?

The question arising after the above studies is: Are these reactions physiologically significant? And if yes, in what situations?

Although not unequivocally proven, it seems likely that at least some of the interactions described in this paper may indeed take place in conditions of oxidative stress in muscle as is for example the reoxygenation after a period of ischemia. It is known that under these conditions an excess of reactive oxygen species are formed among which H_2O_2 is predominant (McCord, 1985). Moreover, hypervalent states of myoglobin have been detected in reperfused hearts and skeletal muscle under oxidative stress (Arduini, Eddy, and Hochstein, 1990; Eddy, Arduini, and Hochstein, 1990a) indicating the interaction of myoglobin with peroxides *in vivo*.

There is no question that the main enzyme responsible for the decomposition of H_2O_2 is glutathione peroxidase. Catalase is believed to participate only in situations where the concentration of H_2O_2 is relatively high. However, the muscle tissue contains very small amounts of the latter enzyme (about 3–5% that of liver tissue). Taking in account that myoglobin is found in high concentrations in muscle (about 0.3 mM) it is plausible to propose that it can react with the H_2O_2 leading to formation of higher valence states. It has to be noted here, that deoxy-myoglobin reacts also with H_2O_2 with a rate constant ten times higher than that of metmyoglobin (Yusa and Shikama, 1987).

The subsequent fate of the hypervalent states of myoglobin depends on whether or not there are available intracellular antioxidants to reduce them. In the presence of antioxidants, the most important of which in our opinion is ascorbic acid, these states are reduced back to metmyoglobin with net H_2O_2 decomposition. Only when the total amount of ascorbic acid is oxidized, Mb(form I) can abstract electrons from neighboring amino acid residues forming globin-centered free radicals. These radicals can react with molecular oxygen and form peroxyl radicals, which have been proposed to be able to initiate the peroxidation of the membrane phospholipids. On the other hand singlet oxygen can also be formed in some cases when excess H_2O_2 is formed as described above.

It is apparent from these studies that the presence of higher concentrations of ascorbic acid might contribute to more effective resistance of the muscle tissue in conditions of oxidative stress as in reoxygenation after a period of ischemia. Such a conclusion is in agreement with the results of Hochstein and his collaborators who increased the plasma concentration of ascorbate in patients undergoing bypass surgery (Eddy, Hurvitz, and Hochstein, 1990b). It was shown that only in cases in which the operation lasted for more than 50 min the addition of ascorbate was beneficial while for shorter operation times results were not significant, indicating that the presence of endogenous antioxidants were sufficient to protect during this time.

3. CONCLUSIONS

The interaction of H_2O_2 with Mb is a multistep process with the transient formation of Mb(form I) which is the counterpart of Compound I of peroxidases. Contrary to the latter, however, Mb(form I) is very unstable leading, in the absence of endogenous antioxidants, to protein-centered free radicals which may initiate deleterious reactions like lipid peroxidation, protein oxidations and DNA damage.

In the presence of antioxidants, like ascorbic acid, Mb(form I) reacts rapidly with them leading to a cyclic process in which H_2O_2 is reduced to H_2O. This process may be beneficial for the cells especially in the case of myocytes which lack catalase.

We feel confident to propose that extra supplementation of ascorbic acid (perhaps together with iron chelators) can contribute to a more efficient resistance capacity of muscle against oxidative stress.

Finally, the proposed mechanism of myoglobin-catalyzed reduction of H_2O_2 by antioxidants may serve as a mechanistic prototype for the interaction of H_2O_2 with other heme-containing proteins like prostaglandin synthase, cyt-P450 and others.

ACKNOWLEDGMENTS

The authors would like to thank Prof. O. Tsolas for his continuing support and comments. This research was supported by grants from the program PENED 95 of General Secretary of Research and Technology, Athens, Greece.

REFERENCES

Allentoff, A.J., Bolton, J.L., Wilks, A., Thompson, J.A., and Ortiz de Montellano, P.R., 1992, Heterolytic versus homolytic peroxide bond cleavage by sperm whale myoglobin and myoglobin mutants, *J. Am. Chem. Soc.* **114**:9744–9749.

Arduini, A., Eddy, L., and Hochstein, P., 1990, Detection of ferrylmyoglobin in the isolated ischemic rat heart, *Free Radic. Biol. Med.* **9**:511–513.

Catalano, C.E., Choe, Y.S., and Ortiz de Montellano, P.R., 1989, Reactions of the protein radical in peroxide-treated myoglobin, *J. Biol. Chem.* **264**:10534–10541.

Chance, B., 1952, The spectra of the enzyme-substrate complexes of catalase and peroxidases, *Arch. Biochem. Biophys.* **41**:404–415.

Eddy, L., Arduini, A., and Hochstein, P., 1990a, Reduction of ferrylmyoglobin in rat diaphragm, *Am. J. Physiol.* **259**:C995–C997.

Eddy, L., Hurvitz, R., and Hochstein, P., 1990b, A protective role for ascorbate in ischemic arrest associated with cardiopulmonary bypass, *J. Appl. Cardiol.* **5**:409–414.

Evans, P.J., and Halliwell, B., 1995, Side-effects of drugs used in the treatment of rheumatoid arthritis, *Biochem. Soc. Symp.* **61**:195–207.

Galaris D., and Korantzopoulos P., 1997, On the molecular mechanism of metmyoglobin-catalyzed reduction of hydrogen peroxide by ascorbate, *Free Radic. Biol. Med.* **22**:657–667.

Galaris, D., Mira, D., Sevanian, A., Cadenas, E., and Hochstein, P., 1988, Co-oxidation of salicylate and cholesterol and generation of electronically-excited states during the oxidation of metmyoglobin by H_2O_2, *Arch. Biochem. Biophys.* **281**:163–169.

Galaris, D., Cadenas, E., and Hochstein, P., 1989a, Glutathione-dependent reduction of peroxides during ferryl- and met-myoglobin interconversion: A potential protective mechanism in muscle, *Free Radic. Biol. Med.* **6**:473–478.

Galaris, D., Cadenas, E., and Hochstein, P., 1989b, Redox-cycling of myoglobin and ascorbate: A potential protective mechanism against oxidative reperfusion injury in muscle, *Arch. Biochem. Biophys.* **273**:497–504.

Galaris, D., Eddy, L. Arduini, A., Cadenas, E., and Hochstein, P., 1989c, Mechanisms of reoxygenation injury in myocardial infarction: Implication of a myoglobin redox cycle, *Biochem. Biophys. Res. Commun.* **160**:1162–1168.

Galaris, D., Kokkoris, S., Toumpoulis., I., and Tsolas, O., 1997, Generation of reactive species and arachidonic acid peroxidation during the oxidation of metmyoglobin by hydrogen peroxide, *Submitted for publication.*

George, P., and Irvine, D.H., 1952, The reaction between metmyoglobin and hydrogen peroxide, *Biochem. J.* **52**:511–517.

Giulivi, C., and Cadenas, E., 1993, The reaction of ascorbic acid with different heme iron states of myoglobin, *FEBS Lett.* **332**:287–290.

Giulivi, C., and Cadenas, E., 1994, Ferrylmyoglobin: Formation and chemical reactivity toward electron-donating compounds, *Meth. Enzymol.* **233**:189–202.

Giulivi, C., Romero, F.J., and Cadenas, E., 1992, The interaction of Trolox C, a water-soluble vitamin E analog with ferrylmyoglobin: Reduction of the oxoferryl moiety, *Arch. Biochem. Biophys.* **299**:302–312.

Grisham, M.B., 1985, Myoglobin-catalyzed hydrogen peroxide dependent arachidonic acid peroxidation, *J. Free Radic. Biol. Med.* **1**:227–232.

Gunther, M.R., Kelman, D.J., Corbett, J.T., and Mason, R.P., 1995, Self-peroxidation of metmyoglobin results in formation of an oxygen reactive tryptophan-centered radical, *J. Biol. Chem.* **270**:16075–16081.

Halliwell, B., and Gutteridge, J.M.C., 1989, *Free Radicals in Biology and Medicine*, 2nd ed., Clarendon Press, Oxford.

Hanan T., and Shaklai, N., 1995, Peroxidative interaction of myoglobin and myosin, *Eur. J. Biochem.* **233**:930–936.

Harel, S., and Kanner, J., 1988, The generation of ferryl or hydroxyl radicals during interaction of haemproteins with hydrogen peroxide, *Free Radic. Res. Commun.* **5**:21–33.

Kanner, J., and Harel, S., 1985, Initiation of membranal lipid peroxidation by activated metmyoglobin and methemoglobin, *Arch. Biochem. Biophys.* **238**:314–321.

Kelman, D.J., DeGray, J.A., and Mason, R., 1994, Reaction of myoglobin with hydrogen peroxide forms a peroxyl radical which oxidizes substrates, *J. Biol. Chem.* **269**:7458–7463.

King, N.K., and Winfield, M.E., 1963, The mechanism of metmyoglobin oxidation, *J. Biol. Chem.* **238**:1520–1528.

Larahjinha, J., Almeida, L., and Madeira, V., 1995, Reduction of ferrylmyoglobin by dietary phenolic acid derivatives of cinamic acid, *Free Radic. Biol. Med.* **19**:329–337.

Larahjinha, J., Vieira, O., Almeida, L., and Madeira, V., 1996, Inhibition of metmyoglobin/H₂O₂-dependent low density lipoprotein lipid peroxidation by naturally occurring phenolic acids, *Biochem. Pharmacol.* **51**:395–402.

Mondal, M.S., and Mitra, S., 1996, Kinetic studies of the two-step reactions of H₂O₂ with manganese-reconstituted myoglobin, *Biochim. Biophys. Acta,* **1296**:174–180.

McCord, J., 1985, Oxygen-derived free radicals in post-ischemic tissue injury, *New Engl. J. Med.* **312:** 159–163.

Newman, E.S.R., Rice-Evans, C.A., and Davies, M., 1991, Identification of initiating agents in myoglobin-induced lipid peroxidation, *Biochem. Biophys. Res. Commun.* **179**:1414–1419.

Osawa, Y., and Korzekwa, K., 1991, Oxidative modification by low levels of HOOH can transform myoglobin to an oxidase, *Proc. Natl. Acad. Sci. USA* **88**:7081–7085.

Osawa, Y., and Williams, M.S., 1996, Covalent crosslinking of the heme prosthetic group to myoglobin by H₂O₂: Toxicological implications, *Free Radic. Biol. Med.* **21**:35–41.

Ostdal, H., Daneshvar, B., and Skibsted, L.H., 1996, Reduction of ferrylmyoglobin by b-lactoglobulin, *Free Radic. Res.* **24**:429–438.

Puppo, A., and Halliwell, B., 1988, Formation of hydroxyl radicals in biological systems. Does myoglobin stimulate hydroxyl radical formation from hydrogen peroxide? *Free Radic. Res. Commun.* **6**:415–422.

Rao, S.I., Wilks, A., and Ortiz de Montellano, P.R., 1993, The role of His-64, Tyr-103, Tyr-146, and Tyr-151 in the epoxidation of styrene and beta-methylstyrene by recombinant sperm whale myoglobin, *J. Biol. Chem.* **268**:803–809.

Rao, S.I., Wilks, A., Hamberg, M., and Ortiz de Montellano, P.R., 1994, The lipoxygenase activity of myoglobin. Oxidation of linoleic acid by the ferryl oxygen rather than protein radical, *J. Biol. Chem.* **269**:7210–7216.

Romero, F.J., Ordonez, I., Arduini, A., and Cadenas, E., 1992, The reactivity of thiols and disulfides with different redox states of myoglobin, *J. Biol. Chem.* **267**:1680–1688.

Sundarsan, M., Yu, Z-X., Ferrans, V.J., Irani, K., and Finkel, T., 1995, Requirement for generation of HO₂ for platelet-derived growth factor signal transduction, *Science,* **270**:296–299.

Turner, J.J.O., Rice-Evans, C.A., Davies, M., and Newman, E.S.R., 1991, The formation of free radicals by cardiac myocytes under oxidative stress and the effects of electron-donating drugs, *Biochem. J.* **277**:833–837.

Wilks A., and Ortiz de Montellano, P.R., 1992, Intramolecular translocation of the protein radical formed in the reaction of recombinant sperm whale myoglobin with H₂O₂, *J. Biol. Chem.* **267**:8827–8833.

Yonetani, T., and Schleyer, H., 1967, Studies on cytochrom c peroxidase. IX. The reaction of ferrylmyoglobin with hydroperoxides and a comparison of peroxide-induced compounds of ferrylmyoglobin and cytochrom c peroxidase, *J. Biol. Chem.* **242**:1974–1979.

Yusa, K., and Shikama, K., 1987, Oxidation of oxymyoglobin to metmyoglobin with hydrogen peroxide: Involvement of ferryl intermediate, *Biochemistry* **26:** 6684–6688.

10

ARE FREE RADICALS INVOLVED IN THE EXPRESSION OF ADHESION MOLECULES?

W. Sluiter, A. Pietersma, and J. F. Koster

Department of Biochemistry
Cardiovascular Research Institute COEUR
School of Medicine and Health Sciences
Erasmus University Rotterdam
P.O. Box 1738, 3000 DR Rotterdam, The Netherlands

It is beyond doubt that the adhesion of leukocytes to the endothelial lining of the vascular wall plays a key role in various biological processes, including inflammation and thrombosis. This phenomenon is the result of a complex interplay among blood flow, cell adhesion and molecular vascular biology factors. Normally, the circulating granulocytes and monocytes in peripheral blood are distributed over a circulating pool and a marginating pool. The marginating pool comprises about 60% of the total number of cells (Athens, Raab, Haab, Aschenbrucker, and Cartwright, 1961; Van Furth and Sluiter, 1986). The leukocytes of the marginating pool roll along the endothelial line of the vessel. This rolling is governed by the shear force of the flowing blood and the strength of the ionic bonds with the vascular endothelium (Tangelden and Afors, 1991). Under inflammatory conditions the number of rolling leukocytes in the postcapillary vesicles increases drastically. Those leukocytes adhere selectively to the endothelium. Next, these cells traverse the vessel wall between adjacent endothelial cells keeping the monolayer intact, pass the subendothelial layer and accumulate at the site of the inflammation. Strinkingly, granulocytes are commonly the first cells here, followed by the monocytes. It is obvious that knowledge of the mechanism(s) by which granulocytes and monocytes adhere to the vessel wall will enlarge therapeutic modalities (Sluiter, Pietersma, Lamers, and Koster, 1993).

The purpose of this chapter is to describe briefly the major endothelial adhesion molecules and their leukocytes counterparts. And to pay attention to the possibility that free radicals act as second messengers determining the expression of the adhesion molecules.

1. ADHESION MOLECULES OF THE ENDOTHELIAL CELL

The major types of adhesion molecules that an endothelial cell can express belong to the immunoglobin superfamily and the sectin family of cellular adhesion receptors. These

Free Radicals, Oxidative Stress, and Antioxidants, edited by Özben.
Plenum Press, New York, 1998.

receptors are glycoprotein in nature and are the anchors that mediate the attachment of leukocytes.

1.1. The Immunoglobulin Superfamily

The members of this family are single-chain glycoproteins with a polypeptide core containing two (intercellular adhesion molecule-2, or ICAM-2 for short), five (ICAM-1) and six (vascular cell adhesion molecule-1, or VCAM-1) extra cellular immunoglobulin-like domains, followed by a transmembrane domain and a short cytosolic tail (Williams and Barclay, 1988; Springer, 1990). These features are shown in Fig. 1.

ICAM-1 is not only expressed on endothelial cells, but also on other cell types, including lymphocytes and monocytes (Dustin, Staunton and Springer, 1988). ICAM-2 is expressed on endothelial cells, lymphocytes and monocytes but not on granulocytes (De Fougerolles, Stacker, Schawatring, and Springer, 1991). VCAM-1 is expressed predominantly by the endothelial cells present in the postcapillary venules (Rice, Munro, Carlos, and Bevilacqa, 1991) and by endothelial cells covering atherosclerotic lesions (Cybulsky, and Gimbrone, 1991). The leukocytic counterparts of these adhesion molecules are summarized in Table 1.

1.2. The Selectin Family

Selectins have an N-terminal lectin followed by an epiderminal growth factor (EGF) motif, six (endothelial leukocyte adhesion molecule-1, ELAM or E-selectin) or nine (granular membrens protein-140, GMP-140 or P-selectin) short consensus repeat motifs, a transmembrane domain and a short cytosolic peptide tail (Springer, 1990). Only endothelial cells express E-selectin which can bind lymphocytes, granulocytes and monocytes. Table 1 shows the counterparts of E- and P-selectin. The ligand for E-selectin is yet not well characterized but the lectin domain of E-selectin recognizes sialated forms of the Lewis X

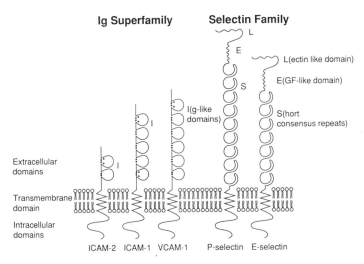

Figure 1. The basic molecular structure of the members of the Ig superfamily and the selectin family of leukocyte adhesion molecules expressed by the vascular endothelium. Black dots represent disulfide bridges within the molecule. Modified from Springer (1990); from Sluiter, W., Pietersma, A., Lamers, J. M. J., and Koster, J. F., 1993, *J. Cardiovasc. Pharmacol.*, **22** (S4): S37–S44.

Table 1. Adhesion molecules on endothelial cells involved in leukocyte adhesion*

Molecule	Family	Basal expression	Stimulators of expression	Minimal time for maximal expression	Ligands on leukocytes
P-selectin (GMP-140)	selectin	absent	histamine, thrombin, ODFR	5–30 min	SLe^a-sugars SLe^x-sugars
E-selectin (ELAM-1)	selectin	absent	IL-1, TNF-alpha, ODFR (?)	2–6 hr	SLe^a-sugars SLe^x-sugars L-selectin
ICAM-1	immunoglobulin	low	IL-1, TNF-alpha, IFN-gamma	4–6 hr	LFA-1, CR3 (CD11a/CD18, CD11b/CD18)
ICAM-2	immunoglobulin	moderate	none, refractory to stimulation	constitutive expression	LFA-1 (CD11a/CD18)
VCAM-1	immunoglobulin	very low	IL-1, TNF-alpha	4–6 hr	VLA-4 (CD49d/CD29)

GMP, granular membrane protein; ODFR, oxygen-derived free radicals; ELAM, endothelial leukocyte adhesion molecule; IL-1, interleukin-1; TNF-alpha, tumor necrosis factor-alpha; ICAM, intercellular adhesion molecule; IFN-gamma, interferon-gamma; VCAM, vascular cell adhesion molecule; LFA, lymphocyte-function-associated antigen; CR, complement receptor.
*From Sluiter, W., Pietersma, A., Lamers, J. M. J., and Koster, J. F., 1993, J. Cardiovasc. Pharmacol., 22 (S4): S37–S44.

and A glycans (SLe-X and Sle-A) of unidentified glycolipids and glycoproteins. A possible candidate is L-selectin (LAM-1), a highly glycosylated selectin family member, which is expressed by all blood leukocytes (Picker, Warnssoek, Burns, Doerschaak, Berg and Butcher, 1991; Tedder, Penta, Levine and Freedman, 1990). P-selectin is found in the Weibel-Palade bodies of the endothelium.

Table 1 summarizes the various mediators that stimulate the expression of the adhesion molecules. It should be realized that there is a time-dependent difference in the expression of these various adhesion molecules. P-selectin is translocated in the first minutes of stimulation and its expression lasts about 5–30 minutes. ICAM-1 is constitutionally expressed but can be upregulated by stimulation with tumor necrosis factor(TNF)-alpha with a maximum at 4 to 6 hours of stimulation. VCAM-1 is normally absent, but its expression reaches a maximum after 4 to 6 hours of stimulation. After 24 hours only ICAM-1 and VCAM remain demonstrable at high levels (Springer, 1990; Osborn, Hession, Tizard, Vassallo, Cuhavsky, Chi-Rosoo, and Labb, 1989).

2. SIGNAL TRANSDUCTION CASCADE OF EXPRESSION

An important event in the signal transduction cascade that leads to the transcription of the genes upon stimulation with cytokines like TNF-alpha is the activation of the transcriptional factor nuclear factor kappaB (NF kappaB). In resting cells NF kappaB is located in the cytosol in an inactive form due to its binding of the NF kappaB inhibitor I kappaB. If upstream signals of the signal transduction cascade (free radicals, an activated protein kinase?) induce a conformational change of I kappaB, NF kappaB is released. While I kappaB will now be degraded by the proteasome, NF kappaB translocates to the nucleus where it binds to NF kappaB regulatory elements in the promotor regions of the VCAM-1, ICAM-1 and E-selectin genes (Neish, Williams, Palmer, Whitley, and Collins, 1992; Voraberger, Schäfer, and Stratowa, 1991; Whelan, Gharsa, Hoeft, Huijsdwijner, Gay, Chandra, Talalat, and Delamarter, 1991). The activation of NF kappaB is essential, but not sufficient for the full assembly of the relevant transcriptional activation complex that induces the cytokine-stimulated gene transcription (Collins, Read, Neish, Whitley, Thonos, and Maniates, 1995).

It has been hypothesized that free radicals function as second messengers in the activation of NF kappaB. This is based upon the observation that NF kappaB can be activated by the addition of hydrogen peroxide and that this activation can be abolished by the addition of radical scavengers (Schreck, Albermann, and Baeuerle, 1992). Furthermore, Matsubara & Ziff (1986) have shown that the administration of TNF-alpha to endothelial cells leads to the generation of superoxide radicals. Recently, the Ras protein has been suggested as the source of this type of free radical (Irani, Xia, Zweier, Sollalt, Der, Fearon, Sunderesan, Finkel, and Goldschmidt-Clermont, 1997). However, a direct role of superoxide in the activation of NF kappaB has been excluded by Schreck *et al.* (1992). Hydrogen peroxide can activate NF kappaB and is generated by the dismutation of superoxide, or directly produced by various enzymes. It has been suggested by Whelan *et al.* (1994) that these sites include the flavoenzymes cytochrome P450 monooxygenase and a phagocyte-type NADPH oxidase, since inhibitors of these enzymes, i.e., SKF525a and apocynin (4-hydroxy-3-methoxy- acetophenone) respectively, also inhibit the TNF-alpha-induced VCAM-1 expression specifically. However, we have shown that the NADPH-oxidase inhibitor apocynin inhibits the cytochrome P450IA1 activity in endothelial cells as well (Pietersma, De Jong, De Wit, Koster, and Sluiter, submitted). Since the phagocyte-type NADPH oxidase remains to be demonstrated in endothelial cells, the inhibitory effect of apocynin on the induction of VCAM-1 may therefor no longer be used to demonstrate a role for the NADPH oxidase in this process. Under hypoxic conditions the expression of VCAM-1 was reduced, whereas the expression of ICAM-1 and E-selectin were not. On this basis, we assume that the oxygen-dependent step in the intracellular signal cascade underlying the TNF-alpha-stimulated transcriptional activation of VCAM-1 resides in the activity of a cytochrome P450-dependent monooxygenase. The flavoprotein inhibitor diphenylene iodonium (DPI) also specifically inhibits the expression of VCAM-1. This theoretically implicates the involvement of the iodonium-sensitive flavoenzymes mitochondrial NADH dehydrogenase, xanthine oxidase and nitric oxide (NO) synthase in the expression of VCAM-1 as well. However, this is not very likely because, 1), inhibition of mitochondrial NADH dehydrogenase by rotenon does not result into a decreased expression of VCAM-1 (Pietersma, De Jong, De Wit, Koster, and Sluiter, submitted), 2), inhibition of xanthine oxidase with allopurinol does not lead to a decreased transcriptional expression of VCAM-1 (Weber, Erl, Pietsch, Ströbel, Ziegel-Heitbroek, and Webe, 1994), and 3), increased delivery of NO as shown by Kahn, Harrison, Olbrych, Alexander, and Medford (1996) suppressed the TNF-alpha-induced expression of VCAM-1 in stead of increasing it. Taken together, this leaves the cytochrome P450 monooxygenase as the sole direct source of hydrogen peroxide involved in the expression of VCAM-1.

We (Pietersma, Koster, and Sluiter, unpublished results) and others (Royall, Gwin, Packs, and Freeman, 1992) have found that the rate of hydrogen peroxide production by endothelial cells did not increase upon stimulation by TNF-alpha. This indicates that hydrogen peroxide does not function as a classical second messenger, but more likely may play a permissive role in the expression of VCAM-1. Then, the hydroxyl radical could be the relevant second messenger. This implies an increase in the cellular free iron content in order to make the Haber Weiss chemistry possible. However, we found no change in the magnitude of the low molecular weight iron pool of the endothelial cells upon stimulation by TNF-alpha (Pietersma, Kraak-Slee, De Jong, De Wit, Van Heugten, Koster, and Sluiter, in preparation), which weakens the hypothesis that free radicals trigger the activation of

NF kappaB in endothelial cells. Surprisingly, we found that the iron chelator desferriox-amine inhibited the expression of VCAM-1 protein, but not of ICAM-1, by endothelial cells. And even more confusing, under this condition the expression of VCAM-1 mRNA was not decreased.

What could be the meaning of these results? We suppose that the oxygen-dependent step in the specific expression of VCAM-1 resides in the cytochrome P450 monooxy-genase. Other products of this enzym complex, but not free radicals, may activate NF kappaB via a presently unknown mechanism. It might well be that these cytochrome P450 products inactivate the NF kappaB inhibitor I kappaB directly or via the activation of some kind of protein kinase. Liu, Goedelel, and Karim (1996) have excluded the contribution of JNK in this latter respect. The activation of NF kappaB then leads to the expression of VCAM-1 transcripts.

Our findings reveal that the TNF-alpha-stimulated expression of VCAM-1 on the endothelial cell membrane is not only specifically regulated at the transcriptional level, but also at the post-transcription level. This latter step is chelatable-iron dependent. We recently have shown (Pietersma *et al.*, 1997) that inhibition of the P38 mitogen activated protein (MAP) kinase suppresses the TNF-alpha-induced VCAM-1 expression, but not ICAM-1 expression, at the post-transcriptional level. Whether infact this MAP kinase is iron dependent remains to be elucidated.

REFERENCES

Athens, J.W., Raab, S.O., Haab, O.P., Mauer, A.M., Aschenbrucker, and Cartwright, G.E., 1961, Leukokinetic studies III. The distribution of granulocytes in the blood of normal subjects, *J. Clin. Invest.* **40:**159–168.

Carlos, T.M., and Harlan, J.M., 1994, Leukocyte-endothelial adhesion molecules. *Blood* **84:**2068–2101.

Collins, T., Read, M.A., Neish, A.S., Whitley, M.Z., Thonos, D., and Maniates, T., 1995, Transcriptional regulation of endothelial cell adhesion molecules:NF kappaB cytokine-induced enhancers, *FASEB J.* **9:**899–909.

Cybulsky, M.I., and Gimbrone, M.A., 1991, Endothelial expression of a mononuclear leukocyte adhesion molecule during atherogenesis, *Science* **251:**788–791.

De Fougerolles, A.R., Stacker, S.A., Schawarting, R., and Springer, T.A., 1991, Characterization of ICAM-2 and evidence for a third counter receptor for LFA-1, *J. Exp. Med.* **174:**253–267.

Dustin, M.L., Staunton, D.E., and Springer, T.A., 1988, Supergene families meet in the immune system, *Immunol. Today* **9:**213–215.

Irani, K., Xia, Y., Zweier, J.L., Sollalt, S.J., Der, C.J., Fearon, E.R., Sunderesan, M., Finkel, T., and Goldschmidt-Clermont, P.J., 1997, Mitogenic signaling mediated by oxidants in Res transformed fibroblasts. *Science* **275:**1649–1652.

Khan, B.V., Harrison, D.G., Oblrych, M.T., Alexander, R.W., and Medford, R.M., 1996, Nitric oxide regulates vascular cell adhesion molecule-1 gene expression and redox sensitive transcriptional events in human vesicles endothelial cells, *Proc. Natl. Acad. Sci. USA* **93:**9114–9119.

Liu, Z-G., Hsu, H., Goedelel, D.V., and Karim, A., 1996, Dissection of TNF receptor 1 effector functions: JNK activation is not linked to apoptosis while NF-kappa-B activation prevents cell death, *Cell* **87:**565–576.

Matsubara, T., and Ziff, M., 1986, Increased superoxide anion release from human endothelial cells in response to cytokines, *J. Immunol.* **137:**3296–3298.

Neish, A.S., Williams, A.J., Palmer, H.J., Whitley, M.J., and Collins, T., 1992, Functional analysis of the humen vascular cell adhesion molecule-1a promotor. J. Exp. Med. **176:**1583–1593.

Osborn, L., Hession, C., Tizard, R., Vasselo, C., Bubewsky, S., Chi-Rosso, G., and Lobb, R., 1989, Direct expression cloning of vascular cell adhesion molecule-1, a cytokine-induced endothelial protein that binds lymphocytes, *Cell* **59:**1203–1211.

Picker, L.J., Warnssoek, R.A., Burns, A.R., Doerschaak, C.M., Berg, E.L., Butcher, E.C., 1991, The neutrophil selectin LECAM-1 presents carbohydrate ligands to the vascular selectins ELAM-1 and GMP-140, *Cell* **66:**921–933.

Pietersma, A., Tilly, B.C., Gaestel, M., De Jong, N., Lee, J.C., Koster, J.F., and Sluiter, W., 1997, P38 mitogen activated protein kinase regulated endothelial VCAM-1 expression at the posttranscriptional level, *Biochem. Biophys. Res. Commun.* **230:**44–48.

Rice, C.E., Munro, J.M., Carlos, C., Bevilacqa, M., 1991, Vascular and non-vascular expression of INCAM-110. A target for mononuclear leukocyte adhesion in normal and inflamed human tissues. *Am. J. Pathol.* **138**:385–393.

Royall, J.A., Gwin, P.D., Packs, D.A., Freeman, B.A., 1992, Responses of vascular endothelial metabolism to lipopolysaccharide and tumor necrosis factor alpha, *Arch. Biochem. Biophys.* **294**:686–699.

Schreck, R., Albermann, K., and Baeuerle, P.A., 1992, Nuclear factor kappa-B: an oxidative stress responsive transcription factor of eukaryotic cells (a review), *Free Rad Res. Commun.* **4**:221–237.

Sluiter, W., Pietersma, A.,Lamers, J.M.J., and Koster, J.F., 1993, Leukocyte adhesion molecules on the vascular endothelium. Their role in the pathogenesis of cardiovascular disease and the mechanism underlying their expression, *J. Cardiol. Pharmacol.* **22**:S37–S44.

Springer, T.A., 1990, Adhesion receptors of the immune systems, *Nature* **346**:425–434.

Springer, T.A., 1994, Traffic signals for lymphocyte recirculation and leukocyte emigration: the multistep paradigm, *Cell* **76**:301–314.

Tedder, T.F.,and Penta, A.C., Levine, H.B., and Freedman, AS., 1990, Expression of the human leukocyte adhesion molecule, LAM-1. Identity with TQ1 and Leu-8 differentiation antigen. *J. Immunol.***144**:532–540.

Tengelder, G.J. Arfors, K.E., 1991, Inhibition of leukocyte rolling in venules by protamine and sulphonated polysaccharides, *Blood* **77**:1565–1571.

Van Furth, R., and Sluiter, W., 1986, Distribution of blood monocytes between a marginating and a circulating pool, *J. Exp. Med.* **163**:474–479.

Voraberger, G., Schafer, R., and Stratowa, C., 1991, Cloning of the human gene for intercellular adhesion molecule 1 and analysis of its 5' regulatory region. Induction by cytokines and phorbol ester, *J. Immunol.* **147**:2777–2786.

Weber, C., Erl, W., Pietsch, A., Ströbel, M., Ziegler-Heitbroek, H.W., and Webe, P.C., 1994, Antioxidants inhibit monocyte adhesion by suppressing nuclear factor kappaB mobilization and induction of vascular cell adhesion molecule-1 in endothelial cells stimulated to generate radicals, *Artheriolscler. Thromb.* **14**:1665–1673.

Williams, A.F., and Barclay, A.N., 1988, The immunoglobulin superfamily-domains for cell surface recognition, *Annu. Rev. Immunol.* **6**:381–405.

Whelan, J. Gharsa, P., Hoeft and Huijsdwijner, R.H., Gray, J., Chandra, G., Talalat, F., and Delamarter, J.F., 1991, An NF kappaB-like factor is essential but not sufficient for cytokine induction of endothelial leukocyte adhesion molecule 1 (ELAM-1) gene transcription, *Nucleic Acids Res.* **19**:2645–2653.

11

REACTIVE OXYGEN SPECIES (ROS) AND ALTERATION OF F_0F_1–ATP SYNTHASE IN AGING AND LIVER REGENERATION

Ferruccio Guerrieri, Giovanna Pellecchia, and Sergio Papa*

Institute of Medical Biochemistry and Chemistry
University of Bari, Italy

1. ABSTRACT

The contribution of oxidative phosphorylation to cellular energy demand changes in the life span, being low in foetal tissues and increasing progressively after the birth until 80% and more of cellular ATP is provided by mitochondrial oxidative phosphorylation in the tissues from adult animals. Aging is characterized by a progressive decline of the oxidative phosphorylation process associated to alterations of respiratory complexes and F_0F_1–ATP synthase. The age dependent changes are tissue specific being more pronounced in the heart (a well differentiated tissue) than in liver. Damage of F_0F_1–ATP synthase has been also observed in the early phase of liver regeneration characterized by retrodifferentation of hepatocytes which change from oxidative to fermentative metabolism. In both cases, aging and liver regeneration, the reactive oxygen species are apparently involved in the damage of mitochondrial F_0F_1–ATP synthase.

2. INTRODUCTION

Under normal conditions, mitochondrial oxidative phosphorylation (OXPHOS) covers more than 80% of the cellular ATP demand, the remaining being met by the anaerobic glycolysis (Papa, 1996). The contribution of OXPHOS to the cellular energy demand changes, however, in the life span. In foetal tissues anaerobic glycolysis is the major source of cellular ATP (Valcarce et al., 1988; Izquierdo et al., 1990). At the birth, following a rapid

---

* Corresponding author: Prof. Sergio Papa, Institute of Medical Biochemistry and Chemistry, University of Bari, Bari, Italy, Piazza G. Cesare, Policlinico, 70124, Bari, Italy. Phone: +39-80-5478428; Fax: +39-80-5478429; E-mail: spauniba@tin.it

Free Radicals, Oxidative Stress, and Antioxidants, edited by Özben.
Plenum Press, New York, 1998.

induction of mitochondrial biogenesis (Valcarce et al., 1988), the OXPHOS process increases progressively reaching the maximum level in the tissues of young and adult animals (Guerrieri et al., 1992a,b). In the tissues of aged animals the OXPHOS capacity decreases (Nohl and Kramer, 1980; Hansford, 1983; Guerrieri et al., 1992a,b). These age dependent changes in OXPHOS capacity are more pronounced in heart and brain cells than in hepatocytes (Guerrieri et al., 1992a,b) which duplicate during the life span.

The central enzyme of OXPHOS is the mitochondrial F_0F_1–ATP synthase of the inner mitochondrial membrane which is made up of a hydrophilic catalytic sector (F_1 moiety), exposed to the mitochondrial matrix and consisting of five nuclear encoded subunits (3α, 3β, γ, δ, and ε), and a proton-translocating membrane integral sector, composed of 10 subunits (Collison et al., 1994) (Fig. 1). Two of the F_0 subunits (ATPase 6 and A6L) are encoded by the mitochondrial genome (Attardi and Schatz, 1988).

To verify if alterations of F_0F_1–ATP synthase contribute to the age dependent changes in OXPHOS, the oligomycin sensitive ATP hydrolase activity in inside-out submitochondrial particles (ESMP) (Lee and Ernster, 1968), isolated from heart and liver of young (3 month old) and senescent (24 month old) rats, was examined (Guerrieri et al., 1992a; Capozza et al., 1994). The activity of the F_0F_1–ATP synthase was found to be depressed in ESMP isolated from senescent rats (Fig. 2A). The decline was much more pronounced in ESMP from heart than in ESMP from liver (Fig. 2A).

In these particles the anaerobic release of transmembrane proton gradient, set up by the respiratory chain, is 70–80% inhibited by the F_0 inhibitor oligomycin (Pansini et al., 1978; Guerrieri et al., 1992a), indicating that this process represents, mainly, the proton permeability through the F_0 sector. In ESMP prepared from mitochondria of aged rats, the anaerobic release of the transmembrane proton gradient was faster than in ESMP from mitochondria of young animals (Fig. 2A). Increase of passive proton conduction in submitochondrial particles is observed after removal of F_1 from the membrane (Pansini et al., 1978; Guerrieri et al., 1993). Immunoblot analysis, using an antiserum against beef-heart isolated F_1 (Guerrieri et al., 1989) showed that the age related decrease in activities of ATP synthase were accompanied by parallel decrease in the amount of immunodetected α/β subunits of the complex (Fig. 2B), which was more pronounced in heart than in liver mitochondria.

These and other observations (Guerrieri et al., 1992a; Guerrieri et al., 1993, Capozza et al., 1994; Guerrieri et al., 1996) thus show that in the inner mitochondrial membrane of tissues of aged rats the decrease in the content and the activity of the F_1 sector of F_0F_1

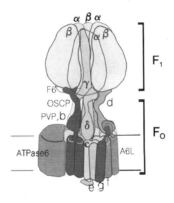

Figure 1. Scheme for the structure of F_0F_1–ATP synthase complex. Reproduced from Papa et al. (1998).

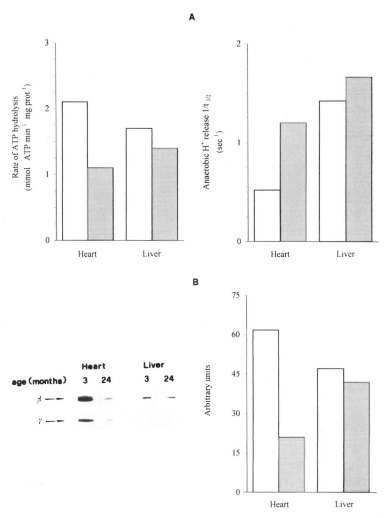

Figure 2. Age dependent changes in activities (A) and immunodetected amount of the the catalytic β subunit (B) of mitochondrial F_0F_1 ATP synthase. Empty columns: 3 month old rats; filled columns: 24 month old rats, The densitometric analysis of immunoblot is reported in right panel B. For experimental details see Guerrieri et al., 1992a.

complex contributes to the decline in the capacity of the oxidative phosphorylation process in mitochondria of associated with aging.

3. ROS DAMAGE OF MITOCHONDRIAL F_0F_1–ATP SYNTHASE

The free radical theory of aging (Harman, 1956) postulates a progressive damage of cellular macromolecules, such as DNA, lipids and proteins, by reactive oxygen species (ROS) which are, essentially, produced at the level of the respiratory chain (Chance et al., 1979). Glutathione (GSH), which represents the main cellular antioxidant system (Viña, 1990) is a tripeptide synthetized in the cytosol and actively imported into the mitochondria (Viña, 1990). Age dependent changes in the intramitochondrial content of GSH have been observed both in

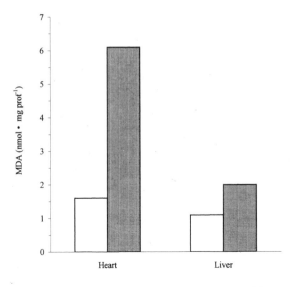

Figure 3. MDA production in isolated mitochondria following exposure to ^{60}Co irradiation. For preparation of mitochondria and its exposure to ^{60}Co irradiation see Guerrieri et al., 1996. MDA production was determined as described by Slater and Sawyer 1971. Empty columns: heart or liver mitochondria from 3 month old rats; filled columns: heart or liver mitochondria from 3 month old rats exposed to ^{60}Co irradiation (3000 rad).

heart and liver (Capozza et al., 1994). However the age dependent decrease of intramitochondrial GSH was much more significant in heart than in liver (Capozza et al., 1994).

"In vitro" exposure of mitochondria, isolated from heart and liver of young rats (3 month old rats), to ROS generated by ^{60}Co irradiation (Zhang et al., 1990; Capozza et al., 1994; Guerrieri et al., 1996) was found to cause an increase of mitochondrial malondialdehyde (MDA) production (Fig. 3), which is an index of increased lipid peroxidation by ROS. The increase of MDA production was higher in ^{60}Co irradiated heart mitochondria than in ^{60}Co irradiated liver mitochondria (Fig. 3). In ^{60}Co irradiated mitochondria a decrease of the mitochondrial content of GSH, which accompanied the decrease in the rate of ATP synthesis (Table 1), was observed. These ^{60}Co induced changes were more pronounced in heart than in liver mitochondria (Table 1).

Oxidative stress of isolated rat liver mitochondria, induced by ter-butyl-hydroperoxide, causes a binding of GSH to free thiol groups of proteins which is related to the impairment of mitochondrial functions (Olafsdottir et al., 1988). Similarly oxidative stress induced by ^{60}Co irradiation of isolated mitochondria was found to increase the level of GSH bound to thiol groups of intramitochondrial membrane proteins. This effect was more pronounced in heart than in liver (Fig. 4).

Table 1. Effect of ^{60}Co irradiation of isolated rat-liver and rat-heart mitochondria

	GSH (nmol·mg prot^{-1})	ATP synthesis (nmol ATP·min^{-1} mg prot^{-1})
Heart		
Control	4.1 ± 0.3	440 ± 13
[60]Co (3000 rad)	1.4 ± 0.2	120 ± 18
Liver		
Control	5.7 ± 0.3	268 ± 11
[60]Co (3000 rad)	4.1 ± 0.3	200 ± 23

For preparation of mitochondria, ^{60}Co irridation, determination of the rate of ATP synthesis and of GSH level see Guerrieri et al., 1996. The data reported in the Table are the means (± SEM) of 6 experiments.

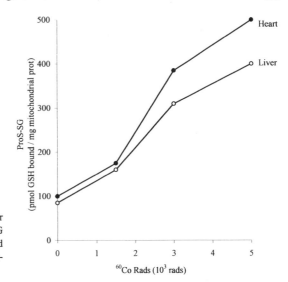

Figure 4. ^{60}Co irradiation of heart (●) or liver (○) mitochondria causes formation of ProS–SG complex. For determination of glutathione bound to mitochondrial proteins (ProS–SG) and ^{60}Co irradiation see Guerrieri et al., 1996.

Direct exposure of sonic submitochondrial particles (ESMP) to ^{60}Co irradiation was found to cause a damage of the structure and function of mitochondrial F_0F_1–ATP synthase (Guerrieri et al., 1993; Capozza et al., 1994; Guerrieri et al., 1996).

In conclusion these "in vitro" experiments demonstrate a direct damage of mitochondrial FoF_1–ATP synthase by ROS.

In agreement with the free radical theory of aging (Harman, 1956) ROS damage of F_0F_1–ATP synthase could be involved in the alteration of OXPHOS observed during aging. In "vivo" this effect can be favoured by the progressive decrease of cellular (Viña, 1990) and mitochondrial GSH (Capozza et al., 1994) observed during aging.

The increase of GSH bound to mitochondrial membrane proteins, following the ^{60}Co irradiation of isolated mitochondria, suggests the possibility that accumulation of ROS in mitochondria causes the formation of protein-S-radicals in membrane proteins directly exposed to the mitochondrial matrix (i.e. subunits of ATP synthase complex) forming denatured proteins (i.e. GS–S protein complexes) which can represent a better substrate for mitochondrial proteases (Goldberg, 1992). The difference between liver and heart mitochondria suggests that replicating liver cell and organelles are better protected from ROS damage (Miquel and Fleming, 1986).

4. CHANGES IN ACTIVITY OF MITOCHONDRIAL F_0F_1–ATP SYNTHASE AND ROS PRODUCTION DURING LIVER REGENERATION

Liver regeneration is an "in vivo" model of cell proliferation which can be induced by surgical removal of a portion of the liver (Bucher and Malt, 1971) until the original mass is restored (Michelopoulos, 1990). The kinetics of the process shows two phases: i) an early prereplicative phase, characterized by a decrease in the mitochondrial content and activity of F_0F_1–ATP synthase and in the capacity of OXPHOS (Buckle et al., 1986; Guerrieri et al., 1995); ii) a proliferative phase characterized by an active DNA synthesis (Michelopoulos, 1990) and recovery of the amount and function of mitochondrial F_0F_1–ATP synthase (Guerrieri et al., 1995). Although the process has been known for a long time, the exact mechanism by which liver regeneration starts and is regulated is, still, not known (Michelopoulos, 1990). It has been suggested (Tsai et al., 1992) that enhanced production of ROS can represent a major event in the very early phase of liver regeneration (see also: Steer, 1995; Rastogi et al., 1995). On the other hand a depression of mitochondrial glutathione has been observed in the early phase of liver regeneration (Fig. 5A) which seems to be chronologically related to structural and functional changes in mitochondrial F_0F_1–ATP synthase (Vendemiale et al., 1995).

The lag phase of liver regeneration is accompanied by a decrease of the rate of mitochondrial ATP synthesis (Fig. 5A) . This decrease of ATP synthesis is linearly related to the decrease of intramitochondrial GSH (Fig. 5B). After the lag phase an immediate complete recovery of the intramitochondrial level of GSH and of the rate of ATP synthesis is

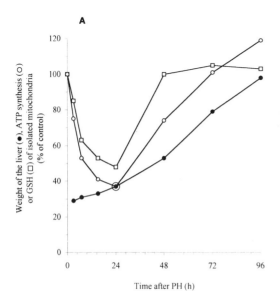

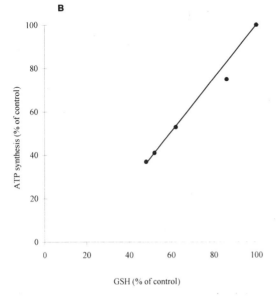

Figure 5. Time course of recovery of liver mass (●), and changes of rate of ATP synthesis (○) and GSH level (□) in mitochondria during liver regeneration (panel A). Linear relationship between changes in GSH content and rate of ATP synthesis in mitochondria during early phase (0–24 h) of liver regeneration (Panel B). For experimental details see Guerrieri et al., 1995 and Vendemiale et al., 1995. The mass of the liver is expressed as percentage of the weight of the liver in sham-operated rats (12.3 ± 1 g). The rate of ATP synthesis is expressed as percentage of rate of ATP synthesis in liver mitochondria isolated from sham operated rats (350 nmol ATP · mg prot^{-1} ± 12) see (Guerrieri et al., 1995). GSH is expressed as percentage of mitochondrial content in sham-operated rats (6.1 ± 0.3 nmol·mg prot^{-1}) (see Vendemiale et al., 1995). The data reported in panel B are obtained from the data shown in panel A.

observed (Fig. 5A), which is associated to the recovery of the liver weight (see also Guerrieri et al., 1995; Vendemiale et al., 1995).

The decrease of intramitochondrial GSH could be related to either a decrease of intramitochondrial import of the tripeptide newly synthetized at the cytosolic level, or to an
intramitochondrial oxidative stress. The fact that cytosolic level of GSH does not decrease
in the early phase of liver regeneration (Vendemiale et al., 1995), but instead increases,
reaching a peak at 48 h after partial hepatectomy (PH) (Vendemiale et al., 1995) (Fig. 5A),
supports the possibility of an extra consumption of intramitochondrial GSH by oxidative
stress in the early phase of liver regeneration. This is confirmed by an increase of intramitochondrial MDA production and of the level of intramitochondrial carbonyl proteins with
a peak at 24 h after PH (Fig. 6A), after which both return to control values in 48 h after
PH (Fig. 6A), when, also the intramitochondrial level of GSH was recovered (cf.Fig. 5A).

The intramitochondrial oxidative stress during the early phase after PH caused, also,
a decrease of free thiol groups of proteins (Fig. 6B). Particularly interesting is the fact
that, in this early phase after PH, the amount of glutathione covalently bound to intramito-

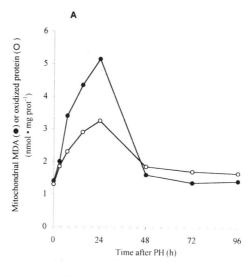

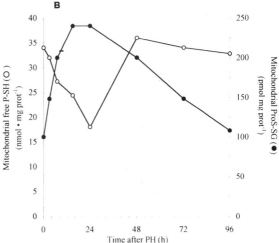

Figure 6. Time course of changes in MDA (●) and carbonyl proteins (○) (panel A) or P–SH (○) and ProS–SG (●) (in panel B) in liver mitochondria during liver regeneration. MDA was determined as reported by Slater and Sawyer, 1971. Carbonyl proteins were determined as reported by Levine et al., 1990. P–SH levels were determined by Elmann procedure and ProS–SG levels were determined as reported by Guerrieri et al., 1996.

chondrial thiol groups of the proteins increased (Fig. 6B). Finally the normal low level of ProSSG is progressively restored (Fig. 6B).

Blue-native PAGE of liver mitochondria (Fig. 7A) identified 5 main protein complexes (see also Schägger and von Jagow, 1991). As for bovine heart mitochondria, the

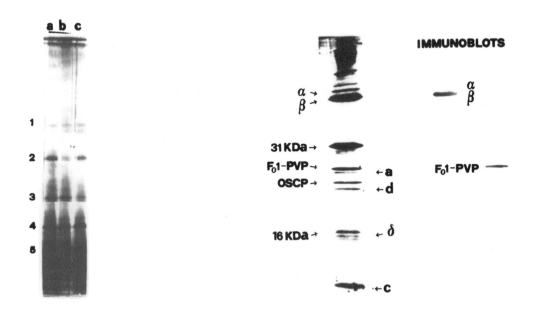

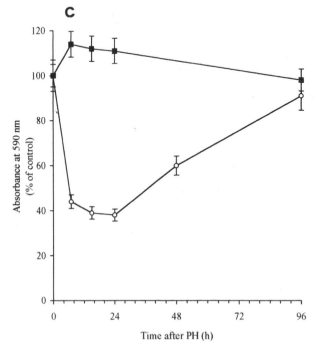

Figure 7. Blue-native polyacrylamide gel electrophoresis (A), identification of band 2 (B) and densitometric analysis of band 2 and band 4 (C). Blue-native gel electrophoresis was performed as described by Schäggher and von Jagow, 1991. Second dimension gel and immunoblot analysis of band 2 (Panel B) were run as described by Guerrieri et al., 1995. Panel A shows the Coomassie blue gels of liver mitochondria isolated from rats at 0 (a), 24 (b), and 96 h (c) after PH. Densitometric analysis at 590 mn of bands 2 (O) and 4 (■) of blue native gels at various time intervals after PH are reported in panel C, taking the area peak of the bands at 0 h after PH as 100%.

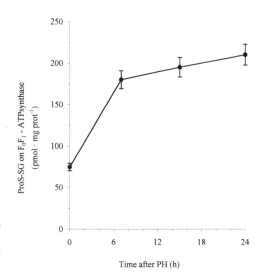

Figure 8. ProS–SG on F_0F_1–ATP synthase isolated by electroelution of band 2 from blue-native gels of liver mitochondria during early phase of liver regeneration (0–24 h after PH). ProS–SG level was determined as described by Guerrieri et al., 1996.

band 2 was identified as F_0F_1–ATP synthase by resolution in the second dimension gel and immunoblot (Fig. 7B). The comparison of blue-native PAGE of mitochondria isolated from rat-liver at different times after PH (Fig. 7A and C), shows a decrease of the density of Coomassie brillant blue stained bands 2, corresponding to F_0F_1–ATP synthase complex, during the early phase of liver regeneration; this was followed by a recovery of the amount of the protein in the second proliferative phase (Fig. 7C). No significant decrease of the density of the other bands was, on the other hand, observed (i.e. band 4 in Fig. 7C) in the early phase of liver regeneration. The F_0F_1–ATP synthase complex, electroeluted from preparative blue-native PAGE, was used for determination of the amount of glutathione bound to the enzyme (ProSSG in Fig. 8). This was found to increase 2–3 times during the early phase after PH (Fig. 8) and then, progressively returned to basal values between 24 and 96 h after PH (not shown).

5. CONCLUDING REMARKS

Common features of various tissues from aged animals and rapidly growing tissues (i.e. foetal tissues, regenerating liver, tumours) are high glycolytic activity and the low mitochondrial oxidative phosphorylation. This appears to be associated to alterations of the structure and function of mitochondrial F_0F_1–ATP synthase complex (Papa, 1996; Papa et al., 1997). The biogenesis of the F_0F_1–ATP synthase depends on the concerted expression of nuclear and mitochondrial genes (Attardi and Schatz, 1988). Synthesis of the nuclear-encoded subunits of F_1 sector of the F_0F_1–ATP synthase seems to control the increased phosphorylation capacity of mitochondria after birth (Valcarce et al., 1988; Izquierdo et al., 1990).

The data reviewed in the present paper show that mitochondria from senescent rats are characterized by a decrease in F_1 content in mitochondria from various tissues. The decrease of F_0F_1–ATP synthase subunits in mitochondria from aged animals appears to be associated with a decrease of intramitochondrial GSH. Both these phenomena, seem to be related to oxidative damage of the mitochondrial proteins which can result in an increase of proteolysis of the damaged proteins (Goldberg, 1992; Stadtman, 1992).

The oxidative damage of F_0F_1–ATP synthase is much more pronounced in heart than in liver mitochondria from aged rats suggesting that liver mitochondria are better protected from ROS damage and aging process (Miquel and Fleming, 1986).

The early prereplicative phase of liver regeneration is characterized by an intramitochondrial oxidative stress which is related to alteration of structure and function of F_0F_1–ATP synthase. The residual ATP synthase complex, isolated from liver mitochondria at 7, 15 and 24 h after PH, shows a 2–3-fold increase in the level of ProSSG. It is conceivable that the formation of ProSSG results in an increased proteolytic cleavage of the enzyme by mitochondrial proteases (Stadtman, 1992; Goldberg, 1992). In line with this hypothesis is the observation that thiol containing compounds, such as GSH (which decreases in mitochondria isolated from livers in the early phase of PH), inhibit a variety of proteases (Ferguson et al., 1993) and that the oxidatively denatured proteins are actively degraded by ATP and Ca^{++} independent proteolytic systems (Davies et al., 1987).

In conclusion, the early pre-replicative phase of liver regeneration after PH is characterized, at mitochondrial level, by a transient increase of ROS accompanied by alteration of mitochondrial oxidative phosphorylation in particular of the F_0F_1–ATP synthase complex. The question remains open if ROS, produced in the early phase after PH, interfere with or even play a direct role in hepatocyte regeneration, possibly acting in cellular signal transduction (Ammendola et al., 1995).

ACKNOWLEDGMENTS

The part of the reviewed work conducted in this laboratory was supported in part by M.U.R.S.T., Italy (40% and 60%) and by C.N.R. and was run collaboration with doctors P. Cantatore, G. Capozza, I. Grattagliano and G. Vendemiale. Finally the help of Angelica Lorusso for computer processing of the manuscript is acknowledged.

REFERENCES

Ammendola, R., Fiore, F., Esposito, F., Caserta, G., Mesuraca, M., Russo, T. and Cimino, F.,1995. Differentially expressed mRNAs as a consequence of oxidative stress in intact cells. *FEBS Lett.* **371**:209–213.

Attardi, G., and Schatz, G., 1988. Biogenesis of mitochondria. *Annu.Rev.Cell.Biol.* **4**:289–333.

Bucher, N.R.L., and Malt, R.A., 1971. Regeneration of liver and kidney. In *Thirthy Years of Liver Regeneration: a Distillate* (N.L.R. Bucker ed.) Little, Brown and Co., Boston, pp. 15–26.

Buckle, M., Guerrieri, F., Pazienza, A., and Papa, S., 1986. Studies on polypeptide composition, hydrolytic activity and proton conduction of mitochondrial F_0F_1 H^+–ATP*ase* in regenerating rat liver. *Eur. J. Biochem.* **155**:439–445.

Capozza, G., Guerrieri, F., Vendemiale, G., Altomare, E., and Papa, S. 1994. Age related changes of the mitochondrial energy metabolism in rat liver and heart. *Arch. Gerontol. Geriatr.* suppl. **4**:31–38.

Collison, I.R., Runswick, M.J., Buchaman, S.K., Fearnley, I.M., Skehel, J.M., van Griffiths, D.E. and Walker, J.E. 1994. F_0 membrane domain of ATP synthase from bovine heart mitochondria: purification, subunit composition and reconstitution with F_1–ATPase. *Biochemistry* **33**, 7971–7978.

Davies K.J.A., Lin S.W., Pacifici R.E. 1987. Protein damage and degradation by oxygen radicals. IV . Degradation of denatured protein. *J. Biol. Chem.* **162**:9914–9920.

Ferguson D.M., Gores, G.J., Bronk S.F., Krom R.A.F., Raaij, M.J. 1993. An increase in cytosolic protease activity during liver preservation. *Transplantation* **55**:627–633.

Goldberg, A.L., 1992, The mechanism and functions of ATP-dependent proteases in bacterial. and animal cells. *Eur. J. Biochem.* **203**:9–23.

Guerrieri, F., Kopecky, J., and Zanotti, F., 1989. Functional and Immunological characterization of mitochondrial F_0F_1–ATP synthase, in: *Organelles in eukaryotic cells: molecular structure and interactions* (J.M. Tager, A. Azzi, and S. Papa, eds.) New York, London: Plenum Co., pp. 197–208.

Guerrieri, F., Capozza, G., Kalous, M., Zanotti, F., Drahota, Z., and Papa, S., 1992a. Age-dependent changes in the mitochondrial F_0F_1–ATP synthase. A*rch. Geront. Geriatr*. **14**:299–308.

Guerrieri, F., Capozza, G., Kalous, M., and Papa, S., 1992b. Age-dependent changes in the mitochondrial F_0F_1–ATP synthase. *Annals of the New York Academy of Sciences*. **671**:395–402.

Guerrieri, F., Capozza, G., Fratello, A., Zanotti, F., and Papa, S., 1993. Functional and molecular changes in F_0F_1 ATP-synthase of cardiac muscle during aging. *Cardioscience* **4**:93–98.

Guerrieri, F., Kalous, M., Capozza, G., Muolo, L., Drahota, Z., and Papa, S., 1994. Age dependent changes in mitochondrial F_0F_1–ATP synthase in regenerating rat-liver. *Biochem. Molec. Biol. Intern*. **33**:117–129.

Guerrieri, F., Muolo, L., Cocco, T., Capozza, G., Turturro, N., Cantatore, P., and Papa, S., 1995. Correlation between rat liver regeneration and mitochondrial energy metabolism. *Biochim. Biophys. Acta* **1272**:95–100.

Guerrieri, F., Vendemiale, G., Turturro, N., Fratello, A., Furio, A., Muolo, L., Grattagliano, I., and Papa, S., 1996. Alteration of mitochondrial F_0F_1–ATP synthase during aging. *Annals of the New York Academy of Sciences,* **786**:62–71.

Hansford, R.G., 1983. Bioenergetics in aging. *Biochim. Biophys. Acta* **726**:41–80.

Harman, D., 1956. Aging: a theory based on free readical and radiation chemistry. *J. Gerontol* **11**:298–300.

Izquierdo, J.M., Luis, A.M., and Cuezva, J.M., 1990. Postnatal mitochondrial differentation in rat-liver. *Jour. Biol. Chem*. **265**:9090–9097.

Lee, C.P., and Ernster, L., 1968. Studies of the energy transfer system of submitochondrial particles. Effects of oligomycin and aurovertin. *Eur. J. Biochem*. **3**:391–409.

Levine R.L., Garland D., Oliver C.N., Amici A. Climent I., Lenz A.G., Ahn B.W. et al., 1990. Determination of carbonyl content in oxidatively modified proteins. *Methods Enzymol* 186: 464–478.

Michelopulos, G.K., 1990. Liver regeneration: molecular mechanisms of growth control. *FASEB J.* **4**:176–187.

Miquel, J., and Fleming, J., 1986. Theoretical and experimental support for an oxygen radical mitochondrial injury. Hypothesis of cell aging in : *Free radicals, aging and degenerative diseases* (J.E. Johnson, R. Walford, D. Harman and J. Miquel).

Nohl, H., and Kramer, R., 1980. Molecular basis of age-dependent changes in the activity of adenine nucleotide translocase. *Mech. Ageing Dev.* **14**:137–144.

Olafsdottir, K., Pascoe, G.A., and Reed, D.J., 1988. Mitochondrial glutathione status during Ca^{2+} ionophore-induced injury to isolated hepatocytes. *Arch.Biochem.Biophys*. **263**:226–235.

Pansini, A., Guerrieri, F., and Papa, S., 1978. Control of proton conduction by the H^+–ATP*ase* in the inner mitochondrial membrane. *Eur. J. Biochem*. **92**:545–551.

Papa, S., 1996. Mitochondrial oxidative phosphorylation changes in the life span. Molecular aspects and physiopathological implications. *Biochim. Biophys. Acta* **1276**:87–105.

Papa, S., Guerrieri, F., Capuano, F., and Zanotti F., 1997. The mitochondrial ATP synthase in normal and neoplastic cell growth, in: Cell growth and oncogenesis (P. Bannash, D. Kanduc, S. Papa, and J. M. Tager, eds.) Basel: Birkhäuser Verlag 1997 in press.

Rastogi, R., Saksena, S., Garg, N.K., and Dhawan, B.N., 1995. Effect of picroliv on antioxidant-system in liver of rats, after partial hepatectomy. *Phytotherapy Research* **9**:364–367.

Slater, T., and Sawyer, B., 1971. The stimulatory effect of carbon tetrachloride and other halogenoalkanes on peroxidative reactions in rat liver fractions in vitro. *Biochem. J.* **123**:805–814.

Schägger H., von Jagow G. 1991. Blue native electrophoresis for isolation of membrane protein complexes in enzymatically active form. *Anal. Biochem.* **199**:223–231.

Stadtman, E.R. 1992. Protein oxidation and aging. *Science* **257**:1220–1224.

Steer, C.J., 1995. Liver regeneration. *The FASEB Journal* **9**:1396–1400.

Tsai, I.L., King, K.L., Chang, C.C., Wei, Y., 1992. Changes of mitochondrial respiratory functions and superoxide dismutase activity during liver regeneration. *Biochem. Int.* **28**:205–217.

Valcarce, C., Navarete, R.M., Encabo, P., Loeches, E., Satrùstegui, J. and Cuezva, J.M., 1988. Postnatal development of rat liver mitochondrial function. The roles of protein synthesis and adenine nucleotides. *Journ. Biol. Chem.* **263**:7767–7775.

Viña, J. (ed), 1990. *Glutathione:Metabolism and physiological function*. CRC Press Boston.

Vendemiale, G., Guerrieri, F., Grattagliano, I., Didonna, D., Muolo, L., and Altomare, E., 1995. Mitochondrial oxidative phosphorylation and intracellular glutathione compartmentation during rat liver regeneration. *Hepatology* **21**:1450–1454.

Vendemiale, G., Grattagliano, I., Altomare, E., Turturro, N. and Guerrieri, F., 1996. Effect of acetaminophen administration on hepatic glutathione compartimentation and mitochondrial energy metabolism in the rat. *Biochemical Pharmacology* **52**:1147–1154.

Zhang, Y., Marcillat, O., Gulivi, C., Ernster, L . and Davies, J.A. 1990. The oxidative inactivation of mitochondrial electron transport chain components and ATP*ase*. *Jour. Biol. Chem.* **265**:16330–16336.

12

FREE RADICALS AND ANTIOXIDANTS IN PHYSICAL EXERCISE

José Viña,* Miguel Asensi, Juan Sastre, José A. Ferrero, Emilio Servera, Amparo Gimeno, and Federico Pallardó

Departamento de Fisiologia
Universidad de Valencia
Valencia, Spain

1. INTRODUCTION

The beneficial effects of exercise are well documented. Indeed, it ameliorates diabetes mellitus, improves the plasma lipid profile, increases bone density and may help to lose weight. However, as stated in the old medical saying "the beneficial effects of exercise are lost with exhaustion". It has been known for some time that exhaustive exercise causes muscle soreness, induces an elevation of cytosolic enzyme activities in blood plasma and may be harmful. In the last decade a considerable amount of information concerning production of free radicals in exhaustive exercise has been obtained. An international symposium took place in Valencia, Spain in 1993 and a book was published on the subject (Sen, Packer, and Hanninen, 1994). This subject has been reviewed by (Bendich, 1991; and Ji, 1995). In this chapter we will summarize the present state of the art on this subject. The reader will find more information in the aforementioned sources.

2. EXERCISE GENERATES FREE RADICALS

The pioneering work by Davies, Quintanilha, Brooks, and Packer (1982) showed that exercise generates free radicals as evidenced by an up to three-fold increase in the signals attributed to free radicals in muscle and in liver. These authors further speculated that the increased production of these radicals may be a stimulus for the increased mitochondrial biogenesis that occurs in physical exercise.

Duthie, Robertson, Maughan, and Morrice (1990) showed that physical exercise causes an elevation in creatin kinase activity in plasma (an index of muscle damage) with-

* Corresponding author: Dr. José Viña, Departamento de Fisiologia, Facultad de Medicina, Avenida Blasco Ibáñez 17, 46010 Valencia, Spain. Tel +6-3864650; Fax +6-3864173.

Free Radicals, Oxidative Stress, and Antioxidants, edited by Özben.
Plenum Press, New York, 1998.

out changes in thiobarbiturate reactive substances (TBARS). However, erythrocyte susceptibility to lipid oxidation was elevated (P < 0.01). Furthermore, these authors observed a decrease in glutathione content in erythrocytes. They did not report on changes in GSSG levels probably due to an inadequate method to determine GSSG (see below).

3. INDICES OF OXIDATIVE STRESS IN EXERCISE: THE QUESTION OF METHODOLOGY

The fact that exercise generates free radicals prompted several investigators to study damage that may occur after exercise and that may be caused by free radical formation. Haramaki and Packer (1994) pointed out the importance of such indices. Reports on the effect of exercise on the oxidation of lipids, protein and nucleic acids show considerable discrepancies. Indeed, while some reports have shown increases in malondialdehyde (MDA) (Haramaki and Packer, 1994), others failed to do so (Duthie *et al.*, 1990). Similar discrepancies have been reported when studying the effects of exercise on glutathione status (Kosower and Kosower, 1978). Indeed while Duthie *et al.* (1990) failed to observe an increase in GSSG levels in blood plasma after exercise, others did find an increase in glutathione disulfide after exhaustion (Gohil, Viguie, Stanley, Brooks, and Packer, 1988; Sastre, Asensi, Gascó, Ferrero, Furukawa, and Viña, 1992). These discrepancies are almost certainly due to the poor methodology used to determine indices of oxidative stress associated with exercise. These shortcomings on the measurement of MDA and glutathione will be discussed below.

MDA is usually estimated by determining TBARS. This method, albeit widely used, is inadequate because it is not specific enough. Indeed, MDA levels are grossly overestimated by this procedure. Recently, Ames and co-workers (Yeo, Helbock, Chyu and Ames, 1994) have reported that "true" MDA levels determined by HPLC-mass spectroscopy are at least an order of magnitude, lower than those found when the TBARS method is used.

Glutathione is another example of a metabolite that can be used to measure oxidative stress (Sies, 1986) that may lead to erroneous conclusions. The hepatic GSH/GSSG ratio is about 100 and thus, a 2% oxidation of GSH will double the concentration of GSSG. Akerboom, Bilzer, and Sies (1982) used N-ethyl maleimide to determine GSH and GSSG in rat liver. We found (Asensi, Sastre, Pallardó, García de la Asunción, Estrela, and Viña, 1994) that currently used h.p.l.c. methods (Reed, Babson, Beatty, Brodie, Ellis, and Potter, 1980) did indeed cause oxidations of glutathione of about 1 to 2% in many tissues. This oxidation was specially marked in blood (Viña, Sastre, Asensi, and Packer, 1995). Gohil *et al.* (1988) did not find a relationship between glutathione oxidation and lactate levels after exhaustive exercise. However, Sastre *et al.* (1992) did find such relationship. The discrepancy can be explained by the inadequate methodology used by Gohil *et al.* (1988) to determine glutathione disulfide in blood.

To sum up, we have attempted to illustrate with two examples, malondialdehyde and glutathione, how inadequate methodology may affect our estimations of oxidative stress associated with physical exercise.

4. EXERCISE AND OXIDATION OF GLUTATHIONE

Glutathione is a major natural antioxidant in the cell (for a review see Viña, 1990). A major advantage of the glutathione system is that there are a number of enzymes which catalyze either the oxidation of GSH i.e. glutathione peroxidase which eliminates hydroperoxides, or the reduction of the disulfide, GSSG. The reducing equivalents are ultimately provided by glucose via the pentose phosphate pathway.

Lew, Pyke, and Quintanilha (1985) were the first to report on changes in glutathione following exhaustive exercise in rats. Gohil *et al.* (1988) reported an oxidation of glutathione in human blood after exhaustion. This was confirmed by Ji (1995) and by Sen, Atalay, and Hanninen (1994).

By using a new h.p.l.c. method to determine glutathione in blood (Asensi *et al.,* 1994; Viña *et al.,* 1995) and also, in some experiments by using the enzymatic method described by Akerboom *et al.* (1982) we observed that blood glutathione is oxidised in humans and in rats after exercise and that oxidation of glutathione is linear with increases in blood lactate (Sastre *et al.,* 1992). This observation led us to conclude that only *exhaustive* exercise causes oxidative stress (Sastre *et al.,* 1992) Indeed, exercise of low intensity (i.e. that does not cause exhaustion) did not cause oxidative stress in normal subjects; this is not the case of patients suffering from chronic obstructive pulmonary disease (see below).

5. ENZYME INDUCTION BY EXERCISE

The fact that exhaustive exercise causes an increase in free radical formation suggested that antioxidant enzymes might be induced by exhaustive exercise. Indeed, Ji, Stratman, and Lardy (1988) and also Sen *et al.* (1994) studied the effect of training on antioxidant enzymes. Exhaustive exercise resulted in an increase in glutathione peroxidase, glutathione reductase, superoxide dismutase and catalase activities. Ji and Fu (1992), studying the effect of physical exercise on the glutathione system hypothesized that both exhaustive exercise and hydroperoxide can cause severe oxidative stress to the body by a similar mechanism.

The original work by Davies *et al.* (1982) suggested that muscle mitochondriogenesis associated with exercise might be caused by the repeated free radical formation. Furthermore they showed that vitamin E deficient animals have 40% lower endurance than controls.

Indeed, the results reported in this section show that cells react against free radical formation associated with exercise by inducing the synthesis of enzymes which tend to detoxify such radicals. This may be very important to determine the possible protective effects of training against damage associated with exercise as discussed in the following section.

6. PROTECTION AGAINST FREE RADICAL DAMAGE BY TRAINING

The majority of persons who practise strenuous exercise do so after training. Thus, an important question was whether training protects against oxidative damage caused by exercise. Shortly after the observation of the free radical formation by exercise, Salminen and Vihko (1983) found that endurance training protected against muscle lipid peroxidation *in vitro*. They also reported muscle catalase activity and the concentration of vitamin E in red muscle fibres is significantly higher after training than in controls. No changes were reported in total sulfhydryl contents.

Ji *et al.* (1988) reported that as little as one hour of treadmill running elevated catalase and mitochondrial superoxide dismutase activity in liver. They also reported that

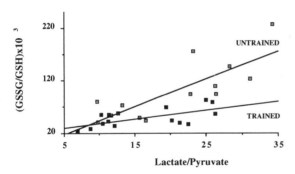

Figure 1. Effect of training on blood glutathione oxidation caused by exhaustive physical exercise in humans. Results are from 7 human volunteers. The training consisted of two bouts of one hour of aerobic exercise twice daily for two months.

liver peroxidation was increased after an acute bout of exercise and concluded that liver and skeletal muscle are capable of adapting to exercise to minimise oxidative injury caused by free radicals.

Laughlin *et al.* (1990) reported that antioxidant enzymes are related to muscle oxidative capacity but that in spite of this, the effect of exercise training on antioxidant enzymes cannot be predicted by measuring changes in oxidative capacity. Recently, Criswell *et al.* (1994) confirmed that indeed high intensity training induces skeletal muscle antioxidant activity.

In spite of much effort devoted to studying the effect of training on enzyme activities, specially on antioxidant enzymes, little work has been dedicated to testing the protective effect of training on metabolites that might be an index on oxidative stress. To our knowledge this question had been tested only on malondialdehyde, measured as TBARS. The possibility that malondialdehyde determination might be flawed by technical errors has already been mentioned above. Thus, we determined the effect of training in changes of glutathione redox status in human blood. For this purpose we used volunteers from the Spanish army. These young recruits were not previously trained. They performed exercise graded to exhaustion, and venous blood was collected before and after the exercise. This was followed by two months of aerobic training in the army and then the exercise test was repeated. Figure 1 shows that indeed training partially protects against exercise-induced oxidation of glutathione.

7. EXERCISE AND AGING

A major feature of aging is that the physiological performance declines as the individual ages. For instance, when comparing values of 20-year-olds versus 80-year-old persons, pulmonary capacity decreases by up to 45%, cardiac output decreases by up to 50% and the kidney mass decreases by up to 35%. The fact that free radicals are involved in aging, by Harman (1956) together with the increased generation of free radicals in exercise, provides with a theoretical frame to study the relationship between exercise and aging.

The idea of a threshold of age in exercise was first stated in the seventies (Edington, Consmas, and McCafferty, 1972). This interesting concept has been clearly stated by Reznick, Witt, Silbermann, and Packer (1992). Briefly, the threshold of age means that for a given exercise of a particular duration and intensity there is a certain age beyond which exercise may not be beneficial. It may even be detrimental to the cell (Golan and Reznick, 1994). Dr. Reznick has shown impressive pictures of histological damage to cells of old animals that had performed physical exercise (Ludatscher *et al.*, 1983). In young animals such histological damage could not be found, although other minor forms of damage, probably caused by oxidative stress, i.e. leakage of cytosolic enzymes could be observed (Sastre *et al.*, 1992). The question of whether exercise increases or decreases life expectancy remains open but some recent experiments in drosophila by Agarwal and Sohal (1994) show that the survival curve of animals who did not fly was shifted to the left, thus indicating that these animals have a shorter average life expectancy. Drosophila who fly have a higher level of a modified DNA base 8-hydroxy-2'deoxy-guanosine (Agarwal and Sohal, 1994). This is an indicator of oxidative stress, specially of damage to mitochondrial DNA (Richter, Park, and Ames, 1988). In other words, evidence is mounting that shows that free radicals generated as a result of *exhaustive* exercise may accelerate the aging process.

8. FATIGUE AND FREE RADICALS

One of the most interesting aspects of exercise physiology is the control of fatigue. Newsholme and his colleagues have studied the metabolic basis of fatigue. In their excellent book on exercise physiology, Newsholme, Leech and Duester (1993) define fatigue as the inability to maintain power output or in more simple words, fatigue is an overwhelming necessity to reduce pace. Many attempts have been made to understand the basis of fatigue. Costill (1985) has underscored the role of glycogen stores in delaying fatigue.

Reid, Haack, Franchek, Valberg, Kobzick, and West (1992) have provided evidence that reactive oxygen radicals are involved in the occurrence of fatigue in skeletal muscle. As a logical consequence, these authors tested the effect of antioxidants as inhibitors of fatigue and found that N-acetyl cysteine, a sulphur-containing amino acid that is a precursor of cysteine, did indeed cause a diminution of fatigue in muscle when subjected to repeated electrical stimulation (Reid, Stokic, Koch, Khawli, and Leis, 1994). These experiments, however, were performed to test fatigue in a strict physiological sense (inability to maintain power output). Of much more interest to both the researcher and the athlete is whether free radicals are involved in the more holistic way defined above, i.e. the overwhelming necessity to reduce pace.

9. EXERCISE, DISEASE, AND FREE RADICALS

Free radicals are thought to be involved in a series of diseases, some of which may be of a general nature like aging and some may be more specific like chronic obstructive pulmonary disease (COPD), atherosclerosis or inflammatory processes. Recently, in an excellent review Halliwell (1994). has summarised the role of free radicals in human disease. Although moderate exercise has beneficial effects on the evolution and outcome of diseases such as diabetes or osteoporosis, exhaustive exercise may be specially damaging in diseases whose pathogenesis involves free radicals.

Free radicals are thought to be involved in the pathogenesis of COPD. Furthermore, we noted that COPD patients become exhausted when they perform the activities that are normally carried out in their ordinary life. Since our experience in healthy people taught us that it is exhaustion and not just exercise that gives rise to free radicals (Sastre *et al.* 1992), we tested the effects of exercise of low intensity on blood GSH/GSSG in COPD patients. Results shown in Figure 2 indicate that indeed, mild exercise which is, however, exhaustive for them, causes an oxidation of blood glutathione in COPD patients. While healthy people become exhausted when they voluntarily perform exercise, COPD patients become exhausted daily as they carry out the low output exercise that is required in the course of their ordinary life.

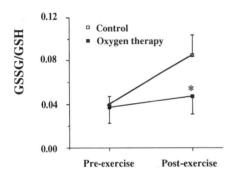

Figure 2. Effect of physical exercise and oxygen therapy on blood glutathione status in chronic obstructive pulmonary disease patients. Results are means from 5 chronic obstructive pulmonary disease patients under two conditions: control (breathing room air) and oxygen therapy (FiO_2 24–26%; flow rate 2–3 liters per minute).

10. ROLE OF ANTIOXIDANTS IN THE PREVENTION OF FREE RADICAL DAMAGE CAUSED BY EXERCISE

The fact that strenuous exercise increases free radical formation and causes a depletion of some antioxidants, indicates that supplementation with specific antioxidants may be beneficial to both the élite athlete and the ordinary person who occasionally performs exhaustive exercise.

Indeed, vitamin E has proved beneficial in protecting against free radical damage caused by exhaustive exercise (Meydani, Evans, Handelman, Biddle, Fielding, Meydani, Burill, Fiatarone, Blumberg and Cannon 1993). After exercise young adults showed an increased urinary excretion of TBARS. All subjects that took a suplement of vitamin E excreted significantly lower amounts of TBARS than those who took placebo (Meydani *et al.*, 1993). Furthermore, tissue damage caused by exhaustion was increased in vitamin E depleted animals. Furthermore, Cannon, Orencole, Fielding, Meydani, Meydani, Fiatarone, Blumberg and Evans (1990) studying the acute phase response in exercise demonstrated the interaction between nutrients and vitamin E on muscle enzyme release after exercise. Administration of coenzyme Q also had a protective effect in exercise induced muscle damage in rats(Shimomura, Suzuki, Sugiyama, Hanaki, and Ozawa (1991).

We found that oral administration of a mixture of glutathione and vitamin C partially protected against oxidation of blood glutathione induced by exercise in humans (see Figure 3). Also, oral administration of either glutathione or N-acetyl cysteine or vitamin C protected, at least in part, against exercise-induced oxidation of blood glutathione (Sastre *et al.*, 1992). Bendich (1991) and Witt, Reznick, Viguie, Starke-Reed, and Packer (1992) reviewed the protective role of antioxidant vitamins on exercise induced cellular damage.

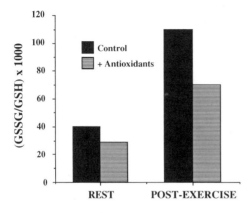

Figure 3. Effect of oral administration of antioxidants on blood glutathione oxidation caused by exhaustive physical exercise in humans. Values are means from 5 human volunteers. They were given a single oral dose of 1 g GSH ± 2 g vitamin C daily during 7 days previous to exercise protocol.

Also, Simon-Schass and Pabst (1988) observed that exhalation of pentane (a product of the peroxidation of polyunsaturated fatty acids), is increased by 100% after exhaustive exercise and that this is partially protected by administration of vitamin E.

As stated before recent evidence suggests that a thiol containing antioxidant, N-acetyl cysteine may inhibit muscle fatigue in humans (Reid *et al.*, 1994).

11. BENEFICIAL EFFECTS OF EXERCISE

This chapter has dealt with the undesirable effects of exercise specially of exhaustive exercise and with the involvement of free radicals in such effects.

However it must be emphasized that physical exercise, specially non-exhaustive exercise, is a healthy practice that has clear benefits in the prevention of many diseases. For instance, exercise attenuates diet-induced atherogenesis (Hasler, Rothenbacher, Mela, and Kris-Etherton, 1987) and can be considered as a cardio-protective factor (Chandrashekhar and Anand, 1991).

The favourable metabolic effects of exercise are well documented. Indeed, it increases high density lipoprotein cholesterol and activates lipoprotein lipase both in untrained and trained persons (Kantor, Culliane, Sady, Herbert, and Thomson, 1987). The importance of moderate exercise in the treatment of diabetes mellitus in undisputed (Felig, 1981).

Exercise together with a low calorie diet is the treatment of choice of most cases of over-weight and indeed it increases oxygen consumption not only during but also after exercise (Bahr, Ingnes, Vaage, Sejersted, and Newsholme, 1987).

The emerging conclusion is that while moderate exercise is a healthy practice, exhaustive exercise causes muscle damage and that such damage is partly due to free radicals. Antioxidants partially prevent exercise-induced oxidative damage.

ACKNOWLEDGMENTS

Work from the authors' laboratory reported here was carried out with the support of Comisión Interministerial de Ciencia y Tecnología Grant no. SAF 95-0558.

REFERENCES

Agarwal, S., Sohal, R.S. (1994). DNA oxidative damage and life expectancy in houseflies. *Proc. Natl. Acad. Sci. USA* **91**, 12332–12335.

Akerboom, Th. P. M., Bilzer, M., Sies, H. (1982). The relationship of biliary glutathione disulfides efflux and intracellular glutathione disulfide content in perfused rat liver. *J. Biol. Chem.* **257**, 4248–4252.

Asensi, M., Sastre, J., Pallardó, F.V., García de la Asunción, J., Estrela, J.M., Viña, J. (1994). A high-performance liquid chromatography method for measurement of oxidized glutathione in biological samples. *Anal Biochem.* **217**, 323–328.

Bahr, R., Ingnes, I., Vaage, O., Sejersted, O.M., Newsholme, E. (1987). Effect of duration of exercise on excess postexercise O_2 consumption. *Am. J. Physiol.* **62**, 485–490.

Bendich, A. (1991) Exercise and Free Radicals—Effects of Antioxidant Vitamins. *Adv Nutr. and Top Sport* **32**, 59–78

Cannon, J.G., Orencole, S.F., Fielding, R.A., Meydani, M, Meydani, S.N., Fiatarone, M.A., Blumberg, J.B., Evans, W.J. (1990) Acute Phase Response in Exercise—Interaction of Age and Vitamin-E on Neutrophils and Muscle Enzyme Release. *Am. J. Physiol.* **259**, R1214–R1219.

Chandrashekhar, Y., Anand, I.S. (1991). Exercise as a Coronary Protective Factor. *Am. Heart Journal* **122**, 1723–1739.

Costill, D.L. (1985). Carbohydrate nutrition before, during and after exercise *Fed. Proc.* 44, 364–368.

Criswell, D., Powers, S., Dodd, S., Lawer, J., Edwards, W., Renshler, K., Grinton, S. (1994). High intensity training induced in skeletal muscle antioxidant activity. *Med Sci. Sports Exerc.* **25**, 1135–1140.

Davies, K.J.A., Quintanilha, A.T., Brooks, G.A., Packer, L. (1982). Free radicals and tissue damage produced by exercise. *Biochem. Biophys. Res. Commun.* **107**, 1198–1205.

Duthie, G.G., Robertson, J.D., Maughan, R.J., Morrice, P.C. (1990). Blood Antioxidant Status and Erythrocyte Lipid Peroxidation Following Distance Running. *Arch. Biochem. Biophys.* **282**, 78–83.

Edington, D.W., Consmas, A.C., McCafferty, W.B. (1972). Exercise and longevity: evidence for a threshold of age. *J. Gerontol.* **27**, 341–343.

Felig, P. (1981). The endocrine pancreas: diabetes mellitus. In "Endocrinology and Metabolism" (Felig *et al* eds.) McGraw-Hill, New York.

Gohil, K., Viguie, C., Stanley, W.C., Brooks, G., Packer, L. (1988). Blood glutathione oxidation during human exercise. *J. Appl. Physiol.* **64**, 115–119

Golan, R., Reznick, A.Z. (1994). Aging and exercise-induced oxidative stress. In "Exercise and oxygen toxicity" (Sen *et al.*, eds) Elsevier, Amsterdam.

Halliwell, B. (1994). Free radicals, antioxidants, and human disease: Curiosity, cause, or consequence? *Lancet* **344**, 721–724.

Haramaki, N., Packer, L. (1994). Oxidative stres indices in exercise. In "Exercise and oxygen toxicity" (Sen *et al.*, eds.) Elsevier Amsterdam.

Harman, D. (1956) Aging: a Theory based on free radical and radiation chemistry. *J. Gerontol.* **11**, 298–300.

Hasler, C.M., Rothenbacher, H., Mela, D.J., Kris-Etherton, P.M. (1987). Exercise attenuates diet- induced arteriosclerosis in the adult rat. *J. Nutr.* **117**, 986–993.

Ji, L.L., Fu, R. (1992). Responses of glutathione system and antioxidant enzymes to exhaustive exercise and hydroperoxide. *J. Appl. Physiol.* **72**, 549–554

Ji, L.L. (1995). Oxidative stress during exercise: implication of antioxidant nutrients. *Free Rad Biol Med* **18**, 1079–1086.

Ji, L.L., Stratman, F.W., Lardy, H.A. (1988). Antioxidant exzyme systems in rat liver and skeletal muscle. Influences of selenium deficiency, cronic training and acute exercise. *Arch. Biochem. Biophys.* **263**, 150–160.

Kantor, M.A., Culliane, E.M., Sady, S.P., Herbert, P.N., Thomson, P.D. (1987). Exercise acutely increases high density lipoprotein-cholesterol and lipoprotein lipase activity in trained and untrained men. *Metabolism* **36**, 188–192.

Kosower, N.S., Kosower, E. M. (1978). The glutathione status of cells. *Int. Rev. Cytol.* **54**, 109–160

Laughlin, M.H., Simpson, T., Sexton, W.L., Brown, O.R., Smith, J.K., Korthuis, R.J. (1990). Skeletal Muscle Oxidative Capacity, Antioxidant Enzymes, and Exercise Training. *J. Appl. Physiol.* **68**, 2337–2343.

Lew, H., Pyke, S., Quintanilha, A. (1985). Changes in the glutathione status of plasma, liver and muscle following exhaustive exercise in rats. *FEBS lett.* **185**, 262–266.

Ludatscher, R., Silbermann, M., Gershon, D., Reznick, A. (1983) The effect of enforced running on the gastrocnemius muscle in aging mice. *Exp. Gerontol.*, **18**, 113–123.

Meydani, M., Evans, W.J., Handelman. G., Biddle, L., Fielding, R.A., Meydani, S.N., Burill, J., Fiatarone, M.A., Blumberg, J.B., Cannon J.G. (1993). Protective effect of vitamin E on exercise induced oxidative damage in young and older adults. *Am. J. Physiol.* **264**, R992–R998.

Newsholme, E., Leech, T., Duester, G. (1993). Keep on running: The science of training and performance. J. Wiley and Sons, Chichester.

Reed, D.J., Babson, J.R., Beatty, P.W., Brodie, A.E., Ellis, W.W., Potter, D.W. (1980). High-performance liquid chromatography analysis of nanomole levels of glutathione, glutathione disulfide, and related thiols and disulfides. *Anal. Biochem.* **106**, 55–62.

Reid, M.B., Haack, K.E., Franchek, K.M., Valberg. P.A., Kobzick, L., West, M. S. (1992). Reactive oxygen in skeletal muscle. I.Intracellular oxidant kinetics and fatigue in vitro. *J. Appl. Physiol.* **73**, 1797–1804.

Reid, M.B., Stokic, D.S., Koch, S.M., Khawli, F.A., Leis, A.A. (1994). N-acetylcysteine inhibits muscle fatigue in humans. *J. Clin. Invest.* **94**, 2468–2474.

Reznick, A.Z., Witt, E.H., Silbermann M., Packer, L. (1992). The threshold of age in exercise and antioxidants action. In "Free radicals and aging". (Chance, B and Emerit, I., eds.) Birkhauser Verlag, Basel,Switzerland.

Richter, C., Park, J.W, Ames, B. (1988). Normal oxidative damage to mitochondrial and nuclear DNA is extensive. *Proc. Natl. Acad. Sci. USA* **85**, 6465–6467.

Salminen, A., Vihko, V. (1983). Endurance training reduces the susceptibility of mouse skeletal muscle to lipid peroxidation in vitro. *Acta Physiol Scand*, **117**, 109–113.

Sastre, J, Asensi, M., Gascó, E., Ferrero, J.A., Furukawa, T., Viña, J. (1992). Exhaustive physical exercise causes and oxidation of glutathione status in blood. Prevention by antioxidant administration. *Am.J.Physiol.* **263**, R992–R995.

Sen, C.K., Atalay, M., Hanninen, O. (1994). Exercise-induced oxidative stress: Glutathione supplementation and deficiency *J. Appl. Physiol.* **77**, 2177–2187

Sen, C.K., Packer, L., Hanninen, O. (1994) Exercise and oxygen toxicity. Elsevier, Amsterdam.

Sies, H. (1986). Biochemistry of oxidative stress *Ang. Chemie* **25**, 1058–1071.

Simon- Schass, I., Pabst, H. (1988). Influence of vitamin E on physical performance *Int. J. Vit. Nutr. Res.* **58**, 49–54.

Shimomura, Y., Suzuki, M., Sugiyama, S. Hanaki,Y., Ozawa, T. (1991) Protective effect of coenzyme Q10 on exercise-induced muscular injury *Biochem. Biophys. Res. Commun.* **176**, 349–355.

Viña. J., (editor) (1990). Glutathione: Metabolism and Physiological Functions. CRC Press, Boston.

Viña, J., Sastre, J., Asensi, M., Packer, L. (1995) Assay of blood glutathione oxidation during physical exercise. *Meth. Enzymol.* **251**, 237–243.

Witt, E.H., Reznick, A., Viguie, C.A., Starke-Reed, P., Packer, L. (1992). Exercise, oxidative damage and effects of antioxidant manipulation. *J. Nutr.* **122**, 766–773.

Yeo, H.C., Helbock, H.J., Chyu, D.W., Ames, B. (1994). Assay of malondialdehyde in biological fluids by gas chromatography-mass spectrometry. *Anal. Biochem.* **220**, 391–396.

13

THE ROLE OF FREE RADICAL MEDIATION OF PROTEIN OXIDATION IN AGING AND DISEASE

Earl R. Stadtman[*]

Laboratory of Biochemistry
National Heart, Lung, and Blood Institute
National Institutes of Health
Building 3, Room 222
Bethesda, Maryland 20892

1. ABSTRACT

A role of protein oxidation in aging is indicated by the following observations: The cellular level of oxidized protein increases with animal age. Age-related changes in enzyme activities can be mimicked by treatment of enzymes from young animals with reactive oxygen species (ROS) *in vitro*. Exposure of animals to conditions of oxidative stress leads to an increase in the intracellular level of oxidized protein. Factors that increase the life span of animals lead also to a decrease in the level of oxidized protein and vice versa. Many age-related diseases are associated with elevated levels of oxidized proteins. Some age-related changes in enzyme activities and cognitive functions can be reversed by exposing old animals to free radical spin traps. The age-related increase in oxidized proteins is a complex function of the balance between a multiplicity of prooxidants, antioxidants, and the activities of proteases that selectively degrade the oxidized forms of proteins.

2. INTRODUCTION

It is well known that animal aging is accompanied by the accumulation of altered, catalytically inactive, or less active forms of many enzymes (Dreyfus, Kahn, and Schapira, 1978; Rothstein, 1977). The possibility that these age-related changes might be due to oxidative damage is suggested by the demonstration that similar changes in enzyme activities are elicited by exposing purified enzyme preparations to oxidation by various mixed-function oxidation (MFO) systems *in vitro* (Fucci, Oliver, Coon, and Stadtman, 1983; Levine, 1983). The subsequent discovery that the MFO-catalyzed oxidation of pro-

[*] Telephone: 301-496-4096. Fax: 301-496-0599.

Free Radicals, Oxidative Stress, and Antioxidants, edited by Özben.
Plenum Press, New York, 1998.

teins leads to conversion of some amino acid side chains to carbonyl derivatives (Levine, 1983a) is the basis of several highly sensitive procedures for assessing the levels of oxidative damage in tissues and cell cultures (Oliver, Ahn, Moerman, Goldstein, and Stadtman, 1987; Levine, Garland, Oliver, Amici, Climent, Lenz, Ahn, Shaltiel, and Stadtman, 1990; Levine, Williams, Stadtman, and Shacter, 1994). In the meantime, it is generally accepted that the carbonyl content of proteins is a reliable indicator of oxidative protein damage and is widely used as a measure of ROS-mediated oxidative damage in studies of aging, oxidative stress, and disease. Using the carbonyl content as a marker of protein oxidation, it has been established that: (a) The level of oxidized protein increases with animal age. (b) Factors that lead to an increase or decrease in the life span of animals lead also to a decrease or increase, respectively, in the intracellular level of oxidized protein (Sohal, Agarwal, Dubey, and Orr, 1993; Gerschman, Gilbert, and Caccamise, 1988; Youngman, Park, and Ames, 1992; Sohal and Dubey, 1994; Sohal, Ku, and Agarwal, 1993; Butterfield, Howard, Yatin, Allen, and Carney, 1997). (c) The sensitivity of animals to oxidative stress increases with animal age and is correlated with an increase in oxidative stress-induced protein oxidation (Agarwal and Sohal, 1993; Sohal, Agarwal, and Sohal, 1995). (d) The age-related loss of various cognitive and motor functions is associated with the accumulation of oxidized forms of proteins in those areas of the brain that are known to control these functions (Carney, Starke-Reed, Oliver, Landum, Cheng, Wu, and Floyd, 1991; Forster, Dubey, Dawson, Stutts, Lal. and Sohal, 1996). (e) The age-related increase in protein carbonyl content and the decrease in the glutamine synthetase and neutral protease activities in the gerbil brain, as well as the age-related loss of gerbil spatial memory function, can all be restored to young animal levels by chronic administration of the free radical spin trap, *tert*-butyl-alpha-phenylnitrone (PBN) (Carney *et al.*, 1991). (f) Aging is associated with an increase in the production of reactive oxygen species (ROS) by mitochondria (discussed below). (g) Elevated levels of oxidatively modified proteins are associated with a number of age-related diseases (see below). In addition, it was found that there is an age-related increase in the surface hydrophobicity of proteins and that this is accompanied by a parallel oxidation of methionine residues to methionine sulfoxide residues (Chao, Ma, and Stadtman, 1997).

3. AGE-RELATED CHANGES IN ENZYME ACTIVITIES AND HEAT STABILITY

Typical patterns of the heat inactivation kinetics of enzymes from young animals ("young" enzyme) and from old animals ("old " enzyme) are illustrated in Fig. 1A. In the case of "young" enzyme, there is a linear relationship between the log of catalytic activity and time of incubation at an elevated temperature, indicative of the fact that the loss of activity is first order with respect to enzyme concentration. In contrast, a similar plot of activity of an "old" enzyme versus time is biphasic, consisting of two linear segments, the first of which is much steeper than that of the "young" enzyme, whereas the slope of the second segment may be either identical or slightly steeper than that of the "young " enzyme (Rothstein, 1977). It follows from this behavior that the enzyme from young animals is fairly homogeneous, whereas enzyme from old animals is a mixture of at least two forms of the enzyme, one or both of which have undergone conformational changes that render them more susceptible to heat denaturation.

It was proposed that the age-related changes in conformation of enzymes reflects a decrease in the rate of enzyme turnover, leading to an increase in the cellular "dwell-time"

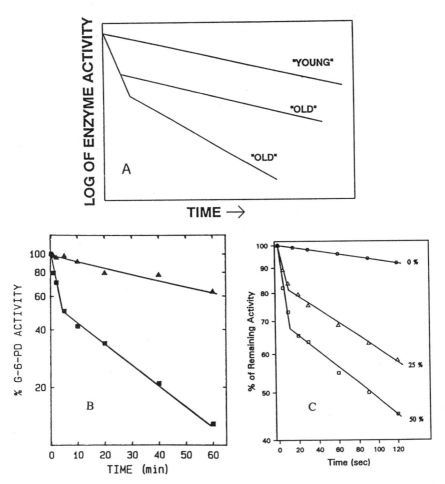

Figure 1. Effect of animal age and metal-catalyzed oxidation on the heat stability of enzymes. (A) Diagrammatic representation of thermal denaturation patterns of enzymes from "young" and "old" animals (after Rothstein, 1977). (B) Heat inactivation profiles of bovine adrenal glucose-6-phosphate dehydrogenase activity at 51°C, before (closed triangles) and after (closed circles) oxidation by the ascorbate/Fe^{+3}/O$_2$ MFO system (see Oliver *et al.*, 1987, for details). (C) Heat inactivation profiles of yeast glucose-6 phosphate dehydrogenase activity, before (0%) and after 25% and 50% inactivation, as indicated, by treatment with the Fe^{+2} and H$_2$O$_2$ at 47°C (see Szweda and Stadtman, 1992, for details).

during which spontaneous conversion of the enzyme to the more heat labile conformation occurs. In the meantime, there is growing evidence that the age-related changes in protein conformation is due, at least in part, to oxidative modifications (Oliver, Fulks, Levine, Fucci, Rivett, Roseman, and Stadtman, 1984; Oliver, Ahn, Wittenberger, Levine, and Stadtman, 1985; Stadtman, 1986). This concept is supported by the demonstration that the intracellular level of oxidized proteins increases with animal age (see below) and the fact that oxidative modification of "young" enzymes converts them to the "old", more heat-labile configuration. For example, (a) as shown in Figs. 1B and 1C, metal-catalyzed oxidation of glucose-6-phosphate dehydrogenase from bovine adrenals (Oliver *et al.*, 1987) or from yeast (Szweda and Stadtman, 1992) converts the enzyme to a form exhibiting heat denaturation kinetics characteristic of the "old" enzyme. (b) The age-related loss of rat liver malic enzyme activity is associated with the loss of one histidine residue per subunit, and a similar loss of activity can be obtained *in vitro* by exposing purified preparations of the "young" enzyme to oxidation by a MFO system, under conditions where only one histidine residue is oxidized (Cordillo, Ayala, F.-Lobato, Bautista, and Muchado, 1988). (c) Oxidation of bovine kidney alkaline phosphatase by an MFO system converts the enzyme to a form exhibiting a biphasic heat inactivation profile similar to that of "old" enzymes (Mordente, Martorana, Miggiano, Meucci, Santini, and Castelli, 1988). (d)The age-related loss of tyrosine hydroxylase activity and concomitant increase in the carbonyl content of the enzyme in *substantia nigra* of rats can be mimicked by treatment of rat striatal homogenates with hydrogen peroxide (P. de la Cruz, Revilla, Venero, Ayala, Cano, and Muchado, 1996). (e) Age-related changes in the heat-lability of rat liver leucyl-tRNA synthetase (Takahashi and Goto, 1987) could be mimicked by exposing partially purified preparations of "young"-type enzyme to the Fe(III)-ascorbate MFO system (Takahashi and Goto, 1990). (f) The age-related changes in catalytic activity, histidine content, and carbonyl content, of rat liver Cu,Zn-superoxide dismutase could be mimicked by treatment of purified preparations of the corresponding bovine erythrocyte enzyme with the Fe(III)-ascorbate MFO system (Maria, Revilla, Ayala, P. de la Cruz, and Machado, 1995). (g) Exposure of rat muscle phosphoglycerate kinase to a nonenzymic MFO system generates the "old" form of the enzyme (Zhou and Gafni, 1991).

4. THE PROTEIN CARBONYL CONTENT IS A MARKER OF OXIDATIVE STRESS-MEDIATED DAMAGE

The oxidative modification of proteins by various reactive oxygen species (ROS) leads to the conversion of some amino acid residues to reactive aldehyde and ketone derivatives (Garrison, 1987; Amici, Levine, Tsai, and Stadtman, 1989; Levine, 1983; Uchida and Stadtman, 1993); therefore, the protein carbonyl content of animal tissues or cell cultures is a widely used measure of protein oxidative damage and oxidative stress. The validity of this criterion is illustrated by the results of studies showing that exposure of animals or cell cultures to a variety of oxidative stress conditions leads to a significant increase in the intracellular levels of protein carbonyl groups. As summarized in Table 1, exposure to x-irradiation, cigarette smoking, hyperoxia, ischemia- reperfusion, rapid correction of hyponatremia, magnesium defficiency, vigorous exercise, ozone, chronic alcohol administration, paraquat toxicity, MFO systems, and induction of oxidative burst in neutrophils and macrophages are among oxidtive stress conditions that have been shown to generate protein carbonyl derivatives.

5. PROTEIN OXIDATION AS A MARKER OF AGING

Studies with a variety of aging models have shown that the protein carbonyl content increases almost exponentially as a function of animal age. Thus, as shown in Fig. 2A, the carbonyl content of protein in cultured human dermal fibroblasts increases exponentially as a function of the age of the fibroblast donor (Oliver *et al.*, 1987), but is independent of the cell passage number (not shown). Moreover, the carbonyl content of protein in cultured fibroblasts from individuals with the premature aging diseases, progeria or Werner's syndrome, is very much higher than that in cultured cells from normal age-matched controls (Oliver *et al.*, 1987); it is, in fact, comparable to the carbonyl content of normal 80-year-old individuals.

Table 1. Oxidative stress-mediated protein carbonyl formation

Oxidative stress	Tissue affected	Reference
Ischemia-reperfusion	Gerbil brain	Oliver, Starke-Reed, Stadtman, Liu, Carney, and Floyd, 1990
Ischemia-reperfusion	Rat heart	Poston and Parenteau, 1992
Ischemia-reperfusion	Dog brain	Liu, Rosenthal, Starke-Reed, and Fiskum, 1993
Ischemia-reperfusion	Rat lung	Ayene, Dodia, and Fisher, 1992
Hyperoxia	Rat hepatocytes	Starke-Reed and Oliver, 1989; Starke, Oliver, and Stadtman, 1987
Hyperoxia	Houseflies	Sohal, Agarwal, Dubey, and Orr, 1993
Hyperoxia	Rat lung	Ayene, Dodia, and Fisher, 1992; Winter and Liehr, 1991
X-irradiation	Houseflies	Agarwal and Sohal, 1993
Exercise	Rat hind leg muscle	Witt, Reznick, Viguie, Starke-Reed, and Packer, 1992
High altitude exercise	Rat hind leg	Radak, Asano, Lee, Ohno, Nakamura, and Goto, 1997
Neutrophil activation	Endogenous protein	Oliver, 1987; Krsek-Staples and Webster, 1993
Neutrophil activation	Endothelial cells	Krsek-Staples and Webster, 1993
Rapid correction of hyponatremia	Rat brain	Mickel, Oliver, and Starke-Reed, 1990
Estrogen administration	Hamster kidney	Winter and Lier, 1991
Paraquat toxicity	Rat lung	Winter and Lier, 1991
Magnesium deficiency	Rat kidney, brain	Stafford, Mak, Kramer, and Weglicki, 1993
Cigarette smoke	Human plasma	Reznick, Cross, Hu, Suzuki, Khwaja, Safadi, Motchnik, Packer, and Halliwell, 1992
Xanthine oxidase/Xanthine	Endothelial cells	Kresk-Staples and Webster, 1993
Ozone	Human plasma	Cross, Reznick, Packer, Davis, Suzuki, and Halliwell, 1992
Metal-catalyzed oxidation	Human plasma	Shacter, Williams, Lim, and Levine, 1994; Shacter, Williams, and Levine, 1995
Alcohol intake	Mouse liver mitochondria	Wielaand and Lauterburg, 1995

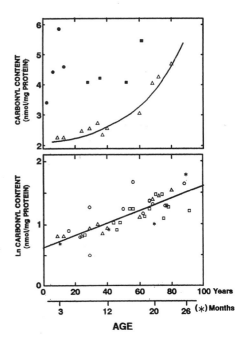

Figure 2. Age-related accumulation of oxidized protein. **A:** Carbonyl content of protein in cultured dermal fibroblasts from normal individuals (open triangles), from patients with progeria (filled circles), from patients with Werner's syndrome (filled squares), adapted from Oliver *et al.* (1987). **B:** Semilog plots of the carbonyl content of proteins versus age for: dermal fibroblasts from normal individuals (open triangles), a replot of the data in upper panel; the occipital lobe of the human brain tissue (Smith *et al.*, 1991); the human eye lens cortex (open squares), replot of data from Garland *et al.* (1988); rat hepatocytes (*), replot of data from Starke-Reed and Oliver (1989). The scale from 0 to 26 months refers to the rat hepatocytes. The scale from 0 to 100 refers to data for human subjects.

The exponential nature of the age-dependent increase in protein carbonyls is illustrated further by the fact that there is a linear relationship between the log of the protein carbonyl content and donor age (Fig. 2B). Furthermore, a similar exponential relationship between protein carbonyl content and animal age has been demonstrated for human brain tissue (Smith, Carney, Starke-Reed, Oliver, Stadtman, and Floyd, 1991), human eye lens proteins (Garland, Russell, and Zigler, 1988), and rat hepatocytes (Starke-Reed and Oliver, 1989) (Fig. 1, lower panel). In addition, an age-dependent increase in protein carbonyl content has been observed to occur in whole body proteins of houseflies (Sohal *et al.*, 1995), in brain and kidney protein of mice (Sohal and Dubey, 1994; Sohal *et al.*, 1993), in brain tissue of gerbils (Carney *et al.*, 1991), and during the aging of human erythrocytes (Oliver *et al.*, 1987).

6. RELATIONSHIP BETWEEN CARBONYL CONTENT AND LIFE SPAN

There is an inverse relationship between animal longevity and the steady-state level of oxidized protein. For example, based on the fact that senility in houseflies is associated with the loss of flight capacity, Sohal *et al.* (1993) were able to separate senescent and presenescent flies of the same chronological age by their ability to fly. They found that, on the average, the "fliers" (presenescent flies) lived about 23% longer than their senescent cohorts ("crawlers"); moreover, the steady-state level of protein carbonyls, measured after 12 days of age, was about 29% higher in the "crawlers" than in the "fliers".

It is well known that rats and mice placed on a caloric restricted diet live much longer than when they are allowed to eat as much as they like. In a study designed to determine if diet restriction had any effect on oxidative protein damage, it was demonstrated that diet restriction led to a 43% increase in life span of mice and that the protein carbonyl content of brain, heart, and kidney tissue in diet-restricted animals is considerably lower than in the*se* from *ad libitum* fed animals (Sohal, Ku, Agarwal, Forster, and Lal, 1994).

In a similar study, the carbonyl contents of two strains of mice, a senescent-resistant strain (SAMR1) and a shorter-lived senescent-prone strain (SAMP8) were compared (Butterfield *et al.*, 1997). They found that the carbonyl content of the SAMP8 mouse was about two times higher than in the SMR1 strain.

In an effort to determine the effect of antioxidant levels on the life span and oxidative stress, Orr and Sohal (1994) prepared transgenic strains of *Drosophila melanogaster* that produce about three times the normal levels of either catalase or superoxide dismutase or three times as much of both enzymes. They found that the life spans of strains producing high levels of only one of these antioxidant enzymes were indistinguishable from the parental strain. However, the strain that overproduced both kinds of enzymes lived 30% longer and had only one-half to two-thirds the level of protein carbonyl groups than did any of the other strains, including the parental strain. These results provide strong support for the oxidative stress hypothesis of aging.

7. PROTEOLYTIC DEGRADATION OF OXIDIZED PROTEINS

In studies designed to determine basic mechanisms that underlie the differential degradation of various enzymes in response to nitrogen or carbon starvation, it was discovered that the activity and immuno-reactive glutamine synthetase (GS) protein in *Klebsiella aerogenes* intact cells declines under conditions of nitrogen starvation (Fulks, 1977). On the basis of this, and the further demonstration that GS in crude extracts of *K. aerogenes* is highly susceptible to oxidative inactivation by an Fe(III)/NAD(P)H/O$_2$ MFO system, it was proposed that the oxidation of proteins targets them for proteolytic degradation (Levine, Oliver, Fulks and Stadtman, 1981; Oliver, Levine, and Stadtman, 1981; Oliver, Fucci, Levine, Wittenberger, and Stadtman, 1982; Oliver, Levine, and Stadtman, 1982; Oliver *et al.*, 1984; Oliver, Ahn, Wittenberger, and Stadtman, 1985). Based on similar considerations, other workers (Wolff, Garner, and Dean, 1986; Wolff and Dean, 1987; Davies, 1986, 1986a) also proposed that free radical-mediated oxidation of proteins targeted them for proteolytic degradation. In the meantime, this proposition is supported by a large body of evidence showing that rat liver hepatocytes (Rivett, 1985, 1985a; Rivett, Roseman, Oliver, Levine, and Stadtman, 1985), erythrocytes and reticulocytes from humans, rabbits, and cows (Davies, 1986a; Davies, Delsignore, and Lin, 1987; Davies and Goldberg, 1987; Salo, Lin, Pacifici, and Davies, 1988; Pacifici, Salo, and Davies, 1989), bovine eye lens epithelium (Murakami, Jahngen, Lin, Davies, and Taylor, 1990), rat liver epithelial cells (Grune, Reinheckel, Joshi, and Davies, 1995), beef and rat liver mitochondria (Grant, Jessup, and Dean, 1993; Marcillat, Zhang, Lin, and Davies, 1988), bacteria (Davies and Lin, 1988, 1988a; Roseman and Levine, 1987), and human hematopoietic cells (Grune, Reinheckel, and Davies, 1996) contain proteases that preferentially degrade the oxidized forms of proteins. These include a Ca(II)-dependent protease, an acid protease (cathepsin B), and the multicatalytic protease (i.e. the 20s proteosome).

8. A QUESTION OF BALANCE

The accumulation of oxidatively damaged proteins during aging and in several diseases reflects the balance between numerous factors that govern the formation and/or destruction of ROS, on the one hand, and numerous factors that govern the concentrations and/or activities of diverse proteases that degrade oxidatively modified forms of proteins, on the other. Evidence that the rate of ROS formation increases with age comes from studies with houseflies (Sohal, 1993; Sohal and Dubey, 1994; Ku and Sohal, 1993) and rodents (Spoerri, 1984; Nohl, Breuninger, and Hegner, 1978; Sawada and Carlson, 1987; Sohal, Arnold, and Sohal, 1990; Muscari, Frascaro, Guarnieri, and Calderara, 1990; Sohal, 1993) showing that the rate of O$_2$ and H$_2$O$_2$ production by mitochondria increases with aging (for review, see also Bandy and Davison, 1990; Ames, Shigenaga, and Hagen, 1993; Benzi and Moretti, 1995). This increase in ROS formation is partly attributable to a decrease in the level of cytochrome oxidase (Sohal, 1993). A loss of cytochrome c oxidase may lead to an increase in the steady-state levels of the reduced forms of electron carriers, which can react directly with O$_2$ to form ROS. However, an age-related increase in the rate of ROS formation could also reflect changes in the levels of endogenously generated antioxidants, such as glutathione, uric acid, bilirubin, or antioxidant enzymes (SOD, glutathione peroxidase, catalase, thiol peroxidase, methionine sulfoxide reductase), which act together with dietary antioxidants (vitamins A, C, E, and many others) to destroy ROS once they are formed. Because the redox active metal ions, Fe(III) and Cu(II), are involved in the formation of the most damaging ROS species, the levels of protein chelators (ferridoxin, transferrin, lactoferrin, ceruloplasmin) and metabolites, such as citrate, as well as many factors that regulate their formation and activities, contribute to the antioxidant function. For example, the ability of ceruloplasmin to scavenge O$_2$ (Goldstein, Kaplan, Edelson, and Weissmann, 1979; Samokyszyn, Miller, Reif, and Aust, 1989) and to catalyze the oxidation of Fe(II) to Fe(III) (Osaki, 1966) serves an antioxidant function by maintaining low levels of Fe(II) and thereby suppresses the reaction of Fe(II) with H_2O_2 to form HO (Krsek-Staples and Webster, 1993). The physiological significance of this function is illustrated by the observation that during aging the ROS-mediated oxidation of ceruloplasmin leads to a loss of its ferroxidase activity and concomitant loss of its ability to protect proteins from oxidative injury (Krsek-Staples and Webster, 1993; Musci, Bonaccorsi de Patti, Fagiolo, and Calabrese, 1993).

Clearly, decreases in the levels of antioxidant defenses could contribute to the accumulation of oxidized proteins during aging. Nevertheless, evidence supporting a loss of antioxidant functions during aging is still contradictory. Some published reports show an age-related decrease, others show no changes, and some even show an increase in the level of particular antioxidant enzymes (see Perez, Lopez, and Barja-de Quiroga, 1991; Matsuo, 1993) for reviews. These conflicting reports may simply reflect the fact that the levels of antioxidant enzymes are tightly regulated. Antioxidant levels are constantly being up-regulated or down-regulated depending on the degree of oxidative stress. Moreover, to compensate for the loss of one form of an antioxidant enzyme activity there is often an increase in the level of another. Such a reciprocal relationship has been observed with respect to the levels of glutathione peroxidase and catalase or superoxide dismutase (Lin, Thomas, and Girotti, 1993; Amstad, Moret, and Cerutti, 1994). However, there is evidence to support the view that an age-related loss in ability to degrade oxidized proteins may be partly responsible for the accumulation oxidized proteins during aging. The neutral protease activities (multicatalytic protease?) of rat liver hepatocytes and gerbil brain tissue have been shown to decrease with animal age (Starke-Reed and Oliver, 1989; Carney et al., 1991). This loss of protease activity could be due to either age-related changes in genetic and regulatory factors that control the intracellular protease levels, the accumulation of oxidized forms of proteins that are resistant to proteolysis (Rivett, 1986: Friguet, Szweda, and Stadtman, 1994; Grant et al., 1993; Grune et al., 1995), or to an increase in the levels of derivatives that inhibit the ability of the proteases to degrade oxidatively modified proteins. The latter possibility is highlighted by the demonstration that cross-linked proteins produced by interactions with HNE are not only resistant to proteolysis by the multicatalytic protease, but they inhibit the ability of the protease to degrade oxidized forms of other proteins (Friguet, Stadtman, and Szweda, 1994).

9. PROTEIN OXIDATION AND DISEASE

It has become increasingly apparent that protein oxidation is associated with the development of various diseases, including Alzheimer's disease (Carney, Smith, Carney, and Butterfield, 1994; Harris, Hensley, Butterfield, Leedle, and Carney, 1995; Chauhan, Chauhan, Brockerhoff, and Wisniewski, 1991; Smith, Perry, Sayre, Anderson, Beal, and Kowall, 1996; Smith, Rudnicka-Nawrot, Richey, Praprotnik, Mulvihill, Miller, Sayre, and Perry, 1995), amyotropic lateral sclerosis (Bowling, Schultz, Brown, and Beal, 1993), rheumatoid arthritis (Chapman, Rubin, and Gracy, 1989), muscular dystrophy (Murphy and Kherer, 1989), cataractogenesis (Garland et al., 1988), respiratory distress syndrome (Gladstone and Levine, 1994), iron-induced carcinogenesis (Toyokuni, Uchida, Okamoto, Hattori-Nakakuki, Hiai, and Stadtman, 1994), cardiovascular disease (Kelly and Birch, 1993; Uchida, Toyokuni, Nishikawa, Kawakishi, Oda, Hiai, and Stadtman, 1994), systemic amyloidosis (Ando, Nyhlin, Sur, Holmgren, Uchida, Sahly, Yamashita, Terasaki, Nakamura, Uchino, and Ando, 1997).

In conclusion, it is abundantly clear that oxidation of protein is implicated in aging, oxidative stress, in the impairment of some biological functions, and in the etiology or progression of several diseases.

REFERENCES

Agarwal, S. and Sohal, R. S., 1993, Relationship between aging and susceptibility to protein oxidative damage, *Biochem. Biophys. Res. Commun.* **194**, 1203–1206.

Ames, B. N., Shigenaga, M. K., and Hagen, T. M., 1993, Oxidants, antioxidants, and the degenerative diseases of aging, *Proc. Natl. Acad. Sci. USA* **90**, 7915–7922.

Amici, A., Levine, R. L., Tsai, L., and Stadtman, E. R., 1989, Conversion of amino acid residues in proteins and amino acid homopolymers to carbonyl derivatives by metal-catalyzed reactions, *J. Biol. Chem.* **264**, 3341–3346.

Amstad, P., Moret, R., and Cerutti, P., 1994, Glutathione peroxidase compensates for the hypersensitivity of Cu,Zn-superoxide dismutase overproducers to oxidant stress, *J. Biol. Chem.* **269**, 1606–1609.

Ando, Y., Nyhlin, N., Suhr, O., Holmgren, G., Uchida, K., Sahly, M. E., Yamashita, T., Terasaki, H., Nakamura, M., Uchino, M., and Ando, M., 1997, Oxidative stress is found in amyloid deposits in systemic amyloidosis, *Biochem. Biophys. Res. Commun.* **232**, 497–502.

Ayene, I. S., Dodia, C., and Fisher, A. B. , 1992, Role of oxygen in oxidation of lipid and protein during ischemia/reperfusion in isolated perfused rat lung, *Arch. Biochem. Biophys.* **296**, 183–189.

Bandy, B. and Davison, A. J., 1990, Mitochondrial mutations may increase oxidative stress: Implications for carcinogenesis and aging, *Free Rad. Biol. Med.* **8**, 523–539.

Benzi, G. and Moretti, A., 1995, Age- and peroxidative stress-related modifications of the cerebral enzymatic activities linked to mitochondria and the glutathione system, *Free Rad. Biol. Med.* **19**, 77–101.

Bowling, A. C., Schultz, J. B., Brown, Jr., R. H., and Beal, M. F., 1993, Superoxide dismutase activity, oxidative damage, and mitochondrial energy metabolism I familial and sporadic amyotrophic lateral sclerosis, *J. Neurochem.* **61**, 2322–2325.

Butterfield, D. A., Howard, B. J., Yatin, S., Allen, K. L., and Carney, J. M., 1997, Free radical oxidation of brain proteins in accelerated senescence and its modulation by N-*tert*-butyl-α-phenylnitrone, *Proc. Natl. Acad. Sci. USA* **94**, 674–678.

Carney, J. M., Smith, C. D., Carney, A. M., and Butterfield, D. A., 1994, Aging- and oxygen-induced modifications in brain biochemistry and behavior, in: *Aging and Cellular Defense Mechanisms*, Volume 63 (Franceschi, C., Crepaldi, G., Cristofalo, V. J., and Vijg, J., eds.), pp. 110–119, New York Academy of Science, New York.

Carney, J. M., Starke-Reed, P. E., Oliver, C. N., Landum, R. W., Cheng, M. S., Wu, J. F., and Floyd, R.A., 1991, Reversal of age-related increase in brain protein oxidation, decrease in enzyme activity loss and loss of temporal and spatial memory by chronic administration of the spin-trapping compound N-*tert*-butyl-α-phenylnitrone, *Proc. Natl. Acad. Sci. USA* **88**, 3633–3636.

Chao, C.-C., Ma, Y.-S., and Stadtman, E. R., 1997, Modification of protein surface hydrophobicity and methionine oxidation by oxidative stress, *Proc. Natl. Acad. Sci. USA* **94**, 2969–2974.

Chapman, M. L., Rubin, B. R., and Gracy, R. W., 1989, Increased carbonyl content of proteins in synovial fluid from patients with rheumatoid arthritis, *J. Rheumatol.* **16**, 15–18.

Chauhan, A., Chauhan, V. P. S., Brockerhoff, H., and Wisniewski, H. M., 1991, Action of amyloid -protein on protein kinase C activity, *Life Sci.* **49**, 1555–1556.

Cordillo, E., Ayala, A., F.-Lobato, M., Bautista, J., and Machada, A., 1988, Possible involvement of histidine residues in loss of enzymatic activity of rat liver malic enzyme during aging, *J. Biol. Chem.* **263**, 8053–8057.

Cross, C. E., Reznick, A. Z., Packer, L., Davis, P. A., Suzuki, Y. J., and Halliwell, B., 1992, Oxidative damage to human plasma proteins by ozone, *Free Rad. Res. Commun.* **15**, 347–352.

Davies, K. J. A., 1986, Intracellular proteolytic systems may function as secondary antioxidant defenses: A Hypothesis, *J. Free Rad. Biol. Med.* **2**, 155–173.

Davies, K. J. A., 1986a, The role of intracellular proteolytic systems in antioxidant defense, In: *Superoxide and superoxide dismutase in chemistry, biology, and medicine* (Rotilio, G., ed.), pp. 443–450, Elsevier Science Publishing, Amsterdam.

Davies, K. J. A. and Goldberg, A. L., 1987, Proteins damaged by oxygen radicals are rapidly degraded in extracts of red blood cells, *J. Biol. Chem.* **262**, 8227–8234.

Davies, K. J. A. and Lin, S. W., 1988, Degradation of oxidatively denatured proteins in *Escherichia coli*, *Free Rad. Biol. Med.* **5**, 215–223.

Davies, K. J. A. and Lin, S. W., 1988a, Oxidatively denatured proteins are degraded by an ATP-independent proteolytic pathway in *Escherichia coli, Free Rad. Biol. Med.* **5**, 225–236.

Davies, K. J. A., Delsignore, M. E., and Lin, S. W., 1987, Protein damage by oxygen radicals. II. Modification of amino acids, *J. Biol. Chem.* **262**, 9902–9907.

Dreyfus, J. C., Kahn, A., and Schapira, F., 1978, Post translational modifications of enzymes, *Curr. Top. Cell. Regul.* **14**, 243–297.

Forster, M. J., Dubey, A., Dawson, K. M., Stutts, W. A., Lal, H., and Sohal, R. S., 1996, Age-related losses of cognitive function and motor skills in mice are associated with oxidative protein damage in the brain, *Proc. Natl. Acad. Sci. USA* **93**, 4765–4769.

Friguet, B., Stadtman, E. R., and Szweda, L., 1994, Modification of glucose-6-phosphate dehydrogenase by 4-hydroxy-2-nonenal, *J. Biol. Chem.* **269**, 21639–21643.

Friguet, B., Szweda, L., and Stadtman, E. R., 1994, Susceptibility of glucose-6-phosphate dehydrogenase modified by 4-hydroxy-2-nonenal and metal-catalyzed oxidation to proteolysis by the multicatalytic protease, *Arch. Biochem. Biophys.* **311**, 168–173.

Fucci, L., Oliver, C. N., Coon, M. J., and Stadtman, E. R., 1983, Inactivation of key metabolic enzymes by mixed-function oxidation reactions: Possible implications in protein turnover and aging, *Proc. Natl. Acad. Sci. USA* **80**, 1521–1525.

Fulks, R. M., 1977, Regulation of glutamine synthetase degradation in *Klebsiella aerogenes, Fed. Proc. Am. Soc. Exptl. Biol.* **36**, 919 (abstr.).

Garland, D., Russell, P., and Zigler, J. S., 1988, The oxidative modification of lens protein, In: *Oxygen Radicals in Biology and Medicine* (Simic, M. G., Taylor, K. S., Ward, J. F., and von Sontag, V., eds.), pp. 347–353, Plenum, New York.

Garrison, W. M., 1987, Reaction mechanisms in the radiolysis of peptides, polypeptides, and proteins, *Chem. Rev.* **87**, 381–398.

Gerschman, R., Gilbert, D. I., and Caccamise, D., 1988, Effect of various substances on survival times of mice exposed to different high oxygen tension, *Am. J. Physiol.* **192**, 563–571.

Gladstone, I. M. and Levine, R. L., 1994, Oxidation of proteins in neonatal lungs, *Pediatrics* **93**, 764–768.

Goldstein, I. M., Kaplan, H. B., Edelson, H. S., and Weissmann, G., 1979, Ceruloplasmin. A scavenger of superoxide anion radicals, *J. Biol. Chem.* **254**, 4040–4045.

Grant, A. J., Jessup, W., and Dean, R. J., 1993, Inefficient degradation of oxidized regions of protein molecules, *Free Rad. Res. Commun.* **18**, 259–267.

Grune, T., Reinheckel, T., and Davies, K. J. A., 1996, Degradation of oxidized proteins in K562 human hematopoietic cells by proteosome, *J. Biol. Chem.* **271**, 15504–15509.

Grune, T., Reinheckel, T., Joshi, M., and Davies, K. J. A., 1995, Proteolysis in cultured liver epithelial cells during oxidative stress, *J. Biol. Chem.* **270**, 2344–2351.

Harris, M., Hensley, K., Butterfield, D. A., Leedle, R. A., and Carney, J. M., 1995, Direct evidence of oxidative injury produced by Alzheimer's -amyloid peptide (1–40) in cultured hippocampal neurons, *Exp. Neurol.* **131**, 193–202.

Krsek-Staples, J. A. and Webster, R. O., 1993, Ceruloplasmin inhibits carbonyl formation in endogenous cell proteins, *Free Rad. Biol. Med.* **14**, 115–125.

Kelley, F. J. and Birch, S., 1993, Ozone exposure inhibits cardiac protein synthesis in the mouse, *Free Rad. Biol. Med.* **14**, 443–446.

Ku, H.-H. and Sohal, R. S., 1993, Comparison of mitochondrial pro-oxidant generation and antioxidant defenses between rat and pigeon: Possible basis of variation in longevity and metabolic potential, *Mech. Ageing Develop.* **72**, 67–76.

Levine, R. L., 1983, Oxidative modification of glutamine synthetase. I. Inactivation is due to loss of one histidine residue, *J. Biol. Chem.* **258**, 11823–11827.

Levine, R. L., 1983a, Oxidative modification of glutamine synthetase. II. Characterization of the ascorbate model system, *J. Biol. Chem.* **258**, 11828–11833.

Levine, R. L., Williams, J. A., Stadtman, E. R., and Schacter, E., 1994, Carbonyl assays for determination of oxidatively modified proteins, *Methods Enzymol.* **233**, 346–357.

Levine, R. L., Garland, D., Oliver, C. N., Amici, A., Climent, I., Lenz, A. G., Ahn, B.-W., Shaltiel, S., and Stadtman, E. R., 1990, Determination of carbonyl groups in oxidatively modified proteins, *Methods Enzymol.* **186**, 464–478.

Levine, R. L., Oliver, C. N., Fulks, R. M., and Stadtman, E. R., 1981, Turnover of bacterial glutamine synthetase: Oxidative inactivation precedes proteolysis, *Proc. Natl. Acad. Sci. USA* **78**, 2120–2124.

Lin, F., Thomas, J. P., and Girotti, A. W., 1993, Hyperexpression of catalase in selenium-deprived murine L1210 cell, *Arch. Biochem. Biophys.* **305**, 176–185.

Liu, Y., Rosenthal, R. E., Starke-Reed, P.E., and Fiskum, G., 1993, Inhibition of postcardiac arrest brain protein oxidation by acetyl-L-carnitine, *Free Rad. Biol. Med.* **15**, 667–670.

Marcillat, O., Zhang, Y., Lin, S. W., and Davies, K. J. A., 1988, Mitochondria contain a proteolytic system which can recognize and degrade oxidatively denatured proteins, *Biochem. J.* **254**, 677–683.

Maria, C. S., Revilla, E., P. de la Cruz, C., and Machado, A., 1995, Cu,Zn-superoxide dismutase during aging, *FEBS Lett.* **347**, 85–88.

Matsuo, M., 1993, Age-related alterations in antioxidant defense, In: *Free Radicals in Aging* (Yu, B.P., ed.), pp. 143–181, CRC Press, Ann Arbor.

Mickel, H. S., Oliver, C. N., and Starke-Reed, P. E., 1990, Protein oxidation and myelinolysis occur in brain following rapid correction of hyponatremia, *Biochem. Biophys. Res. Commun.* **172**, 92–97.

Mordente, A., Martorana, G. E., Miggiano, G. A. D., Meucci, E., Santini, S. A., and Castelli, A., 1988, Mixed function oxidation and enzymes: Kinetic and structural properties of oxidatively modified alkaline phosphatase, *Arch. Biochem. Biophys.* **264**, 502–509.

Murakami, K., Jahnegn, J. H., Li, S. W., Davies, K. J. A., and Taylor, A., 1990, Lens proteosome shows enhanced rates of degradation of hydroxyl radical modified alpha-crystallin, *Free Rad. Biol. Med.* **8**, 217–222.

Murphy, M. E. and Kherer, J. P., 1989, Oxidation state of tissue thiol groups and content of protein carbonyl groups in chickens with inherited muscular dystrophy, *Biochem. J.* **260**, 359–364.

Muscari, C., Frascaro, M., Guarnieri, C., and Calderara, C. M., 1990, Mitochondrial function and superoxide generation from submitochondrial particles of aged rat hearts, *Biochem. Biophys. Acta* **1015**, 200–204.

Musci, G., Bonaccorsi di Patti, M. C., Fagiolo, U., and Calabrese, L., 1993, Age-related changes in human ceruloplasmin, *J. Biol. Chem.* **268**, 13388–13395.

Nohl, H., Breuninger, V., and Hegner, D., 1978, Influence of mitochondrial radical formation on energy-linked respiration, *Eur. J. Biochem.* **90**, 385–390.

Oliver, C. N., 1987, Inactivation of enzymes and oxidative modification of proteins by stimulated neutrophils, *Arch. Biochem. Biophys.* **253**, 62–72.

Oliver, C. N., Starke-Reed, P. E., Stadtman, E. R., Liu, G. J., Carney, J. M., and Floyd, R. A., 1990, Oxidative damage to brain proteins, loss of glutamine synthetase activity, and production of free radicals during ischemia-reperfusion-induced injury to gerbil brain, *Proc. Natl. Acad. Sci. USA* **87**, 5144–5147.

Oliver, C. N., Ahn, B.-W., Moerman, E. J., Goldstein, S., and Stadtman, E. R., 1987, Age-related changes in oxidized proteins, *J. Biol. Chem.* **262**, 5488–5491.

Oliver, C. N., Ahn, B., Wittenberger, M. E., and Stadtman, E. R., 1985, Oxidative inactivation of enzymes: Implication in protein turnover and aging, In: *Cellular Regulation and Malignant Growth* (Ebashi, S., ed.), pp. 320–331, Japan Sci. Soc. Press/Springer-Verlag, Berlin.

Oliver, C. N., Ahn, B., Wittenberger, M. E., Levine, R. L., and Stadtman, E. R., 1985a, Age-related alterations of enzymes may involve mixed-function oxidation reactions, In: *Modification of proteins during aging* (Adelman, R. C. and Dekker, E. E., eds.), pp. 39–52, Alan R. Liss, New York.

Oliver, C. N., Fulks, R., Levine, R. L., Fucci, L., Rivett, A. J., Roseman, J. E., and Stadtman, E. R., 1984, Oxidative inactivation of key metabolic enzymes during aging, In: *Molecular Basis of Aging* (Roy, A. K. and Chatterjee, B., eds.), pp. 235–262, Academic Press, New York.

Oliver, C. N., Fucci, L., Levine, R. L., Wittenberger, M. E., and Stadtman, E. R., 1982, Inactivation of key metabolic enzymes by P450 linked mixed function oxidation systems, In: *Cytochrome P-450, Biochemistry, Biophysics, and Environmental Implications* (Heitanen, E., Laitinen, M., and Hanninen, O., eds.), pp. 531–539, Elsevier Biomedical Press, Amsterdam.

Oliver, C. N., Levine, R. L., and Stadtman, E. R., 1982, Regulation of glutamine synthetase degradation, In: *Experience in Biochemical Perception* (Ornston, L. N. and Sligar, S. G., eds.), pp. 233–249, Academic Press, New York.

Oliver, C. N., Levine, R. L., and Stadtman, E. R., 1981, Regulation of glutamine synthetase degradation, In: *Metabolic interconversion of enzymes* (Holzer, H., ed.), pp. 259–268, Springer-Verlag, Berlin.

Orr, W. C. and Sohal, R. S., 1994, Extension of life-span by over expression of superoxide dismutase and catalase in *Drosophila melanogaster*, *Science* **263**, 1128–1130.

Osaki, S., 1966, Kinetic studies of ferrous ion oxidation with crystalline human ferroxidase (ceruloplasmin), *J. Biol. Chem.* **241**, 5053–5059.

P. de la Cruz, C.P., Revilla, E., Venero, J. L., Ayala, A., Cano, J., and Machado, A., 1996, Oxidative inactivation of tyrosine hydroxylase in *Substantia nigra* of aged rat, *Free Rad. Biol. Med.* **20**, 53–61.

Pacifici, R. E., Salo, D. C., and Davies, K. J. A., 1989, Macroproteinase (M.O.P.): A 670 kDA proteinase complex that degrades oxidatively denatured proteins in red blood cells, *Free Rad. Biol. Med.* **7**, 521–536.

Perez, R., Lopez, M., and Barja-De Quiroga, G., 1991, Aging and lung antioxidant enzymes, glutathione and lipid peroxidation in the rat, *Free Rad. Biol. Med.* **10**, 35–39.

Poston, J. M. and Parenteau, G. L., 1992, Biochemical effects of ischemia on isolated perfused rat heart tissues, *Arch. Biochem. Biophys.* **295**, 35–41.

Radák, Z., Asano, K., Lee, K.-C., Ohno, H., Nakamura, A., Nakamoto, H., and Goto, S., 1997, High altitude training increases reactive carbonyl derivatives but not lipid peroxidation in skeletal muscle of rats, *Free Rad. Biol. Med.* **22**, 1109–1114.

Reznick, A. Z., Cross, C. E., Hu, M.-L., Suzuki, Y. J., Khwaja, S., Safadi, A., Motchnik, P. A., Packer, L., and Halliwell, B., 1992, Modification of plasma proteins by cigarette smoke as measured by protein carbonyl formation, *Biochem. J.* **286**, 607–611.

Rivett, A. J., 1986, Regulation of intracellular protein turnover: Covalent modification as a mechanism of marking proteins for degradation, *Curr. Top. Cell. Regul.* **28**, 291–337.

Rivett, A. J., 1985, Preferential degradation of the oxidatively modified form of glutamine synthetase by intracellular mammalian protease, *J. Biol. Chem.* **260**, 300–305.

Rivett, A. J., 1985a, Purification of a liver alkaline protease which degrades oxidatively modified glutamine synthetase, *J. Biol. Chem.* **260**, 12600–12606.

Rivett, A. J., Roseman, J. E., Oliver, C. N., Levine, R. L., and Stadtman, E. R., 1985, Covalent modification of proteins by mixed-function oxidation: Recognition by intracellular proteases, In: *Intracellular Protein Catabolism* (Khairallan, E. A., Bond, J. S., and Bird, J. W. C., eds.), pp. 317–328, Alan R. Liss, Inc., New York.

Roseman, J. E. and Levine, R. L., 1987, Purification of a protease from *Escherichia coli* with specificity for oxidized glutamine synthetase, *J. Biol.Chem.* **262**, 2101–2110.

Rothstein, M., 1977, Recent developments in age-related alteration of enzymes. *Mech. Aging and Dev.* **6**, 241–257.

Salo, D. C., Lin, S. W., Pacifici, R.E., and Davies, K. J. A., 1988, Superoxide dismutase is preferentially degraded by a proteolytic system from red blood cells following oxidative modification by hydrogen peroxide, *Free Rad. Biol. Med.* **5**, 335–339.

Samokyszyn, V. M., Miller, D. M., Reif, D. W., and Aust, S. D., 1989, Inhibition of superoxide and ferritin-dependent lipid peroxidation by ceruloplasmin, *J. Biol Chem.* **264**, 21–36.

Sawada, M. and Carlson, J. C., 1987, Changes in superoxide radical and lipid peroxide formation in brain, heart, and liver during lifetime of the rat, *Mech. Aging & Dev.* **41**, 125–137.

Shacter, E., Williams, J. A., and Levine, R. L., 1995, Oxidative modification of fibrinogen inhibits thrombin-catalyzed clot formation, *Free Rad. Biol. Med.* **18**, 815–821.

Shacter, E., Williams, J. A., Lim, M., and Levine, R. L., 1994, Differential susceptibility of plasma proteins to oxidative modification: Examination by Western blot immunoassay, *Free Rad. Biol. Med.* **17**, 429–437.

Smith, C. D., Carney, J. M., Starke-Reed, P. E., Oliver, C. N., Stadtman, E. R., and Floyd, R. A., 1991, Excess brain protein oxidation and enzyme dysfunction in normal and Alzheimer's disease, *Proc. Natl. Acad. Sci. USA* **88**, 10540–10543.

Smith, M. A., Perry, P. L., Sayre, L. M., Anderson, V. E., Beal, M. F., and Kowall, N., 1996, Oxidative damage in Alzheimer's disease, *Nature* **382**, 120–121.

Smith, M. A., Rudnicka-Nawrot, M., Richey, P. L., Praprotnik, D., Mulvihill, P., Miller, C. A., Sayre, C. A., and Perry, G., 1995, Carbonyl-related posttranslational modification of neurofilament protein in neurofibrillary pathology of Alzheimer's disease, *J. Neurochem.* **64**, 2660–2666.

Sohal, R. S., 1993, The free radical hypothesis of aging: An appraisal of the current status, *Aging Clin. Exp. Res.* **5**, 3–17.

Sohal, R. S. and Dubey, A., 1994, Mitochondrial oxidative damage, hydrogen peroxide release, and aging, *Free Rad. Biol. Med.* **16**, 621–626.

Sohal, R. S., Agarwal, S., and Sohal, B. H., 1995, Oxidative stress and aging in the Mongolian gerbil (*Meriones unguiculatus*), *Mech. Aging and Development* **81**, 15–25.

Sohal, R. S., Ku, H.-H., Agarwal, S., Forster, M. J., and Lal, H., 1994, *Mech. Aging & Dis.* **79**, 121–133.

Sohal, R. S., Agarwal, S., Dubey, A., and Orr, W. C., 1993, Protein oxidative damage is associated with life expectancy of houseflies, *Proc. Natl. Acad. Sci. USA* **90**, 7255–7259.

Sohal, R. S., Ku, H.-H., and Agarwal, S., 1993, Biochemical correlates of longevity in two closely related rodent species, *Biochem. Biophys. Res. Commun.* **196**, 7–11.

Sohal, R. S., Arnold, L. A., and Sohal, B. H., 1990, Age-related changes in antioxidant enzymes and prooxidant generation in tissues of the rat with special reference to parameters in two insect species, *Free Rad. Biol. Med.* **10**, 495–500.

Spoerri, P. E., 1984, Mitochondrial alterations in aging mouse neuroblastoma cells in culture. *Monogr. Dev. Biol.* **17**, 210–220.

Stadtman, E. R., 1992, Protein oxidation and aging, *Science* **257**, 1220–1224.

Stadtman, E. R., 1986, Oxidation of proteins by mixed-function oxidation systems: Implication in protein turnover, aging, and neutrophil function, *Trends Biochem. Sci.* **11**, 11–12.

Stafford, R. E., Mak, T. M., Kramer, J. H., and Weglicki, W. B., 1993, Protein oxidation in magnesium deficient rat brains and kidneys, *Biochem. Biophys. Res. Commun.* **196**, 596–600.

Starke, P. E., Oliver, C. N., and Stadtman, E. R., 1987, Modification of hepatic proteins in rats exposed to high oxygen concentration, *FASEB J.* **1**, 36–39.

Starke-Reed, P. E. and Oliver, C. N., 1989, Protein oxidation and proteolysis during aging and oxidative stress, *Arch. Biochem. Biophys.* **275**, 559–567.

Szweda, L. I. and Stadtman, E. R., 1992, Iron-catalyzed oxidative modification of glucose-6-phosphate dehydrogenase from *Leuconostoc. mesenteroides, J. Biol. Chem.* **267**, 3096–3100.

Takahashi, R. and Goto, S., 1990, Alteration of aminoacyl-tRNA synthetase with age: Heat labilization of the enzyme by oxidative damage, *Arch. Biochem. Biophys.* **277**, 228–233.

Takahashi, R., and Goto, S., 1987, Influence of dietary restriction on accumulation of heat-labile enzyme molecules in the liver and brain of mice, *Arch. Biochem. Biophys.* **257**, 200–206.

Toyokuni, S. Uchida, K., Okamoto, K., Hattori-Nakakuki, Y., Hiai, H., and Stadtman, E. R., 1994, Formation of 4-hydroxy-2-nonenal-modified proteins in the renal proximal tubules of rats treated with a renal carcinogen ferric nitrilotriacetate, *Proc. Natl. Acad. Sci. USA* **91**, 2616–2620.

Uchida, K. and Stadtman, E. R., 1993, Covalent modification of 4-hydroxynonenal to glyceraldehyde-3-phosphate, *J. Biol. Chem.* **268**, 6388–6393.

Uchida, K., Toyokuni, S., Nishikawa, K., Kawakishi, S., Oda, H., Hiai, H., and Stadtman, E. R., 1994, Michael addition-type 4-hydroxy-2-nonenal adducts in modified low density lipoproteins: Markers for atherosclerosis, *Biochemistry* **33**, 12487–12494.

Wieland, P. and Lauterburg, B. H., 1995, Oxidation of mitochondrial proteins, DNA following administration of ethanol, *Biochem. Biophys. Res. Commun.* **213**, 815–819.

Winter, M. L. and Liehr, J. G., 1991, Free radical-induced carbonyl content in protein of estrogen-treated hamsters assayed by sodium boro[^{3}H]hydride reduction, *J. Biol. Chem.* **266**, 14446–14450.

Witt, E. H., Reznick, A. Z., Viguie, C. A., Starke-Reed, P. E., and Packer, L. (1992) Exercise, oxidative damage, and effects of antioxidant manipulation, *J. Nutr.* **122**, 766–773.

Wolff, S. P. and Dean, R. T., 1987, Glucose autooxidation and protein modification, *Biochem. J.* **245**, 243–250.

Wolff, S. P., Garner, A., and Dean, R. T., 1986, Free radicals, lipids, and protein degradation, *Trends Biochem. Sci.* **11**, 27–31.

Youngman, L. D., Park, J.-Y. K., and Ames, B., 1992, Protein oxidation associated with aging is reduced by dietary restriction of protein calories, *Proc. Natl. Acad. Sci. USA* **89**, 9112–9116.

Zhou, J. Q. and Gafni, A., 1991, Exposure of rat muscle phosphoglycerate kinase to a nonenzymatic MFO system generates the old form of enzyme, *J. Gerontol.* **46**, B217–B221.

14

OXIDATIVE DAMAGE AND FIBROSCLEROSIS IN VARIOUS TISSUES

Antonella Scavazza, Gabriella Leonarduzzi, Simonetta Camandola, Maurizio Parola, and Giuseppe Poli

Department of Experimental Medicine and Oncology
University of Torino, Italy

1. INTRODUCTION

A number of chronic disease processes affecting liver, lung, arteries, kidney and nervous system are characterized by excess deposition of collagen and other extracellular matrix proteins. Such degenerative condition is termed fibrosclerosis and has assumed much importance in human physiopathology.

2. OXIDATIVE THEORY OF FIBROSCLEROSIS

Of interest, with regards to the molecular pathogenesis of fibrosclerosis, is the commonly observed association between increased collagen deposition and enhancement of free radical driven lipid oxidation (Poli and Parola, 1997). Since membrane lipid peroxidation is recognized to exert a variety of biochemical effects through propagation and reactivity of derived products (Poli *et al.*, 1987), e.g. aldehydes (Esterbauer *et al.*, 1991), an oxidative theory to explain overproduction of fibrotic tissue indeed appears fascinating. Actually, a causative role of reactive oxidant species in fibrosclerosis is supported by its prevention or amelioration in the experimental animal following treatment with antioxidants (Mossman *et al.*, 1990; Parola *et al.*, 1992a,b; Parizada *et al.*, 1991).

3. CYTOTYPES MAINLY INVOLVED IN FIBROGENESIS

Following the above reported observations, cellular types and biochemical mechanisms involved in fibrogenesis have been analyzed with the aim of elucidating potential ways in which reactive oxidant species act as modulators.

After careful analysis of the many recent reports on the association between oxidative damage, mainly in terms of lipid peroxidation, and fibrogenesis, it seems that a stereotyped event is involved in the building up of fibrosis at the level of the different tis-

Free Radicals, Oxidative Stress, and Antioxidants, edited by Özben.
Plenum Press, New York, 1998.

Table 1. Stereotyped interaction between macrophages
and fibroblast-like cells in the different tissues

Macrophage	$\longrightarrow$	ECM-producing cell
Alveolar phagocyte		Fibroblast
Kupffer cell	Cross-talk molecules	Stellate cell
Foam cell	(Fibrogenic cytokines)	Smooth muscle c.
Microglial cell	(PGs ?, etc.)	Fibroblast-like c.
Mesangial cell		Fibroblast-like c.

ECM: extracellular matrix; PGs: prostaglandins.

sues and organs: that based on the interaction between cells of the macrophage lineage on one side and extracellular matrix (ECM) producing cells on the other side (Table 1). Interestingly enough, the latter class mainly include cells with a fibroblast-like phenotype rather than true fibroblasts. For instance, stellate cells, also known as Ito cells, play a major role in liver fibrogenesis as well as smooth muscle cells in atherosclerosis. Yet to be defined are the cytotypes producing ECM in kidney, lung and nervous system.

4. PROFIBROGENIC CELLULAR CROSS-TALK MEDIATED BY TGF-β1

In relation to the possible ways of interaction between the recognized key-player cells in fibrosis, a primary mechanism appears that sustained by so called fibrogenic cytokines. In fact, a number of cytokines, among which TGF-β1, PDGF, TNF-α, have been shown potentially able to modulate the metabolism of connective tissue (see for a review Kovacs, 1991) and further overexpressed in various fibrotic diseases (Poli and Parola, 1997). A large body of literature suggests that transforming growth factor β1 plays a key role during tissue repair and fibrogenesis (Poli and Parola, 1997; Gressner, 1991). But an ultimate demonstration of the importance of TGFβ1 in wound healing by fibrosis is given by the evidence that TGFβ1 knock-out mice all die from multifocal inflammatory disease within 3–4 weeks of age (Shull *et al.*, 1992).

Over the last few years we focused our attention on the possible modulation of fibrogenic cytokine gene expression by lipid peroxidation.

5. LIPID PEROXIDATION UP-REGULATES TGF-β1 EXPRESSION

Both acute and chronic rat intoxication with carbon tetrachloride (CCl_4), in parallel to a marked stimulation of membrane lipid peroxidation, produce increased transcription and synthesis of hepatic TGFβ1, processes that are clearly limited to non parenchimal cells (Armendariz-Borunda *et al.*, 1990). We observed that CCl_4-induced overexpression of TGF-β1 was in both treatments paralleled by an increased expression of the procollagen type I gene; further, the prevention of lipid peroxidation by rat pretreatment with vitamin E allowed to abolish CCl_4-induced up-regulation of TGF-β1 and procollagen typeI both after acute (Fig. 1) and chronic poisoning (Parola *et al.*, 1992b). Hence, free radical mediated reactions like those promoted by the hepatic metabolism of the used haloalkane increase the steady-state levels of at least TGF-β1. Unpublished experiments from our laboratory would point CCl_4 acute poisoning as able to induce PDGF and TNFα overexpression as well.

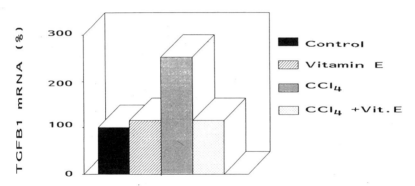

Figure 1. Up-regulation of TGFβ1 mRNA steady-state levels in rat liver after acute CCl₄-poisoning; preventative action of rat preloading with α-tocopherol. Rats were given a single dose of CCl₄ (2.5ml/kg b. wt.). See methods of Parola *et al.*, 1992b, with regard to vitamin E supplementation and TGFβ1 mRNA measurement.

After the acknowledgement of the "in vivo" TGF-β1 up-regulation by the prooxidant drug CCl₄, the scientist's attention then focused on the molecules actually responsible for such effect. So far, most of the related research delt with the possible profibrogenic role played by aldehydic products of lipid peroxidation, in particular those belonging to the hydroxyalkenal class.

6. 4-HYDROXY-2,3-NONENAL INDUCES INCREASE IN MACROPHAGE TGFβ1 mRNA STEADY-STATE LEVELS

Quantitatively and qualitatively relevant aldehydic end-product of lipid peroxidation, 4-hydroxy-2,3-nonenal (HNE) has been deeply characterized over the last years for its biochemical and likely biological effects (Esterbauer *et al.*, 1991). HNE is now considered also for its possible profibrogenic activity because just recently detected in a number of human fibrotic diseases as reported in Table 2.

The treatment of both human and murine macrophagic cell lines with a single HNE amount well within its physiopatological range induced a significant and rapid increase of TGFβ1 mRNA steady-state levels (Leonarduzzi *et al.*, 1997). Consistently, the synthesis of the active polypeptide by HNE-treated macrophages was found higher than in cells untreated or incubated with 2-nonenal or with nonanal (Table 3). The results obtained underline the specific action of HNE with regard to TGFβ1 up-regulation.

Table 2. *In situ* detection of 4-hydroxy-2,3-nonenal in human diseases

Atheromasic lesions of the carotid artery	Bellomo *et al.*, *Proc. Int. Congress*, Pavia, 1996
Neurons of substantia nigra in Parkinson's disease	Yoritaka *et al.*, 1996, *PNAS*
Oculomotor neurons in aged subjects	Yoritaka *et al.*, 1996, *PNAS*
Neurofibrillary tangles in Alzheimer's disease (DNPH labelling)	Smith *et al.*, 1996, *Nature*
Alcohol hepatitis, hemocromatosis, Wilson's disease, primary biliary cirrhosis	Paradis *et al.*, 1996, *Hepatology*, suppl.
Amyloid deposits in systemic amyloidosis	Ando *et al.*, 1997, *Biochem. Biophys. Res. Commun.*

Immunohistochemical detection was performed by using monoclonal antibodies specific for hydroxynonenal-protein adducts, except in the case of carbonyl evaluation by means of dinitrophenylhydrazine (DNPH) labelling.

Table 3. TGFβ1 synthesis by cultivated macrophages
in the presence or absence of different C9 aldehydes

	U937	J774-A1
Control	0.88 ± 0.41	0.94 ± 0.43
HNE 10 μM	1.72 ± 0.85	1.36 ± 0.60
2NE 10 μM	0.64 ± 0.14	0.91 ± 0.67
NA 10 μM	0.87 ± 0.30	1.07 ± 0.48

Data are means ± SE of at least four experiments. See methods in
Leonarduzzi et al., 1997. HNE: 4-hydroxy-2,3-nonenal; 2NE: 2,3-
nonenal; NA: nonanal.

7. LIPID PEROXIDATION-DERIVED ALDEHYDES ENHANCE TYPE I COLLAGEN SYNTHESIS

We observed that HNE treatment of human stellate cells in culture increases both transcription and synthesis of collagen type I (Parola *et al.*, 1993). Other groups confirmed the procollagen type I overexpression by HNE on rat stellate cells (Beno *et al.*, 1994; Tsukamoto, 1994). Actually such effect is shared by other biogenic hydroxy-alkenals, like 4-hydroxyhexenal and 4-hydroxyundecenal, but not by nonanal and 2-nonenal (Parola *et al.*, 1996). Maybe other aldehydic products stemming from membrane oxidative breakdown can theoretically exibit a profibrogenic effect, but one can draw attention to their actual effective concentration, whether within acceptable physio-pathological range or not.

8. CONCLUSIONS

In conclusion, "in vitro" and "in vivo" data are accumulating which markedly support a causative role of tissue damage as induced by reactive oxidant species in fibrotic degeneration of tissues and organs. The involvement of two kinds of cells, macrophages and fibroblasts, or more often fibroblast-like cells, in the fibrogenic processes occurring in the various organs are clearly demonstrated. HNE, definitely one of the major lipid peroxidation products, appears able to upregulate the transcription and the synthesis of TGFβ1 in human and murine macrophages. In parallel, this fairly diffusable aldehyde upregulates expression and synthesis of collagen type I, one of the most represented classes of collagen during progression of fibrotic diseases.

HNE effect on other fibrogenic cytokines as well as on other extracellular matrix proteins, while most likely, remains to be investigated.

ACKNOWLEDGMENTS

The experimental studies were supported by the Ministero dell'Università e della Ricerca Scientifica e Tecnologica, Progetto Nazionale Radicali Liberi ed Equilibrio Redox, Roma, Italy.

Financial support was also provided by the Consiglio Nazionale delle Ricerche, Progetto Strategico Oxidative and Cellular Stress and Progetto Finalizzato ACRO, and by the Associazione Italiana per la Ricerca sul Cancro.

REFERENCES

Ando, Y., Nyhlin, N., Suhr, O., Holmgren, G., Uchida, K., El Sahli, M., Yamashita, T., Terasaki, H., Nakamura,M., Uchino, M., and Ando, M:, 1997, Oxidative stress is found in Amyloid deposits in Systemic Amyloidosis, *Biochem. Biophys. Res. Commun.* **232**:497–502.

Armendariz-Borunda, J., Seyer, J.M., Kang, A.H., and Raghow, R., 1990, Regulation of TGFbeta gene expression in rat liver intoxicated with carbon tetrachloride, *FASEB J.* **4**:215–221.

Bellomo, G., Finardi, G., Maggi, E., Palladini, G., Perugini, C., and Seccia, M., 1996, Antigenic epitopes generated during LDL oxidation and the detection of specific autoantibodies. *Proceedings of the 2nd Int. Conf. on Lipoprotein Oxidation and Atherosclerosis. Biological and Clinical Aspects*, p. L9, September, Pavia.

Beno, D.W.A., Retsky, R.J., and Davis, B.H., 1994, Lipid peroxidation-induced isoprostane independently activates Ito cell MAP kinase and increase type I collagen mRNA abundance, *Hepatology* suppl. **20**, abstr. 776.

Esterbauer, H., Schaur, R.J., and Zollner, H., 1991, Chemistry and Biochemistry of 4-hydroxynonenal, malonaldehyde and related aldehydes, *Free Radic. Biol. Med.* **11**:81–128.

Gressner, A.M., 1991, Liver fibrosis: Perspectives in pathobiochemical research and clinical outlook, *Eur. J. Clin. Chem. Clin. Biochem.* **29**:293–311.

Kovacs, E:J:, 1991, Fibrogenic cytokines: The role of immune mediators in the development of scare tissue, *Immunol. Today* **90**:17–23.

Leonarduzzi, G., Scavazza, A., Biasi, F., Chiarpotto, E., Camandola, S., Vogl, S., Dargel, R., and Poli, G., 1997, The lipid peroxidation end product 4-hydroxy-2,3-nonenal up-regulates transforming growth factor β1 expression and synthesis in the macrophage lineage: a link between oxidative injury and fibrosclerosis, *FASEB J.* **11**, in press.

Mossman, B.T., Marsh, J.P., and Sesko, A., 1990, Inhibition of lung injury, inflammation, and interstitial pulmonary fibrosis by polyethylene glycol-conjugated catalase in a rapid inhalation model of asbestosis, *Am. Rev. Respir. Dis.* **141**:1266–1271.

Paradis, V., Kollinger, M., Fabre, M., Holstege, A., and Bedossa, P., 1996, In situ detection of lipid peroxidation in chronic liver diseases, *Hepatology* suppl. **24**, abstr. 433.

Parizada, B., Weber, M.M., and Nimrod, A., 1991, Protective effect of human recombinant MnSOD in adjuvant arthritis and bleomycin-induced lung fibrosis, *Free Radic. Res. Commun.* **15**:297–301.

Parola, M., Leonarduzzi, G., Biasi, F., Albano, E., Biocca, M.E., Poli, G., and Dianzani, M.U., 1992a, Vitamin E dietary supplementation protects against carbon tetrachloride-induced chronic liver damage and cirrhosis, *Hepatology* **16**:1014–1021.

Parola, M., Muraca, R., Dianzani, I., Barrera, G., Leonarduzzi, G., Bendinelli, P., Piccoletti, R., and Poli, G., 1992b, Vitamin E dietary supplementation inhibits transforming growth factor β1 gene expression in the rat liver, *FEBS Lett.* **308**:267–270.

Parola, M., Pinzani, M., Casini, A., Albano, E., Poli, G., Gentilini, A., Gentilini, P., and Dianzani, M.U., 1993, Stimulation of lipid peroxidation or 4-hydroxynonenal treatment increase procollagen alpha (I) gene expression and synthesis in human liver fat storing cells, *Biochem. Biophys. Res. Commun.* **194**:1044–1050.

Parola, M., Pinzani, M., Casini, A., Leonarduzzi, G., Marra, F., Caligiuri, A., Ceni, E., Biondi, P., Poli, G., and Dianzani, M.U., 1996, Induction of procollagen type I gene expression and synthesis in human hepatic stellate cells by 4-hydroxy-2-3-nonenal and other 4-hydroxy-2,3-alkenals is related to their molecular structure, *Biochem. Biophys. Res.Commun.* **222**:261–264.

Poli, G., Albano, E., and Dianzani, M.U., 1987, The role of lipid peroxidation in liver damage, *Chem. Phys. Lipids* **45**:117–142.

Poli, G., and Parola,M., 1997, Oxidative damage and fibrogenesis, *Free Radic. Biol. Med.* **22**:287–305.

Shull, M.M., Ormsby, I., Kier, A.B., Pawlowski, S., Diebold, R.J., Yin, M., Allen, R., Sidman, C., Proetzel, G., Calvin, D., Annunziata, N., and Doetschman, T., 1992, Targeted disruption of the mouse transforming growth factor-beta1 gene results in multifocal inflammatory disease, *Nature* **359**:693–699.

Smith, M.A., Perry, G., Richey, P.L., Sayre, M.L., Anderson, V.E., Beal, M.F. and Kowall, N., 1996, Oxidative damage in Alzheimer's, *Nature* **382**:120–121.

Tsukamoto, H., 1994, Activation of fat storing cells in alcoholic liver fibrosis: Role of Kupffer cells and lipid peroxidation, in:*Fat Storing Cell and Liver Fibrosis*, (C. Surrenti, A. Casini, S. Milani, and M. Pinzani, eds.), pp. 189–195, Kluwer Academic Publisher, Lancaster.

Yoritaka, A., Hattori, N., Uchida, K., Tanaka, M., Stadtman, E.R., and Mizuno, Y., 1996, Immunoistochemical detection of 4-hydroxynonenal protein adducts in Parkinson disease, *Proc. Natl. Acad. Sci. USA* **93**: 2696–2701.

15

THE ROLE OF FREE RADICALS IN ISCHEMIC INJURY AND THE PRECONDITION STATUS

J. F. Koster and W. Sluiter

Department of Biochemistry
Cardiovascular Research Institute COEUR
School of Medicine and Health Sciences
Erasmus University Rotterdam
P.O. Box 1738, 3000 DR Rotterdam, The Netherlands

This chapter is focused on the role of iron in the mechanism of ischemic injury and the possible involvement of free radicals in preconditioning. Various hypotheses have been proposed for the mechanism of preconditioning.

Ten years after Murry, Jennings and Reimer's publication (1986) despite the huge effort of investigation the mechanism of preconditioning is still illusive.

Preconditioning is the curious phenomenon that four short periods of ischemia (5 min) separated by 5 min reperfusion result in less infarction after a sustained ischemic period than controls. In some way the previous short ischemic periods initiates a protection for the following sustained ischemic period. The protection lasts for about one hour.

It is well established that upon reoxygenation of ischemic tissue free oxygen radicals are generated (Zweier, 1988). However the origin of these radicals is not well known. The damaging action of free oxygen radicals is dependent on the presence of a free transition metal ion, in order to generate via the Haber-Weiss chemistry the hydroxyl radical. The most likely transition metals for the cell is iron or copper. The latter is present in low concentrations and besides bound to enzymes mostly located in ceruplasmin. Iron is abundantly present in transferrin and subcellularly bound in ferritin. However, the free iron concentration is physiologically very low. The latter is often addressed as the low molecular weight (LMW) pool. The iron located in this pool is able to perform the Haber-Weiss chemistry. The role of iron in the postischemic tissue injury has been substantiated by studies in which iron chelators, which inhibit lipid peroxidation, attenuates the reperfusion damage in a variety of experimental animal models (Smith, Gordon, Grisham, Granger and Korthuis, 1989; Xuehun, Prasad, Engelman, Jones and Das, 1991; Van der Kraaij, Van Eijk and Koster (1989). Further substantial evidence is performed with iron overloaded hearts (Van der Kraaij, Mostert, Van Eijk, and Koster, 1988). They show that iron over-

Free Radicals, Oxidative Stress, and Antioxidants, edited by Özben.
Plenum Press, New York, 1998.

151

loading of the hearts results in a dramatically increased reperfusion damage. In contrast to control hearts these iron-loaded hearts are unable to withstand an anoxic period.

In 1991 Nohl, and Stolze show that iron is released in the perfusate upon reperfusion after an ischemic insult, which is confirmed by Prasad, Lim, Ronson, Engelmann, Jones, George, and Das (1992). Voogd, Sluiter, Van Eijk and Koster (1992) have shown that there is already during the ischemic period an increase in the LMW iron pool of the ischemic tissue. Furthermore a period of anoxia does not result in an increase LWM pool. These data are in agreement with the facts that recovery after an anoxic insult is much better than after ischemic insult.

Another explanation of the reperfusion damage is the calcium overload. Due to the acidification of the tissue during ischemia the Na^+ content increased through Na^+/H^+ exchange, which leads to an intracellular calium concentration increase through Na^+/Ca^{2+} exchange. Steenberger, Fralix, and Murphy (1993) show that already during the ischemic period intracellular calcium increases, which decreases upon reperfusion. Toni, and Neely (1990) claim that the cytosolic calcium concentration does not increase during the ischemia but just after the reperfusion. Calcium overloading can lead to a contracture of the heart, but can also be the signal to apoptosis. The latter phenomenon besides necrosis is claimed to occur as a result of an ischemic insult (Gottlieb, Gruel, Zhu, and Engler, 1996). It can be concluded that during ischemia the LMW iron pool increased probably due to an increase in the number of reduction equivalents which are necessary to release iron from ferritin. Also the cytosolic calcium concentration is probably increased due to damaged sarcoplasmatic reticulum or due to the acidification of the cardiomyocyte. It is not unlikely that the damage occurring by reoxygenation is due to the synergistic effect between highly toxic oxygen free radicals generated in the presence of free iron, and calcium overloading. It is also possible that free radicals damage the sarcoplasma first which then leads to a faster increase of calcium ions.

ARE FREE RADICALS AND IRON INVOLVED IN THE MECHANISM OF PRECONDITION?

As stated above precondition (4×5 min ischemia, interfered by 5 min reperfusion) gives a protection against an ischemic insult. This protection lasts for about one hour. It is still unknown which mediator and which mechanism lead to the precondition status. In the paper of Murry et al. (1988) it is suggested that free oxygen radicals can be responsible for triggering the mechanism which leads to preconditioning.

Administration of radical scavengers during the first reperfusion period blocks the beneficial effects of preconditioning. Murry et al. (1988) suggested therefore that already small amounts of free oxygen radicals are sufficient to initiate the cellular activities necessary to induce preconditioning, but not to induce necrosis. Not surprisingly this hypothesis has been confirmed (Tanaka, Fugiwara, Yomasaki and Sasayama, 1994; Osada, Takeda, Sato, Komori and Tamura, 1994) and denied (Iwamoto, Miura, Adachi, Noto, Ogawa, Tsuchida and Imura, 1991; Bossam, Hanson, Bose, McCord, 1991; Richard, Tron and Thuilloz, 1993). If free oxygen radicals are the mediators of preconditioning they should be generated during the short ischemic-reperfusion periods. Bolli, Zughaib, Li, Tang, Sun, Triana, and McCay (1995) have shown that already after the first ischemic period free radicals are formed upon reperfusion. This production of free radicals diminishes as the perfusion is prolonged. Also after the other short ischemic periods free radicals are generated upon reperfusion. This study confirms more or less the findings of Pietri, and Culcasi (1989). These authors also show that after a 5 min ischemic period free radicals are formed. The studies differ in the amount of free radicals formed. Interesting is that the radical detected is the hydroxyl radical. This implicates that in some way free iron has to be present. Voogd et al. (1992) show that already a 5 min ischemic period results in a substantial increased LMW iron pool. Apparently the amount of free radical and the increased LMW iron pool is not enough to result into tissue damage. Voogd *et al.* (1972) measured the LMW iron in preconditioned and control hearts after a sustained period of 15 min ischemia. In the preconditioned heart the amount of LMW iron pool is 23 ± 32 pmole/mg protein while in the control heart 192 ± 45 pmole/mg protein after 15 min ischemia. A substantial decrease in LMW iron pool. It is also possible to mimick preconditioning by applying an anoxic period before the sustained ischemic insult. This treatment results in LMW iron pool of 60 ± 19 pmole/mg protein after 15 min ischemia. Furthermore on administration of glucagon also a preconditioning condition is reached. In this case the LMW iron pool is 74 ± 25 pmole/mg protein. The common features in the various methods (3x ischemia, anoxia, glucagon) to obtain a precondition status are a decrease in glycogen content and a lower lactate formation after the 15 min ischemia. Neely and Grotyohan (1984) show that recovery after ischemia is increasingly related to the glycolytic activity, but independent of tissue ATP just before reperfusion. The involvement of glycogen depletion is substantiated by finding of Wolfe, Sievers, Visseron, and Donnely (1993). They found that the loss of myocardial protection due to preconditioning is correlated with glycogen recovery. However it is also denied that depletion of glycogen is a mediator to preconditioning (Schneider, Nguyen, and Taegtmeyer, 1991; Schneider, and Taegtmeyer, 1991; Lagenstrom, Walker, and Taegtmeyer, 1988). If not the depletion of glycogen is the cause, it is still possible that free oxygen radicals trigger the mechanism leading to preconditioning. It is known that after an anoxic period also free oxygen radicals are generated upon reperfusion (Garlick, Davies, Hearse, and Slater, 1987). There is no reason to think that glucagon induces generation of free radicals. The fact that anoxia also mediates precondition status indicates that an accumulation of an intermediate is probably not necessary.

Besides the low increase in LMW iron pool there is also a diminished increase in cytosolic calcium concentration (Steenbergen, Perlman, London and Murphy, 1993). It is tempting to suggest that this phenomenon is the result of the attenuated increase in LMW iron pool.

From the fact that various stimuli can induce preconditioning it is not very likely that only one pathway will lead to this status. Obviously, different mechanisms underly the same endpoint, the preconditioned status.

Adenosine is advocated for quite a time as mediator for preconditioning. However Lanson and Downey (1993) conclude that adenosine is not involved in preconditioning. Adenosine acts through the A-1 receptor. It has to be realized that A-1 receptor agonism leads to increased glycolysis (Wyatt, Edmonds, Rubio, Berne and Lesley, 1989), which can lead to a depletion of glycogen if administered before the sustained ischemic insult.

It is proposed that the mechanism leading to preconditioning is related with the activation of protein kinase C (Lim, Ythrehus, Downey, 1994). This proposal is mainly based on experiments in which protein kinase C inhibitors are used. It is suggested that the cytosolic protein kinase C is translocated to the plasmamembrane where it can interact with diacylglycerol (DAG). It is stated that by translocation protein kinase C is already in the plasmamembrane when the sustained ischemic insult is applied. During the ischemic insult or upon the reperfusion protein kinase C can exert its beneficial effect on the reperfusion damage. It is also possible that a phosphorylated product of protein kinase C is responsible for the protection. It is unknown by which stimulus during preconditioning the protein kinase C is triggered to translocate. During the short periods of ischemia there is hardly any increase in cytosolic calcium concentration, which otherwise stimulates the activation of protein kinase C. If activated PKC may phosphorylate and activate c-Raf (Daum, Eisenmann-Tappe, Fries, Troppmair and Rapp, 1994). c-Raf is an intermediate in the ERK MAPK cascade. It is rather doubtful if the activation of this cascade leads to protection to ischemic insult. Knight and Buxton (1996) have shown that this cascade is activated not during ischemia but upon reperfusion. More important is that they show that a 5 min ischemia is not enough to activate this ERK MAPK cascade upon reperfusion. Even a previous 10 min ischemia is not enough. Parallel with this ERK MAPK cascade cellular stresses activate two separate MAPK cascades, the JNK/SAPK and the p38/RK cascade (Davies, 1994; Cano, and Mahadevan, 1995). Knight and Buxton (1996) also show that JNK is not activated during ischemia and only upon reperfusion. Also for JNK a 5 min ischemia followed by reperfusion is not enough to activate JNK. Bogoyevitch, Gillespie-Brown, Ketterman, Fuller, Ben-Levy, Ashworth, Marschall and Sugden (1996) confirm that JNK is only activated upon reperfusion and not during the ischemia. The investigators also show that p38/RK is activated already during ischemia. From these data, however, no firm conclusion can be drawn if p38/RK is involved in the preconditioning, because p38/RK activity was studied only at one time-point of 20 min of ischemia. Thus presently, for all MAPK cascades it is not known if they last long enough or if their products are present long enough to protect the heart against a sustained ischemic insult.

REFERENCES

Bogoyevitch, M.A., Gillesie-Brown, J., Ketterman, A.J., Fuller, S.J. Ben-Levy, R., Ashworth, A., Marshall, C.J., and Sugden, P.H., 1996, Stimulation of the stress activated mitogen activated protein kinase subfamilies in perfused heart, *Circ. Res.* **79**:162–173.

Bolli, R., Zughaib, M., Li, X-Y., Tang, X-L., Sun, J.Z., Triana, J.F., and McCay, P.B., 1995, Recurrent ischemia in the canine heart causes recurrent burst of free radical production that have a cumulative effect on contractile functions, *J. Clin. Invest.* **96**:1066–1084.

Bossam, D.A., Hanson, A.K., Bose, S.K., and McCord, J.M., 1991, Ischemic preconditioning is not mediated by free radicals in the isolated rabbit heart, *Free Rad. Biol. Med.* **11**:517–520.

Cano, E., and Mahadevan, L.C. , 1995, Parallel signal processing among mammalian MAPKs, *Trends Biochem. Sci.* **20**: 117–122.

Culcasi, M., Pietri, S., and Cazzone, P.J., 1989, Use of 3,3,5,5-tetramethyl-1-pyrroline-1-oxide spin trap for the continuous flow ESR monitoring of hydroxyl radical concentration in the ischemic and reperfused myocardium, *Biochem. Biophys. Res. Commun.* **164**:1274–1280.

Daum, G., Eisenmann-Tappe, I., Fries, H.-W., Troppmair, J., and Rapp, U.R., 1994, The ins and outs of Ref kinases, *Trends Biochem. Sci.* **19**: 474–480.

Davis, R.J., 1994, MAPKS: new JNK expands the group, *Trends Biochem. Sci.* **19**: 470–473.

Fliss, H., and Gattinger, D., 1996, Apoptosis in ischemic reperfused rat myocardium, *Circ. Res.* **79**:949–956.

Gorlick, P.B., Davies, M.J., Hearse, D.J., and Slater, T.F., 1987, Direct detection of free radicals in the reperfused rat heart using electron spin resonance, *Circ. Res.* **61**: 757–760.

Gottlieb, R.A., Gruel, D.L., Zhu, J.Y., and Engler, B., 1996, Preconditioning in rabbit cardiomyocytes, *J. Clin. Invest.* **97**:2391–2398.

Inamoto, T., Miura, T., Aduski, T., Nolo, T., Ogawa, T., Tsuchida, A., and Iimura, O. 1991, Myocardial infarct size limiting effect of ischemic preconditioning was not alternated by oxygen free radical scavengers in the rabbit, *Circulation* **83**: 1015–1022.

Knight, R.J., and Buxton, D.B., 1996, Stimulation of c-gen kinase and mitogen-activated protein kinase by ischemia and reperfusion in the perfused rat heart, *Biochem. Biophys. Res. Commun.* **218**: 83–88.

Lagerstrom, C.F., Walker, W.E., and Taegtmeyer, H., 1988, Failure of glycogen depletion to improve left ventricular function of the rabbit heart after hypothermic ischemic arrest, *Circ. Res.* **63**:81–86.

Lanson, C.S., and Downy, J.M., 1993, Preconditioning: state of the art myocardial protection, *Cardiovasc. Res.* **27**:542–550.

Lim, Y., Ythrehus, K., and Downey, J.M., 1994, Evidence that translocation of protein kinase C play a key event during ischemic preconditioning of rabbit myocardium, *J. Mol. Cell. Cardiol.* **26**:661–668.

Murry , C.E., Jennings, R.B., and Reimer, K.A., 1986, Preconditioning with ischemia: a delay of lethal cell injury in ischemic myocardium, *Circulation* **74**: 1124–1136.

Murry, C.E., Richard, V.J., Jennings, R.B., and Reimer, K.A., 1988, Precondition with ischemia: Is the productive effect mediated by free radical induced myocardial stunning? (Abstract) *Circulation* **78** (Suppl. 11): 77.

Neely, J.R., and Grotyohan, L.W., 1984, Role of glycolytic products in damage to the ischemic tissue, *Circ. Res.* **55**:816–824.

Nohl, H., Stolze, K., Napetochnig, S., and Ishikawa, T., 1991, Is oxidative stress primarily involved in reperfusion injury of the ischemic heart? *Free Rad. Biol. Med.* **11**:581–588.

Osada, M., Takeda, S., Sato, T., Komori, S., and Tamura, K., 1994, The protective effect of preconditioning on reperfusion arrhythmias is lost by treatment with superoxide dismutase, *Jpn. Cir. J.* **58**: 259–263.

Pietri, S., Culcasi, M., and Cozzone, P.J., 1989, Real-time continuous flow spin trapping of hydroxyl free radical in the ischemic and postischemic myocardium, *Eur. J. Biochem.* **186**:163–173.

Prasad, M.A., Lim, X., Ronson, J.A., Engelman, R.M., Jones, R., George, A., and Das, K.A., 1992, Reduced free radical generation during reperfusion of hypothermically arrested hearts, *Mol. Cell. Biochem.* **111**:97–102.

Richard, V., Tron, C., and Thiullez, C., 1993, Ischaemic preconditioning is not mediated by oxygen free radicals in rats, *Cardiovasc. Res.***27**: 2016–2021.

Schneider, C.A., Nguyen, V.T.B., and Taegtmeyer, H., 1991, Feeding and fastging determing postischemic glucose utilization in isolated working rat hearts, *Am. J. Physiol.* **260**: H542–H548.

Schneider, C.A., and Taegtmeyer, H., 1991, Fasting in vivo delays myocardial cell damage after brief periods of ischemia in the isolatged working rat heart, *Circ. Res.* **68**:1045–1050.

Smith, J.K., Gordon, D.L. Grisham, M.B., Granger, D.N., and Korthuis, R.J., 1989, Role of iron in post ischemic microvascular injury, *Am. J. Physiol.* **256**:H1472–H1477.

Steenbergen, C., Fralix, T.A., and Murphy, E., 1993, Role of increased cytosolic free calcium concentrations in myocardial ischemic injury, *Basic Res. Cardiol.* **88**:456–470.

Steenbergen, C., Perlman, M.E., London, R.E., and Murphy, E., 1993, Mechanism of preconditioning. Ionic alterations, *Circ. Res.* **72**: 112–125.

Tanaka, M., Fujiwara, H., Yomasaki, K., Sasayama, S., 1994, Superoxide dismutase and N-2 mercaptopropionyl-glycine alternate infarction size limitation effect of ischaemia preconditioning in the rabbit, *Cardiovasc. Res.* **28**:980–986.

Toni, M., and Neely, J.R., 1990, Na+ accumulation increases calcium overload and impairs function of an anoxic rat heart, J. Mol. Cell. Cardiol. **22**:57–72.

Van der Kraaij, A.A.M., Van Eijk, H.G., and Koster, J.F., 1989, Prevention of post-ischemic cardiac injury by the orally active chelator 1,2-dimethyl-3-hydroxy-4-pyridone (L1) and the *antioxidant* (+)-cyanidant 3, *Circulation* **80**:158–164.

Van der Kraaij, A.A.M., Mostert, L.J., Van Eijk, H.G., and Koster, J.F., 1988, Iron load increases the susceptitility of rat hearts towards reperfusion damage. Protoection by the anti-oxidant (+)-cyanidanol-3 and desferal. *Circulation* **78**: 442–449.

Voogd, A., Sluiter, W., Van Eijk, H.G. and Koster, J.F., 1992, Low molecular weight iron and the oxygen paradox in isolated rat hearts, *J. Clin. Invest.* **90**: 2050–2055.

Wolfe, C.L., Sievers, R.E., Visseron, F.L.J., and Donnely, T.J., 1993, Loss of myocardial protection after preconditioning correlates with the time course of glycogen recovery within the preconditioned segment, *Circulation* **87**: 881–892.

Wyatt, D.H., Edmonds, M.C., Rubio, R., Berne, R.M., Lasley, R.D., and Mentzner, R.M., 1989, Adenosine stimulated glycolytic flux in isolated perfused rat hearts by A1-adenosine receptors, *Am. J. Physiol.* **257**: H1952–H1957.

Xuehun, L., Prasad, R., Engelman, R., Jones, R.M., and Das, D.K., 1991, Role of iron in membrane phospholipid breakdown in ischemic perfused rat hearts, *Am. J. Physiol.* **259**: H1101–H1107.

Zweier, J.L., 1988, Measurements of superoxide free radicals in perfused heart, *J. Biol. Chem.* **263**: 1353–1357.

16

OXIDATIVE DAMAGE AND REPERFUSION SYNDROME IN HUMAN LIVER TRANSPLANTATION

Fiorella Biasi,[1] Juan Cutrin,[2] Elena Chiarpotto,[2] Mauro Salizzoni,[3]
Alessandro Franchello,[3] Fausto Zamboni,[3] Elisabetta Cerutti,[3]
Roberto Pagni,[4] Giacomo Lanfranco,[4] Isabella Chiappino,[4] and Giuseppe Poli[2]

[1]CNR Centre of Immunogenetics and Experimental Oncology
Torino
[2]Department of Experimental Medicine and Oncology
University of Torino
[3]Liver Transplantation Centre
Azienda Ospedaliera S. Giovanni Battista
Molinette, Torino
[4]Baldi e Riberi Laboratory
Azienda Ospedaliera S. Giovanni Battista
Molinette, Torino, Italy

1. INTRODUCTION

In the transplanted organ, early after reperfusion, a consistent while usually transient tissue damage is detectable which could interfere with appropriate recovery and function. In the case of transplanted human liver, a certain degree of cytolysis as well as post-hepatic cholestasis have actually been monitored early after surgery (Ericzon *et al.*, 1990; Biasi *et al.*, 1995). Then, studies addressed to elucidate the mechanisms underlying post-transplantation liver injury are necessary. The inflammatory process which always follows a cytolytic event generates a variety of molecules actually able to modulate not only cellular function but also its genetic program. Hence, extent and duration of flogistic reactions should be controlled for not being overwhelming tissue and organ homeostasis.

2. ISCHEMIA-REPERFUSION AND OXIDATIVE DAMAGE

Constantly associated with tissue reoxygenation after a defined period of ischemia is an increased steady-state level of reactive oxygen species (ROS) of radical and not radical nature (Omar, McCord, Downey, 1991). They are overproduced because an extensive

Free Radicals, Oxidative Stress, and Antioxidants, edited by Özben.
Plenum Press, New York, 1998.

reduction of molecular oxygen by reducing equivalents accumulated in the cells during the hypoxic state. In turn, ROS are mainly oxidant species, i.e. they abstract one electron to target molecules in order to reach a low energy state. But target molecules of ROS and of secondary derived free radicals can be structural proteins, lipids, amino acids, nucleotides, etc. If such unbalanced ratio between oxidative and reductive reactions is not sufficiently quenched by local antioxidant defense systems, an oxidative damage of the interested tissues then occurs.

Many experimental data are in support of a damaging role of excess of oxidative reactions, also termed oxidative stress, in the ischemia-reperfusion syndrome. However, still we do not have conclusive proofs in humans about the actual role of oxidative damage in the onset and the extent of post-reperfusion damage of transplanted organ, in this case the liver. It first appears essential to characterize and quantitatively analyse oxidative stress in strict comparison with tissue damage in human transplanted liver. Then one should evaluate topology and chronology of the damage, i.e. the cell types involved and the time of their involvement. Eventually, pharmaceutical approach with antioxidants should clearly define the role of ROS-driven reactions in the pathogenesis of reperfusion injury.

3. EVIDENCE OF ENHANCED LIPID PEROXIDATION SOON AFTER LIVER TRANSPLANTATION IN HUMANS

The monitoring of a series of hematic indices of membrane lipid peroxidation, one of the major effect of oxidative stress propagation, led us to demonstrate the early onset of oxidative damage in the transplanted liver as well as its transient occurrence.

Plasma lipid peroxides and red cell malonaldehyde significantly increase within the first two hours from portal vein declamping in almost all patients, and any way their level after 120 min reperfusion statistically correlates, in the single patients, with the extent of cytolysis, recorded at the same time in terms of plasma transaminases. Consistent but negative trend has been observed for plasma vitamin E content as to lipid peroxidation and cytolysis. The two monitored indices of lipid peroxidation showed a peak after one, two days of post-transplantation, quite in correspondence to that of transaminases, then a gradual return to normal values well within the first week. Once again the indices of cytolysis normalized correspondingly (Biasi *et al.*, 1995).

4. CORRELATION BETWEEN LIPID PEROXIDATION AND CYTOLYSIS IN THE REPERFUSED LIVER

The reliable relationship between membrane oxidative breakdown and cytolysis which is achievable 120 min after reperfusion, that is around the end of the intraoperation period, supports the involvement of oxidative stress in the building up of tissue damage early after transplantation surgery.

Still to be defined is the actual onset of lipid peroxidation in the immediate post-reperfusion period and its relationship with the appearance of cytolysis. In Table 1 are reported intra-operation mean values of red cell MDA, plasma LPO and plasma ALT and AST from 19 adults patients of both sexes. MDA level shows an average increase already after 1 min of reperfusion, while the rise of LPO level is evident much later. Plasma transaminases strongly increase already after 1 min of reperfusion and keep rising all through the 120 min period of observation. However, it is quite difficult to achieve unbiased cause-effect relationship between oxidative damage and cytolysis after ischemia-reperfu-

Table 1. Intraoperation levels of red cell malondialdehyde (MDA), plasma lipid peroxides (LPO), and plasma transaminases (AST, ALT) in human liver transplants

Assays	Basal	1 min	10 min	60 min	120 min
MDA	2.6 ± 1.0	3.4 ± 1.3	3.6 ± 1.4	4.2 ± 1.4	3.8 ± 1.3
LPO	3.0 ± 1.4	2.7 ± 1.4	3.1 ± 1.5	3.9 ± 1.4	3.9 ± 1.6
AST	30 ± 10	140 ± 31	410 ± 102	321 ± 83	578 ± 144
ALT	26 ± 7	167 ± 49	351 ± 74	421 ± 85	553 ± 137

Mean values ± SE are from 19 adults patients of both sexes. Blood samples were taken from hepatic veins just after anesthesia (basal) and at different intervals (min) after portal vein declamping. MDA: nmoles/ml packed red cells; LPO: nmoles/ml plasma; AST, ALT: U/l plasma.

sion because of the contribution of hypoxya *per se* as well as of surgical trauma to the cytolytic index.

Only the supplementation of the transplanted patients with suitable mixture of antioxidants could be discriminating. Preliminary data from adults recipients i.v. perfused during the vascular connection of the donated organ (period of warm ischemia) with a mixture containing vitamins C, E and β-carotene, showed a significant reduction of plasma transaminase levels with respect to not supplemented patients (Biasi *et al.*, manuscript in preparation).

5. CHRONOLOGY OF CELL INVOLVEMENT AND DAMAGE DURING THE ISCHEMIA-REPERFUSION SYNDROME AT SINUSOIDAL LEVEL

The first cellular type which appears affected is the endothelial cell. Already at the end of about seven hours of cold ischemia, analysis of bioptic samples by electron microscopy gives evidence of vacuolized and swollen endothelial cells with picnotic nucleus. Such damage is dramatically worsened by reperfusion, with frequent evidence of cariolysis (see Figure 1 a,b,c). Still after 60–90 min from reperfusion, i.e. the time which Fig. 1, c is referring to, hepatocyte mitochondria have normal morphology, glycogen stores are preserved, and only scattered images of parenchimal cell blebbing are detectable (not showed). Of note, already at this early reperfusion time polymorphonuclear cells appear to approach and even to attach to damaged endothelial cells (Fig. 1, c). Kupffer cells have evident morphological signs of activation, e.g. marked reduction of pseudopodes and increased number of intracellular vacuoles. Then, recruitment and activation of phagocytes in the transplanted liver occurs precociously indeed and it likely contributes to the reperfusion damage. Actually, the determination of mieloperoxidase plasma levels at 5 and 15 min of liver reperfusion, so far performed on few patients, would point to a significant rise of this specific polymorphonuclear enzyme after minutes and not hours from reconstitution of normal hepatic flow through the transplanted organ (Table 2).

One could argue, then, that oxidative damage cannot be cause of reperfusion injury but just an epiphenomenon related to marked and almost immediate activation of phagocytes. At present, there are not strong arguments against this hypothesis. However, the patient infusion during the warm ischemia period with a single dose of a registered antioxidant mixture would appear able to prevent the observed precocious increase of mieloperoxidase plasma levels (Table 2).

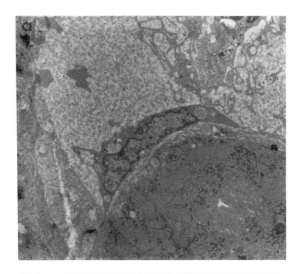

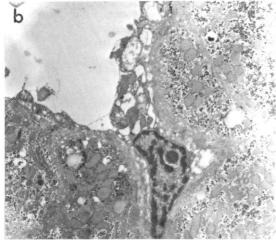

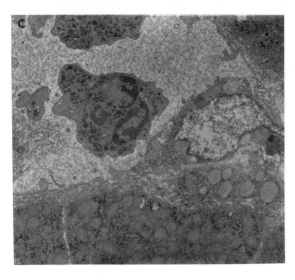

Figure 1. Transmission electron micropho-
tographs of transversal sections of sinusoidal
lumen in liver biopsies taken during the ex-
plantation period (a), after 7 hours of cold ischemia (b), and between 60 and 90 min of
portal vein reperfusion (c). (Uranyl acetate-
lead nitrate; original magnification 3,000×).

Table 2. Myeloperoxidase plasma levels at early intraoperative phases of orthotopic liver transplantation in patients treated or not with a single dose of antioxidant mixture

Time	Myeloperoxidase (ng/ml plasma)	
	W/o antioxidants (4)	Plus antioxidants (3)
Basal	37 ± 6	43 ± 7
5 min post-reperf.	67 ± 6	45 ± 9
15 min post-reperf.	108 ± 14	43 ± 11

Data are means ± SE. The number of patients is in parentheses. In a group of patients a single i.v. administration of one ampoule of Omnibionta antioxidant mixture (Merck, Darmstadt, Germany) was performed during the period of warm ischemia. Blood samples were taken from hepatic veins just after anesthesia and early after portal vein declamping.

On the basis of the available data, a preliminary general consideration regarding cytotypes involved in post-transplant sinusoidal damage and cytolysis could be the following: 1) endothelial cells are undoubtely the first cells to be damaged starting from the cold ischemic period: for this reason, mechanisms of injury other than the oxidative one should mainly act on these cells; 2) immediately after blood reflow in the liver, endothelial cells likely triggers the cooption of phagocytes in bringing about the damage in which hepatocytes are eventually involved; oxidative stress appears to contribute to the early damage of the reperfused sinusoid maybe modulating phagocytic cell activation.

6. REPERFUSION INJURY AT THE LEVEL OF BILE CANALICULI

The intrahepatic biliary tract is one the liver structure most susceptible to ischemia-reperfusion damage. Cholestasis is consistently evident from the second day of surgery, with a peak after approximately 10 to 16 days (Cutrin et al., 1996). With regards to the ultrastructure of the bile canaliculus during orthotopic transplantation, to the end of cold ischemic period the actin cores of microvilli appeared more compact suggesting a change in the functional state of actin filaments; marked and spread alterations of the canaliculi were morphometrically registered after 60–90 min of reperfusion, with loss of microvilli and dilatation of canalicular lumen. Whether free radical-driven reactions are playing a role in the pathogenesis of the observed early impairment of bile canaliculus, is still matter of discussion while the achievement of objective evidence appears experimentally difficult at present at least. The deep disturbance of intracellular calcium homeostasis which occurs during prolonged ischemia has been recognized in other pathological conditions to be responsible for marked microfilamnt derangement (Bennet, Weeds, 1993). It is very likely that such a mechanism of damage is operating during liver transplantation as well. Membrane lipid peroxidation, whose exacerbation takes place soon after organ reperfusion, can contribute in amplification of damage by more than one mechanism. A cause-effect relationship between lipid peroxidation and intracellular calcium derangement has been proved in the isolated hepatocyte model (Albano et al., 1991). Further, monitoring is in progress of possible variations in level and /or activity of gamma-glutamyltranspeptidase during ischemia-reperfusion, based on the possible generation of oxygen reactive species by this enzyme, expressed on the tract of hepatocyte membrane facing the bile canaliculus.

7. CONCLUSION

The tissue injury which usually follows orthotopic human transplantation is now at large characterized. Some mechanisms of damage build up during organ preservation (period of cold ischemia) and organ connection to the recipient's vasculature (period of warm ischemia). After reperfusion, when there is the major expression of tissue injury, the key players appear the rapidly recruited polymorphonuclear cells and the activated Kupffer cells, and among biochemical mechanisms the most likely is oxidative stress. Unbalanced redox equilibrium does not seem merely a consequence of phagocyte's activity. In fact, antioxidant supplementation appears to prevent the rise in myeloperoxidase plasma levels, which would indicate that redox reactions are involved in phagocyte activity. On this line are the several reports on ROS involvement in TNFα gene expression by phagocytic cells and the very recent one on the up-regulation by an aldehydic product of lipid peroxidation of another cytokine gene, coding for transforming growth factor β (Leonarduzzi *et al.*, 1997).

Target of research in progress on oxidative damage in human liver transplantation is the modulatory role of pro- and antioxidant compounds on gene expression related to phagocyte's activity, above all that of inflammatory cytokines and adhesion molecules. In relation to cholestasis, the involvement of ROS produced by hyperactivity of gamma-glutamyltranspeptidase will be checked out soon as well.

ACKNOWLEDGMENTS

The experimental studies were supported by the Ministero dell'Università e della Ricerca Scientifica e Tecnologica, Progetto Nazionale Radicali Liberi ed Equilibrio Redox, Roma, Italy. Financial support was also provided by the Consiglio Nazionale delle Ricerche, Progetto Strategico Oxidative and Cellular Stress and Progetto Finalizzato ACRO, and by the Associazione Italiana per la Ricerca sul Cancro.

REFERENCES

Albano, E., Bellomo, G., Parola, M., Carini, R., Dianzani, M.U., 1991, Stimulation of lipid peroxidation increases the intracellular calcium content of isolated hepatocytes, *Biochim. Biophys. Acta* **1091**:310–316.

Bennet, J., Weeds, A., 1993, Calcium and the cytoskeleton, *Br. Med. Bull.* **42**:385–390.

Biasi, F., Bosco, M., Chiappino, I., Chiarpotto, E., Lanfranco, G., Ottobrelli, A., Massano, G., Donadio, P.P., Vaj, M., Andorno, E., Rizzetto, M., Salizzoni, M., and Poli, G., 1995, Oxidative damage in human liver transplantation, *Free Radic. Biol. Med.* **19**:311–317.

Cutrin, J.C., Cantino, D., Biasi, F., Chiarpotto, E., Salizzoni, M., Andorno, E., Massano, G., Lanfranco, G., Rizzetto, M., Boveris, A., and Poli, G., 1996, Reperfusion damage to the bile canaliculi in transplanted human liver, *Hepatology* **24**:1053–1057.

Ericzon, B.G., Eusufzai, S., Kubota, K., Einarsson, K., Angelin, B., 1990, Characteristics of biliary lipid metabolism after liver transplantation, *Hepatology* **12**:1222–1228.

Leonarduzzi, G., Scavazza, A., Biasi, F., Chiarpotto, E., Camandola, S., Vogl, S., Dargel, R., and Poli, G., 1997, The lipid peroxidation end product 4-hydroxy-2,3-nonenal up-regulates transforming growth factor β1 expression and synthesis in the macrophage lineage: a link between oxidative injury and fibrosclerosis, *FASEB J.* **11**, in press.

Omar, B., McCord, J., Downey, J., 1991, Ischemia-reperfusion, in: *Oxidative Stress: Oxidants and Antioxidants*, (H. Sies, ed.), pp. 493–527, Academic Press, London.

17

PATHOPHYSIOLOGY OF CEREBRAL ISCHEMIA

Mechanisms Involved in Neuronal Damage

Tomris Özben

Akdeniz University Medical Faculty
Department of Biochemistry
07058 Antalya, Turkey

1. INTRODUCTION

Stroke is a common and devastating neurological disorder (Scatton, 1994) which is the third leading cause of death in major industrialized countries and also a major cause of long-lasting disability. Cerebral ischemia is always vascular origin and can be divided into haemorrhagic and thromboembolic, with the latter accounting for approximately 90% of strokes (Scatton, 1994) and results from embolic or thrombotic occlusion of the major cerebral arteries, most often the middle cerebral artery.

Cerebral ischemia can be classed by topography as global or focal and by chronology into reversible and irreversible. Focal hypoxia-ischemia also occurs in such contexts as traumatic insults or cerebral hemorrhages, while global hypoxia-ischemia (Choi, 1990) occurs in cardiac arrest, near drowning and carbon monoxide poisoning.

The brain is critically dependent on its blood flow for a continuous supply of oxygen and glucose. Acute brain damage "stroke" (Choi, 1990) reflects localized tissue hypoxia attributable to reduced blood flow. Cerebral blood flow is normally 50 ml/100 g per min. When CBF is below 20 ml/100 g per min, cerebral infarct develops. In the center, tissue necrosis and cell death occur (Pulsinelli, 1992).

The pathogenesis of brain damage from cerebrovascular occlusion (Pulsinelli, 1992) may be separated into two sequential processes:

1. Vascular and haemotological events (Pulsinelli, 1992) that cause the initial reduction of local cerebral blood flow.
2. Ischemia-induced abnormalities of cellular chemistry that produce necrosis of brain cells.

Heavy studies are being performed to develop drugs that will prevent neurodegeneration following acute ischaemic stroke. For this purpose, animal models have been pro-

Free Radicals, Oxidative Stress, and Antioxidants, edited by Özben.
Plenum Press, New York, 1998.

duced that mimic the neuropathological consequences of stroke. During the past 10–15 years of stroke research, reproducible techniques for the induction of focal and global ischaemia have been developed. These models have several advantages and disadvantages. Reversible or irreversible focal ischaemia models like stroke in humans are useful for investigations of molecular mechanisms of stroke and also for the development of neuroprotective drugs. For studies of the molecular events in cerebral ischemia, an animal model of focal and irreversible ischemia (Chen, Hsu, Hogan, Maricq, and Balentine, 1986) allowing only partial reperfusion to create a reproducible infarction of predictable size and location would be of considerable value. The advantages of using rats for stroke study include the similarity of their intracranial circulation to that of man and the relatively low animal cost which is important for large scale studies for statistical analysis.

In this manuscript, the pathophysiology of lesions caused by focal cerebral ischemia will be discussed.

Both permanent and reversible occlusion of middle cerebral artery develop a reliable and reproducible infarct area in the MCA territory. Ischemia due to middle cerebral artery occlusion encompasses a densely ischemic focus and a less densely ischemic penumbral zone. A rim of moderate ischemia surrounds the severely ischaemic area. This is called penumbra which lies between normally perfused brain and infarct area. Penumbra doesn't exist in global ischemia. The penumbra (Obrenovitch, 1995) defines regions with blood flow below that needed to sustain electrical activity, but above that required to maintain cellular ionic gradiensts, and that lead in time to infarction. In penumbra, pathophysiological mechanisms are dynamic, cell death occurs last and pharmacological intervention has been most successful. Penumbra may extend outside (Pulsinelli, 1992). Cells in the focus are usually doomed unless reperfusion is quickly instituted. In contrast, although the penumbra contains cells "at risk", these may remain viable for at least 4 to 8 hours. Cells in the penumbra may be salvaged by reperfusion or by drugs that prevent an extension of the infarction into the penumbral zone (Siesjö, 1992a). Factors responsible for such an extension, probably include acidosis, edema, dissipative ion fluxes, calcium overload, glutamate toxicity, free radical formation and nitric oxide overproduction. Although, each will be discussed in turn, I wish to emphasize that none works in isolation. For example, effects elicited by a rise in intracellular Ca^{2+} result also from energy failure per se. Furthermore, since calcium triggers several reactions leading to the production of free radicals, the question arises whether the damage is calcium or free radical related. Nonetheless, it (Escuret, 1995; Siesjö, 1992b) appears profitable to discuss the mechanisms separately.

Central to any discussion of the pathophysiology of ischemic lesions is energy depletion. The flow diagram shown in Figure 1 (Pulsinelli, 1992) depicts multiple branching pathways that may be important in ischaemic brain damage. This representation helps one to understand the complexity of the process. There are many potential sites of interaction. It emphasizes the fundamental importance of energy depletion in the genesis of subsequent injurious events. Once initiated, such mechanisms (Pulsinelli, 1992) may no longer require the triggering event. Cell necrosis is secondary to a deranged or perturbed energy metabolism. The brain's ATP supply (White, Grossman, and Krause, 1993) is dependent on continuous perfusion and approaches zero within about 4 minutes of complete ischemia. The following cellular, metabolic and biochemical events (Scatton, 1994; Scheinberg, 1991) occur after vascular occlusion in the brain:

1. There is an immediate reduction of CBF and cerebral oxygen and glucose consumption at the center of the area of ischemia.
2. Cerebral vascular autoregulation is impaired or lost. Brain becomes susceptible to alterations in perfusion pressure. Impaired CBF and cerebral metabolic func-

tion (White *et al.*, 1993) occur in areas of the brain remote from the ischemia, probably as a result of transneuronal inhibition of neuronal activity (diaschisis).

3. There is an immediate initiation of anaerobic glycolysis. Tissue lactate increases and tissue pH decreases. Since CBF and vascular reactivity correlate directly with H^+ ion concentration, these events (Scheinberg, 1991) are responsible for the increased perfusion in the periphery of the ischemic zone.

Brain ischemia and reperfusion (White *et al.*, 1993) elicit the production of both growth factors and proto-oncogene products in at least some neuronal populations. Examples of growth factor responses following ischemia are an increase in basic FGF in the rat hippocampus, caudate, putamen, and cortex, increased production of mRNAs for brain derived neurotrophic factor and NGF in rat dentate gyrus(but a decreased level of neurotrophin-3) and in the rat hippocampus CA_1 and CA_2 regions and increased levels of IGF-1 in the lateral cortex, hippocampus, striatum, thalamus and pyriform cortex. In the case of proto-oncogene products, c-fos (Shimazu, Mizushima, Sasaki, Arai, Matsumoto, Shioda, and Nakai, 1994) and c-jun transcripts are elevated in brain homogenates within 30 minutes of postischemic reperfusion both in rats and gerbils. Stat 1 has a dual function as signal transducer and activator of transcription and is activated in response to growth factors and cytokines. Ischemia (Planas, Justicia, and Ferrer, 1997) caused a strong induction of Stat1. At least one mechanism by which the effects of immediate early transcription regulators are mediated is the induction of stress proteins. Cellular insults elicit the production of heat shock proteins (HSPs). The HSPs appear to facilitate correct folding of proteins, rRNA, and have been proposed to be involved in membrane synthesis. Heat shock messages and proteins are expressed in the brain during reperfusion. There is evidence that the HSP response can be modulated by the presence of free radicals but not simply by increased intracellular calcium. After an ischemic insult the selectively vulnerable neurons in the hippocampus and cortex transcribe large quantities of hsp-70 mRNA, but translation is considerably decreased compared with surviving areas of the hippocampus. Growth factors are involved in basic cellular regulation that affects the mechanisms of cellular repro-

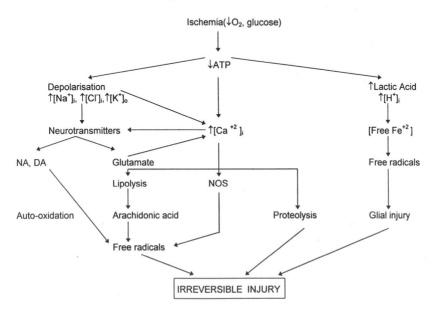

Figure 1. Potential mechanisms of ischemic brain damage. NA: noradrenalin; DA: dopamine; NOS: nitric oxide synthase.

duction, including transcription of the components of membrane synthesis and repair, the administration of growth factors might improve neuronal survival. In some experiments, FGF inhibited glutamate receptor-mediated neuronal damage and NGF reduced the loss of CA_1 neurons in the hippocampus. Treatment with IGF-1 reduced neuronal loss in the cortex and insulin treatment either pre- and post-insult in the rat reduced cortical and striatal neuronal necrosis and reduced infarction rates in the cerebral cortex, thalamus, and substantia nigra pars reticularis. Insulin receptor-mediated protein phosphorylation cascades stimulate transcription of c-fos/c-jun, increase mRNA efflux from the nucleus, induce the enzymatic systems for de novo lipogenesis, activate the mitochondrial pyruvate dehydrogenase complex, modify the phosphatidylinositol transcytoplasmic signaling systems, and stimulate the heat shock response. Insulin's neuroprotective effects may include:

1. Activation of appropriate early transcription of c-fos in the vulnerable zones
2. Phosphorylation of Calmodulin by the insulin receptor and reduction of its ability to bind calcium
3. Stimulation of the lipid synthesis systems vital to membrane repair
4. Insulin dependent promotion of the heat shock response

Catecholamines cause prompt and prolonged inhibition of insulin secretion and reduce the tyrosine kinase activity of the insulin receptor by cAMP induced phosphorylation of serine and threonine residues on the receptor. Moreover, catecholamines promote iron-mediated lipid peroxidation and neuronal death.

A rapid and profound increase in CRH mRNA levels during focal cerebral ischemia is likely to be associated with neurotoxicity, as CRH antagonism (Wong, Loddick, Bongiorno, Gold, Rothwell and Licinio, 1995) has been reported to cause a significant reduction in neuronal loss during ischemia.

2. ACIDOSIS

Following ischemia, oxygen and glucose supplies of brain diminish. ATP decreases rapidly. In 2–3 min, lactic acid formation reaches its maximum level. Lactate and unbuffered hydrogen ions accumulate in tissue in proportion to the carbohydrate stores present at the onset of ischemia.The severity of lactic acidosis is dependent upon the blood and tissue levels of glucose. Hyperglycemia (Pulsinelli, 1992) worsens ischemia at the onset of ischemia. A strong correlation exists between ischemic damage and the glucose content of brain before the ischemia. For example, in normoglycaemic conditions, 10–12 μmol/g lactate is produced by ischemic glycolysis and extracellular pH decreases to 6.8. In hyperglycemic states, lactate formation is 30–40 μmol/g and extracellular pH is 5.5–6.0. Lactate levels over 20 μmol/g lead to necrosis by sudden decrease of pH. Lactic acidosis exerts its lethal effect in several ways. Mitochondrial respiration is depressed at low pH. This inhibits further ATP formation. Low tissue pH (Scheinberg, 1991) enhances free radical formation and lipid peroxidation. In vitro acidosis (Cafe, Torri, and Marzatico, 1993) causes delocalization of protein bound iron, and thereby facilitates iron-catalyzed production of oxygen reactive species. Low pH (Cafe *et al.*, 1993) can enhance free radical reactions also by shifting the unprotonated form of superoxide radical (O^{2-}) to the protonated form (HO^{2-}) which is more lipid soluble and pro-oxidant. In spite of this evidence, it is very difficult to show a correlation between ischemic acidosis and free radicals. Because their production is localized into small subcellular compartments. In addition, acidosis worsens the disturbance occurred during ischemia in calcium ion homeostasis.

3. EDEMA FORMATION

In addition to the rapid change in tissue acid-base status, failure of all energy dependent mechanisms occurs. Failure of ion pumps leads to deterioration of membrane ion gradients, opening of selective and unselective ion channels, and equilibration of most intracellular and extracellular ions(anoxic depolarisation). As a consequence of anoxic depolarisation (Pulsinelli, 1992), potassium ions leave the cell, while, sodium, chloride and calcium ions enter. During the first two hours of ischemia, Na^+ influx reaches its maximum level. 3 Na^+ ions enter the cells and 2 K^+ ions leave the cell. Therefore, the number of ions entering the cell exceeds the number of ions leaving the cell. To maintain electroneutralization, sodium carries chloride ion inside the cells. Cellular osmolarity increases. Depending on this, water shifts from extracellular into intracellular compartment. This causes ischemic neuronal swelling and development of cytotoxic edema. Cytotoxic edema is the first stage of ischemic edema and occurs in the first minutes of ischemia. In addition, increased H^+ ions due to ischemic acidosis are expelled out of the cells in exchange for Na^+ ions. This worsens cytotoxic edema. Cytotoxic edema occurs both in gray and white matter. Blood brain barrier (BBB) is intact. In acute ischemic stroke, edema begins as reversible cytotoxic form, later, changes into vasogenic form. Vasogenic edema follows, secondary to leakiness of the blood brain barrier which is also impaired by energy failure and the hydrostatic force created by the blood pressure. Disruption of BBB leads to extracellular water accumulation and formation of vasogenic edema. The resultant brain edema may compress capillaries, thereby further decreasing regional blood flow into the ischemic zone as well as inhibiting the reperfusion which may follow relief of arterial obstruction or revascularization. Cerebral edema and increased intracranial pressure (Scheinberg, 1991) affect the perfusion of peripheral areas and cell necrose enlarges. Vasogenic edema develops after 4–6 hours of ischemia, reaches maximum at 36–72 hours and lasts 7–14 days. If this edema is severe enough, it causes death in ⅓ of ischemic lesions and ¾ of haemorragics.

During the past several years, it has become increasingly apparent that interleukin-1 (IL-1), particularly IL-1 beta (Betz, Schielke, and Yang, 1996) plays an important role in brain injury during ischemia. Studies from various laboratories have shown that IL-1 beta mRNA and IL-1 beta protein are synthesized early in ischemia and the injection of IL-1 beta into ischemic brain enhances edema formation. The most direct evidence that IL-1 beta contributes to ischemic injury, however, is the demonstration that infarct volume in focal ischemia is reduced following intraventricular injection of an endogenous interleukin-1 receptor antagonist (IL-1 ra), or after IL-1 ra is overexpressed in brain using an adenoviral vector to transfer IL-1 ra cDNA to brain cells. Ischemic injury is also reduced in mice that fail to produce IL-1 beta because of an abnormal interleukin-1 beta converting enzyme gene(ICE knockout mice). At the present time, it is nuclear how IL-1 beta causes brain injury, but several possible mechanisms include 1) stimulation of an inflammatory response through the activation of glia or the induction of other cytokines and/or endothelial adhesion molecules and 2) release of free radicals through stimulation of arachidonic acid metabolism and/or nitric oxide synthase activity.

4. GLUTAMATE NEUROTOXICITY

Glutamate is a major excitatory neurotransmitter (Ozyurt, Graham, Woodruff, and McCulloch, 1988) in the mammalian central nervous system. Extracellular concentration of glutamate (Park, Nehls, Graham, Teasdale, and McCulloch, 1988) are markedly elevated in ischemic brain tissue as a consequence of both enhanced release of the amino acid from neurons and its impaired uptake into glia and neurons. Due to energy failure, presynaptic depolarization causes voltage-sensitive calcium channels (VSCC) open, allowing Ca^{+2} to enter and triggers release of excitatory amino acids (EAA) such as glutamate. VSCC involved (Pulsinelli, 1992; Siesjö, 1992a) is of the N type, while the T and L types could be localized mainly to postsynaptic membranes in dendrites and soma. Synaptic release of EAA is increased also due to the elevated extracellular K^+ concentration. Due to energy failure, glial and neuronal EAA is impaired. The increased EAA (Scatton, 1994) in the extracellular space overactivates post-synaptic EAA receptors. In post-synaptic neurons, there are four distinct glutamate receptors (Hansen, 1995). These are named according to their preferred agonists. Glutamate acts on both N-methyl-D-aspartate (NMDA) and non-NMDA receptors and activates both NMDA and non-NMDA (Kainate and AMPA/Quisqualate) type receptors. Non-NMDA receptors mediate acute neuronal swelling. NMDA receptors are particularly important in mediating subsequent delayed neuronal disintegration. Non-NMDA receptors (Ozyurt et al., 1988; Park et al., 1988; Scatton, 1994) play a more significant role in global ischemia and NMDA receptors in focal ischemia. Furthermore, certain neuronal subpopulations containing the enzyme NADPH-diaphorase or parvalbumin-like immunoreactivity (Choi, 1990) may be vulnerable to non-NMDA receptor mediatad injury.

4.1. Glutamate Receptors

4.1.1. NMDA Receptor (N-Methyl-D-Aspartate). It is activated selectively by glutamate agonist, N-methyl-D-aspartate and named accordingly. NMDA receptor (Choi, 1990) has several distinct sites.

1. The agonist-binding site
2. The glycine-binding site
3. Zinc-binding site
4. Channel site
5. Polyamine binding site
6. Regulatory sites sensitive to changes in pH, phosphorylation or oxidation.

NMDA receptor gates a channel that is permeable to Na^+, K^+ and Ca^{+2}. For the activation of receptor, both glycine and transmitter sites should be activated. NMDA receptor is positively influenced by the modulatory glycine and polyamine. This receptor has a peculiarity. At normal resting potential (which is -65mV) extracellular Mg ions block the channel. Energy failure in ischemia leads to membrane depolarization. An additional 20–30 mV depolarization is sufficient to relieve the Mg^{+2} block of the NMDA receptor. In case of glutamate binding to the receptor, a massive entry of Ca^{+2} ions occur through NMDA receptor gated Ca^{+2} channels.

4.1.2. Kainate Receptor. It is activated by glutamate agonists, Kainate and AMPA (α-amino-3-hydroxy-5-methyl-4-isoxasole propionic acid). NMDA doesn't activate it. It is permeable to Na^+ and K^+ but not to Ca^{+2}.

4.1.3. Quisqualate A Receptor. It is activated by both quisqualate and AMPA. It is also permeable to Na^+ and K^+, but impermeable to Ca^{+2}. Because of their similarities, these two receptors are called AMPA receptors. Although AMPA receptors are impermeable to Ca^{+2}, they indirectly allow entry of Ca^{+2} into neurons by alleviating the Mg^{+2} block of the NMDA receptor and by causing membrane depolarization sufficient to activate voltage operated Ca^{+2} channels.

4.1.4. Quisqualate B Receptor. It is a metabotropic receptor. It doesn't link to a channel. When glutamate binds this receptor, G protein and phospholipase C(PLC) are activated. Activated PLC induces hydrolysis of phosphatidyl inositol-4,5-biphosphate (PIP_2) to generate the second messengers inositol-1,4,5-triphosphate(IP3) and diacylglycerol (DAG). IP3 induce release of Ca^{+2} from endoplosmic reticulum (ER). DAG (Choi, 1990) remains associated with the membrane and activates protein kinase C.

Intense exposure to glutamate (Choi, 1988; Choi, 1994) induce 2 events in neurons:

1. Acute neuronal swelling dependent on extracellular Na^+ and Cl^-.
2. Delayed neuronal disintegration dependent on Ca^{+2}.

Over the past decade, evidence has suggested that glutamate results in neuronal cell death after stroke. More recent evidence (Lipton, 1996) has suggested that in addition to necrotic death in the ischemic core, a number of neurons may also undergo apoptosis. Thus, the hypothesis that intense injury leads to necrosis while mild insult (perhaps in the penumbra) leads to apoptosis may hold in focal cerebral ischemia.

In recent years, several glutamate antagonists and also glutamate release inhibitors (Balkan, Özben, Balkan, Oguz, Serteser and Gümüşlü, 1997a; Kawaguchi and Graham, 1997; Leach, Swan, Eisenthal, Dopson and Nobbs, 1993; Margaill, Parmentier, Callebert, Allix, Boulu and Plotkine, 1996; Meldrum, 1994; Meldrum, Swan, and Leach, 1993; Ozyurt et al., 1988; Park et al., 1988; Shuaib, Mahmood, Wishart, Kanthan, Murabit Ijaz, Miyashita and Howlett, 1995; Swan and Leach, 1992; Umemura, Gemba, Mizuno and Nakashima, 1996; Wiard, Dickerson, Beek, Norton and Cooper, 1995; Winfree, Baker, Connolly, Fiore and Solomon, 1996) were used to prevent glutamate excitotoxicity and promising results were achieved. In models of focal ischemia caused by MCA occlusion, NMDA antagonists (Scatton, 1994) appear to exert their neuroprotective effect in cortical regions but not in the striatum. This may be due to the fact that the striatum is supplied by end-arteries and thus becomes densely ischemic after MCA occlusion while the cerebral cortex of the MCA territory is well endowed with colleteral circulation. Cortical cooling in global transient ischemic models suggests (Baker, Fiore, Frazzini, Choudhri, Zubay, and Solomon, 1995) that hypothermia limits glutamate excitotoxicity by decreasing the release of glutamate during ischemia.

4.2. Excitatory Glutamate Receptor Antagonists

1. NMDA receptor antagonists

   a. Competitive antagonists for the NMDA recognition site (agonist binding site): Linear or cyclic phosphonate analogs of carboxylic acid
   b. Open channel blockers: Ketamine, phencyclidine, dizolcipine (MK-801)
   c. Antagonists of the modulatory glycine site: derivatives of kynurenic acid and indole acetic acid
   d. Antagonists of the polyamine-sensitive modulatory site; Phenyl ethanolamines such as ifenprodil, eliprodil

2. AMPA-kainate receptor antagonists

   Quinoxaline (NBQX)-2,3-benzodiazepine (GYKI52466), (LY-293558).

3. Quisqualate B receptor antagonists

   2-amino-3-phosphonapropionate

Several signaling systems other than those mediated by glutamate may additionally influence hypoxic neuronal injury. Stimulation of adenosine A_1 receptors (Choi, 1990) reduces hypoxic neuronal injury both in vivo and in vitro, whereas adenosine antagonists increase injury. Endogenous adenosine released locally during cerebral ischemia (Jiang, Kowaluk, Lee, Mazdiyasni, and Chopp, 1997) is reported to be neuroprotective. Adenosine kinase inhibition may be a useful approach to the treatment of focal cerebral ischemia. Protection against both glutamate toxicity and ischemic injury (Kogure, 1990) has been reported for gangliosides, which can inhibit the membrane translocation of protein kinase C. GM1 ganglioside (Scheinberg, 1991) stimulates function of Na^+–K^+ATPase, and adenylcyclase and may therefore, promote recovery in ischemic brain. Reduction of ischemic brain injury has also been reported for other protein kinase C inhibitors.

5. CALCIUM OVERLOAD

Calcium ions are among the most powerful intracellular messengers, able to give rise to a great variety of events. In normal conditions, intracellular Ca^{+2} concentration (Cafe *et al.*, 1993) is very low (10^{-7} mol/L) and highly regulated in order to prevent useless activation of Ca^{+2} dependent processes.

Following hypoxia/ischemia, abnormal Ca^{+2} influx into neurons play important role in post-ischemic cell death. Ca^{+2} (Cafe *et al.*, 1993) enters into neurons mainly by ionic channels (voltage-dependent or receptor-operated channels on the plasma membrane. Several mechanisms contribute to the increase in cytosolic Ca^{+2} concentration:

1. Failure of ATP dependent Na^+–K^+ATPase causes rise in extracellular K^+ and depolarization of neuronal membranes. Consequently, voltage sensitive calcium channels are opened and Ca^{+2} enters the cells.
2. Due to insufficient ATP levels, failure of Na^+/Ca^{+2} antiport system and Ca^{+2}ATPase and failure of ATP dependent sequestration of Ca^{+2} by ER contribute to increase in Ca^{+2} influx. Mitochondrial accumulation of Ca^{+2} causes uncoupling of oxidation and phosphorylation.
3. The endoplasmic reticulum membranes of neurons posses other Ca^{+2} channels called Ryanodine receptors (RYRs), because of their sensitivity to the plant alkaloid ryanodine. Membrane depolarization triggers opening of plasma membrane voltage gated Ca^{+2} channels, permitting a small influx of extracellular Ca^{+2} ions. Binding of Ca^{+2} to ryanodine receptors on the ER membrane leads to a massive release of intracellular Ca^{+2} stores.
4. During ischemia, overstimulation of NMDA receptor by excess glutamate leads to a massive entry of Ca^{+2} ions through receptor gated Ca^{+2} channels.
5. Binding of glutamate to Quisqualate B (metabotropic receptor) triggers activation of the G protein which in turn activates phospholipase C. Phospholipase C cleaves phosphatidyl inositol-4,5-biphosphate (PIP_2) to inositol 1,4,5-triphosphate (IP_3) and 1,2-diacylglycerol (DAG). The IP_3 diffuses through the cytosol and binds to a specific IP_3 receptor on the ER membrane. This receptor is a Ca^{+2} channel protein composed of four identical subunits, each containing an IP_3 binding site. Each subunit contains a C-terminal domain that spans the lipid

bilayer twice: together the transmembrane domains of the four monomers form the Ca^{+2} channel. The N-terminal domain is large and contains the IP_3 binding site. IP_3 binding induces opening of the channel allowing Ca^{+2} ions to exit from the ER into the cytosol.

5.1. Effects of Calcium Overload

Intracellular Ca^{+2} overload during ischemia (Siesjö, 1994) has several deleterious consequences:

1. It activates phospholipase A_2 which attacks cellular mebranes liberating fatty acids mainly arachidonic acid (AA) which produces cerebral edema by inhibiting Na^+/K^+–ATPase. It alters membrane permeability and cell function. AA, in turn, (Scatton, 1994) can also amplify NMDA receptor mediated responses by increasing NMDA currents via an action on a site located on the NMDA receptor and by increasing glutamate release from presynaptic terminals and impairing its glial reuptake. Accumulation of free fatty acids and its products have deleterious effects on membranes. At the beginning of ischemia, acidosis is a result of lactic acid overproduction. Afterwards, it is related to the increase in free fatty acids. Intracellular acidosis denatures proteins causing loss of their normal function and also by enhancing glial edema, it causes neuronal death. Oxidation of AA by cyclooxygenase and lipoxygenase produces prostanoids and leukotrienes. These oxidations are the source of oxygen free radicals. Glutamate (Choi, 1990) may also inhibit neuronal cystine uptake, leading to reduced glutathione synthesis and diminished capacity to scavenge free radicals.

2. Activation of phospholipases and resulting phospholipid breakdown triggered by hypoxia-ischemia may have important deleterious effects on subsequent blood flow. Formation of platelet-activating factor (PAF) resulting from phospholipid breakdown (Braquet, Spinnewyn, Demerle, Hosford, Marcheselli, Rossowska, and Bazan, 1989) may be an important cause of post-ischemic hypoperfusion. Pharmacological PAF antagonists, which can attenuate delayed post-ischemic hypoperfusion and other PAF-mediated events such as neutrophil chemotaxis and free radical production show promise as treatments for ischemic insults.

3. Elevated cytosolic Ca^{+2} activates the neutral protease Calpain I. Activated Calpain I, converts xanthin dehydrogenase into xanthine oxidase by cleaving a peptide bond in xanthine dehydrogenase in endothelium of cerebral blood vessels. Elevated xanthine oxidase activity during reperfusion may contribute to increased oxygen radical formation as a result of the metabolism of xanthine to uric acid.

4. Elevated cytosolic Ca^{+2} (Cafe et al., 1993; Scatton, 1994) activates ornitin decarboxylase leading to oversynthesis of polyamines. This causes activation of polyamine site of NMDA receptor amplifying its response, increases Ca^{+2} influx.

5. Trying to buffer Ca^{+2} excess, mitochondria (Cafe et al., 1993; Scatton, 1994) accumulates calcium which uncouples oxidative phosphorylation at the time when ATP production is already reduced.

6. In the presence of Ca^{+2}, DAG activates protein kinase C. Protein kinase C is a soluble protein and it is inactive catalytically in the absence of DAG. Protein kinase C is activated by Ca^{+2} and DAG which remains in the membrane. The activated kinase then phosphorylates several cellular enzymes and receptors, thereby altering their activity. Protein kinase activation depends on both Ca^{+2}

and DAG. This is an indicator of the relationship between the two branches of inozitol-lipid signal pathway. Protein kinase C also phosphorylates NMDA receptor, thereby, (Scatton, 1994) enhances NMDA receptor stimulation via reduction of the Mg^{2+} block.

7. Intracellular Ca^{+2} overload (Scatton, 1994) activates Nitric Oxide Synthase (NOS) leading to the formation of NO from arginine. NO produces cGMP by activating guanylate cyclase. cGMP has anti-oxidant, anti-aggregant, anti-adhesive, anti-inflammatory and dilating effects.

8. Ca^{+2} activates some lytic enzymes (proteases, endonucleases). Endonucleases induce DNA fragmentation, programmed cell death. Proteases degrate protein MAP_2 in microfilaments and microtubulis. This correlates with neuronal degeneration.

5.2. Calcium Antagonists and Pharmacoprotection

Several drugs have been developed to reduce the deleterious transfer of Ca^{+2} ions across the cell membranes. These drugs, termed calcium antagonists (Bowersox, Singh and Luther, 1997; Germano, Bartkowski, Cassel, Lawrence and Pitts, 1987; Meyer, Anderson, Yaksh and Sundt, 1986; Rami and Krieglstein, 1994; Takizawa, Matsushima, Fujita, Nanri, Ogawa and Shinohara, 1995) appear to have some usefulness in experimental and clinical cerebral ischemia. An important group of these calcium antagonists are the dihydropyridines. They block L-type of voltage-sensitive calcium channels present throughout the central nervous system. It has also been shown that a presynaptic selective N-type calcium channel antagonist (SNX-111) (Takizawa *et al.*, 1995) has a protective effect against focal ischemia through the inhibition of glutamate release from presynaptic sites.

6. REACTIVE OXYGEN PARTICLES

One of the mechanisms implicated in the neuronal damage is the formation of oxygen free radicals during ischemia and reperfusion (e.g. reoxygenation). Free radicals formed depending on the duration and severity of ischemia (Cafe *et al.*, 1993; Chan, 1996; Grogaard, Schürer, Gerdin and Arfors, 1989; Schott, Natale, Ressler, Burney and D'alecy, 1989; Traystman, Kirsch and Koehler, 1991) increase much more with lipid peroxidation, cause membrane and neuronal damage with the elevated cytosolic Ca^{2+}.

6.1. Potential Mechanisms for Production of Radicals during Ischemia — Electron Transport Chain

During normal mitochondrial respiration, Cytochrome c is involved in a four-electron transfer to reduce oxygen to water without production of oxygen radicals. During ischemia, when oxygen supply is limited, the electron transport chain of the inner mitochondrial membrane becomes reduced. Oxygen radical production (Traystman *et al.*, 1991) may result. Ubiquinone–cytochrome b region is the major site of radical production.

6.2. Potential Mechanisms for Radical Production during Reperfusion

6.2.1. Free Fatty Acid Metabolites. One source of free radicals in ischemic cells is arachidonic acid, released from membrane phospholipids under the action of Ca^{2+} activated phospholipase A_2. Ischemia leads to a rapid (~30s) increase in arachidonic acid. Arachidonic acid accumulated during ischemia is metabolized upon reperfusion via the lipoxygenase and cyclooxygenase pathways to produce prostaglandins, thromboxanes and superoxide. Oxygen radicals (Traystman *et al.*, 1991) may cause brain injury directly by lipid peroxidation and indirectly by vascular paralysis. Pharmacological agents and hypothermia have been shown to decrease brain free fatty acid accumulation during ischemia. The mechanism for decreased free fatty acids accumulation (Traystman *et al.*, 1991) has been hypothesized to be related to the inhibition of calcium entry into ischemic cells.

6.2.2. Purine Metabolites. With the onset of ischemia, brain adenine nucleotides are metabolized to nucleosides and purine bases. Therefore, cerebral ischemia has been associated with a rapid rise in interstitial concentration of adenosine and hypoxanthine. Brain adenosine concentration increases to 200% of control by 5 s of ischemia and to 500% of control by 60 s of ischemia. Adenosine is metabolised to inosine and hypoxanthine. Hypoxanthine (Traystman *et al.*, 1991) is a substrate for the xanthine oxidase pathway. Under normal contitions, xanthines (Cafe *et al.,* 1993) are oxidized by xanthine dehydrogenase to uric acid and electrons are transferred to NAD^+. Neurons contain high levels of protease Calpain I (Choi, 1990) which undergoes autoproteolytic activation in the presence of high concentrations of free Ca^{2+}. Calpain by breaking down a peptide bond in xanthine dehydrogenase converts it into xanthine oxidase which uses O_2 as electron acceptor during reperfusion liberating superoxide radicals ($O_2^{\cdot-}$). XO/XD ratio in brain microvessels is 4 times greater than in the cortex. For this reason, during reperfusion, superoxide formation by xanthine oxidase (Traystman *et al.*, 1991) is greater in endothelial cells.

Superoxide radicals (Cafe *et al.*, 1993) could be formed also by the co-oxidation of xanthine oxidase with arachidonic acid. It has been demontstrated that linoleic acid, linolenic acid and arachidonic acid (Cafe *et al.*, 1993) cause both significant increases in $O_2^{\cdot-}$ formation and lipid peroxidation.

6.2.3. Nitric Oxide. Ischemia causes an increase in extracellular concentration of excitatory amino acids. Excitatory amino acids bind to and stimulate NMDA receptors within brain and stimulate it. Intracellular Ca^{2+} concentration increased by this and other mechanisms, produces NO from arginine by activating NOS. In normal brain, NO is nontoxic and an important mediator of endothelial-derived relaxation. Howewer, in the presence of superoxide during reperfusion, nitric oxide can lead to the formation peroxynitrite anion. Decomposition of this (Traystman *et al.*, 1991) causes formation of very reactive hydroxyl radical.

6.2.4. Polymorphonuclear Leukocytes. Ischemia leads to activation of leukocytes trapped in the cerebral vasculature which may result in release of chemotactic factors (e.g. leukotrienes) during reperfusion. Leukocytes may interact with platelets to metabolize arachidonic acid and produce lipoxygenase and cyclooxygenase byproducts, including oxygen radicals. Leukocyte accumulation is noted in damaged brain 60 min after cerebral ischemia produced by introducing air emboli into the cerebral circulation and leukocyte accumulation (Traystman *et al.*, 1991) appears to correlate with brain regions having low blood flow.

Activated leukocytes may cause mechanical obstruction of capillaries. However, treatment with antineutrophil serum at the time of reperfusion is not associated with improved postischemic blood flow in rats (Grogaard *et al.*, 1989) or improved neurological recovery in dogs (Schott *et al.*, 1989) despite a reduction in circulating leukocytes. These data suggest that there is an activation of leukocytes trapped in brain during ischemia, wheras postischemic leukocyte depletion has no protective effect. These data also suggest that leukocyte accumulation in brain may impair microvascular circulation, but the presence of leukocyte accumulation is not necessarily the major cause of injury from ischemia and reperfusion. It is also important to recognize that most of the positive evidence of early leukocyte accumulation is based on models using air embolism, which produces significant endothelial damage. The role of leukocyte activation (Traystman *et al.*, 1991) in other models of cerebral ischemia is less clear.

6.2.5. Mononuclear Cells. Two types of mononuclear phagocytes may be important in ischemic injury to brain, the microglia and blood-borne macrophages. Both classes of mononuclear phagocytes have cytotoxic activity and produce oxygen radicals.

6.2.6. Iron and Ascorbic Acid. Areas of the human brain (e.g. globus pallidus and substantra nigra) are rich in iron, yet serebrospinal fluid has no significant iron binding capacity. Depending on the low ferritin content, brain has (Cafe *et al.*, 1993) low iron binding capacity. Most brain iron (Halliwell, 1989) is protein-bound. But little is know of its molecular nature.

An injury to brain cells may result in release of iron ions that can stimulate free radical production. Bleeding and subsequest haemoglobin liberation may have a similar effect. Haemoglobin can be degraded by excess H_2O_2. Catalytic iron ions is released from the haem ring. In addition, haemoglobin reacts with H_2O_2 to form an oxidizing species (probably ferryl) capable of stimulating lipid peroxidation. Thus, haemoglobin. outside the erythrocyte (Halliwell, 1989) is potentially a dangerous protein.

There is a high concentration of ascorbic acid in the grey and white matter of the CNS. Ascorbate concentrations in the cerebrospinal fluid (CSF) is about ten times the plasma level. Neuronal tissue cells have a second transport system that concentrates intracellular ascorbate even more. In the absence of transition metals, ascorbic acid has well established antioxidant properties. But ascorbate/iron and ascorbate/copper mixtures generate free radicals. Most of these events require the presence of O_2. So, it is likely that they will (Cafe *et al.*, 1993; Halliwell, 1989) occur during the reperfusion phase.

Both parenchyma and vascular endothelium produce radicals. Radicals especially superoxide are able to exit cells via anion channels and thereby affect neighboring cell types. Primary damage to endothelium and vascular smooth muscle can result in secondary damage to neurons and astrocytes by impairing oxygen and substrate supply and by producing vasogenic edema. Primary radical damage to neurons and astrocytes can result in cytotoxic edema which can have secondary effects by causing vascular compression. As a result of all these metabolic events, reactive oxygen radicals increase in the lesion sites. This increase cannot be regulated as a result of insufficiency of natural endogen antioxidant defensive molecules. These particules trigger other harmful processes and also accelarate the destruction initiated by lytic enzymes. These events by contributing to the production of free radicals (Traystman *et al.*, 1991) cause a circle.

6.3. The Brain and Nervous System Are Prone to Radical Damage for a Number of Reasons

1. Membrane lipids are very rich in polyunsaturated fatty acid side chains.
2. The brain is poor in catalase, superoxide dismutase and glutathione peroxidase.
3. Some areas of human brain (e.g. globus pallidus and substantia nigra) are rich in iron.

4. Ascorbic acid concentration (Halliwell, 1989) is high in the grey and white matter of the CNS.

6.4. Consequences of Free Radical Induced Damage in Brain

1. Destruction of membranes by lipid peroxidation.
2. Alterations of functiones of membrane proteins (e.g. inhibition of receptors, Na^+/K^+ ATPase).
3. Increase of capillary permeability (formation of vasogenic edema).
4. Destruction of useful cellular enzymes and structures.
5. Activation of lytic enzymes (elastase, protease, phospholipase, lipoxygenase, cyclooxygenase, xanthine oxidase).
6. Damage to mitochondrial oxidative phosphorylation.
7. Destruction of mitochondrial and nuclear DNA.
8. Inhibition of glutamine synthetase. It has been proposed that radicals could inactivate brain glutamine synthetase, the major enzyme responsible for glutamate removal.
9. Activation of poly (ADP–ribose) synthatase. This enzyme polymerises ADP-ribose residues derived from NAD^+ and can lead to depletion of cellular NAD content.
10. Inflammation and leukocyte chemotaxis.
11. Platelet aggregation and adhesion.
12. Break down of extracellular anti-proteolytic enzymes and neurotransmitters.
13. Intracellular K^+ depletion.

Free radicals produced in mitochondria can cause point mutations, DNA cross-link and DNA strand breaks in mitochondrial genes encoding cytochrome b, three large hydrophobic subunits of cytochrome oxidase, two subunits of $F_o–F_1$ ATPase, seven hydrophobic subunits of NADH dehydrogenase and some kinds of mitochondrial RNAs. Damage to mitochondrial genome results in impaired respiration, further increasing the possibility of oxygen radical production. Mitochondrial DNA (Cafe et al., 1993) is particularly susceptible to damage for lack of protective histones or non-histone proteins, limited repair mechanisms. Free radical reactions are a part of normal human metabolism. When produced in excess, radicals can cause tissue injury. However, tissue injury can cause more radical reactions which may contribute to a worsening of the injury. By impairing capillary permeability, they cause penetration of macromolecules and proteins from plasma into interstial fluid and play important role in the development of vasogenic edema.

In addition to free radicals, cytokines, interleukins, TNF, opiates and histamine may contribute to reperfusion injury.

6.5. Struggle against Oxidants

Since, at normal levels, they don't have harmful effects, struggle against reactive oxygen particles is important.

6.6. Stages of Struggle against Oxidants

6.6.1. Avoidance of Events Leading to Free Radical Generation. Determining of the events leading to oxidative stress in the organism and avoidance of these should be the first approach.

6.6.2. Breakdown of Triggered Biochemical Reactions. Second approach is to break-down biochemical reactions triggered by the evenst leading to tissue injury at one or several steps. As a result of biochemical reactions, at the lesion site and surrounding tissue, the amount of oxidants reach to harmful levels. Attempts to breakdown these reaction steps will diminish indirectly the amount of oxidant molecules. Since more than one biochemical reactions are triggered, the attempts to breakdown these reactions will be varied. Approaches for breakdown of triggered metabolic reactions:

1. Inhibition of Na^+ influx
2. Activation of K^+ influx
3. Inhibition of lactic adidosis
4. Inhibition of cytotoxic and vasogenic edema
5. Adenosine/adenosine analogs
6. Accelerators of phospholipid synthesis
7. Ca^{2+} channnel blockers
8. Inhibitors of lipid peroxidation
9. Inhibitors of cyclooxygenase
10. Inhibition of leukotrienes
11. Opiate antagonists
12. PAF inhibitors
13. H_2 receptor antagonists
14. Atrial natiuretic peptide
15. Hypometabolizers
16. Hypothermia

Each one of these approaches is a distinct subject.

6.6.3. Inactivation of Free Radical Generating Cells. Third way of strugggle against oxidants is inhibition of chemotaxy of inflamatuar cells especially neutrophils to lesion site and inhibition of their accumulation. For this purpose antiinflamatuar drugs are proposed to use.

6.6.4. Treatment with Antioxidants. The main struggle against oxidants is antioxidants which inactivate oxidants directly. Antioxidants have four distinct effects:

1. Scavenger effect
2. Quencher effect
3. Repair effect
4. Chain breaking effect

The effect of antioxidants which may result from their reaction with oxidants converting them to new weaker molecules is called their "Scavenging effect". Natural enzymes having scavenger effect exist usually in cytoplasm and mitochondria and extracellular compartment.

The effect of antioxidants which inactivate oxidants by transferring a hydrogen atom is called "Quencher effect". Vitamines, flavonoids, trimetazidine, mannitol and antocyanidins are in this group.

Hemoglobin, ceruloplasmin, heavy minerals bind oxidants, break their chains and exhibit chain breaking effect.

Antioxidants (Halliwell, 1991) are classified into two groups as natural and synthetic (Table 1).

Table 1. Antioxidants

Natural antioxidants	Drugs
Enzymes	Recombinant h-SOD
Superoxide dismutase	Ebselen
Catalase	21-amino steroids (lazaroids)
Glutathione peroxidase	Iron chelators
Hydroperoxidase	Cytokines
Cytochrome-C-oxidase	TNF and interleukin-I
Macromolecules	Xanthine oxidase inhibitors
Ceruloplasmine	Allopurinol
Transferrin	Mannitol
Ferritin	Barbiturates
Hemoglobin	Flavonoids
Myoglobin	
Micromolecules	
Vitamin E and analogues	
Vitamin C	
Thiols	
Glutathione	
N-acetyl-cysteine	
Methionine	
Captopril	
Vitamin A-beta carotene	
Urates-Uric acid	
Ubiquinone	
Glucose	
Bilirubin	

A lot of studies are being performed to diminish and prevent neuronal damage occuring during ischemia by treatment with antioxidants. Some of them (Clemens, Saunders, Ho, Phebus, and Panetta, 1993) yielded positive results. Some of the results are show in Table 2. However, this strategy (Chan, 1996) entails problems(hemodynamic, pharmacokinetic, toxicity, blood-brain barrier permeability, etc.) that may cloud the data interpretation. To eliminate such difficulties, transgenic and knockout mice should be used as an alternative approach. Transgenic and knockout mutants that either overexpress or are deficient in antioxidant enzyme/protein levels have been successfully produced. It has been (Chan, 1996; Chan, Epstein, Li, Huang, Carlson, Kinouchi, Yang, Kamii, Mikawa and Kondo, 1995) shown that an increased level of CuZn-superoxide dismutase (SOD-1) and antiapoptotic protein Bcl-2 in the brains of transgenic mice protects neurons from ischemic/reperfusion injury, whereas a deficiency in CuZn-superoxide dismutase or mitochondrial Mn-superoxide dismutase (SOD-2) exacerbates ischemic brain damage.

7. NITRIC OXIDE

Nitric oxide (NO) which was defined as the molecule of the year in 1992 in Science (Snyder, 1992) has been implicated recently in the pathophysiology of focal cerebral ischemia. Nitrogen monoxide (NO) (Balkan, Balkan, Özben, Serteser, Gümüşlü and Oguz, 1997b; Buisson, Margaill, Callebert, Plotkine and Boulu, 1993; Dalkara and Moskowitz, 1994a; Kader, Frazzini, Solomon and Trifiletti, 1993; Moskowitz and Dalkara, 1996; Samdani, Dawson and Dawson, 1997; Stuhlmiller and Boje, 1995) has recently emerged as an

Table 2. Effects of some antioxidants in various cerebral ischemia models

Antioxidant	Model	Species	Effect	Reference
MCI-186	dMCAO	Rat	↓ Infarct volume	Kawai (1997)
S-allyl cysteine (SAC)	pMCAO	Rat	↓ Infarct volume	Numagami (1996)
α-Lipoic acid	tBCAO	Rat	↑ Survive	Panigrahi (1996)
Tirilazad	pMCAO	Rat	↓ Infarct volume	Park (1994)
LY 231617	Global	Cat	↓ Infarct volume	Clemens (1993)
MCI-186	Global	Rat	Anti-ischemic	Watanabe (1993)
SOD	MCAO+BCAO	Cat	↓ Infarct volume	Matsumiya (1991)
Allopurinol	tBCAO	Rat	↓ Lipid peroxidation	Vanella (1991)
α-Tocopherol	tBCAO	Rat	↓ Lipid peroxidation	Vanella (1991)
Deferoxamine	tBCAO	Rat	↓ Lipid peroxidation	Vanella (1991)
U-78517F	UCAO	Rat	↓ Cortical necrosis	Hall (1991)
Estrogen	UCAO	Gerbil	↓ Cortical necrosis rCBF	Hall (1991)
Ebselen	tMCAO	Cat	↓ Ischemic edema	Johshita (1990)
SOD	MCAO	Rat	↓ Infarct volume	Chan (1990)
Dimethylthiourea	MCAO	Rat	↓ Ischemic edema	Marz (1990)
Ebselen	tMCAO	Cat	↓ Ischemic edema	Johshita (1990)
Ascorbate	MCAO	Rat	↓ Lipid peroxidation	Kinuta (1989)
Catalase	MCAO	Rat	↓ Infarct volume	Liu (1989)
ONO-3144	tMCAO	Cat	↑ rCBF, ↓ Cerebral edema	Johshita (1989)
Tirilazad	tMCAO	Rat	NC in brain edema	Baughman (1989)
Tirilazad	pMCAO	Rat	↓ Brain edema	Young (1988)
Tirilazad	tMCAO	Gerbil	↑ Survive	Hall (1988)
Tirilazad	Global	Dog	↑ rCBF	Natale (1988)
Tirilazad	Global	Cat	↑ pCBF, ↓ Injury volume	Hall (1988)
MCI-186	tMCA	Cat	↓ Cerebral edema	Abe (1988)
DMSO	pMCAO	Rat, Cat	NC in cerebral edema	Little (1981)

important mediator of cellular and molecular events which impact the pathophysiology of cerebral ischemia.

In CNS, NO (Anile, Maira, Iannone, Corte, Mangiola, and Nistico, 1993; Hanbauer, 1993; Moncada, Lekieffre, Arvin, and Meldrum, 1993; Samdani *et al.*, 1997) is a neuronal messenger and mediator having both neurotoxic and neuromodulator effects.

Nitric oxide (NO) was first identified as endothelium derived relaxing factor. It is also (Dalkara and Moskowitz, 1994a) involved in NMDA glutamatergic neurotransmission.

NO is (Bories and Bories, 1995; Dawson, Dawson, London, Bredt and Snyder, 1991; Springall, Suburo, Bishop, Merrett, Moncada, and Polak, 1992) synthesized from the guanido nitrogen of L-arginine and molecular oxygen by nitric oxide synthase (NOS; EC 1,14,13,39). The end products are citrulline and NO. Catalytic conversion of L-arginine to nitric oxide by cytosolic isoform of brain No synthase (Mittal, C. K., 1993) requires superoxide ion, hydrogen peroxide and possibly hydroxyl radical. Two mechanisms (Bredt and Snyder, 1990) have been proposed for NO formation:

1. Hydroxylation of guanido nitrogen (Marletta, 1988)
2. Deimination followed by oxidation of ammonia (Hibbs, 1987)

7.1. Isoforms of Nitric Oxide Synthase

Three isoforms of NOS (Archer, 1993; Bories and Bories, 1995; Dalkara and Moskowitz, 1994a) have been defined and their cDNAs have been isolated in humans.

1. Isozyme I. Constitutive neuronal isoform (cNOS) is present in neurons and epithelial cells and is Ca^{2+}-Calmodulin dependent.
2. Isozyme II. Inducible isoform (iNOS) which is present in macrophages and glial cells.
3. Isozyme III. Endothelial isoform (eNOS) which is present in endothelial cells.

They are coded by three distinct genes located on chromosomes 12, 17 and 7. Amino acid sequence of human isozyms show less than 59% identication. Between species, amino acid sequence of each isoform has been preserved better (more than 90% for isoform I and III and more than 80% for isoform II). All isoforms use L-arginine and molecular oxygen as substrates and NADPH, tetrahydrobiopterin, FAD and FMN as cofactors. All of them (Archer, 1993; Bories and Bories, 1995; Dalkara and Moskowitz, 1994a) bind calmoduline and contain hem.

The constitutive isoform (cNOS) is Ca^{2+}–Calmodulin dependent. It binds calmodulin loosely, When cytosolic Ca^{2+} is increased, it causes firm binding of calmodulin and NOS, thereby producing NO. An increase in intracellular calcium concentration from 100 nM to 500 nM (Dalkara and Moskowitz, 1994a) changes the rate of NO synthesis from < 5% to >95% of maximum. Besides synthesizing nitric oxide (NO), purified neuronal NO synthase (nNOS) (Xia, Dawson, Dawson, Snyder, and Zweier, 1996) can produce superoxide at lower L–Arg concentrations. Regulating arginine levels may provide a therapeutic approach to disorders involving $O_2^{\cdot -}$/NO-mediated cellular injury.

Inducible isoform (iNOS) is calcium insensitive and is stimulated by endotoxins, cytokines, interferon γ and lipopolysaccharides. iNOS binds calmodulin tightly. For this reason, it doesn't require the presence of Ca^{2+}.

By inhibiting iron containing enzymes in parasitic target cells and causing DNA fragmentation, iNOS exerts its cytotoxic effects. iNOS induction occurs in otoimmun diseases and septic shock. iNOS (Bories and Bories, 1995; Dalkara and Moskowitz, 1994a) is present in glial cells in CNS.

Endothelial isoform (eNOS) is present in endothelial cells. It is expressed constitutively. But its expression can be enhanced. Its activity is regulated by Ca^{2+}-Calmodulin. NO originating from endothelium causes dilatation of blood vessels, inhibition of platelet and leucocyte adhesion. It, also, inhibits proliferation of vascular smooth muscle.

NO is produced in neurons, glia cells and vascular endothelium in central nervous system. Depending on their origin, their effects are varied. Neuronal NO increases acute ischemic damage. While, vascular NO, (Dalkara and Moskowitz, 1994a) diminishes ischemic injury by increasing cerebral blood flow. Cerebral ischemia is followed by a local inflammatory response that is thought to participate in the extension of the tissue damage occurring in the postischemic period. It has been reported (Iadecola, Zhang and Xu, 1995b; Iadecola, Zhang, Xu, Casey, and Ross, 1995c) that inhibition of iNOS reduces infarct volume in focal cerebral ischemia and this indicates that NO production may play an important pathogenic role in the progression of the tissue damage that follows cerebral ischemia.

7.2. Properties of Nitric Oxide

NO is lipid soluble. Its half life (Habu, Yokoi, Kabuto, and Mori, 1994; Beckman and Koppenol, 1992) is very short (4 seconds). It (Beckman, Beckman, Chen and Marshall, 1989) reacts rapidly with superoxide. Hemoglobin and other hem proteins inhibit NO by binding it tightly. Rapid metabolism of superoxide by superoxide dismutase and removal o hemoglobin prolong its half life.

Formation of S-nitrosothiol adducts may stabilize the labile NO radical and prolong its biological half life. Sulfhydryl groups in proteins (e.g., albumin) (Dalkara and Moskowitz, 1994a) represent a rich source of reduced thiol and S-nitroso proteins are formed readily under physiological conditions. NO binds to metals (Fe, Cu, Co, Mn, etc.) within cells. These metal ions (Archer, 1993) are required for the activity of cytochrome oxidases.

NO (Dalkara and Moskowitz, 1994a) activates heme containing guanylate cyclase by forming an NO-Heme complex. Elevations in CGMP result. However CGMP dependent mechanisms may only be operational in cells adjacent to those generating NO. Because intracellular calcium levels sufficient to activate NOS (Dalkara and Moskowitz, 1994; Knowles, Palacios, Palmer and Moncada, 1989) inhibit guanylate cyclase. Blood vessels may provide an example. Vasodilatory neuromediators like acetylcholine or bradykinin raise intracellular calcium and stimulate endothelial NOS activitiy to release NO or a related nitrosothiol (EDRF) from endothelial cells. NO, in turn, enhances cGMP synthesis in smooth muscle cells. Nitrovasodilators such as nitroglycerin, nitroprusside or 3-morpholinosynonimine (SIN-1) (Dalkara and Moskowitz, 1994a; Knowles and Moncada, 1992) raise cGMP in vascular smooth muscle by directly liberating or donating NO.

NO (Dalkara and Moskowitz, 1994a; Stamler, Singel and Loscalzo, 1992) can exist in three redox forms:

1. nitrosonium (NO^+)
2. nitric oxide ($NO^.$)
3. nitroxyl anion (NO^-)

Nitric oxide ($NO^.$) activates guanylate cyclase and is neurotoxic. $NO^.$ was proposed as the neurotoxic agent mediating NMDA toxicity. Whereas, Nitrosonium (NO^+) form of NO reacts with the thiol group of NMDA receptor, down regulates NMDA receptor activity and blocks its function. So, it is neuroprotective. NO donors may exert different actions depending upon the redox form of NO generated. Sodium nitroprusside protects neurons because it generates Nitrosonium (NO^+) which can be converted to the toxic $NO^.$ after reduction with cystein. However, whether $NO^.$ or NO^+ (Dalkara and Moskowitz, 1994a) predominates in tissue remains to be determined.

7.3. Nitrogen Monoxide (NO) Mediated Neurotoxicity

NO and its degradation products (Dalkara and Moskowitz, 1994a; Stamler *et al.*, 1992) cause cytotoxicity through formation of Iron–NO complexes with several enzymes including Complex I and II of mitochondrial electron transport, oxidation of protein sulfhydryls and DNA nitration. NO-mediated ADP-ribosylation inhibits glyceraldehyde-3-phosphate dehydrogenase. $NO^.$ contains an unpaired electron and is paramagnetic. It rapidly (Kumura, Yoshimine, Iwatsuki, Yamanaka, Tanaka, Hayakawa, Shiga and Kosaka, 1996) reacts with superoxide ($O_2^{.-}$) to form peroxynitrite anion ($ONOO^-$). Peroxynitrite (Beckman, 1991) can diffuse for several micrometers before decomposing to form the powerful and cytotoxic oxidants hydroxyl radical and nitrogen dioxide.It is stable in alkaline solutions. But, it decays rapidly once protonated.

$$O_2^{.-} + NO^. \longrightarrow ONOO^- + H^+ \longrightarrow ONOOH \longrightarrow OH^. + NO_2^. \longrightarrow NO_3^. + H^+$$

The half life of peroxynitrite at pH 7.4 is 1.9 seconds. SOD, preventing peroxynitrite formation (Beckman *et al.*, 1989; Dalkara and Moskowitz, 1994a) protects tissues.

7.4. Nitrogen Monoxide (NO) and Focal Ischemia

NO has been implicated in the pathophysiology of focal cerebral ischemia. Within 3 to 24 minutes after middle cerebral artery (MCA) occlusion, NO (Dalkara and Moskowitz, 1994a; Kumura, Kosaka, Shiga, Yoshimine and Hayakawa, 1994) increases dramatically from 10 nM to 2.2 μM within cortex as (Rice-Evans, Diplock, and Symons, 1991) detected by a porphyrinic microsensor (Malinski, Patton, Grunfeld, Kubaszewski, Pierchala, Kiechle and Radomski, 1994).

Brain NOS activity increases as well. Nitrite, NO and NOS activity return to baseline within one hour. cNOS activity (Dalkara and Moskowitz, 1994a; Dawson *et al.*, 1992; Kader *et al.*, 1993; Kumura *et al.*, 1994) increases due to a rise in intracellular Ca^{2+}–Calmodulin complex. In ischemic tissues, rapid activation of NOS is followed by its rapid inactivation . The mechanisms responsible for these effects (Dalkara and Moskowitz, 1994a; Kader *et al.*, 1993) may be:

1. Focal ischemia leads to increased synaptic glutamate. Increased synaptic glutamate might increase postsynaptic calcium influx via NMDA receptors. Increased Ca^{2+}–Calmodulin binding (Iadecola, Xu, Zhang, el-Fakahany and Ross, 1995a) activates NOS.
2. Brain NOS (Brüne and Lapetina, 1992) has two distinct serine groups that can be phosphorylated by different protein kinases leading to decreased activity of the enzyme. Rapid depletion of ATP is coupled to NOS dephosphorylation which causes activation of the enzyme.
3. Recently, Ca^{2+}–Calmodulin regulated phosphatase, calcineurin has been demonstrated to dephosphorylate NOS, thereby activating it.
4. The immunomodulatory drug FK-506 inhibits calcineurin and potently inhibits glutamate toxicity in cell culture. This implies that NOS activity in vivo is, at least, in part regulated by phosphorylation and dephosphorylation.
5. NOS produces highly oxidizing NO. NOS requires numerous highly reduced cofactors (NADPH, reduced tetrahydrobiopterin, reduced flavin adenine dinucleotide, reduced flavin mononucleotide). Rapid NO production might inactivate NOS.
6. Another possible inactivation mechanism involves the NMDA receptor. NO oxidizes a critical sulfhydry group in the NMDA receptor which might decrease calcium influx and thereby reduce NOS activity.
7. Finally, activation of a peptidase in ischemic tissue causes NOS deactivation.

Neurons, astrocytes, perivascular nerves and cerebrovascular endothelium (Dalkara and Moskowitz, 1994a) may form NO during cerebral ischemia. Following ischemia, an increased NO production in vascular endothelium or perivascular nerves improves blood flow and is neuroprotective. Infusion of L-arginine (Anile, Maira, Iannone, Della Corte, Mangiola and Nistico, 1993; Dalkara and Moskowitz, 1994a) which dilates pial vessels increases rCBF and reduces infarct size and causes functional recovery. It has been reported (Kader, Frazzini, Baker, Solomon, and Trifiletti, 1994) that mild hypothermia (33 degrees C) modulates the burst of nitric oxide synthesis during cerebral ischemia and may account, at least partially, for its cerebroprotective effects.

NO binding protein, guanylate cyclase, appears more robust and resistant to the effects of ischemia than NOS. Hence, administering NO donors (Dalkara and Moskowitz, 1994a) leads to blood flow increases within the ischemic tissue for at least one hour after arterial occlusion and a decrease in infact size in models of focal ischemia. One may speculate that NO production by perivascular nerves (Anile *et al.*, 1993; Dalkara and Moskowitz, 1994a) improve blood flow in a zone of hypoperfusion. Although the ameliorative effects of raising blood flow within ischemic tissues are known for 10 years, a common treatment strategy hasn't been developed yet. There are several arguments against this. Increasing rCBF promotes edema formation and free radical generation during reperfusion and normal tissue may steal blood from the ischemic zone. Recent findings with NO donors and L-arginine and preliminary data following thrombolytic therapy in stroke patients strongly suggest that early restoration of rCBF can increase tissue survival and restore function. These events (Dalkara and Moskowitz, 1994a) are more critical in ischemic penumbra. Parasympathetic fibers contain NOS. Infarction volume caused by MCA occlusion increases after chronic parasympathectomy.

7.5. L-Arginine Analogs and Nitric Oxide Sythase (NOS) Inhibition

The most widely studied inhibitors are:

1. Nitro-L-arginine methyl ester (L-NAME)
2. Nitro-L-arginine (L-NA)
3. Monomethyl-L-arginine (L-NMMA)

L-arginine analogs (Anderson and Meyer, 1996; Nowicki, Carreau, Duval, Vige and Scatton,1993; Regli, Held, Anderson, Meyer, Thoralf and Sundt, 1996) are competitive inhibitors of NOS.

NOS inhibitors have been administered systemically (intraperitoneal, intravenous), directly into tho brain (microdialysis), topically by superfusion (cranial window preparation) or into the cerebrospinal fluid (intracerebroventricular). There are conflicting results in literature (Dalkara and Moskowitz, 1994a; Dalkara, Yoshida, Irikura and Moskowitz, 1994b; Dawson, 1994; Yamamoto, Golanov, Berger and Reis, 1992) reporting either increases or reductions in infarct volume after permanent or transient MCA occlusion and NOS inhibition (Table 3).

The reasons that may underlie some of the existing controversies:

1. Use of non-selective NOS inhibitors.
2. Choice of species, anesthetics or ischemic models.

Although, each of them may contribute to these contradictory results, none of them is sufficient to explain the observed differences. For example, anesthetics like halothan and isoflurane are reported to increase pial arterial diameter with an NO-mediated, yet unknown mechanism.

The choice of model is also important. For example, injury in the hypoxia-ischemic neonatal rat may be more dependent upon excitotoxic mechanisms and less on reductions in blood flow.

Various species and strains possess different susceptibilities to stroke (level of blood pressure, differences in endothelial NOS activity, extent of collateral blood flow).

The effect of inhibition of nitric oxide synthesis on cerebral ischemic damage (Zhang, Xu, and Iadecola, 1995) may vary depending on the timing of the inhibition relative to the induction of ischemia.

Severe hypertension caused by NOS inhibitors (Dalkara and Moskowitz, 1994a) may also potentiate edema formation and cause bleeding into the core territory. L-NAME has been shown (Anderson and Meyer, 1996; Regli *et al.*, 1996) to prevent intracellular brain acidosis during focal cerebral ischemia independent from regional cortical blood flow changes.

Table 3. Effects of nitric oxide synthase inhibitors on focal cerebral ischemia

Species	Reference	Model	Duration	Inhibitor	Infarct volume
Decreases in infarct volume					
Rat	Matsui, 1997	tMCAO	1,2,4 hour	L-NA	↓
Mice	Kamii, 1996	tMCAO	1 day	7-NI	↓
Rat	Iuliano, 1995	tMCAO	1 day	L-NAME	↓
Rat	Iadecola, 1995	pMCAO	4 days	Aminoguanidin	↓
Rat	Quast, 1995	tMCAO	2 hours	L-NAME	↓
Rat	Ashwal, 1994	tMCAO	5 hour	L-NAME	↓
Cat	Nishikawa, 1994	tMCAO	4 hours	L-NAME	↓
Mouse	Carreau, 1994	pMCAO	6 days	L-NA	↓
Rat	Morikawa, 1994	pMCAO	1 day	L-arginine	↓
Rat	Zhang, 1993	pMCAO	1 hour	SNP	↓
Rat	Ashwal, 1993	tMCAO	5 hour	L-NAME	↓
Cat	Nishikawa, 1993	pMCAO	4 hours	L-NAME	↓
Rat	Ashwal, 1993	tMCAO	3 hours	L-NAME	↓
Rat	Morikawa, 1992	pMCAO	1 day	L-arginine	↓
Rat	Nagafuji, 1992	pMCAO	3 days	L-NA	↓
Rat	Buisson, 1992, 1993	pMCAO	2 days	L-NAME	↓
Rat	Trifiletti, 1992	tUCAO	2.5 hours	L-NA	↓
Mouse	Nowicki, 1991	pMCAO	7 days	L-NA	↓
Increases in infarct volume					
Mice	Kamii, 1996	tMCAO	1 day	L-NAME	↑
Sheep	Marks, 1996	tCI	3 days	L-NNA	↑
Rat	Zhang, 1995	pMCAO	1 day	L-NAME	↑
Rat	Hamada, 1995	pMCAO	7 days	L-NAME	↑
Rat	Kuluz, 1993	tMCAO	3 days	L-NAME	↑
Rat	Zhang, 1993	pMCAO	1 day	L-NAME	↑
Rat	Morikawa, 1993	pMCAO	1 day	L-NA	↑
Rat	Morikawa, 1993	pMCAO	2 days	L-NAME	↑
Rat	Morikawa, 1993	pMCAO	1 day	L-NA	↑
Rat	Kuluz, 1993	tMCAO+UCAO	2 hours	L-NAME	↑
Rat	Zhang, 1993	pMCAO	1 hour	L-NAME	↑
Rat	Dawson, 1992	pMCAO	4 hours	L-NA	↑
Rat	Yamamato, 1992	pMCAO	1 day	L-NA	↑

The effects of NO (Iadecola *et al*, 1995b; Zhang and Iadecola, 1995) may vary with the tissue compartment (e.g., vascular versus parenchymal) or with the time after the onset of ischemia. While endothelial and perivascular NO have neuroprotective roles, parenchymal NO exerts neurotoxic effects. Pharmacological and genetic approaches (Samdani *et al.*, 1997) have significantly advanced our knowledge regarding the role of NO and the different NOS isoforms in focal cerebral ischemia. nNOS and iNOS play key roles in neurodegeneration, while eNOS plays a prominent role in maintaining cerebral blood flow and preventing neuronal injury. Excitotoxic or ischemic conditions excessively activate nNOS, resulting in concentrations of NO that are toxic to surrounding neurons. Conversely, NO generated from eNOS is critical in maintaining cerebral blood flow and reducing infarct volume. iNOS which is not normally present in healthy tissue, is induced shortly after ischemia and contributes to secondary late-phase damage. The protocol chosen for experiments may give rise to contradictory results. Some answers may be forthcoming with the development of selective inhibitors (Dalkara and Moskowitz, 1994a; Hamada, Greenberg, Croul, Dawson and Reivich, 1995; Huang, Huang, Panahian, Dalkara, Fishman and Moskowitz, 1994) for the neuronal or vascular isoforms, or transgenic mice whose neuronal or endothelial NOS are selectively knocked-out. A clearer appreciation of the potential therapeutic utility of NO synthase inhibitors will emerge only when the complexity of their effects on the extent of ischemic damage in vivo is more fully defined and understood.

REFERENCES

Anderson, R. E., and Meyer, F. B., 1996, Nitric oxide synthase inhibition by L-NAME during repetitive focal cerebral ischemia in rabbits, *Am. J. Physiol.* **271**(2 Pt 2): H588–H594.

Anile, C., Maira, G., Iannone, M., Corte, F. D., Mangiola, A., and Nistico, G., 1993, Role of nitric oxide in the regulation of cerebral blood flow under normal conditions and in brain ischaemia in pigs, in *Nitric Oxide: Brain and Immune System,* (S. Moncada, G. Nistico, and E. A. Higgs, eds.), pp. 143–149, Portland Press, London.

Archer, S., 1993, Measurement of nitric oxide in biological models, FASEB J. **7**:349–360.

Baker, C. J., Fiore, A. J., Frazzini, V. I., Choudhri, T. F., Zubay, G. P., and Solomon, R. A., 1995, Intraischemic hypothermia decreases the release of glutamate in the cores of permanent focal cerebral infarcts, *Neurosurgery,***36**(5): 994–1001.

Balkan, E., Balkan, S., Özben, T., Serteser, M., Gümüşlü, S., and Oguz, N., 1997a, The effects of nitric oxide synthase inhibitor, L-NAME on NO production during focal cerebral ischemia in rats: could L-NAME be the future treatment of sudden deafness?, *Inter. J. Neuroscience* **89**: 61–67.

Balkan, S., Özben, T., Balkan, E., Oguz, N., Serteser, M., and Gümüşlü, S., 1997b, Effects of Lamotrigine on brain nitrite and cGMP levels during focal cerebral ischemia in rats, *Acta Neurol. Scand.* **95**:140–146.

Beckman, J. S., 1991, The double-edged role of nitric oxide in brain function and superoxide-mediated injury, *J. Dev. Physiol.* **15**(1): 53–59.

Beckman, J. S., Beckman, T. W., Chen, J., Marshall, P. A., and Freeman, B. A., 1990, Apparent hydroxyl radical production by peroxynitrite: implications for endothelial injury from nitric oxide and superoxide, *Proc. Natl. Acad. Sci., USA* **87**: 1620–1624.

Beckman, J. S., and Koppenol, W., 1992, Why is the half-life of nitric oxide so short?, in: *The Biology of Nitric Oxide, Enzymology, Biochemistry, and Immunology,* (S. Moncada, M. A. Marletta, J. B. Hibbs, and E. A. Higgs, eds.), pp. 131, Portland Press, London.

Betz, A. L., Schielke, G. P., and Yang, G. Y., 1996, Interleukin-1 in cerebral ischemia, *Keio J. Med.* **45**(3): 230–237.

Bories, P. N., and Bories, C., 1995, Nitrate determination in biological fluids by an enzymatic one-step assay with nitrate reductase, *Clin. Chem.* **41**(6): 904–907.

Bowersox, S. S., Singh, T., and Luther, R. R., 1997, Selective blockade of N-type voltage-sensitive calcium channels protects against brain injury after transient focal cerebral ischemia in rats, *Brain Res.* **747**(2): 343–347.

Braquet, P., Spinnewyn, B., Demerle, C., Hosford, D., Marcheselli, V., Rossowska, M., and Bazan, N. G., 1989, The role of platelet-activating factor in cerebral ischemia and releted disorders, *Ann. NY Acad. Sci.,***1559**: 296–312.

Bredt, D. S., and Snyder, S. H., 1990, Isolation of nitric oxide synthetase, a calmoduline-requiring enzyme, *Proc. Natl. Acad. Sci., USA* **87**: 682–685.

Brüne, B., and Lapetina, E. G., 1992, Specific phosphorylation of nitric oxide synthase by protein kinase A, in: *The Biology of Nitric Oxide, Enzymology, Biochemistry, and Immunology,* (S. Moncada, M. A. Marletta, J. B. Hibbs, and E. A. Higgs, eds.), pp. 132, Portland Press, London.

Buisson, A., Margaill, I., Callebert, J., Plotkine, M., and Boulu, R. G., 1993, Mechanisms involved in the neuroprotective activity of a nitric oxide synthase inhibitor during focal cerebral ischemia, *J. Neurochem.* **61**(2): 690–696.

Cafe, C., Torri, C., and Marzatico, F., 1993, Cellular and molecular events of ischemic brain damage, *Funct. Neurol.* **8**(2): 121–133.

Chan, P. H., 1996, Role of oxidants in ischemic brain damage, *Stroke* **27**(6): 1124–1129.

Chan, P. H., Epstein, C. J., Li, Y., Huang, T. T., Carlson, E., Kinouchi, H., Yang, G., Kamii, H., Mikawa, S., and Kondo, T., 1995, Transgenic mice and knockout mutants in the study of oxidative stress in brain injury, *J. Neurotrauma* **12**(5): 815–824.

Chen, S. T., Hsu, C. Y., Hogan, E. L., Maricq, H., and Balentine, J. D., 1986, A model of focal ischemic stroke in the rat: reproducible extensive cortical infarction, *Stroke* **17**(4):738–743.

Choi, D. W., 1988, Calcium-mediated neurotoxicity: relationship to specific channel types and role in ischemic damage, *Trends in Neuroscience* **11**: 465–469.

Choi, D. W., 1990, Cerebral hypoxia: some new approaches and unanswered questions, *J. Neurosci.* **10**(8): 2493–2501.

Choi, D. W., 1994, Calcium and excitotoxic neuronal injury, in: *Calcium Hypothesis of Aging and Dementia*, Volume 747 (J. F. Disterhoft, W. H. Gispen, J. Traber, and Z. S. Khachaturian, eds.), pp. 162–171, Annals of the New York Academy of Sciences, New York.

Clemens, J. A., Saunders, R. D., Ho, P. P., Phebus, L. A., and Panetta, J. A., 1993, The antioxidant LY231617 reduces global ischemic neuronal injury in rats, *Stroke* **24**: 716–723.

Dalkara, T., and Moskowitz, M. A., 1994a, The complex role of nitric oxide in the pathophysiology of focal cerebral ischemia, *Brain Pathol.* **4**: 49–57.

Dalkara, T., Yoshida, T., Irikura, K.,and Moskowitz, M. A., 1994b, Dual role of nitric oxide in focal cerebral ischemia, *Neuropharmacology* **33**(11): 1447–1452.

Dawson, D. A., 1994, Nitric oxide and focal cerebral ischemia: multiplicity of actions and diverse outcome, Cerebrovasc. *Brain Metab. Rev.* **6**(4): 299–324.

Dawson, V. L., Dawson, T. M., London, E. D., Bredt, D. S., and Snyder, S. H., 1991, Nitric oxide mediates glutamate neurotoxicity in primary cortical cultures, *Proc. Natl. Acad. Sci, USA* **88**: 6368–6371.

Dawson, T. M., Dawson, V. L., and Snyder, S. H., 1992, A novel neuronal messenger in brain: The free radical, nitric oxide. *Ann. Neurol.* **32**: 297–311.

Escuret, E., 1995, Cerebral ischemic cascade, *Ann. Fr. Anesth. Reanim.* **14**(1): 103–113.

Garthwaite, J., 1993, Nitric oxide signalling in the nervous system, in *Nitric Oxide: Brain and Immune System*, (S. Moncada, G. Nistico, and E. A. Higgs, eds.), pp. 85–97, Portland Press, London.

Germano, I. M., Bartkowski, H. M., Cassel, M. E., and Pitts, L. H., 1987, The therapeutic value of nimodipine in experimental focal cerebral ischemia, *J. Neurosurg.* **67**: 81–87.

Grogaard, B., Schürer, L., Gerdin, B., and Arfors, K. E., 1989, Delayed hypoperfusion after incomplete forebrain ischemia in the rat. The role of polymorphonuclear leukocytes, *J. Cereb. Blood Flow Metab.* **9**: 500–505.

Habu, H., Yokoi, I., Kabuto, H., and Mori, A., 1994, Application of automated flow injection analysis to determine nitrite and nitrate in mouse brain, *NeuroReport* **5**(13): 1571–1573.

Halliwell, B., 1989, Oxidants and the central nervous system: some fundamental questions, *Acta Neurol. Scand.* **126**: 23–33.

Halliwell, B., 1991, Drug antioxidant effects, *Drugs* **42**(4): 569–605.

Hamada, J., Greenberg, J. H., Croul, S., Dawson, T. M., and Reivich, M., 1995, Effects of central inhibition of nitric oxide synthase on focal cerebral ischemia in rats, *J. Cereb. Blood Flow Metab.* **15**(5): 779–786.

Hanbauer, I., 1993, The role of nitric oxide in neurotransmitter release, in *Nitric Oxide: Brain and Immune System*, (S. Moncada, G. Nistico, and E. A. Higgs, eds.), pp. 135–143, Portland Press, London.

Hansen, A. J., 1995, The importance of glutamate receptors in brain ischemia, in: *Neurochemistry in Clinical Applications*, Volume 363 (L. C. Tang, and S. J. Tang, eds.), pp. 123–131, Plenum Press, New York.

Huang, Z., Huang, P. L., Panahian, N., Dalkara, T., Fishman, M. C., and Moskowitz, M. A., 1994, Effects of cerebral ischemia in mice deficient in neuronal nitric oxide synthase, *Science* **265**(5180): 1883–1885.

Iadecola, C., Xu, S., Zhang, F., el-Fakahany, E. E., and Ross, M. E., 1995a, Marked induction of calcium-independent nitric oxide synthase activity after focal cerebral ischemia, *J. Cereb. Blood Flow Metab.* **15**(1): 52–59.

Iadecola, C., Zhang, F., and Xu, S., 1995b, Inhibition of inducible nitric oxide synthase ameliorates cerebral ischemic damage, *Am. J. Physiol.* **268**(1 Pt 2): R286–292.

Iadecola, C., Zhang, F., Xu, S., Casey, R., and Ross, M. E., 1995c, Inducible nitric oxide synthase gene expression in brain following cerebral ischemia, *J. Cereb. Blood Flow Metab.* **15**(3): 378–384.

Jiang, N., Kowaluk, E. A., Lee, C. H., Mazdiyasni, H., and Chopp, M., 1997, Adenosine kinase inhibition protects brain against transient focal ischemia in rats, *Eur. J. Pharmacol,* **320**(2–3): 131–137.

Kader, A., Frazzini, V. I., Solomon, R. A., and Trifiletti, R. R., 1993, Nitric oxide production during focal cerebral ischemia in rats, *Stroke* **24**:1709–1716.

Kader, A., Frazzini, V. I., Baker, C. J., Solomon, R. A., and Trifiletti, R. R., 1994, Effect of mild hypothermia on nitric oxide synthesis during focal cerebral ischemia, *Neurosurgery*, **35**(2): 272–277.

Kamii, H., Mikawa, S., Murakami, K., Kinouchi, H., Yoshimoto, T., Reola, L., Carlson, E., Epstein, C. J., and Chan, P. H., 1996, Effects of nitric oxide synthase inhibition on brain infarction in SOD-1 transgenic mice following transient focal cerebral ischemia, *J. Cereb. Blood Flow Metab.* **16**(6): 1153–1157.

Kawaguchi, K., and Graham, S. H., 1997, Neuroprotective effects of the glutamate release inhibitor 619C89 in temporary middle cerebral artery occlusion, *Brain Res.* **749**(1): 131–134.

Kawai, H., Nakai, H., Suga, M., Yuki, S., Watanabe, T., and Saito, K. I., 1997, Effects of a novel free radical scavenger, MCI-186, on ischemic brain damage in the rat distal middle cerebral artery occlusion model, *J. Pharmacol. Exp. Ther.* **281**(2): 921–927.

Knowles, R. G., and Moncada, S., 1992, Nitric oxide as a signal in blood vessels, *Trends Biochem. Sci.* **17**: 399–402.

Knowles, R. G., Palacios, M., Palmer, R. M. J., and Moncada, S. 1989, Formation of nitric oxide from L-arginine in the central nervous system: A transduction mechanism for stimulation of the soluble guanylate cyclase, *Proc. Natl. Acad. Sci. USA* **86**: 5159–5162.

Kogure, K., 1990, Trans-NMDA receptor signalling in ischemia-induced brain cell damage, *Trans. Am. Soc. Neurochem* **21**: 185–192.

Kumura, E., Kosaka, H., Shiga, T., Yoshimine, T., and Hayakawa, T.,1994, Elevation of plasma nitric oxide end products during focal cerebral ischemia and reperfusion in the rat, *J. Cereb. Blood Flow Metab.* **14**(3): 487–491.

Kumura, E., Yoshimine, T., Iwatsuki, K. I., Yamanaka, K., Tanaka, S., Hayakawa, T., Shiga, T., and Kosaka, H., 1996, Generation of nitric oxide and superoxide during reperfusion after focal cerebral ischemia in rats, *Am. J. Physiol.* **270**(3 PT 1): C748–C752.

Leach, M. J., Swan, J. H., Eisenthal, D., Dopson, M., and Nobbs, M., 1993, BW619C89, a glutamate release inhibitor, protects against focal cerebral ischemic damage, *Stroke*, **24**: 1063–1067.

Malinski, T., Patton, S., Grunfeld, S., Kubaszewski, E., Pierchala, B., Kiechle, F., and Radomski, M. W., 1994, Determination of nitric oxide in vivo by a porphyrinic microsensor, in: *The Biology of Nitric Oxide, Physiological and Clinical Aspects* (S. Moncada, M. Feelisch, R. Busse, and E. A. Higgs, eds.), pp. 48–53, Portland Press, London.

Margaill, I., Parmentier, S., Callebert, J., Allix, M., Boulu, R. G., and Plotkine, M., 1996, Short therapeutic window for MK-801 in transient focal cerebral ischemia in normotensive rats, *J. Cereb. Blood flow Metab.* **16**(1): 107–113.

Marks, K. A., Mallard, C. E., Roberts, I., Williams, C. E., Gluckman, P. D., and Edwards, A. D., 1996, Nitric oxide synthase inhibition attenuates delayed vasodilation and increases injury after cerebral ischemia in fetal sheep, *Pediatr. Res.* **40**(2): 185–191.

Meldrum, B. S., 1994, Lamotrigine: a novel approach, *Seizure* **3**: 41–45.

Meldrum, B. S., Swan, J. H., and Leach, M. J., 1992, Reduction of glutamate release and protection against ischemic damage by BW1003C87, *Brain Res.* **593**: 1–6.

Meyer, F. B., Anderson, R. E., Yaksh, T. L., and Sundt, T. M., 1986, Effect of nimodipine on intracellular brain pH, cortical blood flow, and EEG in experimental focal cerebral ischemia, *J. Neurosurg.* **64**: 617–626.

Mittal, C. K., 1993, Nitric oxide synthase: involvement of oxygen radicals in conversion of L-arginine to nitric oxide, *Biochem. Biophys. Res. Commun.* **193**(1): 126–132.

Moncada, C., Lekieffre, D., Arvin, B., and Meldrum, B. S., 1993, The involvement of nitric oxide in neuronal damage in a focal model of excitotoxicity in vivo and a global model of ischemia, in *Nitric Oxide: Brain and Immune System*, (S. Moncada, G. Nistico, and E. A. Higgs, eds.), pp. 191–199, Portland Press, London.

Moskowitz, M. A., and Dalkara, T., 1996, Nitric oxide and cerebral ischemia, *Adv. Neurol.* **71**: 365–367.

Nowicki, J. P., Carreau, A., Duval, D., Vige, X., and Scatton, B., 1993, Neuroprotective potential of nitric oxide synthase inhibitors, in *Nitric Oxide: Brain and Immune System*, (S. Moncada, G. Nistico, and E. A. Higgs, eds.), pp. 121–135, Portland Press, London.

Obrenovich, T. P., 1995, The ischaemic penumbra: twenty years on, *Cerebrovasc. Brain Metab. Rev.* **7**(4): 297–323.

Ozyurt, E., Graham, D. I., Woodruff, G. N., and McCulloch, J., 1988, Protective effect of the glutamate antagonist, MK-801 in focal cerebral ischemia in the cat, *J. Cereb. Blood Flow Metab.* **8**(1): 138–143.

Park, C. K., Nehls, D. G., Graham, D. I., Teasdale, G. M., and McCulloch, J., 1988, The glutamate antagonist MK-801 reduces focal ischemic brain damage in the rat, *Ann. Neurol.* **24**: 543–551.

Planas, A. M., Justicia, C., and Ferrer, I., 1997, Stat 1 in developing and adult rat brain. Induction after transient focal ischemia, *Neuroreport* **8**(6): 1359–1362.

Pulsinelli, W., 1992, Pathophysiology of acute ischemic stroke, *The Lancet* **339**:533–537.

Rami, A., and Krieglstein, J., 1994, Neuronal protective effects of calcium antagonists in cerebral ischemia, *Life Sciences* **55**(25/26): 2105–2113.

Regli, L., Held, M. C., Anderson, R. E., Meyer, F. B., Thoralf, M., and Sundt, Jr. M. D., 1996, Nitric oxide synthase inhibition by L-NAME prevents brain acidosis during focal cerebral ischemia in rabbits, *J. Cereb. Blood Flow Metab.* **16**(5): 988–995.

Rice-Evans, C. A., Diplock, A. T., and Symons, M. C. R., 1991, The detection and characterization of free radical species, in: *Techniques in Free Radical Research*, Volume 22 (R. H. Burdon, and P. H. van Knippenberg, eds.), pp. 51–99, Elsevier Science Publishers BV, Amsterdam.

Samdani, A. F., Dawson, T. M., and Dawson, V. L., 1997, Nitric oxide synthase in models of focal ischemia, *Stroke* **28**(6): 1283–1288.

Scatton, B., 1994, Excitatory amino acid receptor antagonists: a novel treatment for ischemic cerebrovascular diseases, *Life Sciences* **55**(25/26):2115–2124.

Scheinberg, P., 1991, The biologic basis for the treatment of acute stroke, *Neurology* **41**: 1867–1873.

Schott, R. J., Natale, J. E., Ressler, S. W., Burney, R. E., and D'Alecy, L. G., 1989, Neutrophil depletion fails to improve neurologic outcome after cardiac arrest in dogs, *Ann. Emerg. Med.* **18**: 517–522.

Shimazu, M., Mizushima, H., Sasaki, K., Arai, Y., Matsumoto, K., Shioda, S., and Nakai, Y., 1994, Expression of c-fos in the rat cerebral cortex after focal ischemia and reperfusion, *Brain Res. Bull.* **33**(6): 689–697.

Shuaib, A., Mahmood, R. H., Wishart, T., Kanthan, R., Murabit, M. A., Ijaz, S., Miyashita, H., and Howlett, W., 1995, Neuroprotective effects of lamotrigine in global ischemia in gerbils. A histological, in vivo microdialysis and behavioral study, *Brain Res.* **702**(1–2): 199–206.

Siesjö, B. K., 1992a, Pathophysiology and treatment of focal cerebral ischemia Part I: Pathophysiology, *J. Neurosurg.* **77**:169–184.

Siesjö, B. K., 1992b, Pathophysiology and treatment of focal cerebral ischemia Part II: Mechanisms of damage and treatment, *J. Neurosurg.* **77**: 337–354.

Siesjö, B. K., 1994, Calcium-mediated processes in neuronal degeneration, in: *Calcium Hypothesis of Aging and Dementia*, Volume 747 (J. F. Disterhoft, W. H. Gispen, J. Traber, and Z. S. Khachaturian, eds.), pp. 140–162, Annals of the New York Academy of Sciences, New York.

Snyder, S. H., 1992, Nitric oxide: first in a new class of neurotransmitters, *Science* **257**: 494–498.

Springall, D. R., Suburo, A., Bishop, A. E., Merrett, M., Moncada, S., and Polak, J. M., 1992, Nitric oxide synthase is present abundantly in human neurons and vascular endothelium, in: *The Biology of Nitric Oxide, Enzymology, Biochemistry, and Immunology,* (S. Moncada, M. A. Marletta, J. B. Hibbs, and E. A. Higgs, eds.), pp. 117–119, Portland Press, London.

Stamler, S. S., Singel, D. J., and Loscalzo, J., 1992, Biochemistry of nitric oxide and its redox-activated forms, *Science* **258**:1898–1902.

Stuhlmiller, D. F., and Boje, K. M., 1995, Characterization of L-arginine and aminoguanidine uptake into isolated rat choroid plexus: differences in uptake mechanisms and inhibition by nitric oxide synthase inhibitors, *J. Neurochem.* **65**(1): 68–74.

Takizawa, S., Matsushima, K., Fujita, H., Nanri, K., Ogawa, S., and Shinora, Y., 1995, A selective N-type calcium channel antagonist reduces extracellular glutamate release and infarct volume in focal cerebral ischemia, *J. Cereb. Blood Flow Metab.* **15**(4): 611–618.

Traystman, R. J., Kirsch, J. R., and Koehler, R. C., 1991, Oxygen radical mechanisms of brain injury following ischemia and reperfusion, *J. Appl. Physiol.* **71**(4): 1185–1191.

Umemura K., Gemba, T., Mizuno, A., and Nakashima, M., 1996, Inhibitory effect of MS-153 on elevated brain glutamate level induced by rat middle cerebral artery occlusion, *Stroke* **27**(9): 1624–1628.

White, B. C., Grossman, L. I., and Krause, G. S., 1993, Brain injury by global ischemia and reperfusion: a theoretical perpective on membrane damage and repair, *Neurology* **43**: 1656–1665.

Wiard, R. P., Dickerson, M. C., Beek, O., Norton, R., and Cooper, B. R., 1995, Neuroprotective properties of the novel antiepileptic lamotrigine in a gerbil model of global cerebral ischemia, *Stroke* **26**(3): 466–472.

Winfree, C. J., Baker, C. J., Connoly, E. S. Jr., Fiore, A. J., and Solomon, R. A., 1996, Mild hypothermia reduces penumbral glutamate levels in the rat permanent focal cerebral ischemia model, *Neurosurgery* **38**(6): 1216–1222.

Wong, M. L., Loddick, S. A., Bongiorno, P. B., Gold, P. W., Rothwell, N. J., and Licinio, J., 1995, Focal cerebral ischemia induces CRH mRNA in rat cerebral cortex and amygdale, *Neuroreport* **6**(13): 1785–1788.

Xia, Y., Dawson, V. L., Dawson, T. M., Snyder, S. H., and Zweier, J. L., 1996, Nitric oxide synthase generates superoxide and nitric oxide in arginine-depleted cells leading to peroxynitrite-mediated cellular injury, *Proc. Natl. Acad. Sci. USA* **93**(13): 6770–6774.

Yamamoto, S., Golanov, E. V., Berger, S. B., and Reis, D. J., 1992, Inhibition of nitric oxide synthesis increases focal ischemic infarction in rat, , *J. Cereb. Blood Flow Metab.* **12**(5): 717–726.

Zhang, F., Xu, S., and Iadecola, C., 1995, Time dependence of effect of nitric oxide synthase inhibition on cerebral ischemic damage, *J. Cereb. Blood Flow Metab.* **15**(4): 595–601.

18

THE ROLE OF FREE RADICALS IN NMDA AND GLUTAMATE EXCITOTOXICITY

A. S. Yalçın,[1] Goncagül Haklar,[1] Belgin Küçükkaya,[1] Meral Yüksel,[1] and Hale Saybaşılı[2]

[1]Department of Biochemistry
Faculty of Medicine
Marmara University
[2]Institute of Biomedical Engineering
Boğaziçi University
Istanbul, Turkey

1. INTRODUCTION

The central nervous system is particularly vulnerable to oxidative stress. Neuronal membranes contain high levels of polyunsaturated fatty acids which are susceptible to free radical attack (Grisham, 1992). Normally, reactive oxygen species (ROS) generated as by-products of cellular metabolism are maintained at very low, non-toxic levels with the aid of free radical scavengers and antioxidants (Sies, 1993). However, in recent years it has been established that oxidative stress is an important causal factor in the pathogenesis of several neurological disorders such as stroke, trauma, seizures and chronic neurodegenerative diseases (Dawson and Dawson, 1996; Simonian and Coyle, 1996).

On the other hand, excessive and persistent activation of glutamate-gated ion channels, in particular NMDA channels, may cause neurotoxicity and neuronal degeneration (Garthwaite, 1994; Lipton and Rosenberg, 1994). Glutamate is the principal excitatory neurotransmitter in the brain. It plays an essential role in several physiological neuronal activities such as learning, memory and neuronal plasticity (Nicholls, 1993). Glutamate acts on three major ionophore-linked receptor families classified according to their most selective agonists: N-methyl-D-aspartate (NMDA), kainate and α-amino-3-hydroxy-5-methyl-4-isoxazole propionic acid (AMPA). NMDA receptors have received the greatest attention with regard to excitotoxicity as they possess high permeability to Ca^{2+} ions besides Na^+ and K^+ ions (Garthwaite, 1994). These receptors have several modulators, i.e., glycine is a co-agonist, Mg^{2+} and Zn^{2+} are voltage dependent competitive and non-compe-tititve antagonists, respectively. The modulators are important tools in investigating NMDA receptor activities. Several studies have pointed out glutamate as one of the major causes in the pathogenesis of different neurological diseases including epilepsy, trauma,

Free Radicals, Oxidative Stress, and Antioxidants, edited by Özben.
Plenum Press, New York, 1998.

domoate poisoning and Alzheimer's disease (Coyle and Putfarcken, 1993; Lipton and Rosenberg, 1994). Thus, a close parallelism exists in the involvement of glutamate and ROS in the mechanism of neuronal cell death.

2. EXCITOTOXICITY AND ROS FORMATION IN DIFFERENT EXPERIMENTAL MODELS

We have studied the relation between glutamate/NMDA excitotoxicity and free radical formation in different experimental models. Initially, we have investigated the effect of excitotoxic concentrations of glutamate and NMDA on the generation of ROS in rat brain homogenates (Küçükkaya, Haklar, and Yalçın, 1996). We have used luminol and lucigenin enhanced chemiluminescence and correlated the results with other parameters of oxidative stress. Excitotoxic concentrations of glutamate and NMDA resulted in increased chemiluminescence formation which correlated well with other parameters of oxidative stress such as conjugated diene formation, MDA accumulation, and protein oxidation. It was concluded that chemiluminescence measurements are suitable for investigating interactions of ROS and excitotoxic amino acids. In the same study, the effects of different NMDA receptor modulators were investigated. Both Mg^{2+} (a voltage-dependent competitive antagonist of NMDA receptor) and Zn^{2+} (a non-competitive antagonist of NMDA receptor) had inhibitory effects whereas glycine, the essential co-agonist of NMDA receptor, increased the NMDA-induced chemiluminescence. On the other hand, addition of sodium nitroprusside, a NO donor, and ascorbate ions significantly increased NMDA-induced chemiluminescence.

Later we have carried our studies on cortical synaptosomes. The synaptosome is the simplest system which retains the full machinery for the uptake, storage and release of neurotransmitters. Moreover, it also involves the presynaptic receptor mediated signal transduction pathways which regulate release (Bradford, 1986). Molecules such as glutathione, NMDA antagonists, Ca^{++} channel blockers and the redox potential of the milieu seem to modulate the end-response of NMDA channel activation (Sucher and Lipton, 1991; Tang and Aizenman, 1993). We have treated freshly prepared crude synaptosomes with NMDA prior to or concomittant with addition of different modulatory molecules. Chemiluminescence measurements were made with lucigenin as enhancer. Synaptosomes preincubated with GSH showed a marked response to NMDA stimulation. Upon the addition of Zn^{++} the response returned to its original level. We have observed that after depolarization, physiological concentrations of Mg^{++} was necessary for the potentiation of the response of synaptosomes to NMDA (Figure 1).

In recent studies we have used hippocampal slices to study the interaction between ROS and glutamate excitotoxicity. Hippocampus is a structural part of the limbic system which participates in long term potentiation, kindling phenomena, and excitotoxicity (Ballyk and Goh, 1992). The excitation of a hippocampal neuron is modulated by voltage and receptor operated channels. Glutamate is released from the presynaptic neuron to the synaptic cleft upon depolarization, and different types of receptor subgroups are activated (Nicholls, 1993). The channel becomes functional under depolarization conditions when the agonist and coagonist (i.e., glutamate/NMDA and glycine) parts of the NMDA receptor is occupied and blockage of the channel by magnesium ions is relieved. NMDA induced increased intracellular calcium leads to nitric oxide (NO) synthesis via Ca^{2+}/calmodulin dependent nitric oxide synthase. However, participation of NO in neurodegenerative phenomena remains contraversial (Iadecola, 1997).

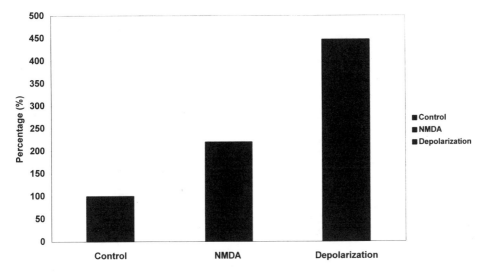

Figure 1. Percentage increase in lucigenin enhanced chemiluminescence in rat cortical synaptosomes after induction with NMDA and KCl.

In our studies, we have used hippocampal slices prepared from 21–25 days old Spraque-Dawley rats. The hippocampi were isolated and then aerated in artificial cerebrospinal fluid (ACSF). Generation of radicals under spontaneous epileptiform activity conditions, in the absence of electrical stimulation, was investigated in ACSF with or without Mg^{2+} ions. Depolarization was maintained by adding potassium to ACSF. Slices were then transferred to counting vials and chemiluminescence was recorded after the addition of luminol or lucigenin. NO detection was based upon the chemiluminescence reaction between NO and purified luminol-H_2O_2 system (Kikuchi *et al.*, 1993). As shown in Figure 2, it was observed that NMDA induced excitotoxicity leads to increase in lucigenin-chemilu-

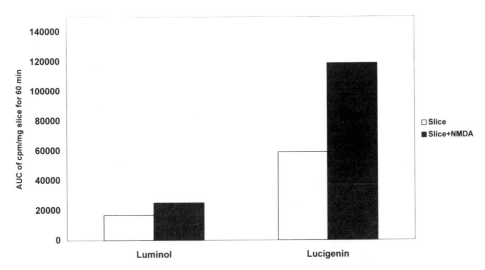

Figure 2. Luminol and lucigenin enhanced chemiluminescence in rat hippocampal slices after induction with NMDA.

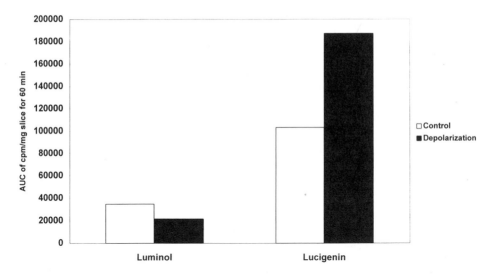

Figure 3. Luminol and lucigenin enhanced chemiluminescence in rat hippocampal slices after induction with KCl.

minescence in hippocampal slices. Luminol-chemiluminescence was also increased, but not significantly. We have also observed that lucigenin chemiluminescence was significantly increased under depolarization conditions (Figure 3). This means that spontaneous epileptiform discharge generated by enhanced NMDA receptor activity with ambient glutamate is sufficient to form ROS. Therefore, depolarization which activates both voltage and receptor operated channels generates ROS and may further cause the release of glutamate to the medium, thus aggravating excitotoxicity. Selective calcium influx to neuron, most probably through NMDA receptor ionophore complex mediates this effect. Consistent with this, addition of AP-5, a competitive inhibitor of NMDA receptor, depressed ROS generation under spontaneous epileptiform discharge conditions in hippocampal

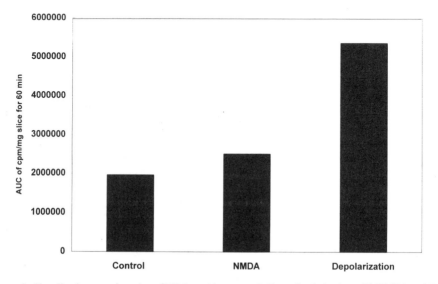

Figure 4. Chemiluminescent detection of NO in rat hippocampal slices after induction with NMDA and KCl.

slices. We have also detected increased NO formation in hippocampal slices under spontaneous epileptiform and depolarization conditions (Figure 4).

3. CONCLUSION

We have shown that enhanced chemiluminescence measurements are valuable for investigating the involvement of ROS in excitotoxicity. This technique can be used in vitro in synaptosomes to correlate glutamate uptake-release and ROS formation under excitotoxic conditions. Finally, our results imply that superoxide and nitric oxide mediated disturbances, via increased NMDA receptor sensitivity under spontaneous epileptiform discharge and depolarization conditions, might be responsible for hippocampal neuronal damage.

REFERENCES

Ballyk, B.A., and Goh, J.W., 1992, Elevation of extracellular potassium facilitates the induction of hippocampal long-term potentiation, *J. Neurosci. Res.* **33**: 596–604.

Bradford, H.F., 1986, The synaptosome, an in vitro model for the synapse, in: *Chemical Neurobiology,* pp. 311–352, Freeman Press, New York.

Coyle, J.T., and Putfarcken, P., 1993, Oxidative stress, glutamate, and neurodegenerative disorders, *Science* 262: 689–695.

Dawson, V.L., and Dawson, T.M., 1996, Free radicals and neuronal cell death, *Death and Differentiation* **3**: 71–78.

Garthwaite, J., 1994, NMDA receptors, neuronal development, and neurodegeneration, in: *The NMDA Receptor,* (G.L. Collingridge and J.C. Watkins, eds.) pp. 428–456, Oxford University Press, New York.

Grisham, M.B., 1992, *Reactive Metabolites of Oxygen and Nitrogen in Biology and Medicine,* pp. 65–66, R.G. Landes Company, Austin.

Iadecola, C., 1997, Bright and dark sides of nitric oxide in ischemic brain injury, *TINS* **20:** 132–139.

Kikuchi, K., Nagano, T., Hayakawa, H., Hirata, Y., and Hirobe, M., 1993, Detection of nitric oxide production from a perfused organ by a luminol-H_2O_2 system, *Anal. Chem.* **65**: 1794–1799.

Küçükkaya, B., Haklar, G., and Yalçın, A.S., 1996, NMDA excitotoxicity and free radical generation in rat brain homogenates: application of a chemiluminescence assay, *Neurochem. Res.* **21**: 1533–1536.

Lipton, S.A., and Rosenberg, P.A., 1994, Excitatory amino acids as a final common pathway for neurologic disorders, *Mech. Dis.* **330**: 613–622.

Nicholls, D.G., 1993, The glutamatergic nerve terminal, *Eur. J. Biochem.* **212**: 613–631.

Sies, H., 1993, Strategies of antioxidant defence, *Eur. J. Biochem.* **215**: 213–219.

Simonian, N.A., and Coyle, J.T., 1996, Oxidative stress in neurodegenerative diseases, *Annu. Rev. Pharmacol. Toxicol.* **36**: 83–106.

Sucher, N.J., and Lipton, A., 1991, Redox modulatory site of the NMDA receptor channel complex: regulation by oxidized glutathione, *J. Neurosci. Res.* **30**: 582–591.

Tang, L.H., and Aizenman, E., 1993, Long-lasting modification of the N-Methyl-D-Aspartate receptor channel by a voltage-dependent sulfhydryl redox process, *Molec. Pharmacol.* **44**: 473–478.

19

THE ROLE OF OXIDATIVE STRESS IN THE PATHOLOGICAL SEQUELAE OF ALZHEIMER DISEASE

Mark A. Smith and George Perry

Institute of Pathology
Case Western Reserve University
2085 Adelbert Road, Cleveland, Ohio 44106

1. INTRODUCTION

Oxidative stress and damage are accepted features of degenerating and at risk neuronal populations in Alzheimer disease (Smith et al., 1994a,b, 1995a, 1996a, 1997a,b; Sayre et al., 1997a). Nonetheless, there is controversy concerning how oxidative stress meshes with currently accepted disease hypotheses as well as whether oxidative stress is an initiator of the disease or is instead a result of the disease process (Figure 1).

In this review, we present evidence showing that oxidative stress is involved in all of the known pathogenic mechanisms governing the disease including cytoskeletal phosphorylation (reviewed in Trojanowski et al., 1993), apolipoprotein E genotype (reviewed in Roses, 1995) and amyloid cascade (reviewed in Selkoe, 1997). Furthermore, recent findings show that oxidative stress is a very early indicator of neuronal dysfunction in Alzheimer disease that occurs concurrently, if not earlier, than the abnormal phosphorylation of cytoskeletal elements. Therefore, it seems that oxidative stress is responsible for many, if not all, of the pathological sequelae of Alzheimer disease.

1.1. Current Theories

The pathological presentation of Alzheimer disease, the leading cause of senile dementia, involves regionalized neuronal death and an accumulation of intraneuronal and extracellular lesions termed neurofibrillary tangles and senile plaques, respectively (reviewed in Smith, 1997). Several independent hypotheses have been proposed to link the pathological lesions and neuronal cytopathology with, among others, apolipoprotein E genotype (Corder et al., 1993; Roses, 1995); hyperphosphorylation of cytoskeletal proteins (Trojanowski et al., 1993), and amyloid-β metabolism (Selkoe, 1997). However, no

Free Radicals, Oxidative Stress, and Antioxidants, edited by Özben.
Plenum Press, New York, 1998.

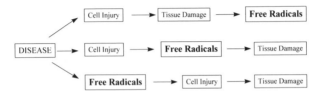

Figure 1. Free radicals are associated with almost all disease processes. Indeed, both cell injury and tissue damage liberate free radicals which could then exacerbate a disease process. However in some cases including, it seems, Alzheimer disease, free radicals are a primary and early factor in the disease process.

one of these theories is alone sufficient to explain the diversity of biochemical and pathological abnormalities that are documented in the disease. Furthermore, attempts to mimic the disease by a perturbation of one of these elements using cell or animal models, including transgenic animals, also do not result in the a spectrum of pathological alterations.

2. OLD AGE IS ESSENTIAL FOR ALZHEIMER DISEASE

The most important etiological aspect of Alzheimer disease is that the disease is an age-related condition (Katzman, 1986) (Figure 2). Importantly, this holds true even in individuals with a genetic predisposition, i. e., those individuals with autosomal dominant inheritance of Alzheimer disease or individuals with Down syndrome who also develop the lesions of Alzheimer disease. Aging is associated with an increase in the adventitious production of oxygen-derived radicals, i.e., reactive oxygen species (ROS), and a concurrent decrease in the ability to defend against such ROS. This decrease in reserve capacity also leads to a compromised ability to deal with abnormal sources of ROS with aging, genetic predisposition and/or disease status.

3. OXYGEN RADICAL GENERATION AND OXIDATIVE STRESS IN ALZHEIMER DISEASE

ROS production occurs as a ubiquitous byproduct of both oxidative phosphorylation and the myriad of oxidases necessary to support aerobic metabolism. In Alzheimer disease, in addition to this background level of ROS, there are a number of additional con-

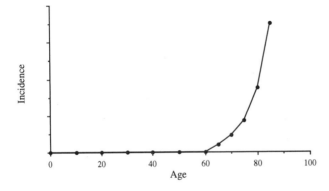

Figure 2. The prevalence of Alzheimer disease increases sharply with age. Adapted from statistics from the National Institute of Aging and the Alzheimer Association.

tributory sources that are thought to play an important role in the disease process: (1) The increase in redox-active iron in neurofibrillary tangle-bearing neurons and senile plaques (Good et al., 1992; Smith et al., 1997a) is of particular importance since transition metals, through Fenton chemistry, are potent catalysts for the generation of ROS. Furthermore, aluminum, which also accumulates in NFT-containing neurons (Good et al., 1992), is able to stimulate iron-induced lipid peroxidation (Oteiza, 1994). (2) Activated microglia, surrounding senile plaques, are a source of NO and O_2^-, the precursors of peroxynitrite which has ·OH-like activity and is involved in the pathological spectrum of Alzheimer disease (Colton and Gilbert, 1987; Smith et al., 1997b). (3) Amyloid-β has been reported to spontaneously fragment to form free radical peptides (Butterfield et al., 1994; Hensley et al., 1994; Sayre et al., 1997b). (4) Advanced glycation end products in concert with transition metals promote redox cycling to produce ROS (Baynes, 1991; Yan et al., 1994, 1995). Furthermore, advanced glycation end products and amyloid-β activate specific receptors such as the receptor for advanced glycation end products (RAGE) (Yan et al., 1996) and the class A scavenger-receptor resulting in increases in ROS (Yan et al., 1996; El Khoury et al., 1996). (5) Genetic abnormalities in mitochondrial metabolism which act to increase ROS production are found in Alzheimer disease (Davis et al., 1997).

The relative importance and temporal sequence of involvement of these or other sources of ROS in the pathogenesis of the disease remains to be determined. Nonetheless, in all cases, the gross result is oxidative damage of neurons, neurofibrillary tangles and senile plaques. Identified types of damage include advanced glycation end products (Smith et al., 1994a), nitration (Smith et al., 1997b), advanced lipid peroxidation end products (Sayre et al., 1997a) as well as carbonyl-modified neurofilament protein and free carbonyls (Smith et al., 1991; Smith et al., 1994a, 1995b, 1996a; Ledesma et al., 1994; Vitek et al., 1994; Yan et al., 1994; Montine et al., 1996a; Sayre et al., 1997a). Modifications, by indicating imbalance in oxidative metabolism, provide important evidence of biologically significant oxidative stress. Furthermore, these modifications provide a biochemical explanation, through protein crosslinking, for the insolubility of the pathologic lesions (reviewed in Smith et al., 1995a; Smith et al., 1996b) as well as their resistance to proteolytic removal (Cras et al., 1995) since, while oxidative modifications target proteins for removal, oxidatively crosslinked proteins inhibit the proteosome (Friguet et al., 1994).

Attesting to the biological significance of oxidative damage in the pathogenesis of Alzheimer disease, there is a close association between increased levels of the antioxidant enzymes superoxide dismutase (Pappolla et al., 1992) and heme oxygenase-1 (Smith et al., 1994b; Premkumar et al., 1995) and the cytoskeletal abnormalities found in Alzheimer disease. Indeed, in quantitative immunocytochemical studies, there is a complete overlap between heme oxygenase-1 and Alz50, an early marker of τ abnormalities, indicating that cytoskeletal abnormalities are associated with increased oxidative stress or vice versa (Smith and Perry, unpublished observation).

4. RELATIONSHIP OF OXIDATIVE STRESS TO OTHER HYPOTHESES OF ALZHEIMER DISEASE

Age-related increases in oxidative stress, in concert with other parameters identified in Alzheimer disease, may be the major contributor toward all aspects of the disease. In other words, oxidative stress may provide a link between all of the currently accepted views regarding disease pathogenesis (Figure 3).

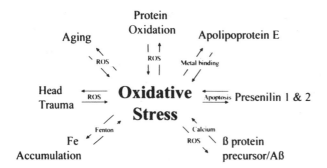

Figure 3. All of the known risk factors for Alzheimer disease likely operate through a common mechanism involving oxidative stress.

4.1. Oxidative Stress and Cytoskeletal Phosphorylation

Biochemical studies have been severely hampered by the extreme insolubility of the paired helical filaments that comprise neurofibrillary tangles (Selkoe et al., 1982; Smith et al., 1996b). Indeed, a great deal of research has focused on the mechanisms by which normal soluble cytoskeletal elements, such as τ and neurofilaments, are transformed into the insoluble paired helical filaments. One suggestion is that 'abnormal phosphorylation' of τ is the first step towards paired helical filament formation (Goedert et al., 1991; Greenberg et al., 1992) and, in this regard, a number of specific kinases and phosphatases have been implicated to play a role in the disease (reviewed in Trojanowski et al., 1993). However, while increased phosphorylation decreases microtubule and intermediate filament stability, salient features of the pathology of Alzheimer disease (Perry et al., 1991; Alonso et al., 1994, 1996; Iqbal et al., 1994), it is quite apparent that the formation of paired helical filaments is not mediated solely by phosphorylation. Indeed, *in vitro* phosphorylation of normal τ or complete dephosphorylation of paired helical filaments has no effect on solubility (Goedert et al., 1991; Greenberg et al., 1992; Gustke et al., 1992; Smith et al., 1996b). Moreover, τ phosphorylation in Alzheimer disease is similar to that seen during mitosis where paired helical filaments are not found (Pope et al., 1994; Preuss et al., 1995) and led to the suggestion that senescent neurons in Alzheimer disease might be undergoing an abortive mitotic event (Vincent et al., 1996; McShea et al., 1997).

Phosphorylation is intimately tied to oxidative stress by the mitogen activated phosphorylation kinase (MAP) pathway (Guyton et al., 1996) as well as through activation of transcription factor NFκB (Schreck et al., 1991), and, while there is controversy concerning the kinases involved in the phosphorylation of τ in Alzheimer disease, it is probable that the MAP kinase pathway is involved (Ledesma et al., 1992). If this linkage is confirmed, it would provide a connection between abnormal phosphorylation of proteins in the cell bodies and increased oxidative stress and furthermore would also provide additional evidence that oxidative stress occurs first. Consistent with this hypothesis, all pyramidal neurons of the hippocampus show increased free carbonyls (Smith et al., 1996a), lipid peroxide adduction products (Sayre et al., 1997a) and nitrotyrosine (Smith et al., 1997b) while only those displaying neurofibrillary pathology show increased phosphorylation of neurofilament and τ proteins (Sternberger et al., 1985; Grundke-Iqbal et al., 1986). These findings suggest that a critical level of oxidative damage must occur prior to increased phosphorylation and, by implication, activation of signaling cascades. Heme oxygenase-1, an inducible antioxidant enzyme, also shows the same pattern of immunocytochemical involvement since it accumulates only in neurons displaying phosphorylated τ (Smith and Perry, unpublished data). Therefore, like the kinase response, it seems that a critical level of oxidative damage is necessary to initiate an antioxidant response, including induction of heme oxygenase-1 (Premkumar et al., 1995). Future studies addressing the connections between signaling cascades that regulate both heme oxygenase and increased phosphorylation are clearly necessary.

4.2. Oxidative Stress and the Amyloid Hypothesis

Amyloid-β is toxic to cultured cells (Yankner et al., 1990; reviewed in Iversen et al., 1995) and a number of mechanisms have been invoked for this neurotoxicity including membrane depolarization (Carette et al., 1993), increased sensitivity to excitotoxins (Koh et al., 1990), and alterations in calcium homeostasis (Mattson et al., 1992). However, the leading hypothesis is that amyloid-β-mediated neuronal damage is mediated by free radicals and, as such, can be attenuated using antioxidants such as vitamin E (Behl et al., 1992, 1994) or catalase (Zhang et al., 1996) although there has been some controversy regarding whether the protection from the latter involves its antioxidant activity (Lockhart et al., 1994). Nonetheless, increased amyloid-β production may be the link between the senile plaque pathology of Alzheimer disease and oxidative stress. While critically addressing this issue requires careful study of the relationship of amyloid-β deposition and increased oxidative markers, such a study is difficult to accomplish with cases of Alzheimer disease alone and is best studied with animal models with known chronology of amyloid-β deposition. In this regard, it will be extremely interesting to determine the oxidative status of the recently reported transgenic rodent models of Alzheimer disease (Games et al., 1995; Hsiao et al., 1996).

One of the main caveats related to the amyloid-β hypothesis is that there is not a strong correlation between the amount of amyloid-β *in vivo* and the degree of dementia (Mann et al., 1988; Bennett et al., 1993; Masliah and Terry, 1993). Nonetheless, albeit in a small number of cases, mutations in βPP clearly precipitate the disease in an autosomal dominant fashion (reviewed in Selkoe, 1996, 1997) and βPP-transgenic animal models display many of the neuropathological correlates of Alzheimer disease (Games et al., 1995; Hsiao et al., 1996). An interesting hypothesis is that βPP/amyloid-β may be intimately involved in free radical homeostasis. Indeed, amyloid-β is reported to spontaneously generate peptidyl radicals (Butterfield et al., 1994; Goodman et al., 1994; Harris et al., 1995; Prehn et al., 1996) but also mutations in βPP are associated with increased DNA fragmentation, possibly involving oxidative mechanisms (see below).

4.3. Oxidative Stress and ApoE, Presenilins and Other Genetic Factors

The utility of genetically assigning presenilins 1 and 2 (Sherrington et al., 1995; Selkoe, 1997) and apolipoprotein E (Corder et al., 1993) as biologically significant factors in the pathogenesis and possibly etiology of Alzheimer disease has done little to resolve their mechanistic role. Nevertheless, intense investigation has revealed putative roles that involve oxidative damage. In the case of presenilin 2, increased expression increase DNA fragmentation and apoptotic changes (Wolozin et al., 1996), both important consequences of oxidative damage. Apolipoprotein E, in brains and cerebrospinal fluid, is found adducted with the highly reactive lipid peroxidation product, hydroxynonenal (Montine et al., 1996b). Furthermore, apolipoprotein E is a strong chelator of copper and iron, important redox-active transition metals, and as such can protect cells from oxidative insult (Miyata and Smith, 1996). Finally, interaction of apolipoprotein E with amyloid-β only occurs in the presence of oxygen (Strittmatter et al., 1993).

4.4. Oxidative Stress and Apoptosis

Apoptosis is the programmed morphological and biochemical changes cells undergo within hours of cell death. In this process cells are digested within their own membrane by proteases and nucleases as well as by increased ROS. The availability of the TUNEL technique to detect DNA fragmentation has greatly advanced detection of apoptotic cells. Nonetheless, without morphological changes, it is unclear whether DNA fragmentation is apoptotic or is, instead, mediated by oxidative stress (Berlin and Haseltine, 1981). This issue is particularly germane for Alzheimer disease where fragmentation is found in many neurons yet neurons live for years following the onset of disease (Su et al., 1994). Indeed, apoptotic changes including nuclear and other morphologies are rarely noted in Alzheimer disease (Cotman and Su, 1996). On the other hand, oxidative damage, including nuclear damage, is increased in all cells of the brain in areas affected in Alzheimer disease (Smith et al., 1996a).

5. IS OXIDATIVE STRESS AN EARLY EVENT IN THE PATHOGENESIS OF ALZHEIMER DISEASE?

One important question potential caveat regarding the role of oxidative stress in Alzheimer disease pathogenesis is whether it represents a primary or secondary event (Mattson et al., 1995; Smith et al., 1995c). However, recent evidence clearly indicates that oxidative damage is associated with the very earliest neuronal and pathological changes characteristic of Alzheimer disease (Ledesma et al., 1994; Smith et al., 1994a, 1995a, 1996a, 1997b; Vitek et al., 1994; Yan et al., 1994; Sayre et al., 1997a). Therefore, it is now apparent that oxidative stress represents a very early contributor to the disease and this early role is borne out by clinical management of oxidative stress which both reduces the incidence and severity of Alzheimer disease (Rogers et al., 1993; Rich et al., 1995; Stewart et al., 1997).

6. OXIDATIVE STRESS AND TREATMENT STRATEGIES

Considering the heterogeneity of etiological factors responsible for Alzheimer disease and the lack of an animal model displaying the full spectrum of pathological changes, it is not surprising that successful pharmacological interventions have yet to be established. To date, most therapeutic regiments have been aimed at replacing the deficient neurotransmitter systems in an attempt to alleviate the symptoms of the disease. However, though potentially useful in providing short term relief, such approaches do not solve the fundamental problem of neuronal degeneration and death.

Given that oxidative stress may be the initiator of the pathological sequelae of Alzheimer disease, it is becoming increasingly apparent that therapeutic reduction of oxidative stress might prove most efficacious. Indeed, this appears to be the case, in that agents that inhibit free radical formation, reduce both the incidence and the progression of Alzheimer disease (McGeer and Rogers, 1992; Rogers et al., 1993; Rich et al., 1995; Stewart et al., 1997). For the latter, several studies have found epidemiological relationships (Breitner et al., 1994; McGeer and Rogers, 1992) that, together with the results of metal chelation treatment (McLachlan et al., 1991), strongly suggest that agents that prevent oxidative damage show the most promise in forming a successful therapeutic approach for the treatment of Alzheimer disease.

7. CONCLUSIONS

In summary, we have presented evidence that stochastic mechanisms of protein damage by oxidative stress may underline all of the commonly accepted notions on Alzheimer disease pathogenesis, including hyperphosphorylation, apolipoprotein E genotype, and mutations of the β-protein precursor. Further studies, to examine the types and extent of oxidative damage, should identify which antioxidant agents will prove most efficacious in the treatment of the disease.

ACKNOWLEDGMENTS

This work was supported through grants from the National Institutes of Health (AG09287) and the American Health Assistance Foundation (AHAF). M.A.S. is a Daland fellow of the American Philosophical Society.

REFERENCES

Alonso, A. C., Zaidi, T., Grundke-Iqbal, I., and Iqbal, K., 1994, Role of abnormally phosphorylated tau in the breakdown of microtubules in Alzheimer disease, *Proc. Natl. Acad. Sci. USA* **91**:5562–5566.

Alonso, A. C., Grundke-Iqbal, I., and Iqbal, K., 1996, Alzheimer's disease hyperphosphorylated tau sequesters normal tau into tangles of filaments and disassembles microtubules, *Nature Med.* **2**:783–787.

Baynes, J. W., 1991, Role of oxidative stress in development of complications in diabetes, *Diabetes* **40**, 405–412.

Behl, C., Davis, J., Cole, G. M., and Schubert, D., 1992, Vitamin E protects nerve cells from amyloid-β protein toxicity, *Biochem. Biophys. Res. Commun.* **186**:944–950.

Behl, C., Davis, J. B., Lesley, R., and Schubert, D., 1994, Hydrogen peroxide mediates amyloid β protein toxicity, *Cell* **77**:817–827.

Bennett, D. A., Cochran, E. J., Saper, C. B., Leverenz, J. B., Gilley, D. W., and Wilson, R. S., 1993, Pathological changes in frontal cortex from biopsy to autopsy in Alzheimer's disease, *Neurobiol. Aging* **14**:589–596.

Berlin, V. and Haseltine, W. A., 1981, Reduction of adriamycin to a semiquinone-free radical by NADPH cytochrome P-450 reductase produces DNA cleavage in a reaction mediated by molecular oxygen, *J. Biol. Chem.* **256**:4747–4756.

Breitner, J. C. S., Gau, B. A., Welsh, K. A., Plassman, B. L., McDonald, W. M., Helms, M. J., and Anthony, J. C., 1994, Inverse association of anti-inflammatory treatments and Alzheimer's disease: initial results of a co-twin control study, *Neurology* **44**:227–232.

Butterfield, D. A., Hensley, K., Harris, M., Mattson, M., and Carney, J., 1994, β-amyloid peptide free radical fragments initiate synaptosomal lipoperoxidation in a sequence-specific fashion: implications to Alzheimer's disease, *Biochem. Biophys. Res. Commun.* **200**:710–715.

Carette, B., Poulain, P., and Delacourte, A., 1993, Electrophysiological effects of 25–35 amyloid-β-protein on guinea-pig lateral septal neurons, *Neurosci. Lett.* **151**:111–114.

Colton, C. A., and Gilbert, D. L., 1987, Production of superoxide anions by a CNS macrophage, the microglia, *FEBS Lett.* **223**:284–288.

Corder, E. H., Saunders, A. M., Strittmatter, W. J., Schmechel, D. E., Gaskell, P. C., Small, G. W., Roses, A. D., Haines, J. L., and Pericak-Vance, M. A., 1993, Gene dose of apolipoprotein E type 4 allele and the risk of Alzheimer's disease in late onset families, *Science* **261**:921–923.

Cotman, C. W., and Su, J. H., 1996, Mechanisms of neuronal death in Alzheimer's disease. *Brain Pathol.* **6**:493–506.

Cras, P., Smith, M. A., Richey, P. L., Siedlak, S. L., Mulvihill, P., and Perry, G., 1995, Extracellular neurofibrillary tangles reflect neuronal loss and provide further evidence of extensive protein cross-linking in Alzheimer disease, *Acta Neuropathol.* **89**:291–295.

Davis, R. E., Miller, S., Herrnstadt, C., Ghosh, S. S., Fahy, E., Shinobu, L. A., Galasko, D., Thal, L. J., Beal, M. F., Howell, N., and Parker, W. D. Jr., 1997, Mutations in mitochondrial cytochrome c oxidase genes segregate with late-onset Alzheimer disease, *Proc. Natl. Acad. Sci. USA* **94**:4526–4531.

El Khoury, J., Hickman, S. E., Thomas, C. A., Cao, L., Silverstein, S. C., and Loike, J. D., 1996, Scavenger receptor-mediated adhesion of microglia to β-amyloid fibrils, *Nature* **382**:716–719.

Friguet, B., Stadtman, E. R., and Szweda, L. I., 1994, Modification of glucose-6-phosphate dehydrogenase by 4-hydroxy-2-nonenal. Formation of cross-linked protein that inhibits the multicatalytic protease. *J. Biol. Chem.* **269**:21639–21643.

Games, D., Adams, D., Alessandrini, R., Barbour, R., Berthelette, P., Blackwell, C., Carr, T., Clemens, J., Donaldson, T., Gillespie, F., Guido, T., Hagopian, S., Johnson-Wood, K., Khan, K., Lee, M., Leibowitz, P., Lieberburg, I., Little, S., Masliah, E., McConlogue, L., Montoya-Zavala, M., Mucke, L., Paganini, L., Penniman, E., Power, M., Schenk, D., Seubert, P., Snyer, B., Soriano, F., Tan, H., Vitale, J., Wadsworth, S., Wolozin, B., and Zhao, J., 1995, Alzheimer-type neuropathology in transgenic mice overexpressing V717F β-amyloid precursor protein, *Nature* **373**:523–527.

Goedert, M., Sisodia, S. S., and Price, D. L., 1991, Neurofibrillary tangles and beta-amyloid deposits in Alzheimer's disease, *Curr. Opin. Neurobiol.* **1**:441–447.

Good, P. F., Perl, D. P., Bierer, L. M., and Schmeidler, J., 1992, Selective accumulation of aluminum and iron in the neurofibrillary tangles of Alzheimer's disease: a laser microprobe (LAMMA) study, *Ann. Neurol.* **31**:286–292.

Goodman, Y., Steiner, M. R., Steiner, S. M., and Mattson, M. P., 1994, Nordihydroguaiaretic acid protects hippocampal neurons against amyloid beta-peptide toxicity, and attenuates free radical and calcium accumulation, *Brain Res.* **654**:171–176.

Greenberg, S. G., Davies, P., Schein, J. D., and Binder, L. I., 1992, Hydrofluoric acid-treated tau PHF proteins display the same biochemical properties as normal tau, *J. Biol. Chem.* **267**:564–569.

Grundke-Iqbal, I., Iqbal, K., Tung, Y. C., Quinlan, M., Wisniewski, H. M., Binder, L. I., 1986, Abnormal phosphorylation of the microtubule-associated protein τ (tau) in Alzheimer cytoskeletal pathology, *Proc. Natl. Acad. Sci. USA* **83**: 4913–4917.

Gustke, N., Steiner, B., Mandelkow, E. M., Biernat, J., Meyer, H. E., Goedert, M., and Mandelkow, E., 1992, The Alzheimer-like phosphorylation of tau protein reduces microtubule binding and involves Ser-Pro and Thr-Pro motifs, *FEBS Lett.* **307**:199–205.

Guyton, K. Z., Liu, Y., Gorospe, M., Xu, Q., and Holbrook, N. J., 1996, Activation of mitogen-activated protein kinase by H_2O_2. Role in cell survival following oxidant injury, *J. Biol. Chem.* **271**:4138–4142.

Harris, M. E., Carney, J. M., Cole, P. S., Hensley, K., Howard, B. J., Martin, L., Bummer, P., Wang, Y., Pedigo, N. W. J., and Butterfield, D. A., 1995, β-amyloid peptide-derived, oxygen-dependent free radicals inhibit glutamate uptake in cultured astrocytes: implications for Alzheimer's disease, *Neuroreport* **6**:1875–1879.

Hensley, K., Carney, J. M., Mattson, M. P., Aksenova, M., Harris, M., Wu, J. F., Floyd, R. A., and Butterfield, D. A., 1994, A model for β-amyloid aggregation and neurotoxicity based on free radical generation by the peptide: relevance to Alzheimer disease, *Proc. Natl. Acad. Sci. USA* **91**:3270–3274.

Hsiao, K., Chapman, P., Nilsen, S., Eckman, C., Harigaya, Y., Younkin, S., Yang, F., and Cole, G., 1996, Correlative memory deficits, Aβ elevation, and amyloid plaques in transgenic mice, *Science* **274**:99–102.

Iqbal, K., Zaidi, T., Bancher, C., and Grundke-Iqbal, I., 1994, Alzheimer paired helical filaments. Restoration of the biological activity by dephosphorylation, *FEBS Lett.* **349**:104–108.

Iversen, L. L., Mortishire-Smith, R. J., Pollack, S. J., and Shearman, M. S., 1995, The toxicity in vitro of β-amyloid protein, *Biochem. J.* **331**:1–16.

Katzman, R., 1986, Alzheimer's disease, *N. Engl. J. Med.* **314**:964–973.

Koh, J. Y., Yang, L. L., and Cotman, C. W., 1990, β-amyloid protein increases the vulnerability of cultured cortical neurons to excitotoxic damage, *Brain Res.* **533**:315–320.

Ledesma, M. D., Correas, I., Avila, J., and Diaz-Nido, J., 1992, Implication of brain cdc2 and MAP2 kinases in the phosphorylation of tau protein in Alzheimer's disease, *FEBS Lett.* **308**:218–224.

Ledesma, M. D., Bonay, P., Colaco, C., and Avila, J., 1994, Analysis of microtubule-associated protein tau glycation in paired helical filaments, *J. Biol. Chem.* **269**:21614–21619.

Lockhart, B. P., Benicourt, C., Junien, J. L., and Privat, A., 1994, Inhibitors of free radical formation fail to attenuate direct β-amyloid$_{25-35}$ peptide-mediated neurotoxicity in rat hippocampal cultures, *J. Neurosci. Res.* **39**: 494–505.

Mann, D. M. A., Marcyniuk, B., Yates, P. O., Neary, D., and Snowden, J. S., 1988, The progression of the pathological changes of Alzheimer's disease in frontal and temporal neocortex examined both at biopsy and at autopsy, *Neuropathol. Appl. Neurobiol.* **14**:177–195.

Masliah, E. and Terry, R. D., 1993, Role of synaptic pathology in the mechanisms of dementia in Alzheimer's disease, *Clin. Neurosci.* **1**:192–198.

Mattson, M. P., Cheng, B., Davis, D., Bryant, K., Lieberburg, I., and Rydel, R. E., 1992, β-amyloid peptides destabilize calcium homeostasis and render human cortical neurons vulnerable to excitotoxicity, *J. Neurosci.* **12**:376–389.

Mattson, M. P., Carney, J. W., and Butterfield, D. A., 1995, A tombstone in Alzheimer's?, *Nature* **373**:481.

McGeer, P. L., and Rogers, J., 1992, Anti-inflammatory agents as a therapeutic approach to Alzheimer's disease, *Neurology* 42:447–449.

McLachlan, D. R., Kruck, T. P., Lukiw, W. J., and Krishnan, S. S., 1991, Would decreased aluminum ingestion reduce the incidence of Alzheimer's disease?, *Can. Med. Assoc. J.* 145:793–804.

McShea, A., Harris, P. L. R., Webster, K. R., Wahl, A., and Smith, M. A., 1997, Abnormal expression of the cell cycle regulators P16 and CDK4 in Alzheimer's disease., *Am. J. Pathol.* 150 (in press).

Miyata, M. and Smith, J. D., 1996, Apolipoprotein E allelle-specific antioxidant activity and effects on cytotoxicity by oxidative insults and β-amyloid peptides, *Nature Genetics* 14:55–61.

Montine, T. J., Amarnath, V., Martin, M. E., Strittmatter, W. J., and Graham, D. G., 1996a, E-4-hydroxy-2-nonenal is cytotoxic and cross-links cytoskeletal proteins in P19 neuroglial cultures, *Am. J. Pathol.* 148:89–93.

Montine, T. J., Huang, D. Y., Valentine, W. M., Amarnath, V., Saunders, A., Weisgraber, K. H., Graham, D. G., and Strittmatter, W. J., 1996b, Crosslinking of apolipoprotein E by products of lipid peroxidation, *J. Neuropathol. Exp. Neurol.* 55: 202–210.

Oteiza, P. I., 1994, A mechanism for the stimulatory effect of aluminum on iron-induced lipid peroxidation, *Arch. Biochem. Biophys.* 308:374–379.

Pappolla, M. A., Omar, R. A., Kim, K. S., and Robakis, N. K., 1992, Immunohistochemical evidence of antioxidant stress in Alzheimer's disease, *Am. J. Pathol.* 140:621–628.

Perry, G., Kawai, M., Tabaton, M., Onorato, M., Mulvihill, P., Richey, P., Morandi, A., Connolly, J. A., and Gambetti, P., 1991, Neuropil threads of Alzheimer's disease show a marked alteration of the normal cytoskeleton, *J. Neurosci.* 11:1748–1755.

Pope, W. B., Lambert, M. P., Leypold, B., Seupaul, R., Sletten, L., Krafft, G., and Klein, W. L., 1994, Microtubule-associated protein tau is hyperphosphorylated during mitosis in the human neuroblastoma cell line SH-SY5Y, *Exp. Neurol.* 126:185–194.

Prehn, J. H., Bindokas, V. P., Jordan, J., Galindo, M. F., Ghadge, G. D., Roos, R. P., Boise, L. H., Thompson, C. B., Krajewski, S., Reed, J. C., and Miller, R. J., 1996, Protective effect of transforming growth factor-β 1 on β-amyloid neurotoxicity in rat hippocampal neurons, *Mol. Pharmacol.* 49:319–328.

Premkumar, D. R. D., Smith, M. A., Richey, P. L., Petersen, R. B., Castellani, R., Kutty, R. K., Wiggert, B., Perry, G., and Kalaria, R. N., 1995, Induction of heme oxygenase-1 mRNA and protein in neocortex and cerebral vessels in Alzheimer's disease, *J. Neurochem.* 65:1399–1402.

Preuss, U., Doring, F., Illenberger, S., and Mandelkow, E. M., 1995, Cell cycle-dependent phosphorylation and microtubule binding of tau protein stably transfected into Chinese hamster ovary cells, *Mol. Biol. Cell* 6:1397–1410.

Rich, J. B., Rasmusson, D. X., Folstein, M. F., Carson, K. A., Kawas, C., and Brandt, J., 1995, Nonsteroidal anti-inflammatory drugs in Alzheimer's disease, *Neurology* 45:51–55.

Rogers, J., Kirby, L. C., Hempelman, S. R., Berry, D. L., McGeer, P. L., Kaszniak, A. W., Zalinski, J., Cofield, M., Mansukhani, L., Willson, P., and Kogan, F., 1993, Clinical trial of indomethacin in Alzheimer's disease, *Neurology* 43:1609–1611.

Roses, A. D., 1995, On the metabolism of apolipoprotein E and the Alzheimer diseases, *Exp. Neurol.* 132:149–156.

Sayre, L. M., Zelasko, D. A., Harris, P. L. R., Perry, G., Salomon, R. G., and Smith, M. A., 1997a, 4-Hydroxynonenal-derived advanced lipid peroxidation end products are increased in Alzheimer's disease, *J. Neurochem.* 68:2092–2097.

Sayre, L. M., Zagorski, M. G., Surewicz, W. K., Krafft, G. A., and Perry, G., 1997b, Mechanisms of neurotoxicity associated with amyloid β deposition and the role of free radicals in the pathogenesis of Alzheimer's disease. A critical appraisal. *Chem. Res. Toxicol.* (in press).

Schreck, R., Rieber, P., Baeuerle, P. A., 1991, Reactive oxygen intermediates as apparently widely used messengers in the activation of the NF-κB transcription factor and HIV-1, *EMBO J.* 10, 2247–2258.

Selkoe, D. J., Ihara, Y., and Salazar, F. J., 1992, Alzheimer's disease: insolubility of partially purified paired helical filaments in sodium dodecyl sulfate and urea, *Science* 215:1243–1245.

Selkoe, D. J., 1996, Amyloid beta-protein and the genetics of Alzheimer's disease, *J. Biol. Chem.* 271: 18295–18298.

Selkoe, D. J., 1997, Alzheimer's disease: genotypes, phenotypes, and treatments, *Science* 275:630–631.

Sherrington, R., Rogaev, E. I., Liang, Y., Rogaeva, E. A., Levesque, G., Ikeda, M., Chi, H., Lin, C., Li, G., Holman, K., Tsuda, T., Mar, L., Foncin, J. -F., Bruni, A. C., Montesi, M. P., Sorbi, S., Rainero, I., Pinessi, L., Polinsky, R. J., Wasco, W., Da Silva, H. A. R., Haines, J. L., Pericak-Vance, M. A., Tanzi, R. E., Roses, A. D., Fraser, P. E., Rommens, J. M., St. George-Hyslop, P. H., 1995, Cloning of a gene bearing missense mutations in early-onset familial Alzheimer's disease, *Nature* 375: 754–760.

Smith, C. D., Carney, J. M., Starke-Reed, P. E., Oliver, C. N., Stadtman, E. R., Floyd, R. A., and Marksberry, W. R., 1991, Excess brain protein oxidation and enzyme dysfunction in normal aging and in Alzheimer disease, *Proc. Natl. Acad. Sci. USA* 88:10540–10543.

Smith, M. A., Taneda, S., Richey, P. L., Miyata, S., Yan, S.-D., Stern, D., Sayre, L. M., Monnier, V. M., and Perry, G., 1994a, Advanced Maillard reaction products are associated with Alzheimer disease pathology, *Proc. Natl. Acad. Sci. USA* **91**:5710–5714.

Smith, M. A., Kutty, R. K., Richey, P. L., Yan, S.-D., Stern, D., Chader, G. J., Wiggert, B., Petersen, R. B., and Perry, G., 1994b, Heme oxygenase-1 is associated with the neurofibrillary pathology of Alzheimer's disease, *Am. J. Pathol.* **145**:42–47.

Smith, M. A., Sayre, L. M., Monnier, V. M., and Perry, G., 1995a, Radical AGEing in Alzheimer's disease, *Trends Neurosci.* **18**:172–176.

Smith, M. A., Rudnicka-Nawrot, M., Richey, P. L., Praprotnik, D., Mulvihill, P., Miller, C. A., Sayre, L. M., and Perry, G., 1995b, Carbonyl-related posttranslational modification of neurofilament protein in the neurofibrillary pathology of Alzheimer's disease, *J. Neurochem.* **64**:2660–2666.

Smith, M. A., Sayre, L. M., Vitek, M. P., Monnier, V. M., and Perry, G., 1995c, Early AGEing and Alzheimer's, *Nature* **374**:316.

Smith, M. A., Perry, G., Richey, P. L., Sayre, L. M., Anderson, V. E., Beal, M. F., and Kowall, N., 1996a, Oxidative damage in Alzheimer's, *Nature* **382**:120–121.

Smith, M. A., Siedlak, S. L., Richey, P. L., Nagaraj, R. H., Elhammer, A., and Perry, G., 1996b, Quantitative solubilization and analysis of insoluble paired helical filaments from Alzheimer disease, *Brain Res.* **717**:99–108.

Smith, M. A., Harris, P.L.R., Sayre, L.M. and Perry, G., 1997a, Redox-active iron is associated with the pathological lesions in Alzheimer disease, *J. Neuropath. Exp. Neurol.* **56**:608.

Smith, M. A., Harris, P. L. R., Sayre, L. M., Beckman, J. S., and Perry, G., 1997b, Widespread peroxynitrite-mediated damage in Alzheimer's disease, *J. Neurosci.* **17**:2653–2657.

Smith, M. A., 1997, Alzheimer Disease, in: *International Review of Neurobiology*, (R. J. Bradley and R. A. Harris, eds.), in press, Academic Press, San Diego.

Sternberger, N. H., Sternberger, L. A., and Ulrich, J., 1985, Aberrant neurofilament phosphorylation in Alzheimer disease, *Proc. Natl. Acad. Sci. USA* **82**: 4274–4276.

Stewart, W. F., Kawas, C., Corrada, M., and Metter, E. J., 1997, *Neurology* **48**:626–631.

Strittmatter, W. J., Weisgraber, K. H., Huang, D. Y., Dong, L. M., Salvesen, G. S., Pericak-Vance, M., Schmechel, D., Saunders, A. M., Goldgaber, D., and Roses, A. D., 1993, Binding of human apolipoprotein E to synthetic amyloid β peptide: isoform-specific effects and implications for late-onset Alzheimer disease, *Proc. Natl. Acad. Sci. USA* **90**:8098–8102.

Su, J. H., Anderson, A. J., Cummings, B. J., and Cotman, C. W., 1994, Immunohistochemical evidence for apoptosis in Alzheimer's disease, *Neuroreport* **5**:2529–2533.

Trojanowski, J. Q., Schmidt, M. L., Shin, R.-W., Bramblett, G. T., Goedert, M., and Lee, V. M.-Y., 1993, PHF-τ (A68): From pathological marker to potential mediator of neuronal dysfunction and degeneration in Alzheimer's disease, *Clin. Neurosci.* **1**:184–191.

Vincent, I., Rosado, M., and Davies, P., 1996, Mitotic mechanisms in Alzheimer's disease?, *J. Cell Biol.* **132**:413–425.

Vitek, M. P., Bhattacharya, K., Glendening, J. M., Stopa, E., Vlassara, H., Bucala, R., Manogue, K., and Cerami, A., 1994, Advanced glycation end products contribute to amyloidosis in Alzheimer disease, *Proc. Natl. Acad. Sci. USA* **91**:4766–4770.

Wolozin, B., Iwasaki, K., Vito, P., Ganjei, J. K., Lacana, E., Sunderland, T., Zhao, B., Kusiak, J. W., Wasco, W., and D'Adamio, L., 1996, Participation of presenilin 2 in apoptosis: enhanced basal activity conferred by an Alzheimer mutation, *Science* **274**:1710–1713.

Yan, S.-D., Chen, X., Schmidt, A.-M., Brett, J., Godman, G., Zou, Y.-S., Scott, C. W., Caputo, C., Frappier, T., Smith, M. A., Perry, G., Yen, S.-H., and Stern, D., 1994, Glycated tau protein in Alzheimer disease: a mechanism for induction of oxidant stress, *Proc. Natl. Acad. Sci. USA* **91**:7787–7791.

Yan, S.-D., Yan, S. F., Chen, X., Fu, J., Chen, M., Kuppusamy, P., Smith, M.A., Perry, G., Godman, G.C., Nawroth, P., Zweier, J.L. and Stern, D., 1995, Non-enzymatically glycated tau in Alzheimer's disease induces neuronal oxidant stress resulting in cytokine gene expression and release of amyloid β-peptide, *Nature Medicine* **1**:693–699.

Yan, S.-D., Chen, X., Fu, J., Chen, M., Zhu, H., Roher, A., Slattery, T., Zhao, L., Nagashima, M., Morser, J., Migheli, A., Nawroth, P., Stern, D., Schmidt, A. M., 1996, RAGE and amyloid-β peptide neurotoxicity in Alzheimer's disease, *Nature* **382**:685–691.

Yankner, B. A., Duffy, L. K., and Kirschner, D. A., 1990, Neurotrophic and neurotoxic effects of amyloid beta protein: reversal by tachykinin neuropeptides, *Science* **250**:279–282.

Zhang, Z., Rydel, R. E., Drzewiecki, G. J., Fuson, K., Wright, S., Wogulis, M., Audia, J. E., May, P. C., and Hyslop, P. A., 1996, Amyloid beta-mediated oxidative and metabolic stress in rat cortical neurons: no direct evidence for a role for H_2O_2 generation, *J. Neurochem.* **67**:1595–1606.

20

BRAIN CYTOCHROME OXIDASE ACTIVITY AND HOW IT RELATES TO THE PATHOPHYSIOLOGY OF MEMORY AND ALZHEIMER'S DISEASE

F. Gonzalez-Lima, J. Valla, and A. Čada

Department of Psychology and Institute for Neuroscience
The University of Texas at Austin
Mezes 330, Austin, Texas 78712

1. ABSTRACT

Recent studies indicate that mitochondrial electron transport dysfunction is involved in various neurodegenerative diseases, including Alzheimer's disease. Although much attention has been devoted to Alzheimer's disease, relatively little has been devoted to the role of oxidative energy metabolism in this disease. Whether genetic or environmental, the pathogenesis of Alzheimer's disease involves a cascade of multiple intracellular events, eventually resulting in failure of oxidative energy metabolism. It is proposed that impairment of cytochrome oxidase activity in energy metabolism initiates the degenerative process by the formation of reactive oxygen species and free radicals. Brain energy failure was produced in an animal pharmacological model based on cytochrome oxidase enzymatic inhibition with sodium azide. Quantitative cytochemical methods for assessing cytochrome oxidase activity in individual neurons are discussed in relationship to the selective vulnerability of larger projection neurons in Alzheimer's brains. Cytochrome oxidase activity is discussed in relationship to neuronal oxidative metabolism, memory and Alzheimer's disease.

2. INTRODUCTION

Cytochrome c oxidase (C.O., also known as cytochrome aa_3, ferrocytochrome c, oxygen oxidoreductase, EC 1.9.3.1) is the mitochondrial enzyme responsible for the utilization of oxygen for aerobic energy metabolism. In the final step of the cellular electron transport chain (respiratory chain), molecules of C.O. pass their electrons to molecular oxygen. An estimated 95% of the oxygen used by cells reacts in this single process (Wikstrom, Krab & Saraste, 1981).

Free Radicals, Oxidative Stress, and Antioxidants, edited by Özben.
Plenum Press, New York, 1998.

Among the initial events in Alzheimer's disease (A.D.), patients exhibit decreased C.O. activity as well as spatial memory deficits. Therefore, we modeled these particular aspects of A.D. in animals by experimentally compromising brain energy metabolism in ways expected to decrease C.O. activity and affect spatial memory. We then assessed cellular changes in C.O. activity in the experimental brains and in A.D. brains with a quantitative C.O. histochemical method developed in our laboratory (Čada, Gonzalez-Lima, Rose & Bennett, 1995; Gonzalez-Lima & Čada, 1994; Gonzalez-Lima, Valla & Matos-Collazo, 1997). We propose that C.O. inhibition is the major source of oxygen radicals that lead to oxidative stress in A.D. The proposed role of C.O. in A.D. is based on the following three lines of scientific evidence.

1.1. Cytochrome Oxidase Deficiency in Alzheimer's Disease

A specific deficiency of C.O. activity in late-onset Alzheimer's disease brain tissue has been confirmed independently by several laboratories in the U.S. and Canada, including our lab (Chagnon, Betard, Robitaille, Cholette, 1995; Gonzalez-Lima *et al.*, 1997; Kish, Bergeron, Rajput, Dozic, Mastrogiacomo, Chang, Wilson, DiStefano, & Nobrega, 1992; Mutisya, Bowling, & Beal, 1994; Parker Jr., Parks, Filley, & Kleinschmidt-DeMasters, 1994b). Since the brain relies almost exclusively on the aerobic metabolism of glucose for its energy, C.O. function is essential for normal brain function. Parker and coworkers found C.O. activity decreases of 17.2% to 50% in platelet mitochondria isolated from patients with A.D (Parker Jr., Filley, & Parks, 1990; Parker Jr., Mahr, Filley, Parks, Hughes, Young, & Cullum, 1994). This systemic deficiency is also supported by the finding of mutations in mitochondrial C.O. genes found in A.D. (Davis, Miller, Herrnstadt, Ghosh, Fahy, Shinobu, Galasko, Thal, Beal, Howell, & Parker, 1997), and the mounting evidence cited below that a C.O. catalytic defect is a primary event in A.D. leading to cellular oxidative damage.

1.2. Oxidative Damage in Alzheimer's Disease

A chronic defect in C.O. activity in A.D. leads to oxidative damage to mitochondrial DNA and cell death in the brain (Mecocci, MacGarvey, & Beal, 1994). Increasing evidence indicates that oxidative damage is one of the important changes in the development of cell injury in A.D. (Sayre, Zelasko, Harris, Perry, Salomon, & Smith, 1997; Smith, Lawerence, Monnier, & Perry, 1995a; Smith, Rudnicka-Nawrot, Richey, Praprotnik, Mulvihill, Miller, Sayre, & Perry, 1995b; Smith, Perry, Richey, Sayre, Anderson, Beal, & Kowall, 1996a; Smith, Sayre & Perry, 1996b; Smith, Richey-Harris, Sayre, Beckman, & Perry, 1997; Yan, Chen, Schmidt, Brett, Godman, Zou, Scott, Caputo, Frappier, Smith, Perry, Yen, & Stern, 1994; Yan, Yan, Chen, Fu, Chen, Kuppusamy, Smith, Perry, Godman, Nawroth, Zweier, & Stern, 1995). Measurement of the degree of DNA damage is important for assessing toxicity of a C.O. defect in A.D. or the effectiveness of a treatment as an inhibitor of oxidative stress. Mitochondria are the major intracellular source of reactive oxygen species. Hence, mitochondrial DNA is subject to severe oxidative damage, much more than nuclear DNA. Damage is assessed by the detection of various base modifications, particularly 8-hydroxy-deoxy-guanosine, which can lead to point mutations because of mispairing (reviewed by Richter, 1995).

1.3. Quantitative C.O. Histochemistry in Alzheimer's Disease

Seligman, Karnovsky, Wasserkroug and Hanker (1968) developed a histochemical technique based on the oxidative polymerization of diaminobenzidine to a reaction product that chromatically labeled C.O. in heart, liver and kidney. Wong-Riley (1979) modified this technique, applying it to the nervous system and later demonstrating that the optical density of histochemically stained sections is closely correlated with the concentration of C.O. in the tissue (Hevner & Wong-Riley, 1989). Histochemistry has become the tool of choice for displaying visually the regional oxidative metabolic capacity of the nervous system. In 1991 Gonzalez-Lima and Garrosa introduced a more sensitive method to quantify changes in regional cerebral C.O. activity within histochemically stained sections, utilizing internal tissue standards of known C.O. activity together with computerized image analysis. This new method has been further validated, refined and applied to various tissues and animal species in recent years (Coomber, Crews, & Gonzalez-Lima, 1997; Gonzalez-Lima, 1992; Gonzalez-Lima & Čada, 1994; Gonzalez-Lima & Jones, 1994). In 1997 Gonzalez-Lima et al. reported the first successful quantitative cytochemical study of C.O. activity in normal humans and A.D. patients. This more sensitive histochemical method used in postmortem brains was able to detect C.O. activity decreases of 17.7% in overall cell bodies and neuropil, and of 10.3% in peak neuropil activity in mesencephalic neurons (Gonzalez-Lima et al., 1997).

2. GENERAL METHODS

2.1. Quantitative C.O. Histochemistry

Tissues were stained for C.O. activity using the quantitative C.O. cytochemical technique of Gonzalez-Lima et al. (1997) which has the advantage that it provides cellular spatial resolution at the light microscope level.

Figure 1 is a schematic of the model of the diaminobenzidine (DAB) histochemical reaction adapted from Wong-Riley (1989). C.O. is an integral transmembrane protein of the inner mitochondrial membrane. Electrons donated from DAB reduce cytochrome c (Cyt C). C.O. then (Cyt a-a$_3$) catalyzes the transfer of electrons from Cyt c to molecular oxygen to form water. The staining in the histochemical reaction is produced when DAB is oxidized to an indamine polymer (OXID DAB). Since continuous reoxidation of Cyt C by C.O. is required for the accumulation of the visible OXID DAB, this reaction serves to visualize C.O. activity. The staining of the reaction product is further intensified by the addition of cobalt to the preincubation solution (Gonzalez-Lima & Jones, 1994).

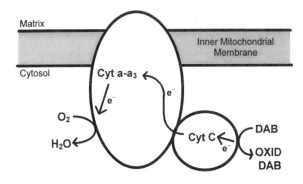

Figure 1.

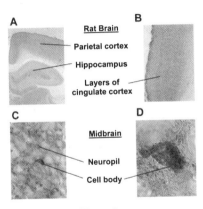

Figure 2.

Figure 2 shows examples of nervous tissue stained with our C.O. method. Panels A and B correspond to sections of the rat brain at the regional macroscopic level. Panel C and D show human midbrain tissue at low and high magnifications at the light microscope level. The grains of dark stain in the cell body and neuropil correspond to the mitochondria stained with the oxidized DAB reaction product visualized cytochemically.

Briefly, our current method producing results illustrated in the Figure 2 plates involved 3 steps:

1. *Tissue sectioning and staining.* Frozen tissue was sectioned at 40 μm and picked up on clean slides in a Frigocut 2800 cryostat at −15°C. Slides were processed for C.O. quantitative histochemistry as described in Gonzalez-Lima and Čada (1994). Preincubation was followed by incubation at 37°C for 60 min.

2. *Spectrophotometry.* C.O. activity standards were made from the brains of 12 adult male rats. The brains were homogenized at 4°C and aliquots were frozen and sectioned at varying thickness to develop a gradient of C.O. activity in the sections used as standards. Standards were stained together with tissue sections to generate a calibration curve between standard C.O. activity and optical density in the tissue. C.O. activity in the standards was spectrophotometrically measured with methods adapted from Wharton & Tzagoloff (1967) and Hevner, Liu & Wong-Riley (1993) as reported in Čada *et al.* (1995). Activity units were defined at pH 7 and 37°C as 1 unit oxidizes 1 μmol of reduced cytochrome c per min (μmol/min/g tissue wet weight).

3. *Microscopic Image Analysis.* Stained tissue slides were mounted on a light microscope or DC-powered light box connected to an imaging system. A high resolution CCD video camera captured the images and transmitted them to a frame grabber in a computer where the image was digitized. Analysis was completed with JAVA (Jandel Scientific) imaging software. A calibration strip containing known optical densities (Kodak) was imaged at the beginning of each session and was used to construct a calibration curve for the conversion of gray levels to optical density for that session. Each imaging session was thereby independently calibrated to optical density. Thickness standards of known C.O. activity (measured spectrophotometrically) were included in each staining batch and were imaged on the DC-powered light box using JAVA. The change in optical density showed linear relationships with respect to tissue activity, section thickness, and incubation time in each of the staining batches. The optical density and activity

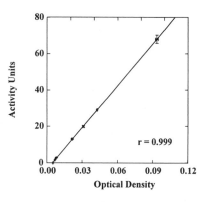

Figure 3. Calibration curve showing the linear relationship between C.O. activity and optical density in the histochemical tissue standards. Activity units were measured spectrophotometrically and are expressed as µmol/min/g tissue wet weight. Optical density of both brain and heart paste standards was measured with an imaging system calibrated with an optical density step tablet. Standard error bars of mean measures are shown, but in most cases are smaller than the size of the symbols.

measurements of these standards were then used to construct a regression equation. Optical density measures of the sections were thereby independently converted to C.O. activity units using the calibration curve generated with the standards' optical density and activity units as illustrated in Figure 3.

2.2. Experimental C.O. Inhibition with Sodium Azide

Azide was administered using the same methods described previously (Čada et al., 1995). Briefly, each rat is anesthetized with secobarbital (40 mg/kg) and implanted subcutaneously with an Alzet 2ML4 osmotic minipump containing either sodium azide solution (160 g/L) or the 0.9% saline vehicle. The Alzet 2ML4 minipump has a 2 ml reservoir and provides a constant infusion rate of 2.5 µl/hr for 28 days; therefore, the dose of sodium azide delivered is 400 µg/hr. This treatment regimen has been shown to decrease C.O. activity by 35% in mitochondria extracted from rat forebrain homogenates (Bennett et al., 1992b).

Brain energy failure based on C.O. enzymatic inhibition was produced by this pharmacological model. Thus we could address three fundamental questions: 1) What regional changes in C.O. activity are found in azide-treated rats with compromised brain energy metabolism? 2) What behavioral and brain deficits are common among this model and A.D. patients? and 3) Do C.O. deficits in this model parallel cellular oxidative metabolism changes in Alzheimer's patients with spatial memory deficits?

3. EFFECTS OF EXPERIMENTAL CYTOCHROME OXIDASE INHIBITION

In order to test the hypothesis that direct C.O. inhibition will result in similar neural metabolic and behavioral deficits found in naturally aged and chronic ischemic rats, sodium azide, which specifically inhibits C.O. activity, was administered to 4 month old rats. The brains of rats treated for 2 weeks with sodium azide or saline vehicle, via subcutaneously implanted osmotic minipumps, were analyzed for C.O. histochemistry as previously described. We found that this treatment regimen decreases overall brain C.O. activity by about 30%, with significant variability in degree between some brain regions (Čada et al., 1995). Since this model selectively impairs spatial memory and the expression of hippocampal long-term potentiation, we could relate C.O. deficits with behavioral deficits. Commonly affected brain regions where the amount of C.O. inhibition was consistently correlated with the degree of spatial memory deficits were identified for future detailed cellular comparisons with the A.D. brains.

3.1. Background and Rationale for Using Sodium Azide to Inhibit C.O. Activity

Analyses of C.O. histochemical measures across many species have shown that there is heterogeneity in the distribution of regional metabolic capacity (Wong-Riley, 1989 for review). In addition to its interregional heterogeneity, C.O. activity can be altered by manipulating neuronal activity. For example, C.O. activity in the auditory nuclei can be inhibited by blocking auditory input for a period of days (Wong-Riley *et al.*, 1978). A similar inhibition of C.O. activity has been demonstrated in the visual system of cats by a variety of techniques that block neural input from one eye (Wong-Riley & Riley, 1983). Nobrega, Raymond, DiStefano, & Burnham, (1993) induced an increase in C.O. activity restricted to limbic areas of rat brains measured 28 days after a series of electroconvulsive treatments. Taken together, these results imply that C.O. activity is positively correlated with neuronal activity.

Bennett and coworkers developed a rat model of persistent systemic C.O. inhibition (Bennett, Diamond, Parker, Stryker, & Rose, 1992a; Bennett, Diamond, Stryker, Parks, & Parker, Jr., 1992b). Continuous subcutaneous infusion of sodium azide (NaN_3) in rats produced a chronic partial inhibition (35–39%) of cytochrome oxidase activity in mitochondrial fractions of whole brain homogenates, without significantly affecting other enzymes of the respiratory chain (complexes I, II, and III). This azide treatment impaired spatial learning and memory and the expression of hippocampal long-term potentiation (LTP), a physiological model of long-lasting synaptic plasticity (Bennett *et al.*, 1992a,b; Bennett & Rose, 1992). Thus, there is a priori evidence to assume that particular regions related to spatial memory functions will be more sensitive to C.O. inhibition than others; for example the hippocampal formation, the amygdala and the frontal cortex have been implicated in numerous studies (reviewed by Aggleton, 1992; Olton, Wible, Pang, & Sakurai, 1989). Azide-induced C.O. inhibition also leads to formation of reactive free radicals and oxidative damage (Partridge, Monroe, Parks, Johnson, Parker Jr., Eaton, & Eaton, 1994).

It has been shown that C.O. is more vulnerable to insult in Alzheimer's disease than are other enzymes of the respiratory chain (Beal, Hyman, & Koroshetz, 1993; Kish *et al.*, 1992; Mecocci *et al.*, 1994; Mutisya *et al.*, 1994; Parker *et al.*, 1990, 1994ab). The clinical finding that a C.O. defect in Alzheimer's disease occurs in the periphery as well as in the brain suggests that a defect in metabolism in Alzheimer's disease might be widespread (Blass, Sheu, & Cedarbaum, 1988; Blass, 1993). Yet the genetic studies on differences between cytochrome oxidase subunits in different brain regions suggest that systemic inhibition of C.O. activity may still be associated with selective regional vulnerability within the brain (Chandrasekaran, Stoll, Giordano, Atack. Matocha, Brady, & Rapoport, 1992).

The goals of our experiment were to quantify histochemically the effects of sodium azide treatment on C.O. activity in rat brain and to determine whether there is differential vulnerability among brain regions in response to this treatment. The goal was to select a sampling of regions from telencephalic, diencephalic, and mesencephalic levels to test the hypothesis that regional effects of systemic azide are heterogeneous in the brain. The main findings are discussed in the context of how preferential sodium azide inhibition of C.O. in some brain regions could contribute to the learning and memory deficits that are induced by this treatment. These results were reported in more detail in Čada *et al.* (1995).

3.2. Methodological Histochemical Considerations

Calibration curves were generated using histochemically measured O.D. of the brain paste standards and the spectrophotometrically determined C.O. activity units of the brain paste thickness standards. The three sets of standards, though reacted in separate incubation media, showed no significant differences (t-tests, p > 0.21) between their optical densities. The interassay coefficient of variation (standard deviation × 100 / mean) was 0.82% for the three staining batches. For each one of the paste standards, its average value showed a standard error (S.E.) below 0.5% of the mean measured across the batches. A regression equation generated using the activity of the thickness standards from separate incubations and the O.D. of the brain paste standards resulted in a linear function: r = 0.978, $y = 6377.24x - 29.89$. Using this regression equation, the O.D. obtained from each structure was converted to activity units (μmol/min/g tissue wet weight).

In the image analysis of regions, the central part of each region was located and measured by two experimenters. The intra- and inter-rater reliability for this sampling technique was assessed by comparing 15 measures made twice from the same 15 loci, separated by a three-month interval. For the intra-rater reliability, the same experimenter made the measures, and this resulted in a mean coefficient of variation of 5.41% and a correlation of r = 0.93. For the inter-rater reliability, a second experimenter made the second measures, and this resulted in a coefficient of variation of 5.81% and a correlation of r = 0.87 between measures of the same regions.

Using calibration standards of known C.O. activity together with quantitative image analysis of histochemical sections formed the basis for the applied quantitative approach (Gonzalez-Lima & Garrosa, 1991). Both of these tools are commonly used in 2-deoxyglucose autoradiography (Gonzalez-Lima, 1992; Gonzalez-Lima et al., 1993) and other metabolic mapping techniques (Biegon & Wolff, 1986; Nobrega, 1992; Nobrega et al., 1993). Histochemical techniques such as the one used here contain some variability between tissue processed in different incubation reactions. However, if tissue processing procedures are strictly reproduced and complete sets of standards of known C.O. activity are included with each incubation medium—as was done in the present study—the problem of interassay variability is largely resolved. Thus the percent differences found between the various brain regions in response to azide treatment could not be accounted for simply by variability between assays or paste standards. The important point is not that the tissue level of C.O. activity is different among brain regions (Wong-Riley, 1989, for review); but rather that the between-group percent decreases produced by azide treatment are significantly larger in some regions.

3.3. Regional Effects of Sodium Azide on Brain C.O. Activity

The total C.O. activity was 170.43 ± 0.1 units (mean ± S.E.) for the control brains, and 121.06 ± 0.1 units for the azide-treated brains. This corresponded to an overall –29% difference between the groups. The maximal activity among the regions sampled was 228 units in the occipital cortex (Area 18) of the control brains. The maximal activity in the azide-treated brains was below 180 units, found in the basolateral amygdaloid nucleus (BlA). The total average for C.O. activity units in the control brains (170.43 ± 0.1) was very similar to the total content of C.O. activity in the whole rat brain (158 ± 5) reported by Hevner et al. (1993). Our somewhat higher value may be attributed to the fact that it represents a mean from a sample of 22 regions rather than a whole brain homogenate that may contain more white matter (with lower activity). In addition, our C.O. units were defined at pH 7 and 37°C, as in our original method (Gonzalez-Lima & Garrosa, 1991), as opposed to pH 6 and 30°C as done by Hevner et al. (1993). The agreement between these control brain C.O. values suggests that both assays were optimal in unmasking enzyme activity to maximal or near maximal levels (Hevner et al., 1993). The definition of C.O.

activity units using optimal conditions of measurement may be preferable for studies evaluating absolute C.O. activity, rather than simpler routine assays (Hess & Pope, 1953) done at room temperature in which C.O. units can be defined reliably but at below maximal levels (Gonzalez-Lima & Čada, 1994).

An expected finding was that all brain regions examined showed a reliable decrement in C.O. activity after sodium azide treatment. The regional evaluation of C.O. activity was performed in two steps, from a more general evaluation of integrated activity at three rostrocaudal levels to a more specific evaluation of a sample of 22 separate regions. In both evaluations, the C.O. activities measured in the sodium azide treated rats were significantly lower than those found in the control rats ($p < 0.01$, ANOVA followed by subsequent comparisons with corrected t-tests).

The first regional analysis showed that in addition to the between-groups reduction in C.O. activity at each level, there was indication of the mesencephalic level being more strongly affected than the others. The mean percent reductions ($\pm$ S.E.) for each level were: 27.31 ± 0.77 for the telencephalic, 29.27 ± 1.18 for the diencephalic, and 35.70 ± 0.33 for the mesencephalic levels. In the sodium azide treated rats, the percentages of C.O. activity reduction in the sections (n = 12/level) at the mesencephalic levels were significantly greater than those at the telencephalic levels when compared using nonparametric tests (Mann-Whitney U – Wilcoxon Rank Sum W tests: U=10, W=88, $p < 0.01$). But no significant differences in percent reduction were found between telencephalic and diencephalic levels ($p > 0.9$).

The more specific regional analysis involved measuring C.O. activities in 22 brain regions. This evaluation served to confirm the inhibitory effect of azide treatment on a region-by-region basis and allowed the identification of the regions with the highest decrement in activity, corresponding to the deep mesencephalic reticular area (-37.11%) and the central amygdala (-37.02%). These regions showed decrements which were significantly greater than the other decrements found in the telencephalic (U=28, W=106, $p < 0.05$) and diencephalic (U=28, W=106, $p < 0.05$) levels, but not significantly greater than mesencephalic levels (U=53, W=131, $p > 0.29$). Other activity decrements at the regional telencephalic and diencephalic levels showed statistically comparable means.

3.4. Differential Vulnerability of Brain Regions to C.O. Inhibition

We made three main observations. First, there was a general decrement in C.O. activity following sodium azide treatment. This decrement ranged from 27% to 35% in the integrated activities measured at telencephalic, diencephalic and mesencephalic levels, and between 25% and 37% in the individual regions analyzed. This decrement measured histochemically is consistent with a previous report of 35–39% decrease in biochemically assayed mitochondrial C.O. activity in brain homogenates from rats given the same azide treatment (Bennett *et al.*, 1992b). Second, midbrain reticular formation and central amygdala appeared more vulnerable than other regions to the sodium azide effects. Third, differences in regional vulnerability were manifested as different patterns of interregional activity correlations found in control and treated brains (Čada *et al.*, 1995).

The brain is especially vulnerable to sodium azide treatment because of its disproportionate aerobic energy requirement, thus making it strongly dependent on oxidative phosphorylation (Wong-Riley, 1989). Impairments of aerobic respiration have the potential for limiting the activity of the brain (Wikström *et al.*, 1981). Decreases of 25–37% in C.O. activity may be of potential functional significance for brain processes affected by aging such as learning and memory (Bennett *et al.*, 1992a,b). For example, in a comparison of C.O. activity in mitochondria isolated from brains of 4-month-old and 30-month-old rats, a significant age-related decrease of 25% in C.O. activity was observed in the parietotemporal cortex (Curti, Giangare, Redolfi, Fugaccia & Benzi, 1990). A C.O. activity decrease of 25% or more may be associated with a reduced capacity for ATP production, as suggested by a parallel regulation of C.O. and Na^+, K^+-ATPase activities in brain (Hevner, Duff & Wong-Riley, 1992).

A possible explanation for the regional effects of azide on C.O. activity may be differences in access of azide to different regions, but there is presently no experimental support for this mechanism. Systemic infusion of sodium azide produced heterogeneity both in the degree of C.O. inhibition and in the correlations of C.O. activity among brain regions. Whether all brain regions presumably had equal access to this highly diffusible compound is unknown. Another possible explanation for the selective vulnerability of some brain regions may be related to the differential expression of C.O. genes in different brain regions. Chandrasekaran et al. (1992) have provided evidence that this is the case in monkey brain. For example, cDNA clones for the three C.O. subunits encoded by mitochondrial DNA showed higher levels of mRNA in frontal pole, dorsal lateral frontal cortex, and hippocampus than in the primary visual or somatosensory cortices, in agreement with heterogeneous C.O. histochemistry in these regions. Chandrasekaran et al. (1992) concluded that such differences may be related to differences in the distribution of neuropil versus cell bodies in the brain regions investigated; and they further suggested that these genetically-mediated regional differences may be relevant to selective regional vulnerability in Alzheimer's disease.

In a subsequent study, Chandrasekaran, Giordano, Brady, Stoll, Martin, & Rapoport, (1994) investigated the expression of C.O.-related genes in the temporal cortex and motor cortex of Alzheimer's brains, and found significantly more decreased mitochondrial RNA levels in the temporal region as compared to the motor region or the same regions in healthy age-matched controls. Therefore, the studies of Chandrasekaran et al. (1992, 1994) provide evidence to suggest that the selective vulnerability of C.O. inhibition may be linked to genetically-inherent capabilities of some regions particularly relevant for memory. This evidence may be relevant for the observed regional heterogeneity in C.O. inhibition after azide treatment.

To our knowledge, no studies on C.O. activity in Alzheimer's brains have been done in the mesencephalic reticular formation or the central amygdala where C.O. inhibition was greatest in the present study after azide treatment. The fact that histopathological plaques and tangles may or may not be abundant in these regions in humans, is not a sufficient argument to discount the potential role of mitochondrial pathophysiological events in these regions in relation to some memory deficits of Alzheimer's patients (Mecocci et al., 1994). The deep mesencephalic reticular formation and the central amygdaloid nucleus exhibited the largest overall decreases in activity in response to the azide treatment. The ascending reticular input from the mesencephalic reticular formation represents an important pathway in the reticular activating system linked to behavioral arousal (Gonzalez-Lima & Scheich, 1985) and memory consolidation (Bloch, 1976); and the central amygdala has a well-established role in memory modulation and behavioral dysfunction (Aggleton, 1992).

3.5. Effects of C.O. Inhibition on Memory Functions

The hippocampal formation as well as the deep mesencephalic reticular formation and the central amygdala are disproportionately affected by sodium azide. The hippocampus is a preferential target of damage in A.D. (Ball et al., 1985) and is typically associated with memory functions in humans and other mammals. Čada *et al.* (1995) found that the interregional positive correlations between C.O. activity in the hippocampal formation and other brain regions, which existed in control brains, disappeared after sodium azide treatment. This functional uncoupling may affect primarily associative memory functions which are the product of interactions between different brain regions (Gonzalez-Lima & McIntosh, 1996).

This azide treatment has been shown to produce learning and memory deficits on appetitively- and aversively-motivated tasks that are highly impaired by hippocampal dysfunction (Bennett *et al.*, 1992a,b; Bennett & Rose, 1992). Indeed, this azide treatment impairs the functional organization of long-term potentiation in the hippocampal formation (Bennett *et al.*, 1992a). Inhibition of C.O. activity can produce some of the symptoms commonly associated with A.D., such as deficits in memory and orientation. Bennett *et al.* (1992) chronically administered sodium azide to rats and found that the subjects showed difficulty with generalized learning tasks (two-way shuttle box and radial maze). These memory impairments did not appear to be secondary to any significant sensory or motor impairments.

Further examination by Bennett and Rose (1992) found that sodium azide administration impairs performance in the Morris water maze task in rats, another spatial learning task. Again without showing motor impairment, the subjects demonstrated difficulty in both acquisition and retention of the task, strengthening the hypothesis that azide infusion produces an A.D.-like learning and memory deficit. This evidence seems to indicate that sodium azide treatment could be a useful animal model of A.D. memory impairment, and it also supports the theory advocating the C.O. defect as the producer of A.D. symptoms.

In addition, ischemic cerebrovascular disease can produce vascular dementia and is a significant risk factor for A.D. A surgical model of chronic cerebrovascular insufficiency that demonstrates consistent spatial memory deficits in aged rats was used by us to study C.O. inhibition (de la Torre, Čada, Nelson, Davis, Sutherland & Gonzalez-Lima, 1997). The objective was to compare 19-month-old rats subjected to bilateral ligation of the carotid arteries with sham-operated controls. Subjects were tested in the Morris water maze memory tasks weekly for four weeks following surgery. Brains were analyzed with C.O. histochemistry. Our data suggest that chronic cerebrovascular insufficiency leads to metabolic and memory impairments in these rats; hippocampal and posterior parietal regions particularly showed C.O. inhibition in the absence of the morphological histopathology that characterizes more severe ischemia (de la Torre *et al.*, 1997).

4. QUANTITATIVE CYTOCHEMISTRY OF CYTOCHROME OXIDASE IN ALZHEIMER'S BRAINS

The clinical significance of the above experimental models was assessed by considering them in relation to C.O. changes in Alzheimer's brains. Frozen tissue samples from A.D. cases, confirmed histopathologically and clinically diagnosed as demented, were compared to control cases matched by age, sex and postmortem time. Although not every regional effect found in the animal models may have a homologous counterpart in A.D. brains, it is hypothesized that C.O. deficits will involve similar cellular groups and cellular compartments that are more vulnerable to A.D. histopathology. Based on our first study, limited to nuclei of the inferior colliculus, we anticipate cellular alterations in C.O. metabolism in specific groups of larger projection neurons that are predominantly affected by A.D., while neighboring neurons are spared (Gonzalez-Lima *et al.*, 1997).

4.1. Background and Rationale for Analysis of C.O. Activity in A.D. Brains

Since C.O. is a mitochondrial enzyme responsible for the activation of oxygen for aerobic metabolism in all animal cells, and the brain relies almost exclusively on the aerobic metabolism of glucose for its energy, C.O. is an essential enzyme for brain function. Furthermore, since C.O. is intimately tied to the production of adenosine triphosphate (ATP), the energy molecule of cells, sustained changes in the demand for ATP energy are reflected in histochemical changes in C.O. activity (Hevner *et al.*, 1992; Wong-Riley, 1979; Wong-Riley, 1989). Therefore, C.O. activity assessed postmortem can be used as an endogenous marker of long-term change in oxidative metabolic energy capacity (Nobrega *et al.*, 1993). Gonzalez-Lima and Garrosa (1991) developed a histochemical method for quantification of C.O. enzymatic activity using calibrated C.O. activity standards measured spectrophotometrically in combination with tissue densitometry. A cytochemical extension of this C.O. quantitative method was implemented using human tissue and cellular microimaging at the light microscope level.

Alzheimer's disease (A.D.) is a pathophysiological state which seems to have a C.O. metabolic alteration. Various researchers have described a C.O. deficit in blood platelets (Parker *et al.* 1994a; Parker, 1991; Parker *et al.*, 1990) as well as in brain homogenates (Kish *et al.*, 1992; Mutisya, Bowling, & Beal, 1994; Parker, Jr., *et al.*, 1994b) from A.D. patients. Administration of substances known to antagonize C.O. activity, such as sodium azide, have been shown to produce behavioral deficits in animals, such as spatial learning deficits in the Morris water maze, similar to some symptoms of A.D. patients (Bennett *et al.*, 1992a,b).

Although extensively mapped and relatively well-defined, the auditory system has been subjected to little scrutiny in regard to A.D. Our laboratory specialization in auditory learning and memory (Gonzalez-Lima, 1992) led us to begin examining the auditory system of A.D. patients. Ohm and Braak (1989) examined the auditory brainstem nuclei of A.D. patients and in 3 out of 7 cases found "considerable neuritic plaque formation" in the inferior colliculus (IC), whereas no plaques were found in the other auditory brainstem nuclei. Sinha and colleagues examined the auditory system in A.D. more thoroughly in 9 A.D. cases and 8 age-matched controls (Sinha, Hollen, Rodriguez, & Miller, 1993). They located numerous plaques and neurofibrillary tangles in the IC, all contained within the central nucleus of A.D. patients, and no changes in the controls. Therefore, it was anticipated based on these studies that C.O. activity differences between A.D. patients and age-matched controls would be found in the central nucleus of the IC rather than in the other IC nuclei (dorsal and external). It is also possible that initial alterations in the IC may contribute to functional impairments found early in the auditory system of A.D. patients (Grimes, Grady, Foster, Sunderland, & Patronas, 1985).

Thus we investigated C.O. activity patterns in the auditory system, specifically in the IC, to quantify cellular metabolic changes which may occur as part of the pathophysiology of A.D. In particular, we tested the hypothesis that C.O. metabolic alterations in auditory neurons in A.D. would be specific to the larger projection neurons in the central nucleus. This specific prediction is in line with the greater vulnerability of larger projection neurons found in A.D., for example in cortical layers III and V (Van Hoesen, 1990) and in nucleus basalis (Arendt, Bigl, Arendt, & Tennstedt, 1983) and the hypothesis that A.D. affects predominantly specific groups of larger projection neurons while neighboring neurons are spared (Sinha *et al.*, 1993).

4.2. Methodological Cytochemical Considerations in A.D. Brains

Frozen tissue samples were received from the Alzheimer's Disease Research Center Neuropathology Core at the University of Southern California School of Medicine and from the Brain Bank of the Michigan Alzheimer's Disease Research Center at the University of Michigan Medical Center. Tissue samples were stored at $-40°C$ until processing. The non-A.D., non-demented control group consisted of 5 males and 3 females with a mean age of 79.6 ± 3.1 years, a mean post-mortem interval of 6.9 ± 1.6 hours, and a mean brain weight of 1287.5 ± 39.8 grams. The A.D. group consisted of 7 males and 1 female with a mean age of 78.3 ± 2.9 years, a mean post-mortem interval of 6.5 ± 1.3 hours, and a mean brain weight of 1175.0 ± 50.9 grams. Patients' reports included years since diagnosis of A.D. dementia (mean $= 8.6 \pm 1.1$ years) and confirmation of A.D. histopathology. The available drug histories showed no neuroleptic use. Student's two-tailed t tests demonstrated no significant differences between the control group and the A.D. group in age (p > 0.75), post-mortem interval (p > 0.84), and brain weight (p > 0.10).

Tissue was sectioned at 40 µm in the transverse plane and picked up on clean slides in a Frigocut 2800 cryostat at $-15°C$. Slides were processed for C.O. quantitative cytochemistry as in our previously described histochemical procedures (Gonzalez-Lima, 1992; Gonzalez-Lima & Čada, 1994; Gonzalez-Lima & Garrosa, 1991; Gonzalez-Lima & Jones, 1994). Preincubation was followed by incubation at 37°C for 120 min. Adjacent sections were stained with Cresyl violet to delineate the cytoarchitecture of the IC.

Standards of C.O. activity were made from the brains of 12 adult male rats. Each rat was decapitated, its brain rapidly removed and stored in 4°C sodium phosphate buffer (pH 7.4) until all 12 brains were collected. The brains were homogenized at 4°C and aliquots were frozen in 1.5 mL microtubes and sectioned at varying thicknesses to develop a gradient of C.O. activity in the sections used as standards. Sections of the standards were stained together with the IC sections for the generation of a regression equation between standard C.O. activity and optical density in the tissue in each staining batch [16]. C.O. enzyme activity was spectrophotometrically measured using a method adapted from Wharton and Tzagoloff (1967) and Hevner *et al.* (1993) as reported in Čada *et al.* (1995). Activity units were defined at pH 7 and 37°C as in our original quantitative method [14] where 1 unit oxidizes 1 µmol of reduced cytochrome c per min (µmol/min/g tissue wet weight). Activity units can also be expressed in terms of grams of protein content (Lowry, Rosebrough, Farr, & Randall (1951) by multiplying the reported values by 10 since our brain standards contained an average of 10% protein (Gonzalez-Lima & Čada, 1994).

Stained slides were mounted on an Olympus light microscope connected to an image processing system. A 40x objective was used and sections were stained lightly to avoid distributional error, glare, and diffraction errors. Pixel spacing was calibrated with a stage micrometer separately for vertical and horizontal dimensions. For spatial calibration, the JAVA software was used in the computer to load conversion values into look-up tables. The optical density and activity measurements of these standards were then used to construct a regression equation. The change in optical density showed linear relationships with respect to tissue activity, section thickness, and incubation time in each of the three staining batches done (r = 0.96, 0.97, and 0.97). Optical density measures of the IC sections were thereby independently converted to C.O. activity units using the calibration curves generated with the standards' optical density and activity units.

A total of 480 cells were sampled, consisting of 30 cells per subject for 16 subjects. The inferior colliculus was subdivided into three separate nuclei: the central (ICC), dorsal (ICD) and external (ICE). To avoid mistakenly sampling from outside the intended nucleus, observations were restricted to the central part of the IC subdivisions. Ten cells were sampled randomly per nuclear subdivision per subject from up to 8 sections, with sections sampled separated by at least a 120 μm interval. Additional selection criteria for sampling were that tissue was free from artifacts of staining and tissue processing (such as cracks, folds, and foreign particles) and that cells showed no obvious signs of morphological abnormalities (such as vacuolations, dendritic swellings, eccentric nuclei, etc.). A total of eight densitometric and six morphometric measures were taken from each cell; for these measurements, the cell body was oriented in the center of the image area (Figure 2, panel C). All measurements were taken using JAVA software.

4.3. Results of Quantitative C.O. Cytochemistry in Control and A.D. Brains

Comparisons between large and small cell subgroups across the entire inferior colliculus showed that the larger than average cells (> 12.1 μm in diameter) contained higher C.O. activity as revealed in their perikaryon average ($p < 0.001$), perikaryon peak ($p < 0.002$), and primary branch peak ($p < 0.02$) measures (one-way ANOVA). This is consistent with the view that large cell bodies tend to be of projection neurons supporting greater processes that require higher metabolic demands.

Comparisons of the same morphometric and C.O. activity measurements done with the 8 controls (n=240 cells) were made with the 8 A.D. subjects (n=240 cells), matched by age and postmortem time, and the results were statistically analyzed for each measure. No morphometric differences were found between A.D. and control neuronal measures. The activity measurements from the A.D. tissue did not differ significantly from the controls when large and small cells were combined in the analysis (one-way ANOVA). However, the large A.D. cells in the ICC (n=35), as compared to the control cells (n=37), were deficient in C.O. activity in the overall average ($p < 0.032$) and neuropil peaks ($p < 0.012$) measures (Figure 4). This corresponded to a 17.7% decrement in overall average activity and a 10.3% decrement in peak neuropil activity. No significant activity differences were found in the smaller than average cell subgroup.

4.4. Discussion of the C.O. Activity Defect Found in A.D. Neurons

To our knowledge, this study used the first quantitative cytochemical method developed which allows microdensitometric cellular measures to be expressed as actual C.O. enzyme activity units. Quantitative C.O. cytochemistry may be one of the most appropriate metabolic mapping techniques for examining postmortem neural tissue from human subjects. C.O. is an endogenous respiratory enzyme which, when assessed cytochemically, can illustrate the effect of heightened or lessened metabolic demands on individual neurons over an extended period of time. Therefore, cytochemical examination of C.O. activity levels, and thereby, long-term neuronal activity patterns, provides a method for quantifying cellular metabolic differences across groups of subjects and various pathophysiological states.

The localized C.O. deficit found could not be accounted for simply by a nonspecific effect, such as loss of neurons, as may be the case of C.O. activity biochemical measures in tissue homogenates. The larger than average cells of the ICC suffered a C.O. decrement in A.D., but there were no other C.O. alterations or morphometric differences in the inferior colliculus of A.D. patients that would indicate any generalized metabolic change. Hence, this finding supports the hypothesis that A.D. affects mainly specific groups of

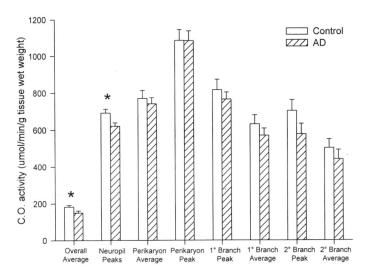

Figure 4. Comparison of the larger than average cells of the ICC between A.D. and control. The controls showed significantly more C.O. activity in the overall average and neuropil peaks measures (*p < 0.05). This may reflect the heightened vulnerability to A.D. of the larger projection neurons. All measurements were taken using JAVA (Jandel) software, as follows: *Overall Average*: With the cell body in the center of the image area, the C.O. activity was averaged across the entire rectangular image area (165 × 130 μm). This area included primarily the neuropil that surrounded each single cell sampled, the cell body, and capillary space. *Neuropil Peaks*: The three highest points of peak activity in the neuropil of each rectangular image area of 165 × 130 μm were selected and measured. Each peak was measured by averaging an area contained within a 13 × 13 pixel (5 × 5 μm) window. *Perikaryon Average*: The average activity of the cell body, excluding the nucleus and any processes. This was measured by outlining the perimeter of the cell body and nucleus and averaging the interior of that outline. *Perikaryon Peak*: The point of highest activity in the perikaryon; the average of a 5 × 5 pixel (2.4 × 2.4 μm) window. *Primary Branch Peak*: The point of highest activity in the primary branch of the largest arborizing process; the average of a 5 × 5 pixel window. *Primary Branch Average*: The average activity of the above primary branch, measured by outlining the branch from the cell body to the first visible secondary branch and averaging the activity within. *Secondary Branch Peak*: The point of highest activity in a secondary branch of the largest arborizing process; the average of a 3 × 3 pixel (1.7 × 1.7 μm) window. *Secondary Branch Average*: The average activity of the above secondary branch, measured by outlining the branch from the primary branch to the first visible tertiary branch and averaging the activity within.

larger projection neurons while neighboring neurons are spared (Sinha *et al.*, 1993). A greater abundance of neurofibrillary tangles and neuritic plaques in A.D. is found in cortical layers III and V, which contain larger projection neurons than the other layers (Van Hoesen, 1990).

But not all the larger neurons in the inferior colliculus were affected. When the larger cells were examined across all three inferior colliculus nuclei, no significant effects were found as the group differences were specific to the ICC sample. This specific metabolic effect is consistent with the A.D. study of Sinha *et al.* (1993), which showed that the anatomic histopathology in the inferior colliculus was all contained within the ICC. But since the C.O. deficit in A.D. precedes the anatomic histopathology, it is of little value to try to correlate the number of tangles and plaques with the C.O. activity in the same tissue sample. For example, a good correlation between both measures may simply mean that a nonspecific decrease in C.O. activity may result from loss of neurons or damage in the sample. Conversely, one may find a poor correlation between C.O. and A.D. histopathology measures when there are C.O. deficits in areas in which anatomic changes have not yet developed. Therefore, the important correlation is not that they coexist at the same time in a tissue sample, rather it is that the same regions and cell types selectively impaired in C.O. metabolism match those that are more vulnerable to develop A.D. histopathology.

Chandrasekaran *et al.* (1992), have proposed that the selective vulnerability to A.D. histopathology of some brain regions is related to their differential expression of C.O. genes. They also concluded that differences in the distribution of neuropil versus cell bodies in these regions may be related to genetically-mediated C.O. differences relevant to selective histopathological vulnerability in A.D. (Chandrasekaran *et al.*, 1992). This conclusion was supported by another study of the expression of C.O. genes in the temporal and motor cortex of A.D. patients. Chandrasekaran *et al.* (1994) found significantly more decreased C.O. mitochondrial RNA levels in the temporal cortex as compared to the motor cortex, or in A.D. as compared to the same regions in age-matched controls. These studies suggest that the selective histopathological vulnerability of some regions in A.D. may be linked to genetically-inherent C.O. differences in these regions. This evidence may be relevant for the observed C.O. deficit in the ICC.

In addition, a metabolic role of a C.O. mitochondrial defect in A.D. may take place even in the absence of subsequent anatomic histopathology. It has been shown consistently that C.O. activity is more vulnerable in A.D. than are other enzymes of the respiratory chain (Beal *et al.*, 1993; Kish *et al.*, 1992; Mecocci *et al.*, 1994; Mutisya *et al.*, 1994; Parker Jr. *et al.*, 1990; Parker Jr. *et al.*, 1994a,b). The evidence that a C.O. defect in A.D. occurs in the periphery, as well as in the brain, suggests that a C.O. defect in A.D. may be widespread. Yet our findings, together with studies showing regional brain differences in C.O. activity and gene expression, suggest that systemic impairment of C.O. activity in A.D. may still be associated with selective vulnerability of certain regions and cellular compartments, such as neuropil adjacent to larger neurons in some cortical layers and subcortical nuclei.

Neurons which are subjected to greater metabolic demands seem to develop more C.O. in order to produce and maintain heightened levels of activity; and this development can be visualized for examination and quantification through our staining and image processing procedures. Reductions in C.O. activity due to pathology can also be measured, making these techniques especially relevant to study the pathophysiology of A.D.

5. CYTOCHROME OXIDASE INHIBITION IN ALZHEIMER'S PATHOPHYSIOLOGY

The etiology of sporadic A.D. is as yet unknown, but the evidence reviewed here suggests that late-onset A.D. patients share a common pathophysiology: a systemic mitochondrial C.O. inhibition affecting the electron transport chain and producing reactive oxygen species that lead to oxidative damage and neurodegeneration. This defect seems to be selective to C.O., the fourth and final complex of the electron transport chain to oxygen.

It should also be emphasized that *C.O. activity inhibition* is the proper measure in these studies, not simply the loss of enzymes due to neuronal death. Interestingly, Parker and his group (1994b) point out that, in a study utilizing A.D. patients in conjunction with aged and diseased controls, assays of brain mitochondria for measurement of peptide concentrations showed that the amounts of C.O. (cytochrome aa_3) were not significantly different between groups, indicating that C.O. was present in normal levels in the A.D. brain mitochondria, but in a catalytically abnormal state. This C.O. defect provides the foundation for the proposed pathophysiologic model of this disease.

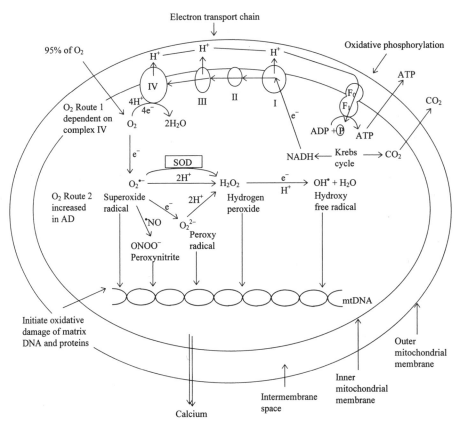

Figure 5. Schematic diagram of the proposed pathophysiologic model of Alzheimer's disease based on mitochondrial C.O. inhibition leading to oxidative damage (Route 2) and loss of calcium homeostasis.

The literature showing the existence of a C.O. defect in A.D., together with our findings, indicate generally that a mitochondrial defect is present during the course of this disorder. Our goal is to integrate this defect into a pathophysiologic model of the disease based on C.O. inhibition leading to oxidative damage (Figure 5). There are a number of very plausible arguments being put forth which can implicate a C.O. defect with A.D. A discussion of some of these follows, beginning with the reduction of cellular energy metabolism, and following with our emphasis on the oxidative damage effects of the increased generation of free radicals due to a C.O. defect. Finally, there will be a discussion linking the C.O. defect to other A.D. features, namely the amyloid plaque and neurofibrillary tangle, as well as neuronal death in A.D.

5.1. Reduction of Cellular Metabolic Energy Due to C.O. Inhibition

Probably the most obvious effect which follows from a C.O. defect is that cellular energy output will be reduced due to C.O.'s placement in the electron transport chain that produces ATP by oxidative phosphorylation. Some researchers have suggested that A.D. is a hypometabolic brain disease (Landin, Blennow, Wallin & Gottfries, 1993). However, this energy use factor is difficult to tease out in the laboratory through the typical whole-brain blood flow and glucose use methods due to the confounding loss of brain cells and correspondingly lowered glucose uptake. One could predict the effects of reduced cellular energy: decreased ability to control ionic flow and resting potential (Beal *et al.*, 1993), increased vulnerability to insult (Bennett & Rose, 1992), and other impairments of cellular function, many of which could be additionally detrimental to the health of the cell and lead to its eventual demise. Again, the energy decline model is likely an effect of C.O. inhibition, but there are other features of A.D. which can be more directly related to the pathogenic effects of a C.O. defect.

5.2. Increased Production of Free Radicals and Reactive Oxygen Species in Neurons Due to C.O. Inhibition

Free radicals are atoms or molecules which contain one or more unpaired electrons in their outer orbitals instead of the usual two. These species often are highly reactive as they "steal" electrons from other molecules. Free radicals have been implicated in a number of biological conditions from inflammation to normal aging (Harman, 1988) and are part of the body's normal physiological processing. In the mitochondria, free radicals are continually formed within the electron transport chain in a process driven by donated electrons. Typically, the body uses several mechanisms to ensure that free radical levels remain controlled: free radical "scavengers" roam the system and tightly controlled reactions usually prevent the release of radicals before they are neutralized.

In the mitochondria during oxidative phosphorylation, electron transport complexes I, III and IV pump protons (H+) for the production of ATP through the reduction of molecular oxygen to water controlled by cytochrome oxidase (complex IV), a process involving four electrons (Figure 5, Route 1). This is the major pathway for most oxygen consumption in mammals (Partridge *et al.*, 1994). However, another route exists, which reduces oxygen to superoxide ($O_2^{\cdot-}$), a potent free radical, in a one-electron process without the involvement of cytochrome oxidase in electron transport (Figure 5, Route 2). It has been estimated that 2% of oxygen utilized is converted to this or other free radicals during oxidative phosphorylation (Boveris & Chance, 1973). But any oxidative stress in neurons is likely to be greater than in other cells because the proportion of oxygen consumption by the brain is over tenfold higher than in the rest of the body. For example, although the adult brain has 2% of the total body weight at least 20% of the total oxygen intake is consumed by the brain at rest (Sokoloff, 1989).

If the other respiratory chain complexes remain active in A.D. (as is suggested by Parker *et al.*, 1994a,b), impairment of cytochrome oxidase could increase levels of $O_2^{\cdot-}$ by forcing more O_2 to be reduced through the secondary, one-electron route. Most of this $O_2^{\cdot-}$ is efficiently converted to hydrogen peroxide (H_2O_2) by membrane-bound superoxide dismutase (SOD). The level of H_2O_2 subsequently increases, and in most cells it is largely deactivated by glutathione peroxidase. However, neurons are particularly vulnerable to oxidative stress because they contain low levels of glutathione (Smith *et al.*, 1995). For this reason glutathione peroxidase is not shown in Figure 5 as an important antioxidant defense that could prevent accumulation of hydrogen peroxide in neurons. Thus more H_2O_2 can be a substrate which leads to the formation of a highly reactive hydroxyl radical in neurons.

In the presence of ferrous or cupric ions, H_2O_2 can participate in the Fenton reaction which results in the formation of the hydroxyl radical (OH). Hydroxyl radical can also be formed in a Haber-Weiss type of reaction when $O_2^{\cdot-}$ and H_2O_2 interact in the presence of an iron catalyst. Hydroxyl radical is probably the most damaging radical in that it will react with whatever biologic molecule is in its vicinity—proteins, lipids, etc.—and its substrate, H_2O_2, can cross cellular membranes into the extracellular space, where few antioxidant defense mechanisms exist (Southorn & Powis, 1988). Neuronal membranes are likely to suffer more peroxidative damage than other cellular membranes because they contain a higher amount of polyunsaturated fatty acids susceptible to free radical damage (Smith *et al.*, 1995).

Partridge and coworkers (1994) demonstrated this sequence of events in rat brain sub-mitochondrial particles. Using the spin-trapping method of radical detection, they found that sodium azide inhibition of cytochrome oxidase does indeed lead to the increased production of H_2O_2 and, subsequently, the hydroxyl radical.

The hydroxyl radical, being intensely reactive, has a very short life span and mean effective radius; however, upon reacting with a stable molecule, hydroxyl can form other free radicals, thus initiating a chain reaction that could be thousands of events long. This seems to be the case in lipid membrane peroxidation, a common free radical effect which leads to a loss of fluidity, breakdown of membrane secretory functions, and loss of trans-membrane ionic gradients. Hydroxyl radicals can also damage proteins, carbohydrates, and nucleic acids, both in the mitochondria and the nucleus (Southorn & Powis, 1988).

The defect in C.O. in A.D. can lead to increased free radical production, and free radicals may severely damage or kill a neuron via lipid peroxidation and by attacking mitochondrial and nuclear DNA. Mecocci *et al.*, (1994) found significant threefold increases in oxidative damage to mitochondrial DNA in A.D. brains relative to age-matched controls. Neurons in A.D. also show widespread oxidative damage mediated by peroxynitrite (Smith *et al.*, 1997). Peroxynitrite ($ONOO^-$) is a strong oxidant formed from the reaction of super-oxide with nitric oxide (NO). Unrestrained free radical production due to C.O. inhibition is sufficient to cause the neuronal degeneration and synaptic loss which in turn may cause the dementia and memory loss of A.D. patients.

Free radical damage may also mediate the toxicity implicating protein aggregates, such as paired helical filament tau and amyloid beta proteins, in A.D. and neuronal cell death (Smith *et al.*, 1995). Neurofilament and cytoskeletal proteins with high proportion of lysine residues, such as tau, are especially vulnerable to oxidative damage and the formation of advanced glycation end products that generate reactive oxygen intermediates (Yan *et al.*, 1994). Furthermore, the oxygen free radicals generated by C.O. inhibition and by subsequent products such as glycated tau may induce the release of amyloid beta-peptide (Yan *et al.*, 1995).

It is plausible that amyloid may have other roles in the pathogenesis of A.D., such as that of a trophic factor responding to neural degeneration (Ihara, 1988; Saitoh, Sundsmo, Roch, Kimura, Cole, Schubert, Oltersdorf, & Schenk, 1989; Uchida, Ihara, & Tomonaga, 1988). This is a reasonable conclusion given that beta-amyloid precursor protein (APP) is used for cell membrane repair in many body regions, including the brain (Meier-Ruge, Bertoni-Freddari, & Iwangoff, 1994). Meier-Ruge and colleagues (1994) take this supposition further by claiming that the reduction in glucose turnover and ATP in the cells (possibly as a result of C.O. inhibition and free radical damage) prevents the proper utilization of APP (for membrane repair), leading to the buildup and subsequent senile plaques.

5.3. A C.O. Defect Can Produce Disruptions of Intracellular Calcium Homeostasis

One important function of mitochondria is to maintain the homeostasis of calcium levels within the cell, a task accomplished by sequestration of excess calcium (Beal, 1992a; Gunther & Pfeiffer, 1990). When this ability is impaired due to C.O. inhibition, i.e., by free radical damage to the mitochondrial membrane or decreased ATP production, the cytosolic calcium level may rise (Richter & Kass, 1991). A potential result of this increased calcium level is calcium-dependent activation of calpain, an intracellular protease known to be involved in the degradation of cytoskeletal proteins (Schlaepfer, Lee, Lee, & Zimmerman, 1985).

Calpain has been found to be involved with the A.D.-characteristic amyloid plaques, perhaps facilitating their formation by improperly breaking down amyloid precursor protein, preventing its normal function and causing it to build up into the insoluble plaques (Iwamoto, Thangipon, Crawford, & Emson, 1991; Shimohama, Suenaga, Araki, Yamaoaka, Shimizu, & Kimura, 1991; Siman, Card, & Davis, 1990). Neurofibrillary tangles may also be related to calpain-induced degradation of or interference with the fibril proteins (Iwamoto *et al.*, 1991; Nixon & Cataldo, 1994).

Impairment of neuronal C.O. metabolism resulting in decreased ATP production can have varied effects, such as decreasing the ability of the cell to maintain ionic potentials, as mentioned above. If the repolarization of a neuronal membrane is inhibited, prolonged or inappropriate opening of some voltage-gated calcium channels may result, increasing the cytosolic calcium level—this increase can cause a dysfunction of NMDA channels (Beal *et al.*, 1993). NMDA channels typically regulate calcium and sodium influx, are gated by magnesium and glutamate-activated NMDA receptors, and have been linked to synaptic plasticity and memory formation (Beal, 1992b).

Decreasing the ability of magnesium to gate NMDA channels could result in the opening of the channels by endogenous glutamate—thus increasing the flow of both sodium and calcium into the cell (Beal, 1992a; Beal *et al.*, 1993). Further attempts by the cell to mediate this influx of positively-charged ions in order to regain the resting potential will increase the demand for ATP and quickly deplete the cell of energy. Also, the mitochondria will preferentially absorb the excess calcium instead of produce ATP (Beal *et al.*, 1993), further reducing cellular energy output while also increasing the possibility of further mitochondrial damage due to mitochondrial calcium overload (Nicotera *et al.*, 1990). Neuronal damage resulting from a number of subsequent excitotoxic mechanisms can also kill the neurons (see Beal, 1992a,b; Beal *et al.*, 1993, for a review).

6. CONCLUSIONS

There is a compelling need to develop biological markers to unequivocally identify A.D. patients for emerging therapeutic interventions (Arai, 1996). A.D. is characterized by initial memory loss, followed by progressive loss of neurons leading to dementia and loss of all nervous functions, and eventually death. A.D. is now the fourth-largest killer of adults 65 and older, and this disease may be affecting one of every three families in the United States (E. Gonzalez-Lima & F. Gonzalez-Lima, 1987). Although rare familial types of A.D. follow traditional Mendelian genetics, the great majority of A.D. cases appear late in life and have no clearly identifiable nuclear genetic defects. It is in this majority of late-onset cases that defects in C.O. activity have been linked to mitochondrial DNA mutations (Davis *et al.*, 1997). This project suggests four conclusions:

1. Compromised oxidative energy metabolism and spatial memory resulting from aging, chronic cerebrovascular insufficiency, and C.O. enzymatic inhibition have similar brain regions and cell types affected in common with Alzheimer's brains, as determined by quantitative C.O. histochemistry.

2. A C.O. inhibition in the mitochondrial electron transport chain is sufficient to initiate the pathophysiology associated with Alzheimer's disease. Memory defi-

cits in A.D. may result from the metabolic effects of C.O. inhibition on brain function before the accumulation of neuromorphologic pathology.

3. Neuronal damage and death in A.D. occur as a result of increased free radical production stemming from the decreased cytochrome oxidase activity. The proposed pathophysiology of Alzheimer's disease will lead to a cascade of multiple intracellular events, resulting from failure of oxidative energy metabolism. For example, proteins such as tau will be particularly vulnerable to oxidative damage. Increases in senile plaques in A.D. may be promoted by an energy-underfunded attempt to repair free radical damage and by a calcium-mediated calpain interaction due to the instability of the mitochondrial membrane. Neurofibrillary tangles could also be promoted by calpain activation, and the vulnerability of the neuron to glutamate excitotoxicity may increase.

4. Monitoring of C.O. activity and oxidative DNA damage may serve to evaluate the success of early treatments in reversing this enzymatic defect to prevent oxidative stress leading to neurodegeneration. This project will serve as the basis for testing pharmacological agents for improving C.O. metabolism aimed at reversing brain energy failure and oxidative damage in animal models and ultimately in demented patients.

ACKNOWLEDGMENTS

Supported by NIH grant RO1 MH43353 to FGL and by NIH grant T32 MH18837 to AC. We thank the Alzheimer's Disease Research Center Neuropathology Core, USC School of Medicine, Los Angeles, California 90033, which is funded by AG05142, National Institute for Aging; and the Brain Bank of the Michigan Alzheimer's Disease Research Center at the University of Michigan, Ann Arbor, Michigan 48104, which is funded by AG08671, National Institute for Aging, for the tissue used in this study.

REFERENCES

Aggleton, J.P.(ed.), 1992, *The Amygdala: Neurobiological Aspects of Emotion, Memory, and Mental Dysfunction,* Wiley-Liss, New York.

Arai, H., 1996, Biological markers for the clinical diagnosis of Alzheimer's disease, *Tohoku Journal of Experimental Medicine* **179**: 65–79.

Arendt, T., Bigl, V., Arendt, A., and Tennstedt, A., 1983, Loss of neurons in the nucleus basalis of Meynert in Alzheimer's disease, paralysis agitans and Korsakoff's disease, *Acta Neuropathologica* **61**: 101–108.

Ball, M.J., Fisman, M., Hachinski, V., Blume, W., Fox, A., Kral, V.A., Kirshen, A.J., Fox, H., and Merskey, H., 1985, A new definition of Alzheimer's disease: A hippocampal dementia, *Lancet* **1**: 14–16.

Beal, M.F., 1992a, Does impairment of energy metabolism result in excitotoxic neuronal death in neurodegenerative illnesses? *Annals of Neurology* **31**: 119–130.

Beal, M.F., 1992b, Mechanisms of excitotoxicity in neurologic diseases, *The FASEB Journal* **6**: 3338–3342.

Beal, M.F., Hyman, B.T., and Koroshetz, W., 1993, Do defects in mitochondrial energy metabolism underlie the pathology of neurodegenerative disorders? *Trends in Neurosciences* **16**: 125–131.

Bennett, M.C., Diamond, D.M., Parker, W.D., Jr., Stryker, S.L., and Rose, G.M., 1992a, Inhibition of cytochrome oxidase impairs learning and hippocampal plasticity: a novel animal model of Alzheimer's disease, in: *Alzheimer's Disease Therapy: A New Generation of Progress,* (J. Simpkins, F.T. Crews and E.M. Meyer, eds.), pp. 485–501, Plenum Press, New York.

Bennett, M.C., Diamond, D.M., Stryker, S.L., Parks, J.K., and Parker, W.D., Jr., 1992b, Cytochrome oxidase inhibition: a novel animal model of Alzheimer's disease, *Journal of Geriatric Psychiatry & Neurology* **5**: 93–101.

Bennett, M.C. and Rose, G.M., 1992, Chronic sodium azide treatment impairs learning of the Morris water maze task, *Behavioral and Neural Biology* **58**: 72–75.

Biegon, A. and Wolff, M., 1986, Quantitative histochemistry of acetylcholinesterase in rat and human brain post-mortem, *Journal of Neuroscience Methods* 16: 39–45.

Blass, J.P., Sheu, R.K.-F., and Cedarbaum, J.M., 1988, Energy metabolism in disorders of the nervous system, *Revue Neurologique* 144: 543–563.

Blass, J.P., 1993, Metabolic alterations common to neural and non-neural cells in Alzheimer's disease, *Hippocampus* 3 Spec No: 45–53.

Bloch, V., 1976, Brain activation and memory consolidation, in: *Neural Mechanisms of Learning and Memory*, (M.R. Rosenzweig and E.L. Bennett, eds.), pp. 582–590, MIT Press, Cambridge, Massachusetts.

Boveris, A. and Chance, B., 1973, The mitochondrial generation of hydrogen peroxide: General properties and the effect of hyperbaric oxygen, *Biochemical Journal* 134: 707–716.

Čada, A., Gonzalez-Lima, F., Rose, G.M., and Bennett, M.C., 1995, Regional brain effects of sodium azide treatment on cytochrome oxidase activity: A quantitative histochemical study, *Metabolic Brain Disease* 10: 303–319.

Chagnon, P., Betard, C., Robitaille, Y., Cholette, A., Gauvreau, and D., 1995, Distribution of brain cytochrome oxidase activity in various neurodegenerative diseases, *Neuroreport* 6: 711–715.

Chandrasekaran, K., Stoll, J., Giordano, T., Atack, J.R., Matocha, M.F., Brady, D.R., and Rapoport, S.I., 1992, Differential expression of cytochrome oxidase (COX) genes in different regions of monkey brain, *Journal of Neuroscience Research* 32: 415–423.

Chandrasekaran, K., Giordano, T., Brady, D.R., Stoll, J., Martin, L.J., and Rapoport, S.I., 1994, Impairment in mitochondrial cytochrome oxidase gene expression in Alzheimer's disease, *Molecular Brain Research* 24: 336–340.

Coomber, P., Crews, D., and Gonzalez-Lima, F., 1997, Independent effects of incubation temperature and gonadal sex on the volume and metabolic capacity of brain nuclei in the leopard gecko (*Eublepharis macularius*), a lizard with temperature-dependent sex determination, *Journal of Comparative Neurology* 380: 409–421.

Curti, D., Giangare, M.C., Redolfi, M.E., Fugaccia, I., and Benzi, G., 1990, Age-related modifications of cytochrome *c* oxidase activity in discrete brain regions, *Mechanisms of Ageing and Development* 55: 171–180.

Davis, R.E., Miller, S., Herrnstadt, C., Ghosh, S.S., Fahy, E., Shinobu, L.A., Galasko, D., Thal, L.J., Beal, M.F., Howell, N., and Parker, W.D., 1997, Mutations in mitochondrial cytochrome *c* oxidase genes segregate with late-onset Alzheimer disease, *Proceedings of the National Academy of Sciences* 94: 4526–4531.

de la Torre, J.C., Čada, A., Nelson, N., Davis, G., Sutherland, R.J., and Gonzalez-Lima, F., 1997, Reduced cytochrome oxidase and memory dysfunction after chronic brain ischemia in aged rats, *Neuroscience Letters* 223: 165–168.

Gonzalez-Lima, E.M. and Gonzalez-Lima, F., 1987, Sources of stress affecting caregivers of Alzheimer's disease patients, *Health Values* 11: 3–10.

Gonzalez-Lima, F., 1992, Brain imaging of auditory learning functions in rats: Studies with fluorodeoxyglucose autoradiography and cytochrome oxidase histochemistry, in: *Advances in Metabolic Mapping Techniques for Brain Imaging of Behavioral and Learning Functions*, (F. Gonzalez-Lima, T. Finkenstadt and H. Scheich, eds.), pp. 39–109, Kluwer Academic Publishers, The Netherlands.

Gonzalez-Lima, F., Helmstetter, F.J., and Agudo, J., 1993, Functional mapping of the rat brain during drinking behavior: A fluorodeoxyglucose study, *Physiology & Behavior* 54: 605–612.

Gonzalez-Lima, F., Valla, J., and Matos-Collazo, S., 1997, Quantitative cytochemistry of cytochrome oxidase and cellular morphometry of the human inferior colliculus in control and Alzheimer's patients, *Brain Research* 752: 117–126.

Gonzalez-Lima, F. and Čada, A., 1994, Cytochrome oxidase activity in the auditory system of the mouse: A qualitative and quantitative histochemical study, *Neuroscience* 63: 559–578.

Gonzalez-Lima, F. and Garrosa, M., 1991, Quantitative histochemistry of cytochrome oxidase in rat brain, *Neuroscience Letters* 123: 251–253.

Gonzalez-Lima, F. and Jones, D., 1994, Quantitative mapping of cytochrome oxidase activity in the central auditory system of the gerbil: A study with calibrated activity standards and metal-intensified histochemistry, *Brain Research* 660: 34–49.

Gonzalez-Lima, F. and McIntosh, A.R., 1996, Conceptual and methodological issues in the interpretation of brain-behavior relationships, in: *Developmental Neuroimaging: Mapping the Development of Brain and Behavior*, (R.W. Thatcher, G.R. Lyon, J. Ramsey and N. Krasnegor, eds.), pp. 235–253, Academic Press, Orlando, Florida.

Gonzalez-Lima, F. and Scheich, H., 1985, Ascending reticular activating system in the rat: A 2-deoxyglucose study, *Brain Research* 344: 70–88.

Grimes, A.M., Grady, C.L., Foster, N.L., Sunderland, T., and Patronas, N.J., 1985, Central auditory function in Alzheimer's disease, *Neurology* 35: 352–358.

Gunther, T.E. and Pfeiffer, D.R., 1990, Mechanisms by which mitochondria transport calcium, *American Journal of Physiology* 258: c755–c786.

Harman, D., 1988, Free radicals in aging, *Molecular and Cellular Biochemistry* **84**: 155–161.

Hess, H.H. and Pope, A., 1953, Ultramicrospectrophotometric determination of cytochrome oxidase for quantitative histochemistry, *Journal of Biological Chemistry* **204**: 295–306.

Hevner, R.F., Duff, R.S., and Wong-Riley, M.T.T., 1992, Coordination of ATP production and consumption in brain: parallel regulation of cytochrome oxidase and Na$^+$, K$^+$-ATPase, *Neuroscience Letters* **138**: 188–192.

Hevner, R.F., Liu, S., and Wong-Riley, M.T.T., 1993, An optimized method for determining cytochrome oxidase activity in brain tissue homogenates, *Journal of Neuroscience Methods* **50**: 309–319.

Hevner, R.F. and Wong-Riley, M.T.T., 1989, Brain cytochrome oxidase: purification, antibody production, and immunohistochemical/histochemical correlations in the CNS, *Journal of Neuroscience* **9**: 3884–3898.

Ihara, Y., 1988, Massive somatodendritic sprouting of cortical neurons in Alzheimer's disease, *Brain Research* **459**: 138–144.

Iwamoto, N., Thangnipon, W., Crawford, C., and Emson, P.C., 1991, Localization of calpain immunoreactivity in senile plaques and in neurones undergoing neurofibrillary degeneration in Alzheimer's disease, *Brain Research* **561**: 177–180.

Kish, S.J., Bergeron, C., Rajput, A., Dozic, S., Mastrogiacomo, F., Chang, L., Wilson, J.M., DiStefano, L.M., and Nobrega, J.N., 1992, Brain cytochrome oxidase in Alzheimer's disease, *Journal of Neurochemistry* **59**: 776–779.

Landin, K., Blennow, K., Wallin, A., and Gottfries, C.G., 1993, Low blood pressure and blood glucose levels in Alzheimer's disease: Evidence for a hypometabolic disorder? *Journal of Internal Medicine* **233**: 257–363.

Lowry, O.H., Rosebrough, N.J., Farr, A.L., and Randall, R.J., 1951, Protein measurement with the folin phenol reagent, *Journal of Biological Chemistry* **193**: 295–275.

Mecocci, P., MacGarvey, U., and Beal, M.F., 1994, Oxidative damage to mitochondrial DNA is increased in Alzheimer's disease, *Annals of Neurology* **36**: 747–751.

Meier-Ruge, W., Bertoni-Freddari, C., and Iwangoff, P., 1994, Changes in brain glucose metabolism as a key to the pathogenesis of Alzheimer's disease, *Gerontology* **40**: 246–252.

Mutisya, E.M., Bowling, A.C., and Beal, M.F., 1994, Cortical cytochrome oxidase activity is reduced in Alzheimer's disease, *Journal of Neurochemistry* **63**: 2179–2184.

Nicotera, P., Bellomo, G., and Orrenius, S., 1990, The role of Ca^{2+} in cell killing, *Chemical Research in Toxicology* **3**: 484–494.

Nixon, R.A. and Cataldo, A.M., 1994, Free radicals, proteolysis, and the degeneration of neurons in Alzheimer disease: How essential is the b-amyloid link? *Neurobiology of Aging* **15**: 463–469.

Nobrega, J.N., 1992, Brain metabolic mapping and behavior: Assessing the effects of early developmental experiences in adult animals, in: *Advances in Metabolic Mapping Techniques for Brain Imaging of Behavioral and Learning Functions. NATO ASI Series, Vol. D68*, (F. Gonzalez-Lima, T. FinkenstSdt and H. Scheich, eds.), pp. 125–149, Kluwer Academic Publishers, Dordrecht/Boston/London.

Nobrega, J.N., Raymond, R., DiStefano, L., and Burnham, W.M., 1993, Long-term changes in regional brain cytochrome oxidase activity induced by electroconvulsive treatment in rats, *Brain Research* **605**: 1–8.

Ohm, T.G. and Braak, H., 1989, Auditory brainstem nuclei in Alzheimer's disease, *Neuroscience Letters* **96**: 60–63.

Olton, D.S., Wible, C.G., Pang, K., and Sakurai, Y., 1989, Hippocampal calls have mnemonic correlates as well as spatial ones, *Psychobiology* **17**: 228–229.

Parker, W.D., Jr., Filley, C.M., and Parks, J.K., 1990, Cytochrome oxidase deficiency in Alzheimer's disease, *Neurology* **40**: 1302–1303.

Parker, W.D., Jr., 1991, Cytochrome oxidase deficiency in Alzheimer's disease, *Annals of the New York Academy of Sciences* **640**: 59–64.

Parker, W.D., Jr., Mahr, N.J., Filley, C.M., Parks, J.K., Hughes, M.A., Young, D.A., and Cullum, C.M., 1994a, Reduced platelet cytochrome c oxidase activity in Alzheimer's disease, *Neurology* **44**: 1086–1090.

Parker, W.D., Jr., Parks, J., Filley, C.M., and Kleinschmidt-DeMasters, B.K., 1994b, Electron transport chain defects in Alzheimer's disease brain, *Neurology* **44**: 1090–1096.

Partridge, R.S., Monroe, S.M., Parks, J.K., Johnson, K., Parker, W.D., Jr., Eaton, G.R., and Eaton, S.S., 1994, Spin trapping of azidyl and hydroxyl radicals in azide-inhibited submitochondrial particles, *Archives of Biochemistry and Biophysics* **310**: 210–217.

Richter, C., 1995, Oxidative damage to mitochondrial DNA and its relationship with aging, *International Journal of Biochemistry and Cell Biology* **27**: 647–653.

Richter, C. and Kass, G.E.N., 1991, Oxidative stress in mitochondria: its relationship to cellular Ca^{2+} homeostasis, cell death, proliferation, and differentiation, *Chemico-Biological Interactions* **77**: 1–23.

Saitoh, T., Sundsmo, M., Roch, J., Kimura, N., Cole, G., Schubert, D., Oltersdorf, T., and Schenk, D.B., 1989, Secreted form of amyloid protein precursor is involved in the growth regulation of fibroblasts, *Cell* **58**: 615–622.

Sayre, L.M., Zelasko, D.A., Harris, P.L., Perry, G., Salomon, R.G., and Smith, M.A., 1997, 4-Hydroxynonenal-derived advanced lipid peroxidation end products are increased in Alzheimer's disease, *Journal of Neurochemistry* **68**: 2092–2097.

Schlaepfer, W.W., Lee, C., Lee, V., and Zimmerman, U.J., 1985, An immunoblot study of neurofilament degradation in situ and during calcium-activated proteolysis, *Journal of Neurochemistry* **44**: 502–509.

Seligman, A.M., Karnovsky, M.J., Wasserkrug, H.L., and Hanker, J.S., 1968, Nondroplet ultrastructural demonstration of cytochrome oxidase activity with a polymerizing osmiophilic reagent, diaminobenzidine (DAB), *Journal of Cell Biology* **38**: 1–14.

Shimohama, S., Suenaga, T., Araki, W., Yamaoaka, Y., Shimizu, K., and Kimura, J., 1991, Presence of calpain II immunoreactivity in senile plaques in Alzheimer's disease, *Brain Research* **558**: 105–108.

Siman, R., Card, J.P., and Davis, L.G., 1990, Proteolytic processing of B-amyloid precursor by calpain I, *The Journal of Neuroscience* **10**: 2400–2411.

Sinha, U.K., Hollen, K.M., Rodriguez, R., and Miller, C.A., 1993, Auditory system degeneration in Alzheimer's disease, *Neurology* **43**: 779–785.

Smith, M.A., Lawrence, S.M., Monnier, V.M., and Perry, G., 1995a, Radical ageing in Alzheimer's disease, *Trends in Neuroscience* **18**: 172–176.

Smith, M.A., Rudnicka-Nawrot, M., Richey, P.L., Prapotnik, D., Mulvihill, P., Miller, C.A., Sayre, L.M., and Perry, G., 1995b, Carbonyl-related posttranslational modification of neurofilament protein in the neurofibrillary pathology of Alzheimer's disease, *Journal of Neurochemistry* **64**: 2660–2666.

Smith, M.A., Perry, G., Richey, P.L., Sayre, L.M., Anderson, V.E., Beal, M.F., and Kowall, N., 1996a, Oxidative damage in Alzheimer's, *Nature* **382**: 120–121.

Smith, M.A., Sayre, L., and Perry, G., 1996b, Is Alzheimer's a disease of oxidative stress? *Alzheimer's Disease Review* **1**: 63–67.

Smith, M.A., Richey-Harris, P.L., Sayre, L.M., Beckman, J.S., and Perry, G., 1997, Widespread peroxynitrite-mediated damage in Alzheimer's disease, *Journal of Neuroscience* **17**: 2653–2657.

Sokoloff, L., 1989, Circulation and energy metabolism of the brain, in: *Basic Neurochemistry*, (G.J. Siegel, B.W. Agranoff, R.W. Albers and P. Molinoff, eds.), pp. 471–495, Little Brown, Boston.

Southorn, P.A. and Powis, G., 1988, Free radicals in medicine, *Mayo Clinic Proceedings* **63**: 381–408.

Uchida, Y., Ihara, Y., and Tomonaga, M., 1988, Alzheimer's disease brain extract stimulates the survival of cerebral cortical neurons from neonatal rats, *Biochemical and Biophysical Research Communications* **150**: 1263–1267.

Van Hoesen, G.W., 1990, The dissection by Alzheimer's disease of cortical and limbic neural systems relevant to memory, in: *Brain Organization and Memory: Cells, Systems, and Circuits*, (J.L. McGaugh, N.M. Weinberger and G. Lynch, eds.), pp. 234–261, Oxford University Press, New York.

Wharton, D.C. and Tzagoloff, A., 1967, Cytochrome oxidase from beef heart mitochondria, *Methods of Enzymology* **10**: 245–250.

Wikstrom, M., Krab, K., and Saraste, M., 1981, *Cytochrome Oxidase: A Synthesis*, Academic Press, New York.

Wong-Riley, M.T.T., Merzenich, M.M., and Leake, P.A., 1978, Changes in endogenous reactivity to DAB induced by neuronal inactivity, *Brain Research* **141**: 185–192.

Wong-Riley, M.T.T., 1979, Changes in the visual system of monocularly sutured or enucleated cats demonstrated with cytochrome oxidase histochemistry, *Brain Research* **171**: 11–28.

Wong-Riley, M.T.T., 1989, Cytochrome oxidase: an endogenous metabolic marker for neuronal activity, *Trends in Neurosciences* **12**: 94–101.

Wong-Riley, M.T.T. and Riley, D.A., 1983, The effect of impulse blockage on cytochrome oxidase activity in the cat visual system, *Brain Research* **261**: 185–193.

Yan, S.D., Chen, X., Schmidt, A.M., Brett, J., Godman, G., Zou, Y.S., Scott, C.W., Caputo, C., Frappier, T., Smith, M.A., Perry, G., Yen, S.H., and Stern, D., 1994, Glycated tau protein in Alzheimer's disease: a mechanism for induction of oxidant stress, *Proceedings of the National Academy of Sciences* **91**: 7787–7791.

Yan, S.D., Yan, S.F., Chen, X., Fu, J., Chen, M., Kuppusamy, P., Smith, M.A., Perry, G., Godman, G.C., Nawroth, P., Zweier, J.L., and Stern, D., 1995, Non-enzymatically glycated tau in Alzheimer's disease induces neuronal oxidant stress resulting in cytokine gene expression and release of amyloid beta-peptide, *Nature Medicine* **1**: 693–699.

21

THE ROLE OF FREE RADICALS DAMAGE TO THE OPTIC NERVE AND VITREOUS AFTER ISCHEMIA*

E. Berenshtein,[1] E. Banin,[2] N. Kitrossky,[1] M. Chevion,[1] and J. Pe'er[2]

[1]Department of Cellular Biochemistry
[2]Department of Ophthalmology
The Hebrew University—Hadassah Medical School
P.O.B. 12272, Jerusalem 91120, Israel

1. INTRODUCTION

Glaucoma is one of the major causes of blindness worldwide, affecting the retina and the optic nerve through injury inflicted by the elevated intra-ocular pressure (IOP). It is postulated that the pathological changes are a result of repeated episodes of ischemia and reperfusion in the ocular tissues, caused by the fluctuating IOP (Kalvin et al., 1966; Winterkorn and Beckman, 1995). In previous investigations, by us as well as by others, it was shown that transient ischemia most probably causes tissue damage through a burst of free radicals and non-radical oxygen derived active species (ROS) (Chevion et al., 1993; Ophir et al., 1993, 1994) and the combined action of other mechanisms, including: loss of energy stores, subsequent impairment of membrane transport processes, an accumulation of intracellular calcium followed by deregulation of calcium homeosatsis, and the release of potentially toxic substances such as excitatory amino acids. ROS, and particularly the highly reactive hydroxyl radical, could cause cellular damage by attacking a number of cellular components, including peroxidation of polyunsaturated fatty acids in membranes, inactivation of enzymes, and fragmentation of DNA molecules.

Transition metal ions, such as iron and copper, play an essential mediatory role in the manifestation of the injury to the tissue following ischemia/reperfusion. During the ischemic phase, there is an increase in the content of the labile and redox-active pools of iron and copper (Chevion et al., 1993; Voogd et al., 1992, 1994; Berenshtein et al., 1997),

* Address for correspondence: Prof. Mordechai Chevion, The Dr. W. Ganz Chair of Heart Studies, Hebrew University—Hadassah Medical School, P.O.Box 12272, Jerusalem 91120, Israel, Telephone number: 972-2-6758158; 972-2-6758160, Fax number: 972-2-6415848; 972-2-6784010

Free Radicals, Oxidative Stress, and Antioxidants, edited by Özben.
Plenum Press, New York, 1998.

that renders the tissue more susceptible to the free radicals produced during the reperfusion phase.

Many investigations, both clinical and with experimental animals, have shown that the damage to the optic nerve is the important factor in the pathophysiology of the loss of vision in glaucoma (Gutman et al., 1993; Bunt-Milam et al., 1987). However, the role of free radicals in this optic nerve damage has not been investigated. Also, the possible roles of the vitreous in such damage have not been clearly determined. They could include:

1. The vitreous tissue contains glutathione, superoxide dismutase (SOD), ascorbate, urate and other compounds which protect against free radicals.
2. Lipid peroxidation of retina has led to the development of cataract (Micelli-Ferrari et al., 1996). Thus, the vitreous could act as a means for ROS communication among ocular tissues.
3. It is postulated that the vitreous contains labile pools of iron and copper which could participate in the catalysis of free-radical induced tissue injury. Of particular importance for this aspect are eye injuries caused by irradiation or by foreign bodies (brass, iron, copper).

In the present study we examined the role of free radicals in the optic nerve and vitreous after ischemic damage by elevated IOP. The efficacy of another complex, the gallium–desferrioxamine (Ga/DFO) to curb the free radical formation, was evaluated by extensive monitoring of the biochemical profile of both the vitreous and the optic nerve from cat eyes subjected to ischemia/reperfusion.

2. MATERIALS AND METHODS

2.1. Experimental Protocols

Cats weighing 2.5 to 3.5 kg were anesthetized with ketamine HCl (Ketalar, Parke Davis, UK) 10 mg/kg injected intramuscularly. Additional ketamine, if needed, was administered intramuscularly in combination with the relaxing agent Xylazine (0.1 ml/kg). All animal experiments were conducted in compliance with the ARVO Resolution on the Humane Use of Animals in Research.

Seventy eyes (of 35 cats) were studied according to the following protocol: thirty minutes before each experiment, 100 mg/kg of sodium salicylate (Aldrich Chemicals, Milwaukee, WI), dissolved in 5 ml of normal saline, were injected into the femoral vein of each cat. Salicylate was used as a hydroxyl radical trap, which, upon scavenging ·OH radicals, forms two relatively stable hydroxylation adducts, 2,3- and 2,5-dihydroxybenzoic acids (DHBA).

Ischemia was induced by two protocols: first for the experiments without protection and second for the experiments that provided protection against ischemia/reperfusion injury by Ga/DFO.

In the first protocol the intraocular pressure in one eye of each cat was increased for generation of ischemia for 60 or 90 min, followed by 5 min of reperfusion (18 cats), and for 90 min without reperfusion (6 cats), followed by enucleation and separation of the optic nerve and vitreous.

In the second protocol of the experiment, both eyes of each cat underwent ischemia for 90 minutes followed by 5 minutes of reperfusion as detailed below (5 cats). One eye served as an internal, untreated control, and the fellow eye was the treated eye in which reperfusion was preceded by intravenous injection of Ga/DFO. Ischemia and reperfusion were produced using a pressurized infusion system. A sterile needle was inserted into the anterior chamber of each eye, and infusion tubings were attached to two separate flexible plastic bags, surrounded by a pressure cuff, containing eye irrigation solution. The IOP was gradually increased until blood flow in the retinal vessels ceased, as seen by indirect ophthalmoscopy.

Consequently, ischemia was initially induced in one eye of each cat, randomly chosen to be the control eye. Fifteen minutes later (to retain the same duration of ischemia at 90 minutes in each eye), the fellow eye underwent the same procedure. At the end of 90 minutes of ischemia in the first control eye, the cuff was fully deflated and blood circulation resumed, as was verified ophthalmoscopically. After 5 minutes of reperfusion, the control eye was enucleated and the optic nerve (about 2 mm) and part of the vitreous were separated and immediately placed in 0.3 ml of ice-cold phosphate-buffered saline (pH 7.4) for measurements of the tissue levels of DHBA, salicylate, and protein.

Regarding the fellow eye, 5 minutes before completion of ischemia (90 min), 2.5 mg/kg of Ga/DFO (stochiometric ratio 1:1) in 10 ml saline was injected into the femoral vein of the cats. Five minutes later, the cuff of the manometer was deflated and reperfusion was then evident. Five minutes after that, the experimental eye was enucleated and the neural retina dissected and prepared for chemical analysis as described above. Cats were then sacrificed by IV injection of KCl.

A series of control experiments were conducted on 6 eyes (3 cats) which were subjected to ischemia (90 min) of both eyes, in an identical protocol as the Ga/DFO treated cats, except that these cats were "treated" with saline. The comparison between the two eyes of each cat did not show any significant difference in the free radicals/ oxidant stress parameters.

2.2. Quantitation of DHBA, Salicylate, and Protein

The optic nerve and vitreous from each eye was homogenized, and 50 µl was set aside for the determinaton of protein level according to Bradford (1976), following which the centrifugation supernatants were kept on ice for immediate analysis or stored at −80°C for future quantitation.

The two major products of salicylate hydroxylation, 2,3- and 2,5-DHBA (Fig. 1), were identified and quantitated by high-pressure liquid chromatography coupled with electrochemical detection (HPLC–ECP), using a Varian 5000 Liquid Chromatograph equipped with a Rheodyne 7125 sample injector (20 and 10 µl loops). The column used for separation of salicylate and DHBA was a 25 cm × 4 mm LiChrospher 100RP-18, 5 µm (E. Merck, Darmstadt, Germany). Salicylate was identified and quantitated fluorimetrically, using a FD-300 detector (SpectroVision Inc., Chelmsford, MA, USA).

2.3. Analysis of Results

The results of the experimental eyes were compared to those of the control eye for each cat according to the protocol groups, and analyzed statistically using the Mann-Whitney test. $P < 0.05$ was considered to be significant.

3. RESULTS

After 60 min of ischemia followed by reperfusion, normalized 2,3-DHBA and 2,5-DHBA levels were non-significantly higher in the optic nerve than in the nonischemic

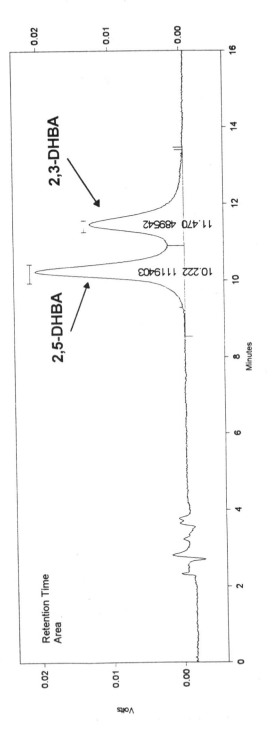

Figure 1. The HPLC analysis of the 2,5- and 2,3-DHBA.

Table 1. Parameters of oxidative stress in the early reperfusion phase of the optic nerve in the cat (ratio between non-ischemic and ischemic/reperfused eyes)

Ratio	2,5-DHBA	2,3-DHBA
Non-ischemic control/ ischemia (90 min) and reperfusion (5 min)	0.72	0.31*

n = 12/12; DHBA = dihydroxybenzoic acid derivative

*Denotes that the difference between the control group and the Ga/DFO-treated group is statistically significant ($p < 0.05$, Mann-Whitney test).

controls. The level of 2,5-DHBA also did not change significantly following a longer period of ischemia; in contrast, following 90 min of ischemia and reperfusion, levels of 2,3-DHBA were higher than in those of the fellow controls, indicating that the mean generation of ·OH radicals was on average approximately 3.2 times higher in experimental eyes than in control eyes (P = 0.0134, Table 1).

The results of the experiments without reperfusion (Ischemia alone) showed that, following 90 min of complete ischemia, the changes in the levels of 2,3- and 2,5-DHBA were not significant between the experimental (ischemic) and control eyes (Table 2).

In the second part of experiment, protection against optic nerve damage by free radicals was provided by using the new iron chelator complex gallium–desferrioxamine (Ga/DFO). As can be clearly seen, levels of 2,3-DHBA (Table 3) were markedly lower in the optic nerve of eyes subjected to ischemia and treated with Ga/DFO, as compared to fellow control, ischemic/reperfused-untreated eyes (0.17 and 0.53, respectively) ($p < 0.05$, Mann-Whitney test). The mean normalized level of 2,3-DHBA, which is considered the more specific marker of hydroxyl radicals, was approximately 3.1 times higher in the un-

Table 2. Levels of 2,5- and 2,3-DHBA in the optic nerve after 90 min of ischemia without reperfusion (compared to non-ischemic control eye)

	2,5-DHBA	2,3-DHBA
Non-ischemic control	1.50 ± 0.33	0.212 ± 0.094
Only ischemia (90 min)	1.55 ± 0.32	0.233 ± 0.084

n = 5/5; DHBA = dihydroxybenzoic acid derivative

(DHBA level ng/ml normalized to Salicylate mg/ml and Protein mg/ml)

Table 3. Levels of 2,5- and 2,3-DHBA in the optic nerve following 90 min of ischemia and 5 min of reperfusion treated with Ga/DFO (compared to non-treated control eye)

	2,5-DHBA	2,3-DHBA
Ischemia (90 min); reperfusion (5 min)	3.62 ± 1.37	0.53 ± 0.11
Ischemia (90 min); reperfusion (5 min) + Ga/DFO	2.70 ± 0.88	0.17 ± 0.06*

n = 4/4; DHBA = dihydroxybenzoic acid derivative

(DHBA level ng/ml normalized to Salicylate mg/ml and Protein mg/ml)

*Denotes that the difference between the control group and the Ga/DFO-treated group is statistically significant ($p < 0.05$, Mann-Whitney test)

Table 4. Levels of 2,5- and 2,3-DHBA in the vitreous after 90 min of ischemia without reperfusion (compared to non-ischemic control eye)

	2,5-DHBA	2,3-DHBA
Ischemia (90 min); reperfusion (5 min)	8.78 ± 1.63	0.83 ± 0.24
Ischemia (90 min); reperfusion (5 min) + Ga/DFO	3.96 ± 0.64*	0.22 ± 0.06*

n = 5/5; DHBA = dihydroxybenzoic acid derivative
(DHBA level ng/ml normalized to Salicylate mg/ml and Protein mg/ml)
*Denotes that the difference between the control group and the Ga/DFO-treated group is statistically significant (p < 0.05, Mann-Whitney test)

treated eyes, as compared to those treated by Ga/DFO. The level of 2,5-DHBA was lower but not to a significant degree (Table 3).

In addition, the levels of DHBA in the vitreous were also studied in the second part of the investigation. As shown in Table 4, the levels of 2,5- and 2,3-DHBA in the vitreous were higher than in the optic nerve and even than in the retina (Ophir et al., 1994). In the treated group, the levels of both 2,5- and 2,3-DHBA were significantly lower (approximately 2–4 times) than in the control (non-treated, ischemic-reperfused eye) group (Table 4).

4. DISCUSSION

Ischemia, followed by reperfusion, has been shown to cause damage in a variety of tissues and organs. In the retina, for example, previous studies showed histologic and functional injury following ischemia in monkeys (Hayreh and Weingeist, 1980), rats (Hughes, 1991) and cats (Reinecke et al., 1962). A previous study from our group has suggested the possible involvement of free radicals in causing this damage and the rapid time course of formation of these radicals following only minutes of reperfusion (Ophir et al., 1993).

According to our hypothesis, the protection afforded by Ga/DFO is due to the combined effects of the "push–pull" mechanisms. The proposed site of its action is the interference with the metal ion-dependent phase of hydroxyl radical generation. It has been suggested (Chevion, 1988, 1991) that redox-active ions of transition metals, mainly of Fe and Cu, which are present in trace amounts in tissue and which appear in higher levels in tissues and body fluids immediately following ischemia (Chevion et al., 1993; Voogd et al., 1992; Nohl et al., 1991), serve as catalytic sites in the critical step of transformation of superoxide radicals into the highly reactive hydroxyl radicals during the early phase of reperfusion. The "pull" component of Ga/DFO action is performed by desferrioxamine, by virtue of its ability to chelate the ferric iron and other ions, and thus reduce their availability to serve as catalytic sites of repeated production of hydroxyl radicales. The "push" action is supplied by Gallium which may displace Fe^{3+} or Cu^{2+} in their binding sites. Since Gallium is not redox-active under physiological conditions, this will also serve to reduce hydroxyl radical formation.

The combination of Gallium and desferrioxamine has other potential advantages. DFO, in itself, is a relatively large and randomly oriented molecule with low ability to penetrate cells. This may be the reason that it was found capable of entering most cells and provides only minimal protection against free radical damage in experimental models. When combined with DFO, however, the Ga/DFO complex assumes a well defined compact structure which probably enhances its potential to infiltrate into cells.

In conclusion, optic nerve and vitreous reperfusion injury following 90 minutes of ischemia was markedly reduced by injection of Ga/DFO. Our studies substantiate the conclusion that injury is caused following formation of highly reactive hydroxyl radicals. We propose that Ga/DFO reduces tissue injury via a combined "push–pull" mechanism, interfering with the availability of redox-active transition metal ions which catalyze the formation of hydroxyl radicals. It is proposed that further studies with this class of drugs may assist in reducing reperfusion injury in ocular disorders such as acute retinal artery occlusion and glaucoma, and in conditions of ischemia reperfusion in other organs.

REFERENCES

Berenshtein, E., Goldberg, C., Kitrossky, N., and Chevion, M., 1997, Pattern of mobilization of copper and iron following myocardial ischemia: possible criteria for tissue injury, *J. Mol. Cel. Biol.,* in press.

Berenshtein, E., Banin, E., Pe'er, J., Kitrossky, N. and Chevion, M., 1996, Ga/DFO protect retina against reperfusion injury, *VIII Biennial Meeting International Society for Free Radical Research,* Barcelona, Spain. p.157.

Bradford, M., 1976, A rapid and sensitive method for the quantitation of microgram quantities of protein utilizing the principle of protein-dye binding. *Anal. Biochem.* **72**: 248–254.

Bunt-Milam, A.H., Dennis, M.B., Jr., and Bensinger, R.E., 1987, Optic nerve head axonal transport in rabbit with hereditary glaucoma, *Exp. Eye Res.* **44**: 537–551.

Chevion, M., 1988, A site specific mechanism for free radical induced biological damage: the essential role of redox-active transition metals, *Free Rad. Biol. Med.* **5**: 27–37.

Chevion, M., 1991, Protection against free radical-induced and transition metal-mediated damage: The use of "pull" and "push" mechanisms, *Free Rad. Res. Comms.* **12–13**: 691–696.

Chevion, M., Jiang, Y., Har-El, R., Berenshtein, E., Uretzky, G. and Kitrossky, N., 1993, Copper and iron are mobilized following myocardial ischemia: possible criteria for tissue injury, *Proc. Natl. Acad. Sci. USA* **90**: 1102–1106.

Gutman, I., Melamed, S., Ashkenazi, I., and Blumenthal, M., 1993, Optic nerve compression by carotid arteries in low-tension glaucoma, *Graefes Arch. Clin. Exp. Ophthalmol.* **231**: 711–717.

Hayreh, S.S., and Weingeist, T.A., 1980, Experimental occlusion of the central retinal artery of the retina: IV: retinal tolerance time to acute ischemia, *Br. J Ophtalmol.* **64**: 818–825.

Hughes, W.F., 1991, Quantitation of ischemic damage in the rat retina, *Exp. Eye Res.* **53**: 573–582.

Kalvin, N.H., Hamasaki, D.I., Glass, J.D., Experimental glaucoma in monkeys. II. Studies of intraocular vascularity during glaucoma, 1966, *Arch. Ophthalmol.* **76**: 94–103.

Micelli-Ferrari, T., Vendemiale, G., Grattagiano, I., Boscia, F., Arnese, L., Altomare, E., and Cardia, L., 1996, Role of lipid peroxidation in the pathogenesis of myotopic and senile cataract, *Br. J. Ophthalmol.* **80**: 840–843.

Nohl, H., Stolze, K., Napetschnig, S. and Ishikawa, T., 1991, Is oxidative stress primarily involved in reperfusion injury of the ischemic heart? *Free Rad. Biol. Med.* **11**: 581–588.

Ophir, A., Berenshtein, E., Kitrossky, N., Berman, E. R., Photiou, S., Rothman, Z. and Chevion, M., 1993, Hydroxyl radical generation in the cat retina during reperfusion following ischemia, *Exp. Eye Res.* **57**: 351–357.

Ophir, A., Berenshtein, E., Kitrossky, N., and Averbukh, E., 1994, Protection of the transiently ischemic cat retina by zinc-desferrioxamine. *Invest. Ophthalmol, Vis. Sci.* **35**: 1212–1222.

Reinecke, R.D., Kuwabara, T., Cogan, D.C., and Weiss, D.R., 1962, Retinal vascular patterns: V: Experimental ischemia of the cat eye, *Arch. Ophthalmol.* **67**: 470–475.

Voogd, A., Sluiter, W., Eijk, H. G. v. and Koster, J. F., 1992, Low molecular weight iron and the oxygen paradox in isolated rat hearts, *J. Clin. Invest.* **90**: 2050–2055.

Winterkorn, J.M., Beckman, R.L., 1995, Recovery from ocular ischemic syndrome after treatment with verapamil, *J. Neuroophthalmol.* **15**: 209–211.

22

MECHANISMS OF ANTIOXIDANT ACTION

Enrique Cadenas

Department of Molecular Pharmacology and Toxicology
School of Pharmacy
University of Southern California
Los Angeles, California 90033

1. INTRODUCTION

Oxidative stress can be viewed as the disturbance in the oxidant—antioxidant balance in favor of the former (Sies, 1985). Over the years, the research disciplines interested in oxidative stress have been growing steadfastedly, thus increasing our knowledge of the importance of the cell redox status and aiding at the recognition of oxidative stress as a process with implications for a large number of pathophysiological states. From this multi- and interdisciplinary interest in oxidative stress emerges a picture that attest to the vast consequences of the complex and dynamic interplay of oxidants and antioxidants in a cellular setting.

Consequently, our view of oxidative stress is growing in scope and new future directions. Likewise, the term 'reactive oxygen species'—adopted at some stage in order to include nonradical oxidants such as H_2O_2 and $^1O_2$—fails nowadays to reflect the rich variety of reactive species in free radical biology, encompassed by nitrogen-, sulfur-, oxygen-, and carbon-centered radicals. Consideration and measurement of cellular steady-state level of reactive species—determined by their production and removal—is of utmost important in the study of oxidative stress. This is in line with the concept advanced by Imlay and Fridovich (1991) that disease states linked to oxidative stress should be confirmed quantitatively by changes in the concentrations of oxidative species. For example, it is well established that subtle changes in the cellular redox status mark the initial events of signaling cascades deciding the fate of the cell. Major sources of reactive oxygen species are encompassed by aerobic metabolism, specialized physiological functions, and xenobiotic metabolism. Extensive research in the last two decades has provided compelling evidence that oxygen dependence imposes universal toxicity to all aerobic life processes; accordingly, the formation of reactive oxygen species ($O_2^{\cdot-}$ and H_2O_2) seems to be commonplace in aerobically metabolizing cells (Boveris and Cadenas, 1997). The formation of oxyradicals during the respiratory burst and the release of the endothelium-derived releasing factor (nitric oxide) represent specialized physiological functions also contributing to the cellular steady-state level of oxidants.

Free Radicals, Oxidative Stress, and Antioxidants, edited by Özben.
Plenum Press, New York, 1998.

As a corollary of the widespread participation of oxidative stress in pathophysiological situations, a vast number of antioxidant therapies have been suggested or implemented. A review of the current status of antioxidant therapy (Rice, Evans, and Diplock, 1993) has focused on the antioxidant profile of individuals and its role in protection against amplification of certain disease process, which are known to be associated with oxidative stress. Although this concept is gaining significance in coronary heart disease, inflammation, and atherosclerosis, and there is some evidence for the beneficial effects of free radical scavenging drugs, the authors conclude that the implementation of antioxidant therapies requires a better understanding of the involvement of free radicals and the molecular mechanisms by which they exert cytotoxicity in disease states. It may also be surmised that knowledge is required on the concerted activity of antioxidant molecules in a cellular or extracellular setting as well as identification of potential specific sites for antioxidant action.

This chapter examines some of the reactions of free radicals with a number of small molecule antioxidants, with emphasis on the decay pathways of the antioxidant radical species ensuing from these interactions. Specific molecular mechanisms of action of natural and synthetic antioxidants, their involvement in signal transduction pathways and gene expression, and their evaluation in antioxidant therapies by clinical trials are covered by recent comprehensive treatises (Cadenas and Packer, 1996; Clerch and Massaro, 1997; Forman and Cadenas, 1997; Packer & Cadenas, 1997; Sies, 1985, 1997). This overview is meant for the students who attended the NATO Advance Study course in Antalya, Turkey, in May–June, 1997 and it is expected to serve as a background for the mechanisms of antioxidant action described in other chapters and that address more specialized areas of free radical biology and medicine.

2. SOME MECHANISMS OF ANTIOXIDANT ACTION

For the sake of convenience, three different mechanisms of antioxidant action for small molecules are described here: (I) transfer of the radical character with formation of a reactive antioxidant-derived radical, (II) trapping of free radicals with formation of a stable or inert free radical trap, and (III) molecules which mimic antioxidant enzyme activities. The first category applies mainly to natural antioxidants and some selected synthetic antioxidants, whereas the second and third are of pharmacological interest and describe the mechanisms inherent in the antioxidant action of some synthetic compounds.

2.1. Transfer of the Radical Character and Formation of a Reactive Antioxidant-Derived Radical

A basic concept in antioxidant activity is summarized in the electron- or hydrogen transfer illustrated in the general equation below: by means of this reaction, the radical character (initially centered on a strong oxidant ($R^{\bullet}$), for example, hydroxyl radical) is transferred to an "antioxidant" molecule (AH^{-}). The formation of an antioxidant-derived radical is a process inherent in the oxidant scavenging activity and understanding the decay pathways of the *antioxidant-derived radical* ($A^{\bullet -}$) is critical for the evaluation of the functions of these compounds in a biological milieu.

$$AH^{-} + R^{\bullet} \rightarrow A^{\bullet -} + RH \qquad [1]$$

Fig. 1 lists the reduction potentials of the redox couples potentially involved in Eq. 1. Following a concept advanced by Buettner and Jurkiewicz (1993) and inferred from the

Redox Couple	$E^{\circ\prime}$/Volts
$HO^{\cdot},H^{+}/H_2O$	+2.18
$RO^{\cdot},H^{+}/ROH$	+1.60
$ONOO^{-}/N^{\cdot}O_2$	+1.40
$HO_2^{\cdot},H^{+/H}_2O_2$	+1.07
$ROO^{\cdot},H^{+}/ROOH$	+1.00
$Fe^{IV}=O/Fe^{III}$	+0.99
NO_2/NO_2^{-}	+0.87
$GS^{\cdot}/GS^{-}$	+0.85
$PUFA^{\cdot},H^{+}/PUFA-H$	+0.60
$U^{\cdot-}/UH^{-}$	+0.56
$\alpha\text{-}TO^{\cdot}/\alpha\text{-}TOH$	+0.48
$NO^{\cdot}/NO^{-}$	+0.39
$H_2O_2,H^{+}/H_2O,HO^{\cdot}$	+0.32
$A^{\cdot-}/AH^{-}$	+0.28

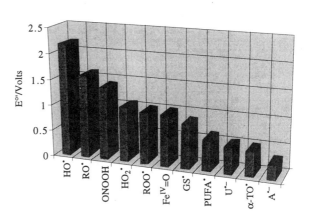

Figure 1. One-electron reduction potentials of some biologically relevant redox couples. Abbreviations: $Fe^{IV}=O$, oxoferryl complex in myoglobin; Fe^{III}, metmyoglobin; PUFA, polyunsaturated fatty acids. U, uric acid. α-TOH, α-tocopherol, A, ascorbic acid.

data in Fig. 1, a distinction between oxidants and antioxidants is not necessarily correct; instead, the terms strong oxidants and weak oxidants appear more suitable, for they permit to consider the chemical reactivity inherent in the antioxidant-derived radical. The latter, with a reduction potential and a chemical reactivity lower—and a life time usually longer—than that of the oxidant initially scavenged, is not inert and can have toxic implications provided the proper cellular settings are given. The chemical reactivity of the antioxidant- derived radical is documented by examples such as the inhibition of α_1-antiproteinase by the radical form of uric acid (Aruoma and Halliwell, 1989) and the propagation of lipid peroxidation within the LDL particle by the α-tocopheroxyl radical (Ingold et al., 1993).

This section surveys the main biochemical properties of major 'antioxidant' molecules, discussed in terms of an electron-transfer reaction involving the removal of the primary radical species and of the decay pathways of the antioxidant derived radical.

2.1.1. Vitamin E. Vitamin E or α-tocopherol, a chain-breaking antioxidant, is known to react with a variety of free radicals at fairly high rates: it reacts readily with secondary peroxyl radicals (Eq. 2) and with $HO_2^{\cdot-}$ but not $O_2^{\cdot-}$—at a rate of 2×10^5 $M^{-1}s^{-1}$ (Bielski & Cabelli, 1991).

$$\alpha T{-}OH + ROO^{\cdot} \rightarrow \alpha T{-}O^{\cdot} + ROOH \qquad [2]$$

The reductive and oxidative decay of the antioxidant-derived radical (chromanoxyl radical) formed in reactions of the type of Eq. 2, are illustrated in the scheme in Fig. 2. The reductive pathways, encompassing the $\alpha T{-}O^{\cdot} \rightarrow \alpha T{-}OH$ transition, may be considered antioxidant or prooxidant. The antioxidant character is illustrated by the efficient recovery of vitamin E by ascorbate. Other antioxidants, such as urate and ubiquinol, may also accomplish this reaction (see below). GSH cannot recover the vitamin E chromanoxyl radical and thiyl radicals do not react with Trolox C, the water-soluble analog of α-tocopherol. $^{\cdot}O_2^{\cdot-}$ may theoretically support a reductive decay pathway of the chromanoxyl radical as inferred from pulse radiolysis experiments carried out with Trolox C ($k_3 = 4.5 \times 10^8$ $M^{-1}s^{-1}$) (Cadenas et al., 1989).

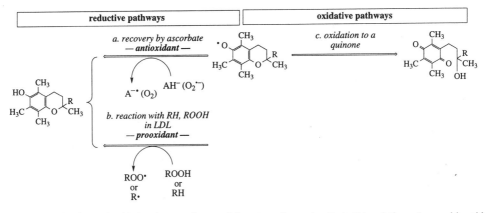

Figure 2. Reductive and oxidative decay pathways of the α-tocopheroxyl radical. Abbreviations: A, ascorbic acid; ROO˙, lipid peroxyl radical; R˙, lipid alkyl radical.

$$\alpha T\text{–}O^\bullet + O_2^{\bullet -} + H^+ \rightarrow \alpha T\text{–}OH + O_2 \qquad [3]$$

The prooxidant character of the reductive pathway is shown in Fig. 2, pathway b: the $\alpha T\text{–}O^\bullet \rightarrow \alpha T\text{–}OH$ transition may be coupled to propagation of lipid peroxidation within lipoprotein particles (LDL) upon reaction of this radical with PUFA moieties, that is, H abstraction from a bisallylic methylene group of PUFAs or with lipid hydroperoxides in oxidized LDL. The former reaction proceeds slowly (0.1 ± 0.005 $M^{-1}s^{-1}$) and the latter at a 10-fold higher rate (Bowry et al., 1992; Bowry and Stocker, 1993). The fact that tocopheroxyl radicals in LDL may act as chain-transfer agents illustrates the chemical reactivity of the antioxidant-derived radical. Some physico-chemical features inherent in the antioxidant status of LDL particles were summoned to account for the different activities of vitamin E (Ingold et al., 1993) as a chain-breaking antioxidant in biological membranes (Eq. 2) and as a chain-transfer agent in LDL (Fig. 2, pathway b).

The oxidative decay of the tocopheroxyl radical is exemplified by reactions involving cleavage of the C–O bond in the pyrane ring of chromane and leading to the formation of tocopherylquinone (Fig. 2, pathway c). This reaction is apparently irreversible, although the quinone may be reduced back to vitamin E in vitro with high concentrations of ascorbate and at low pH (Liebler et al., 1989). That the oxidative decay of the tocopheroxyl radical to the tocopherylquinone takes place in a biological setting is supported by the occurrence in rate live of small amounts of tocopherylquinone (10–12% of the total tocopherol) (Bieri & Tolliver, 1981).

2.1.2. Ubiquinol. In addition to its role as redox component of the mitochondrial electron-transport chain, ubiquinone may function in its reduced form as an antioxidant (Beyer & Ernster, 1990) and membrane labilizer. A large body of circumstantial evidence points to the function of ubiquinol as an antioxidant: experiments with reconstituted membrane systems, mitochondrial membranes, and low density lipoproteins and observations on intact animals as well as in the clinical setting. Unfortunately, there is little kinetic data on the mechanistic aspects by which ubiquinol exerts its antioxidant activity and the precise molecular and cellular mechanism(s) underlying the antioxidant functions of ubiquinol remain to be elucidated.

It could be speculated that the hydrogen-donating activity by which ubiquinol presumably acts as an antioxidant (Cadenas et al., 1992) resembles the reaction of $O_2^{\bullet -}$ with hydroquinones, a facile electron/hydride transfer. Accordingly, peroxyl radicals may be suggested to undergo an analogous reactions, in addition to a rapid reaction of ubiquinol with perferryl complexes of the type $ADP-Fe^{III}-O_2^{\cdot-}$, used in lipid peroxidation experimental models (Forsmark et al., 1991). These pathways leading to the formation of the antioxidant-derived radical, ubisemiquinone, are illustrated in Fig. 3. Another source of ubisemiquinone (or antioxidant-derived radical) may result from its ability to maintain a more efficient recycling of vitamin E (Eq. 4) than ascorbate does (Kagan et al., 1990a,b).

$$UQH^- + \alpha T-O^\cdot \rightarrow UQ^{\cdot-} + \alpha T-OH \qquad [4]$$

Similar to the case of the α-tocopheroxyl radical, the decay of the antioxidant-derived radical, ubisemiquinone, is encompassed by reductive and oxidative decay pathways (Fig. 3). The former may be visualized to occur via the cytochrome b_{562} component of the mitochondiral "Q" cycle. The latter involves autoxidation of the ubisemiquinone, a well documented process, which proceeds with second order rate constants varying between $1-100$ $M^{-1}s^{-1}$ (Cadenas et al., 1977).

The reductive and oxidative decay pathways of ubisemiquinone are important when addressing the antioxidant properties of ubiquinol in connection with the antioxidant status of LDL. Despite the lack of kinetic information on the reactions leading to ubisemiquinone formation, it was proposed that the $UQH^- \rightarrow UQ^{\cdot-}$ transition coupled to peroxyl radical reduction in conjunction with the oxidative decay of the ubisemiquinone (as shown in Fig. 3) served to export the radical character in the form of $O_2^{\cdot-}$ from the LDL particle into the aqueous phase (Ingold et al., 1993). $O_2^{\cdot-}$ formed during this reaction was proposed to react with the vitamin E radical (Ingold et al., 1993) (Eq. 3 above), a reaction already described

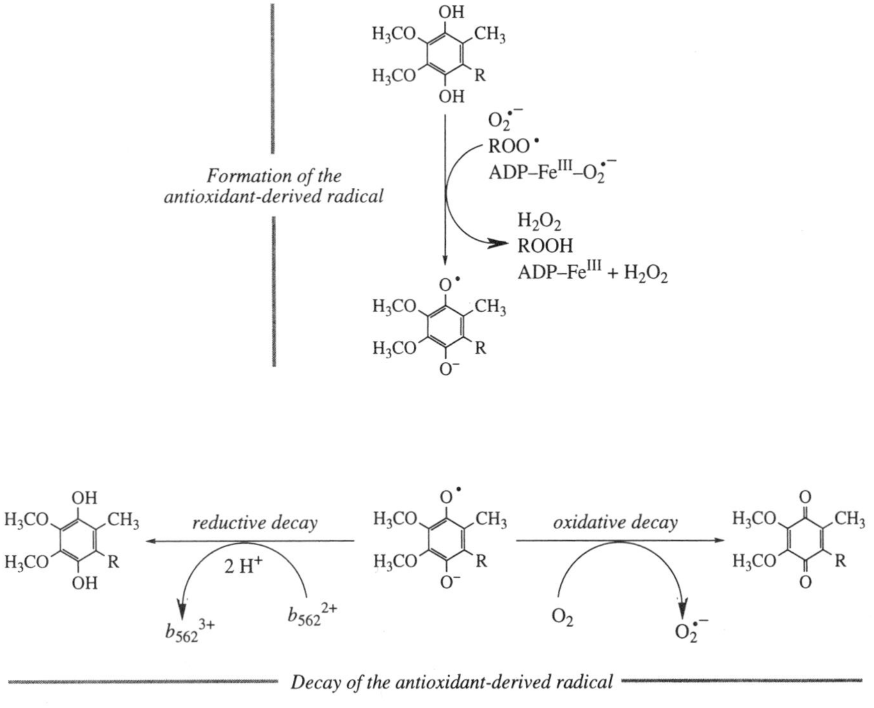

Figure 3. Formation and decay of the antioxidant-derived radical of ubiquinol.

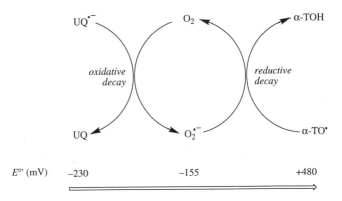

Figure 4. Ubisemiquinone-dependent, superoxide anion-mediated recovery of vitamin E.

for the Trolox radical in a pulse radiolysis study (k_3 = 4.5 × 10^8 M^{-1}s^{-1} (Cadenas et al., 1989). The proposed recovery of the vitamin E radical by O$_2^{\cdot-}$ is expected to take place in the aqueous phase. In summary, the oxidative decay of the semiquinone is coupled to the reductive decay of the tocopheroxyl radical by means of the O$_2$ ⇔ O$_2^{\cdot-}$ transition (Fig. 4).

2.1.3. Uric Acid. Pulse radiolysis studies provided information on the rate constants for the reactions of uric acid with HO$^\cdot$, CCl$_3$OO$^\cdot$, NO$_2^\cdot$, guanyl radicals and for that involved in an efficient repair of glutathionyl radicals (GS$^\cdot$) (1.4 × 10^7 M^{-1}s^{-1}) (Willson et al., 1985; Simic and Jovanovic, 1989). In addition, urate was described to react with hypochlorous acid, singlet oxygen, ozone, and nitrogen dioxide (and possibly peroxynitrite). The radicl character in the antioxidant-derived radical may be placed at O^8 (being the hydroxy group the strongest acid) (Simic and Jovanovic, 1989) or at N^9/N^7 (Maples and Mason, 1988) (Eq. 5):

The antioxidant properties of uric acid were extensively described (see Becker, 1993; Halliwell, 1997); however, the antioxidant-derived radical, with a reduction potential of +0.51 V, inhibits yeast alcohol dehydrogenase (Kittridge and Willson, 1984) and human α$_1$-antiproteinase (Aruoma and Halliwell, 1989).

$$\text{[structure]} + R^\cdot \longrightarrow \text{[structure]}^\cdot \quad \text{or} \quad \text{[structure]} + RH \qquad [5]$$

Ascorbic acid appears to be at the end of the recovery process, for it repairs efficiently urate-derived radicals (10^6 M^{-1}s^{-1}), thus providing another example of transfer of the radical character to a less reactive species, which is less likely to have toxicological implications in a biological milieu. Scavenging of HO$^\cdot$ by urate, followed by recovery of the urate radical by ascorbic acid, involves the sequential formation of species with decreasing reduction potentials: [$E_{HO^\cdot/HO^-}$ = +2.31 V] > [$E_{U^\cdot/UH^-}$ = +0.51 V] > [$E_{A^{\cdot-}/AH^-}$ = +0.28 V].

2.1.4. Thiols. The reaction of thiols with different radicals yields mainly thiyl radicals (Eq. 6), processes extensively characterized by the classical pulse radiolysis work by Asmus (1990), von Sonntag (1987), Wardman (1988, 1990), and Willson (1985). Thiols react with oxygen-, carbon-, and nitrogen-centered radicals as well as with the high oxidation state of hemoproteins at different rates (Eq. 6).

$$RS^- + R^\bullet + H^+ \longrightarrow RS^\bullet + RH$$

$R^\bullet$	$M^{-1}s^{-1}$
$O_2^{\bullet-}$	15
$HO_2^\bullet$	6×10^2
$HO^\bullet$	10^{10}
$NPh{-}O^\bullet$	3×10^5
$Fe^{IV}{=}O$	
$Ar{-}NH^\bullet$	

[6]

The routes for decay of the thiyl radical thus formed (Eq. 6) were elegantly described by Wardman (1988, 1990), who considers two main pathways which exert a kinetic control of and are critical to the free radical biochemistry of thiols: conjugation reactions (with thiolate (RS^-) or oxygen) and dimerization reactions (Fig. 5A). Conjugation reactions, especially with thiolate, are expected to prevail in a biological setting, given the high concentration of GSH (1% of which is as GS^- at the cell's pH). Conjugation reactions with thiolate yield the disulfide anion radical ($[GSSG]^{\bullet-}$), a powerful reductant with a reduction potential of -1.6 V (Surdhar and Armstrong, 1986). Those with O_2 yield the corresponding peroxyl radical. Dimerization reactions may contribute little to the decay of thiyl radicals given the required high steady-state level of $GS^\bullet$ for the bimolecular collision.

The importance of conjugation reactions in thiyl radical biochemistry is illustrated by the following: first, by means of conjugation, a strong oxidant ($GS^\bullet$) with a reduction potential of ~ 0.86 V is transformed into a powerful reductant ($[GSSG]^{\bullet-}$) ($E^{\circ\prime} = -1.6$ V) (Wardman, 1988, 1990). Second, conjugation reactions are also important to evaluate the kinetic regulation of reactions otherwise thermodynamically unfavorable. For example, the reactions of GSH with aminopyrine (Wilson et al., 1986), 1-naphthoxyl radicals (D'Arcy-Doherty et al., 1986), alloxan radicals (Winterbourn and Munday, 1990), and diaziquone radicals (Ordoñez and Cadenas, 1992) are thermodynamically unfavorable; however, efficient removal of the thiyl radical by conjugation drives the equilibrium of these reactions towards the right favoring repair of these organic radicals by GSH.

Third, the strong antioxidant properties of dithiols, such as dihydrolipoic acid, may be partly explained in terms of the intramolecular conjugation between the thiyl radical and the adjacent thiolate (Eq. 7). In this case, a prevalence of conjugation of the thiyl radical with thiolate relative to that with O_2 may be a feature inherent in the antioxidant properties of these compounds. The lipoic acid radical (cyclic disulfide anion radical), for example, reduces FAD to FADH at a rate of $2.4 \times 10^8 M^{-1}s^{-1}$.

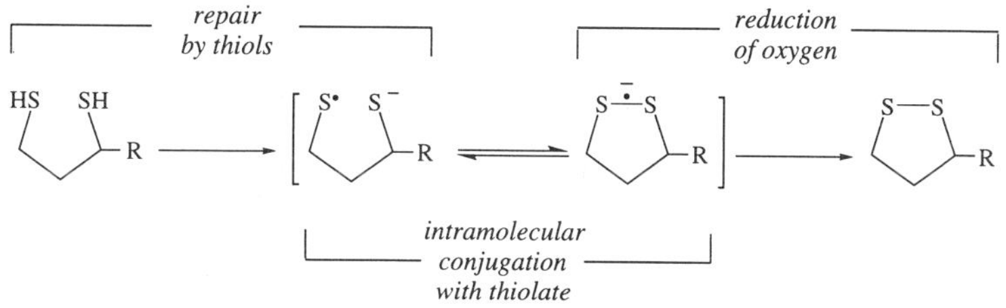

[7]

The antioxidant properties of dihydrolipoate have been examined in a large number of experimental models encompassing the reduction of peroxyl, ascorbyl, and chromanoxyl radicals (Packer et al., 1997), protection against microsomal lipid peroxidation, and

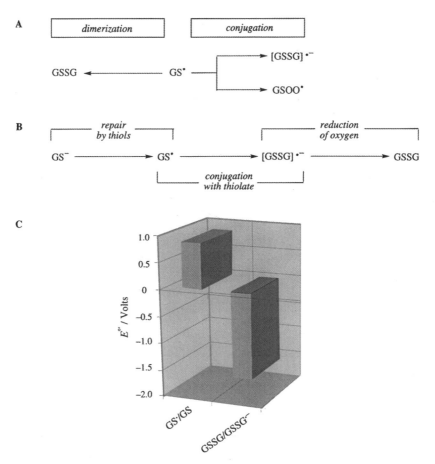

Figure 5. Decay of thiyl radicals. (A) Dimerization and conjugation. (B) Role of conjugation in thiyl radical decay. (C) Reduction potentials of the redox couples before (GS·/GS⁻) and after (GSSG/GSSG·⁻) thiyl radical conjugation with thiolate.

cofactor for the peroxidase activity of ebselen (Haennen and Bast, 1991). These properties are of pharmacological interest, because lipoic acid may be used as a drug to lessen or prevent oxidative stress conditions; the concentration of lipoate in cells is in the low μM range and it does not occur free but bound to protein complexes; uptake of lipoate by hepatic tissue occurs by a carrier-mediated process at low concentrations of the disulfide and by diffusion at high concentration (Peinado et al., 1989).

2.1.5. Vitamin C. One of the most important features of vitamin C in connection with its antioxidant activity is the electron transfer-mediated recovery of the vitamin E radical (Fig. 1, pathway a). Ascorbate is otherwise an excellent electron donor and it reacts with a vast group of free radicals following the general Eq. 8. In addition to these free radical species, ascorbate also reacts at high rates with triplet carbonyls (1.2×10^9 $M^{-1}s^{-1}$) and singlet oxygen (8×10^6 $M^{-1}s^{-1}$), a fact that should be contemplated when evaluating the effect(s) of ascorbate on lipid peroxidation: recombination of secondary lipid peroxyl radicals is known to yield substantial amounts of triplet carbonyls and singlet oxygen.

$$AH^- + R^\bullet \longrightarrow A^{\bullet-} + RH$$

$R^\bullet$	$M^{-1}s^{-1}$
$HO^\bullet$	1.1×10^{10}
$RO^\bullet$	1.4×10^9
$O_2^{\bullet-}/HO_2^\bullet$	1.2×10^7
$RS^\bullet$	6.0×10^8
$^3RO^*$	1.2×10^9
$U^\bullet$	1.0×10^6
$ROO^\bullet$	1.5×10^6
$A^{\bullet-}$	2.0×10^5
$\alpha\text{-}TO^\bullet$	2.0×10^5
$^1O_2$	8.0×10^6

[8]

Fig. 6 illustrates the following concepts: first, the transfer of the radical character to 'less reactive' species established by a sequence determined by the reduction potentials of the radicals involved. For example, the reduction potential of secondary lipid peroxyl radicals is +1.00 V and those of the α-tocopheroxyl- and ascorbyl radicals +0.48 V and +0.28 V, respectively. Hence, the radical character, originally in the form of a strong oxidant (ROO·) is now in the form of a weak oxidant (A·⁻). Second, that the chemical reactivity of the antioxidant-derived radical may have toxicological implications as documented for the radical forms of urate and vitamin E. Third, the sequences drawn in Fig. 6 suggest a role for ascorbate as the ultimate reductant within an antioxidant concerted mechanism.

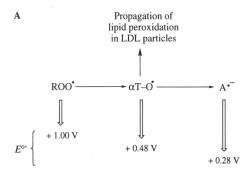

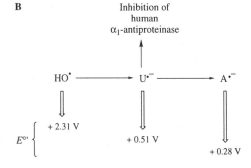

Figure 6. Transfer of the radical character and chemical reactivity of the antioxidant-derived radical. (A) Example with the sequence peroxyl radical → α-tocopheroxyl radical → ascorbyl radical. (B) Example with the sequence hydroxyl radical → urate radical → ascorbyl radical.

The reduction potentials listed in Fig. 1 and the notions summarized in Fig. 6 suggest that the concerted activity of ascorbic acid with other 'antioxidant' molecules would result in ultimate transfer of the radical character to a weak oxidant, the ascorbyl radical. It may be inferred that this radical is nontoxic and it decays readily by a second-order disproportionation process (Eq. 9; $k_9 = 2 \times 10^5$ $M^{-1}s^{-1}$) yielding the ascorbate monoanion and dehydroascorbate (Bielski, 1982). The enzymes in charge of regenerating ascorbate are less well characterized: a monodehydroascorbate reductase and a dehydroascorbate reductase transfer one and two electrons to the ascorbyl radical and to dehydroascorbate, respectively, to generate ascorbic acid. The former enzyme—better characterized in mammalian tissues—reduces $A^{\cdot -}$ at expense of NADH or NADPH, whereas the latter reduces dehydroascorbate at expense of GSH. GSH can also reduce dehydroascorbate nonenzymically. These issues were recently addressed by Buettner and Jurkiewicz (1997), who also suggested that the intensity of the ascorbyl radical ESR signal could serve as a marker of oxidative stress.

$$A^{\cdot -} + A^{\cdot -} + 2H^+ \rightarrow A + AH_2 \qquad [9]$$

Hence, it may be surmised that ascorbic acid, placed at the end of an electron transfer process and—according to its reduction potential—a weak oxidant, decays by disproportionation to nonradical products. It is worth mentioning, that the EPR signal observed in various experimental models of oxidative stress is that of the ascorbyl radical, thus strengthening the notion that vitamin C may be the ultimate antioxidant in a free radical chain. The tissue content of ascorbic acid is relatively high ranging from 125–140 mg/kg in brain and liver to lower levels in adipose tissue and blood (~10 mg/kg).

The transfer of the radical character described in this section was exmplified with natural antioxidant; these mechanisms are also expected to underlie the action of selected synthetic derivatives of natural antioxidants, such asα-tocopherol analogs and anionic tocopherol esters, among others.

2.2. Trapping of Free Radicals with Formation of a Stable or Inert Free Radical Trap

Nitrone radical traps were introduced as valuable tool for EPR spectroscopy and, indeed, the use of spin traps has advanced significantly the field of free radical biology. Beyond analytical chemistry, a pharmacological aspect of nitrones was recently introduced by Floyd and his colleagues (1997). The protective effect of these compounds against oxidative stress in several experimental models is expected to be a consequence of a primary trapping of free radicals as shown in Eq. 10. At variance with the transfer of the radical character with formation of an antioxidant-derived radical discussed in the previous section, the reaction of nitrones (α-phenyl-*t*-butyl nitrone (PBN) in Eq. 10) with free radicals yields a stable radical adduct (Floyd et al., 1997). Because the latter is virtually inert, by trapping the radical R·, it prevents a series of free radical propagation reactions.

$$[10]$$

The concept brought forward in Eq. 10 is central to the understanding of the protective effects of nitrones on experimental models entailing brain oxidative stress and neurodegeneration as well as the effects of nitrones on the cell's redox state (Hensley et al., 1997). Floyd et al. (1997) has described several experimental neuroedegenerative models (such as brain stroke, aging, concussion, and excitotoxicity) where nitrones are protective against the underlying oxidative stress-mediated damage. PBN, together with 5,5'-dimethyl-1-pyrroline-N-oxide (DMPO), a water-soluble radical trap, protected both glutathione peroxidase and glutathione reductase against oxidative damage elicited by UV, H_2O_2 and UV, and O_3 (Tabatabaie and Floyd, 1994). The free radical-scavenging activity of PBN, DMPO, and α-(4-pyridyl-1-oxide)-N-t-butylnitrone (POBN) prevented DNA base oxidation (as 8-hydroxy-2'-deoxyguanosine) elicited by γ-irradiation (Young et al., 1996).

The protective effects of PBN extend to damage elicited by cytokines and nitric oxide (NO$^\bullet$), which are involved in the selective destruction of pancreatic β-cells in insulin-dependent diabetes mellitus (Tabatabaie et al., 1997). Likewise, PBN was shown to prevent NO$^\bullet$ formation caused by gp120 (an HIV-1 envelope glycoprotein implicated in the development of AIDS dementia) and to protect against gp120-induced behavioral impairment (Tabatabaie et al., 1996).

The interruption of a free radical chain by means of the reaction in Eq. 10 is critical to the 'antioxidant' or protective function of these compounds. Although nitrones protect against oxidative stress in the above experimental models by mechanisms other than those ascribed to conventional antioxidants (see section 2.1 above), a precise molecular mechanism is not available yet. The fate of the radical adduct, in terms of its metabolism and excretion as well as its involvement in other reactions encompassed by these oxidative stress models remain to be studied. A more recent view on the 'antioxidant' function of nitrones was provided by the finding that they suppress gene transcriptional events associated with pathophysiological states (NFκB-regulated cytokines and inducible nitric oxide synthase) (Hensley et al., 1997). It may be surmised that an effect of nitrones at a proximal level to oxidatively susceptible signal cascades is unlikely.

2.3. Molecules which Mimic Antioxidant Enzyme Activities

This section addresses briefly the antioxidant activity displayed by synthetic compounds, which mimic the specific reactions catalyzed by antioxidant enzymes, such as glutathione peroxidase and superoxide dismutase.

2.3.1. Glutathione Peroxidase Mimics. Ebselen (2-phenyl-1,2-benzisoselenazol-3(2H)-one is a synthetic selenium-containing heterocycle extensively studied as a glutathione peroxidase mimic (Sies, 1993). The mechanism of hydroperoxide reduction by ebselen in the presence of GSH appears to be kinetically identical to the glutathione peroxidase-catalyzed reaction (Maiorino et al., 1988). The activities of ebselen in biological model systems are numerous and have been recently listed by Noguchi and Niki (1997). A key for the glutathione peroxidase-like activity of ebselen is the formation of a highly reactive selenol intermediate formed during the reaction of ebselen with GSH and which carries the majority of the catalytic activity of the selenoorganic compound. It is outside the scope of this chapter to discuss the active species of ebselen involved in its function as a glutathione peroxidase mimic; the reader is referred to the review by Sies and Masumoto (1997), which addresses these issues in detail. Other synthetic compounds also display a glutathione peroxidase-like activity, such as α-(phenylselenenyl) ketones and diaryl tellurides (Cotgreave and Engman, 1997) (Fig. 7).

In addition to being glutathione peroxidase mimics, the above compounds possess an antioxidant activity on their own: diaryl selenides bearing electron-donating substituents inhibit efficiently microsomal lipid peroxidation triggered by ADP–Fe^{2+}/ascorbate

(Anderson et al., 1994); this effect may be ascribed to the quenching of peroxyl radicals by diarylselenides to yield a selenium-centered radical cation. Likewise, ebselen scavenges trichloromethyl peroxyl radicals (Schöneich et al., 1990) and somewhat less efficiently singlet oxygen (Scurlock et al., 1991).

A novel aspect of ebselen is its scavenging of peroxynitrite (Masumoto and Sies, 1996) entailing a bimolecular reaction of the selenoorganic compound with peroxynitrite ($k = 2 \times 10^6 \, M^{-1} s^{-1}$) to yield the selenoxide of the parent molecule; subsequently, the selenoxide is readily reduced by GSH back to ebselen. This catalytic cycle encompassed by the reduction of peroxynitrite by GSH may represent a defense mechanism agains peroxynitrite. In this context, ebselen was shown to protect DNA against singlet strand breakage elicited by peroxynitrite (Roussyn et al., 1996).

2.3.2. Superoxide Dismutase Mimics. The antioxidant activity of nitroxides has been examined in experimental models ranging from molecular levels to whole-body levels. Samuni and Krishna (1997) described the mechanisms underlying the antioxidant properties of nitroxides in terms of superoxide dismutase mimic activities encompassing oxidative and reductive modes. The former involves a reversible oxidation of the nitroxide to an oxoammonium cation (Eqs. 11–12), whereas the latter involves primary reduction of the nitroxide by superoxide anion (Eqs. 13–14). The overall balance entailed by the oxidative and reductive modes is that of the superoxide dismutase-catalyzed disproportionation of superoxide.

$$\text{N}-\text{O} + \text{O}_2^- + 2\text{H}^+ \longrightarrow \text{} ^+\text{N}=\text{O} + \text{H}_2\text{O}_2 \qquad [11]$$

$$^+\text{N}=\text{O} + \text{O}_2^- \longrightarrow \text{N}-\text{O} + \text{O}_2 \qquad [12]$$

$$\text{N}-\text{O} + \text{O}_2^- + \text{H}^+ \longrightarrow \text{N}-\text{OH} + \text{O}_2 \qquad [13]$$

$$\text{N}-\text{OH} + \text{O}_2^- + \text{H}^+ \longrightarrow \text{N}-\text{O} + \text{H}_2\text{O}_2 \qquad [14]$$

In addition for superoxide anion, nitroxides can react with semiquinones, metal ions (Samuni and Krishna, 1997), and the high oxidation state of myoglobin and hemoglobin. The involvement of nitroxides in the latter process, i.e., the detoxification of high oxidation states of myoglobin or hemoglobin, progresses without changes in the concentration of the nitroxide (Mehlhorn and Swanson, 1992). This is of interest, because, conventionally, the oxidation of myoglobin by H_2O_2 to ferrylmyoglobin could be coupled to a pseudo-catalytic activity entailing oxidation of known electron donors (Giulivi and Cadenas, 1995). Furthermore, in the case of nitroxides, the presence of another reductant may reduce the oxoammonium cation described above to its hydroxylamine (Mehlhorn and Gomez, 1993). The lack of change in nitroxide concentration when coupled to a pseudoperoxidatic acitivity was accounted for by the initial reaction of the nitroxide with a globin radical, rather than with the oxoferryl complex (Mehlhorn and Swanson, 1992).

3. CONCLUDING REMARKS

This chapter surveys some thermodynamic and chemical aspects of the electron-transfer reactions inherent in the mechanism of action of antioxidants (that is, processes leading to the formation of an antioxidant-derived radical) and provides a brief description of the 'antioxidant' properties of radical traps and compounds which mimic antioxidant enzymes. The process of radical transfer has been described numerous times in the pulse radiolysis literature (see for example Willson et al., 1985) and it is expected to be important for the concerted activity of various antioxidant molecules in a biological setting. The latter aspect has been emphasized by Buettner and Jurkiewicz (1993, 1997) who consider ascorbic acid—at the end of the free radical transfer chain (Fig. 6)—a biomarker of oxidative stress, which may be used as a noninvasive indicator of oxidative stress. The 'constitutive' ascorbyl radical EPR signal in different tissues increases substantially in conditions of oxidative stress.

It is clear that ascorbic acid, a good electron donor, may behave as a primary antioxidant—by reacting directly with different free radicals species (Eq. 8)—or as a secondary antioxidant, in a concerted manner with other 'antioxidant' molecules and in charge of recovering the so-called antioxidant-derived radical (Fig. 6). These antioxidant functions affect the cell's level of ascorbate and, accordingly, critical metabolic reactions which require ascorbate, such as collagen production. Brown and Jones (1997) discuss the effects of a decreased ascorbate concentration in cells during oxidative stress on modulation of gene expression triggered by a deficiency in collagen production and alterations in the extracellular matrix. Likewise, a decrease in cellular ascorbate during oxidative stress correlates with an increase in lipid peroxidation products resulting in an enhanced gene expression. Another example is set by the effects of antioxidants such as tocopherol and tocopherol analogs on cell signal transduction, such as inhibition of protein kinase C activity, MAP kinase activities (albeit less well characterized), and of NFκB activation (Azzi et al., 1997).

REFERENCES

Anderson, C.M., Hallberg, A., Linden, M., Brattsand, R., Moldéus, P., and Cotgreave, I.A., 1994, Antioxidant activity of some diarylselenides in biological systems. *Free Rad. Biol. Med.* **16**:17–28.

Aruoma, O.I., and Halliwell, B., 1989, Inactivation of α_1-antiproteinase by hydroxyl radicals. The effect of uric acid. *FEBS Lett.* **244**:76–80.

Asmus, K.-D., 1990, Sulfur-centered free radicals. *Meth. Enzymol.* **186**:168–180.

Azzi, A., Boscoboinik, D., Cantoni, O., Fazzio, A., Marilley, D., O'Donnel, V., Özer, N.K., Spycher, S., Tabataba-Vakili, S., and Tasinato, A., 1997, Modulation by oxidants and antioxidants of signal transduction and smooth muscle cell proliferation, In *Oxidative Stress and Signal Transduction*, (Forman, H.J., and Cadenas, E., eds.), pp. 323–342, Chapman & Hall, New York.

Becker, B.H., 1993, Towards the physiological function of uric acid. *Free Rad. Biol. Med.* **14**:615–631.

Beyer, R.E., and Ernster, L., 1990, The antioxidant role of coenzyme Q. In *Highlights in Ubiquinone Research* (Lenazz, G., Barnabei, O., Rabbi, A., and Battino, M., eds.), pp. 191–213, Taylor and Francis, London.

Bielski, B.H.J., 1982, Chemistry of ascorbic acid radicals. In *Ascorbic Acid: Chemistry, Metabolism, and Uses* (Seib, P.A., and Tolbeert, B.M., eds.), pp. 81–100, American Chemical Society, Washington.

Bielski, B.H.J., and Cabelli, D.E., 1991, Highlights of current research involving superoxide and perhydroxyl radicals in aqueous solutions. *Int. J. Radiat. Biol.* **59**:291–319.

Bieri, J.G., and Tolliver, T.J., 1981, On the occurrence of a-tocopherylquinone in rat tissue. *Lipids* **16**:777–789.

Boveris, A., and Cadenas, E., 1997, Cellular sources and steady-state levels of reactive oxygen species. In *Oxygen, Gene Expression, and Cellular Function* (Clerch, L.B., and Massaro, D.J., eds.), pp. 1–25, Marcel Dekker Inc., New York.

Bowry, V.W., Ingold, K.U., and Stocker, R., 1992, Vitamin E in human low-density lipoprotein. When and how this antioxidant becomes a prooxidant. *Biochem. J.* **288**:341–344.

Bowry, V.W., and Stocker, R., 1993, Tocopherol-mediated peroxidation. The prooxidant effect of vitamin E on the radical-initiated oxidation of human low-density lipoprotein. *J. Am. Chem. Soc.* **115**:6029–6044.

Brown, L.A.S., and Jones, D.P., 1997, The biology of ascorbic acid, In *Handbook of Synthetic Antioxidants*, (Packer, L., and Cadenas, E., eds.), pp. 117–154, Marcel Dekker Inc., New York.

Buettner, G.R., and Jurkiewicz, B.A., 1993, Ascorbate free radical as a marker of oxidative stress: An EPR study. *Free Rad. Biol. Med.* **14**:49–55.

Buettner, G.R., and Jurkiewicz, B.A., 1997, Chemistry and Biochemistry of Ascorbic acid, In *Handbook of Synthetic Antioxidants*, (Packer, L., and Cadenas, E., eds.), pp. 91–115, Marcel Dekker Inc., New York.

Cadenas, E., Boveris, A., Ragan, C.I., and Stoppani, A.O.M., 1977, Production of superoxide radicals and hydrogen peroxide by NADH-ubiquinone reductase and ubiquinol-cytochrome c reductase from beef heart mitochondria. *Arch. Biochem. Biophys.* **180**:248–257.

Cadenas, E., Merenyi, G., and Lind, J., 1989, Pulse radiolysis study on the reactivity of trolox C phenoxyl radical with superoxide anion. *FEBS Lett.* **253**:235–238.

Cadenas, E., Hochstein, P., and Ernster, L, 1992, Pro- and antioxidant functions of quinones and quinone reductases in mammalian cells. *Adv. Enzymol.* **65**:97–146.

Cadenas, E., and Packer, L. (eds.), 1996, *Handbook of Natural Antioxidants*, Marcel Dekker Inc., New York.

Clerch, L.B., and Massaro, D.J. (eds.), 1997, *Oxygen, Gene Expression, and Cellular Function*, Marcel Dekker Inc., New York.

Cotgreave, I.A., and Engman, L., 1997, The development of diaryl chalcogenides and a-(phenylselenyl) ketones with antioxidant and glutathione peroxidase-mimetic properties, In *Handbook of Synthetic Antioxidants* (Packer, L., and Cadenas, E., eds.), pp. 305–320, Marcel Dekker Inc., New York.

D'Arcy-Dohert, M., Wilson, I., Wardman, P., Basra, J., Patterson, L.H., and Cohen, G.M., 1986, Peroxidase activation of 1-naphthol to pahthoxy or naphthoxy-derived radicals and their reactions with glutathione, *Chem.–Biol. Interact.* **58**:199–215.

Floyd, R.A., Liu, G.-J., and Wong, P.K., 1997, Nitrone radical traps as protectors of oxidative damage in the central nervous system, In *Handbook of Synthetic Antioxidants* (Packer, L., and Cadenas, E., eds.), pp. 339–350, Marcel Dekker Inc., New York.

Forman, H.J., and Cadenas, E. (eds.), 1997, *Oxidative Stress and Signal Transduction*, Chapman and Hall, New York.

Forsmark, P., Åberg, F., Norling, B., Nordenbrand, K., Dallner, G., and Ernster, L., 1991, Vitamin E and ubiquinol as inhibitors of lipid peroxidation in biological membranes. *FEBS Lett.* **285**:39–43.

Giulivi, C., and Cadenas, E., 1994, Ferrylmyoglobin: Formation and chemical reactivity toward electron-donating compounds. *Meth. Enzymol.* **233**:189–202.

Haennen, G.R.M.M., and Bast, A., 1991, Scavenging of hypochlorous acid by lipoic acid. *Biochem. Pharmacol.* **42**:2244–2246.

Halliwell, B., 1997, Uric acid: An example of antioxidant evaluation, In *Handbook of Antioxidants* (Cadenas, E., and Packer, L., eds.), pp. 243–258, Marcel Dekker Inc., New York.

Hensley, K., Carney, J.M., Stewart, C.A., Tabatabaie, T., Pye, Q., and Floyd, R.A., 1997, Nitrone-based free radical traps as neuroprotective agents in cerebral ischaemia and other pathologies. *Int. Rev. Neurobiol.* **40**:299–317.

Imlay, J.A., and Fridovich, I., 1991, Assay of metabolic superoxide production in Escherichia Coli. *J. Biol. Chem.* **266**:6957–6965.

Ingold, K.U., Bowry, V.S., Stocker, R., and Wallilng, C., 1993, Autoxidation of lipids and antioxidation by a-tocopherol and ubiquinol in homogeneous solution and in aqueous dispersions of lipids: Unrecognized consequences of lipid particle size as exemplified by oxidation of human low density lipoprotein, *Proc. Natl. Acad. Sci. USA* **90**:45–49.

Kittridge, K., and Willson, R.L., 1984, Uric acid substantially enhances the free radical inactivation of alcohol dehydrogenase. *FEBS Lett.* **170**:162–164.

Liebler, D.C., Kaysen, K.L., and Kennedy, T.A.S., 1989, Redox cycles of vitamin E: Hydrolysis and ascorbic acid dependent reduction of 8a-(alkyldioxyl)tocopherones. *Biochemistry* **28**:9772–9777.

Maiorino, M., Roveri, A., Coassin, M., and Ursini, F. (1988), Kinetic mechanism and substrate specificity of glutathione peroxidase activity of ebselen (PZ51). *Biochem. Pharmacol.* **37**:2267–2271.

Maples, K.R., and Mason, R.P., 1988, Free radical metabolite of uric acid. *J. Biol. Chem.* **263**:1709–1712.

Masumoto, H., and Sies, H., 1996, The reaction of ebselen with peroxynitrite. *Chem. Res. Toxicol.* **9**:262–267.

Mehlhorn, R.J., and Swanson, C.E., 1992, Nitroxide-stimulated H_2O_2 decomposition by peroxidases and pseudoperoxidases. *Free Radic. Res. Comms.* **17**:157–175.

Mehlhorn, R.J., and Gomez, J., 1993, Hydroxyl and alkoxyl radical production by oxidation products of metmyoglobin. *Free Radic. Res. Comms.* **18**:29–41.

Noguchi, N., and Niki, E., 1997, Antioxidant properties of Ebselen, In *Handbook of Synthetic Antioxidants* (Packer, L., and Cadenas, E., eds.), pp. 285–304, Marcel Dekker Inc., New York.

Ordoñez, I.D., and Cadenas, E., 1992, Thiol oxidation coupled to DT-diaphorase-catalyzed reduction of diaziquone. Reductive and oxidative pathways of diaziquone semiquinone modulated by glutathione and superoxide dismutase. *Biochem. J.* **286**:481–490.

Packer, L., and Cadenas, E. (eds.), 1997, *Handbook of Synthetic Antioxidants*, Marcel Dekker Inc., New York

Packer, L., Witt, E.H., and Tritschler, H.J., 1997, Antioxidant properties and clinical applications of alpha-lipoic acid and dihydrolipoic acid. In *Handbook of Antioxidants* (Cadenas, E., and Packer, L., eds.), pp. 545–591, Marcel Dekker Inc., New York.

Peinado, J., Sies, H., and Akerboom, T.P.M., 1989, Hepatic lipoate uptake. *Arch. Biochem. Biophys.* **273**:389–395.

Rice-Evans, C.A., and Diplock, A.T., 1993, Current status of antioxidant therapy. *Free Radical Biol. Med.* **15**:77–96.

Roussyn, I., Briviba, K., Masumoto, H., and Sies, H., 1996, Selenium-containing compounds protect DNA from damage caused by peroxynitrite. *Arch. Biochem. Biophys.* **330**:216–218.

Samuni, A., and Krishna, M.C., 1997, Antioxidant properties of nitroxides and nitroxide SOD mimics, In *Handbook of Synthetic Antioxidants* (Packer, L., and Cadenas, E., eds.), pp. 351–373, Marcel Dekker Inc., New York.

Schöneich, C., Narayanaswami, V., Asmus, K.-D., and Sies, H., 1990, Reactivity of ebselen and related selenoorganic compounds with 1,2-dichloroethane radical cations and halogenated peroxyl radicals. *Arch. Biochem. Biophys.* **282**:18–25.

Scurlock, R., Rougee, M., Bensasson, R.V., Evers, M., and Dereu, N., 1991, Deactivation of singlet molecular oxygen by organ-selenium compounds exhibiting glutathione peroxidase activity and by sulfur-containing homologs. *Photochem. Photobiol.* **54**:733–736.

Sies, H. (ed.), 1985, *Oxidative Stress*, Academic Press, London.

Sies, H., 1993, Ebselen, a selenoorganic compound as glutathione peroxidase mimic. *Free Rad. Biol. Med.* **14**:313–323.

Sies, H. (ed.), 1997, *Antioxidants in Disease Mechanisms and Therapy*, Academic Press, San Diego.

Sies, H., and Masumoto, H., 1997, Ebselen as a glutathione peroxidase mimic and as a scavenger of peroxynitrite, In *Antioxidants in Disease Mechanisms and Therapy* (Sies, H., ed.), pp. 229–246, Academic Press, London.

Simic, M.G., and Jovanovic, S.V., 1989, Antioxidation mechanisms of uric acid. *J. Am. Chem. Soc.* **111**:5778–5782.

Surdhar, P.S., and Armstrong, D.A., 1986, Redox potential of some sulfur-containing radicals, *J. Phys. Chem.* **90**:5915–5917.

Tabatabaie, T., and Floyd, R.A., 1994, Susceptibility of glutathione peroxidase and glutathione reductase to oxidative damage and the protective effect of spin trapping agents. *Arch. Biochem. Biophys.* **314**:112–119.

Tabatabaie, T., Kotake, Y., Wallis, G., Jacob, J.M., and Floyd, R.A., 1997, Spin trapping agent phenyl N-tert-butylnitrone protects against the onset of drug-induced insulin-dependent diabetes mellitus. *FEBS Lett.* **407**:148–152.

Tabatabaie, T., Stewart, C., Pye, Q., Kotake, Y., and Floyd, R.A., 1996, *In vivo* trapping of nitric oxide in the brain of neonatal rats treated with the HIV-1 envelope protein gp120: Protective effectgs of α-phenyl-*tert*-butylnitrone. *Biochem. Biophys. Res. Commun.* **221**:386–390.

von Sonntag, C., 1987, *The Chemical Basis of Radiation Biology*, Taylor and Francis, London.

Wardman, P., 1988, Conjugation and oxidation of glutathione via thiyl free radicals. In *Glutathione Conjugation, Mechanisms, and Biological Significance* (Sies, H., and Ketterer, B., eds.), pp. 44–72, Academic Press, London.

Wardman, P., 1990, Thiol reactivity towards towards drugs and radicals: Some implications in the radiotherapy and chemotherapy of cancer. In *Sulfur-centered Reactive Intermediates in Chemistry and Biology* (Chatgilialoglu, C., and Asmus, K.-D., eds.), pp. 415–427, Plenum Press, New York.

Wilson, I., Wardman, P., Cohen, G.M., and D'Arcy-Doherty, M., 1986, Reductive role of glutathione in the redox cycling of oxidizable drugs. *Biochem. Pharmacol.* **35**:21–22.

Willson, R.L., Dunster, C.A., Forni, L.G., Gee, C.A., and Kittridge, K.J., 1985, Organic free radicals and proteins in biochemical injury: Electron- or hydrogen-transfer reactions? *Phil. Trans. R. Soc. Lond. B* **311**:545–563.

Winterbourn, C.C., and Munday, R., 1990, Concerted action of reduced glutathione and superoxide dismutase in preventing redox cycling of dihydropyrimidines, and their role in atnioxidant defence, *Free Rad. Res. Commun.* **8**:287–293.

Young, H.K., Floyd, R.A., Maidt, M.L., and Dynlacht, J.R., 1996, Evaluation of nitrone spin-trapping as radioprotectors. *Radiat. Res.* **146**:227–231.

23

REPAIR SYSTEMS AND INDUCIBLE DEFENSES AGAINST OXIDANT STRESS

Kelvin J. A. Davies*

Ethel Percy Andrus Gerontology Center
The University of Southern California
Los Angeles, California 90089-0191

1. ANTIOXIDANT DEFENSES

Thankfully we are not defenseless against oxygen radicals, and other activated oxygen species, to which we are constantly exposed. All aerobic organisms, including human beings, utilize a series of primary antioxidant defenses in an attempt to protect against oxidant damage, and numerous damage removal and repair enzymes to remove and/or repair molecules that do get damaged (Davies, 1986; Davies, 1993). This chapter will concentrate on damage removal/repair systems and on genes that are inducible during adaptation to oxidative stress.

2. DIRECT REPAIR SYSTEMS

Damage removal and/or repair systems may be classified as either direct or indirect (Davies, 1986; Davies, 1993) as shown in Fig. 1. Direct repair, about which we know only a little, has so far only been demonstrated for a few classes of oxidized molecules. One important direct repair process is the re-reduction of oxidized sulfhydryl groups on proteins. Cysteine residues in proteins are highly susceptible to autooxidation and/or metal-catalyzed oxidation. When two nearby cysteine residues within a protein oxidize they often form a disulfide bond, producing a more rigid protein. Disulfide bonds can also form between two proteins promoting the formation of large supramolecular assemblies of inactivated enzymes and proteins; this is called inter-molecular cross-linking. Both intra-molecular disulfide cross-links and inter-molecular disulfide cross-links can be reversed to

* Mailing address: Professor Kelvin J. A. Davies, Associate Dean for Research and James E. Birren Chair of Gerontology, Ethel Percy Andrus Gerontology Center, The University of Southern California, 3715 McClintock Avenue, Room 306, Los Angeles, California 90089-0191. Telephone: (213)740-8959, Fax No.: (213)740-6462, E-mail: "Kelvin@usc.edu"

Free Radicals, Oxidative Stress, and Antioxidants, edited by Özben.
Plenum Press, New York, 1998.

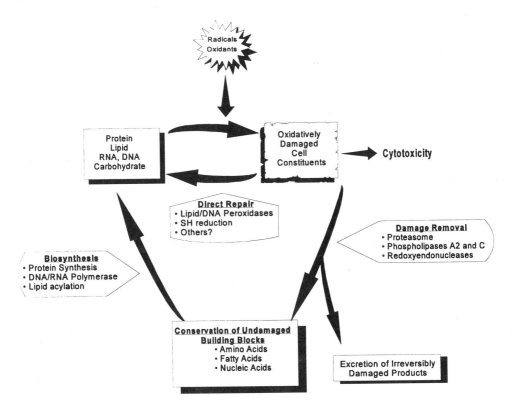

Figure 1. Damage removal and repair systems.

some extent by disulfide reductases within cells (Davies, 1993). Our understanding of such enzymatic reactions is still at an early stage. Another important sulfhydryl oxidation process is the oxidation of methionine residues to methionine sulfoxide, typically causing loss of enzyme/protein function. The enzyme methionine sulfoxide reductase can regenerate methionine residues within such oxidized proteins and restore function (Brot and Weissbach, 1991). As with disulfide reductases, our understanding of methionine sulfoxide reductases is still in its infancy.

Direct repair of DNA hydroperoxides by glutathione peroxidase has been reported from *in vitro* studies (Ketterer and Meyer, 1989). The extent to which DNA peroxides are actually formed *in vivo*, however, is not completely clear. Also not yet studied is the extent to which DNA peroxides may be directly repaired by glutathione peroxidases *in vivo*. Other, relatively straightforward, mechanisms of DNA repair are also being explored (Demple and Harrison, 1994).

3. DAMAGE REMOVAL AND REPAIR SYSTEMS (INDIRECT "REPAIR")

Although our knowledge of direct repair systems, as outlined above, is still rather rudimentary a great deal more is known about indirect repair systems (Fig. 1). Indirect repair involves two distinct steps (Davies, 1986; Davies, 1993); first the damaged molecule (or the damaged part of a molecule) must be recognized and excised, removed, or degraded. Next, a replacement of the entire damaged molecule must be synthesized, or the excised portion of the damaged molecule must be made and inserted.

3.1. Degradation and Replacement of Oxidized Proteins

Extensive studies have revealed that oxidized proteins are recognized by proteases and completely degraded (to amino acids), entirely new replacement protein molecules are then synthesized *de novo* (Davies, 1986; Davies, 1987a;Davies, 1989; Davies and Delsignore, 1987b; Davies, Delsignore, and Lin, 1987c; Davies and Goldberg, 1987d; Davies and Goldberg, 1987e; Davies, Lin, and Pacifici, 1987f; Davies and Lin, 1988a; Davies and Lin, 1988b; Giulivi and Davies, 1993; Giulivi, Pacifici, and Davies, 1994; Grune, Reinheckel, Joshi, and Davies, 1995; Grune, Reinheckel, Talbot, and Davies, 1996; Marcillat, Zhang, Lin, and Davies, 1988; Murakami, Jahngen, Lin, Davies, and Taylor, 1990; Pacifici and Davies, 1991; Pacifici, Kono, and Davies, 1994; Pacifici, Salo, Lin, and Davies, 1989; Salo, Pacifici, and Davies, 1990; Stadtman, 1986; Stadtman, 1993; Starke-Reed, Oliver, and Stadtman, 1989; Taylor, Daims, Lee, and Surgenor, 1982; Taylor and Davies, 1987). It appears that oxidized amino acids within oxidatively modified proteins are eliminated, or used as carbon sources for ATP synthesis. Since an oxidatively modified protein may contain only two or three oxidized amino acids it appears probable that most of the amino acids from an oxidized and degraded protein are reutilized for protein synthesis. Thus, during oxidative stress many proteins synthesized as damage replacements are likely to contain a high percentage of recycled amino acids. During periods of particularly high oxidative stress the proteolytic capacity of cells may not be sufficient to cope with the number of oxidized protein molecules being generated. A similar problem may occur in aging, or with certain disease states, when proteolytic capacity may decline below a critical threshold of activity required to cope with normal oxidative stress levels. Under such circumstances oxidized proteins may not undergo appropriate proteolytic digestion, and may, instead cross-link with one another or form extensive hydrophobic bonds. Such aggregates of damaged proteins are detrimental to normal cell functions and lead to further problems. A summary of protein oxidative damage, recognition and degradation by proteases, or cross-linking and aggregation is presented in Fig. 2 and represents a pictorial synthesis of many detailed studies (Davies, 1989; Davies and Lin, 1988a; Davies and Lin, 1988b; Giulivi and Davies, 1993; Giulivi *et al.*, 1994; Grune *et al.*, 1995; Grune *et al.*, 1996; Marcillat *et al.*, 1988; Murakami *et al.*, 1990; Pacifici and Davies, 1991; Pacifici *et al.*, 1994; Pacifici *et al.*, 1989; Salo *et al.*, 1990; Stadtman, 1986; Stadtman, 1993; Starke-Reed *et al.*, 1989; Taylor *et al.*, 1982; Taylor and Davies, 1987).

In bacteria such as *Escherichia coli* a series of proteolytic enzymes act cooperatively in the recognition and degradation of oxidatively modified soluble proteins (Davies and Lin, 1988a; Davies and Lin, 1988b). A similar series of proteolytic enzymes appear to conduct the degradation of oxidatively modified soluble proteins in mammalian mitochondria (Marcillat *et al.*, 1988). In bacteria and in mitochondria, therefore, the proteolytic role shown for proteasome in Fig. 2 is actually played by a series of cooperative proteases. In the cytoplasm and nucleus of eucaryotic cells, however, oxidized soluble proteins largely appear to be recognized and degraded by the proteasome complex (Giulivi and Davies, 1993; Giulivi *et al.*, 1994; Grune *et al.*, 1995; Grune *et al.*, 1996; Pacifici *et al.*, 1994; Pacifici *et al.*, 1989; Salo *et al.*, 1990) as shown in Fig. 2. Proteasome is a 670 kDa multi-enzyme complex that appears to be ubiquitously expressed in the cytoplasm and nuclei of all eucaryotic cells. More than 15 individual polypeptides, each present in multiple copy, with molecular weights ranging from 20,000 daltons to 35,000 daltons make up the pro-

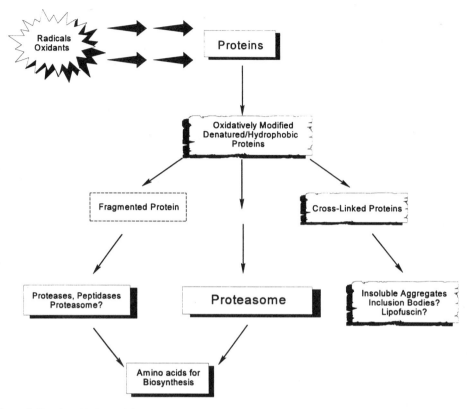

Figure 2. Protein oxidation and the degradation of oxidatively modified proteins by proteasome in eucaryotic cells.

teasome complex; the exact composition varies with species and cell type. Each of the component proteasome polypeptides is encoded by a separate gene and many of these genes have now been cloned and sequenced (Coux, Tanaka, and Goldberg, 1996; Rivet, 1993). The results of such cloning and sequencing studies reveal that proteasome is a completely non-classical protease complex. Indeed, the proteasome subunits have no discernable sequence identity to any known proteins, except for a small degree of sequence overlaps with some of the heat shock proteins (Coux et al., 1996; Rivet, 1993).

The core 670 kDa proteasome can combine with other, ubiquitin conjugating and ATPase, subunits to form a 1,50 kDa proteolytic complex (sometimes called Ubiquitin Conjugate Degrading Enzyme or U.C.D.E.N.). This 1,500 kDa form of proteasome is responsible for the ATP- and ubiquitin-dependent or stimulated proteolysis within eucaryotic cells, and probably plays an important role in antigen processing and cell differentiation (Coux et al., 1996; Rivet, 1993). The 1,500 kDa proteasome form, however, appears not to recognize oxidized proteins (Pacifici et al., 1989). It is in fact the 670 kDa "core" proteasome complex that recognizes oxidatively modified proteins and selectively degrades them in an ATP- and ubiquitin-independent manner (Giulivi and Davies, 1993; Giulivi et al., 1994; Grune et al., 1995; Grune et al., 1996; Pacifici et al., 1994; Pacifici et al., 1989; Salo et al., 1990). In this regard it is interesting that although reticulocytes contain both the 1,500 kDa and the 670 kDa forms of proteasome, mature erythrocytes contain only the 670 kDa core proteasome (Pacifici et al., 1989). Presumably the terminally differentiated erythrocyte can survive perfectly well without ATP/ubiquitin-stimulated proteolysis, and has no need for antigen processing or further differentiation. The housekeeping function of recognizing and degrading oxidatively damaged proteins, however, conducted by the 670 kDa core proteasome, appears to be required throughout the entire life of the red blood cell (Davies, 1986; Davies, !993; Pacifici *et al.*, 1989).

Recognition of oxidized soluble proteins in the cell cytoplasm and nucleus by proteasome appears to occur via binding to exposed hydrophobic patches on the damaged proteins (Giulivi and Davies, 1993; Giulivi *et al.*, 1994; Grune *et al.*, 1995; Grune *et al.*, 1996; Pacifici, Kono and Davies, 1994; Pacifici *et al.*, 1989; Salo *et al.*, 1990). Although the process of protein oxidation (which of course means oxidation of consistent amino acids) often involves changes that make some amino acid residues more hydrophilic, changes in charge relationships on a protein can cause significant unfolding or partial denaturation. Such partial denaturation exposes previously shielded stretches of primary sequence that are hydrophobic in nature. Exposed hydrophobic patches on the surface of oxidized proteins appear to act as recognition and binding sequences for the 670 kDa core proteasome (Giulivi and Davies, 1993; Giulivi *et al.*, 1994; Grune *et al.*, 1995; Grune *et al.*, 1996; Pacifici *et al.*, 1994; Pacifici *et al.*, 1989; Salo *et al.*, 1990).

Proteasome appears equally able to recognize and degrade damaged protein substrates generated by a wide variety of oxidant exposures. Thus, hemoglobin damaged by the indiscriminate hydroxyl radical is selectively degraded by proteasome at about the same efficiency as hemoglobin damaged by metal-catalyzed oxidation with hydrogen peroxide at the heme-moiety (Pacifici *et al.*, 1989; Salo *et al.*, 1990).

Similarly, Cu,Zn superoxide dismutase is recognized and degraded by proteasome equally well following either ˙OH exposure or specific copper-catalyzed active-site inactivation by the enzymes product, H_2O_2 (Pacifici *et al.*, 1989; Salo *et al.*, 1990). Again, the common link between the non-specific damage caused by ˙OH, and the site-specific oxidation caused by H_2O_2 reacting with a protein-bound transition metal catalyst, is the partial denaturation and exposure of hydrophobic patches caused by both. Hydrophobic and bulky residues are the preferred substrates for proteolytic cleavage by the core proteasome (Davies, 1986; Davies, 1993; Grune *et al.*, 1996; Taylor *et al.*, 1982). Recently this laboratory has shown that treatment of rat liver Clone 9 epithelial (Grune *et al.*, 1995) cells and K562 human hematopoietic cells (Grune *et al.*, 1996) in culture with an antisense oligonucleotide directed against the proteasome C-2 subunit gene, drastically diminishes C-2 expression, overall proteasome activity, and the ability of these mammalian cells to successfully degrade oxidatively modified proteins following an imposed oxidative stress. Furthermore, protein oxidation has been shown to occur in disease states such as cataract formation (Davies, 1989; Murakami *et al.*, 1990; Taylor *et al.*, 1982; Taylor and Davies, 1987) and proteasome appears to have an important role in limiting cataracts in healthy individuals (Davies, 1989; Murakami *et al.*, 1990; Taylor *et al.*, 1982; Taylor and Davies, 1987). Finally, several lines of research now indicate that proteasome may be a significant factor in the overall aging process (Pacifici and Davies, 1991; Stadtman, 1986; Stadtman, 1993; Starke-Reed *et al.*, 1989).

3.2. Degradation and Replacement/Repair of Oxidized Membrane Lipids

Lipid peroxidation was the first type of oxidative damage to be studied. Membrane phospholipids are continually subjected to oxidant challenges. The process of lipid peroxidation is comprised of a set of chain reactions which are initiated by the abstraction of a hydrogen atom (from carbon) in an unsaturated fatty acyl chain (Mead, 1976; Sevanian and Hochstein, 1985). In an aerobic environment, oxygen will add to the fatty acid at the carbon centered lipid radical (L˙) to give rise to a lipid peroxyl radical (LOO˙). Once initiated, LOO˙ can further propagate the peroxidation chain reaction by abstracting a hydrogen atom from other vicinal unsaturated fatty acids (Mead, 1976; Sevanian and Hochstein, 1985). The resulting lipid hydroperoxide (LOOH) can easily decompose into several reactive species including: lipid alkoxyl radicals (LO˙), aldehydes (*e.g.*, malonyldialdehyde), alkanes, lipid epoxides, and alcohols (Mead, 1976; Sevanian and Hochstein, 1985). Cholesterol has also been shown to undergo oxidation, to give rise to a variety of epoxides and alcohols.

Peroxidized membranes become rigid, lose selective permeability and, under extreme conditions, can lose their integrity. Water-soluble lipid peroxidation products (most notably the aldehydes) have been shown to diffuse from membranes into other sub-cellular compartments (Comporti, 1986; Sevanian and Hochstein, 1985a; Sevanian, Wratten, McLeod, and Kim, 1988; Vaca, Wilhelm, and Harms-Ringdahl, 1988). Dialdehydes can act as crosslinking reagents, and are thought to play a role in the protein aggregation which forms the age pigment lipofuscin (Davies, 1988). Several laboratories are investigating the possibility that lipid peroxidation products may form DNA adducts, thus giving rise to mutations and altered patterns of gene expression (Vaca *et al.*, 1988). Others have noted inhibition of enzyme function by lipid peroxidation products. It is very clear that the process of lipid peroxidation, and its products, can be detrimental to cell viability. Cumulative effects of lipid peroxidation have been implicated as underlying mechanisms in numerous pathological conditions including: atherosclerosis, hemolytic anemias, and ischemia reperfusion injuries (Sevanian and Hochstein, 1985a).

Lipid bilayers which have been oxidized become better substrates for phospholipase enzymes (Fig. 1). Phospholipase A_2 acts at the sn-2 position of the phospholipid glycerol backbone to generate a free fatty acid and a lysophospholipid. Phospholipase A_2 has been shown to preferentially hydrolyze fatty acids from oxidized liposomes (Sevanian and Kim, 1985b). Structural perturbations due to changes in membrane microviscosity, and the increased hydrophilic nature of oxidized lipids may be responsible for the increased susceptibility to Phospholipase A_2 action (Sevanian *et al.*, 1988; Van Kuijk, Sevanian, Handelman, and Dratz, 1987). Removing fatty acid hydroperoxides from the membrane compartment will help prevent further propagation reactions. Additionally, it has been demonstrated that fatty acid hydroperoxides released into the cytosol are substrates for glutathione peroxidase. Glutathione peroxidase detoxifies fatty acid hydroperoxides by reducing them to their corresponding hydroxy-fatty acids (Van Kuijk *et al.*,1987). Lysophospholipids left in the membrane posses potential detergent properties which have been shown to disrupt membrane structure and function. Lysophospholipids can serve as substrates for reacylation reactions (readdition of fatty acids to the sn-2 position˙) to regenerate intact phospholipids (Lubin, Shohet and Nathan, 1972; Zimmerman and Keys, 1988).

Recent work suggests that it is possible to reduce fatty acid hydroperoxides (to their corresponding alcohols) without hydrolysis and release from the membrane compartment (Ketterer and Meyer, 1989; Thomas, Maiorino, Ursini, and Girotti, 1990; Ursini, Maiorino, Valente, Ferri, and Gregolin, 1982; Zhang, Maiorino, Roveri, and Ursini, 1989). A member of the glutathione peroxidase family, phospholipid hydroperoxide glutathione peroxidase, which acts preferentially on phospholipid hydroperoxides has been characterized by Ursini, Maiorino, and their co-workers (Maiorino, Chu, Ursini, Davies, Doroshow, and Esworthy, 1991; Thomas *et al.*, 1990; Ursini *et al.*, 1982; Zhang *et al.*, 1989). A glutathione transferase with activity towards lipid hydroperoxides has also been extracted from nuclei (Ketterer and Meyer, 1989).

Peroxidized membranes and lipid oxidation products represent a constant threat to aerobic cells. It is now widely held that in addition to preventing initiation of peroxidation (with compounds like vitamin E), cells have also developed a variety of mechanisms for maintaining membrane integrity and homeostasis by repairing oxidatively damaged lipid components (Fig. 1).

3.3. Repair of Oxidized DNA

Ribo- and deoxyribo-nucleic acids are also vulnerable to oxidative damage and, perhaps most importantly, DNA has been shown to incur oxidative damage *in vivo* (Adelman, Saul, and Ames, 1988; Fridovich, 1978; Kasai, Crain, Kuchino, Nishimura, Ootsuyama, and Tanooka, 1986; Povirk and Steighner, 1989; Richter, 1988a; Richter, Park, and Ames, 1988b; Simic, Bergtold, and Karam, 1989). Although DNA is a relatively simple biopolymer, made up of only four different nucleic acids, its integrity is vital to cell division and survival. Oxidative alterations to nucleic acid polymers has been shown to disrupt transcription, translation, and DNA replication, and to give rise to mutations and (ultimately) cell senescence or death (Ames, 1989; Harman, 1956; Harman, 1981; Schraufstatter, Hyslop, Jackson, and Cochrane, 1988; Simic *et al.*,1989; Spector, Kleiman, Huang, and Wang, 1989; Tice and Setlow, 1985). Despite the precious nature of the genetic code, it too appears to be a target for oxidative damage. The amount of oxidative damage, even under normal physiological conditions, may be quite extensive with estimates as high as 1 base modification per 130,000 bases in nuclear DNA (Richter *et al.*, 1988b). Damage to mitochondrial DNA is estimated to be even higher at 1 per 8,000 bases. Fragments of oxidatively modified mitochondrial DNA have been implicated in cancer and aging (Richter *et al.*, 1988a).

Oxidants can elicit a wide variety of DNA damage products several of which have been carefully characterized (Simic *et al.*, 1989; Teebor, Boorstein, and Cadet; 1988). The types of DNA damage can be grouped into: strand breaks (single and double), sister chromatid exchange, DNA–DNA and DNA–protein cross links, and base modifications. All four DNA bases can be oxidatively modified, however, the pyrimidines (cytosine and, especially, thymidine) appear to be most susceptible (Tice and Setlow, 1985). Bases undergo ring saturation, ring opening, ring contraction, and hydroxylation. These types of alteration usually result in a loss of aromaticity and planarity, which can cause local distortions in the double helix (Tice and Setlow, 1985). Depending on the type and extent of damage, the altered bases can be found either attached to, or dissociated from the DNA molecule to generate apurinic/apyrimidinic (AP) sites (Johnson and Demple, 1988; Povirk and Steighner, 1989; Tice and Setlow, 1985).

Radical interactions with DNA appear to be fairly nonspecific; hence, the phosphodiester backbone may also be oxidatively damaged. Damage to the sugar and phosphate moieties, which form the backbone, may result in strand breaks (Simic *et al.*, 1989; *Teebor et al.*, 1988). Depending on the site of radical attack, unusual 3' and 5' ends (*i.e.*, non 3'–OH, non 5'–PO_4) can be generated. These abnormal ends are not substrates for DNA polymerases, and must be removed before any repair can occur (Johnson and Demple, 1988).

Reports of glutathione transferases (Johnson and Demple, 1988) and peroxidases using thymidine hydroperoxide as a substrate have been published (Fig. 1). DNA may also undergo oxidative demethylation (Teebor *et al.*, 1988; Tice and Setlow, 1985). DNA methylases may play an important role in restoring methylation patterns and maintaining epigenetic phenomena. Inhibition of poly (ADP–ribose) polymerase has been shown to exacerbate H_2O_2 genotoxicity, although the mechanism for this is not yet clear (Spector *et al.*, 1989). There is ample evidence that several procaryotic and eucaryotic enzymes repair oxidatively damaged DNA by both direct (Ketterer and Meyer, 1989; Teebor *et al.*, 1988; Tice and Setlow, 1985), and excision repair mechanisms (Chan and Weiss, 1987; Doetsch, Helland, and Haseltine, 1986; Doetsch, Henner, Cunningham, Toney, and Helland, 1987; Greenberg and Demple, 1989; Johnson and Demple, 1988; Kasai *et al.*, 1986; Levin, Johnson, and Demple, 1988; Povirk and Steighner, 1989; Spector *et al.*, 1989; Teebor *et al.*, 1988; Tice and Setlow, 1985; Wallace, 1988) as shown in Fig. 1. Pilot studies using cell-free extracts *in vitro* indicated that some endo- and exo-nucleases preferentially cleave oxidized DNA. Bacterial mutants deficient in these putative DNA oxy-repair enzymes: *nth-* (endonuclease III), *xth-* (exonuclease III), and *nfo-* (endonuclease IV) were used to continue these studies *in vivo* (Cunningham, Saporito, Spitzer, and Weiss, 1986). It was found that strains deficient in exonuclease III (*xth-*) are hypersensitive to H_2O_2 (Cunningham *et al.*, 1986). Endonuclease IV deficiency (*nfo-*) also produced a hypersensitivity to organic hydroperoxides, and to oxidants generated by bleomycin (Cunningham *et al.*, 1986). When the *nfo* and *xth* mutations were combined, lethal oxidant effects are drastically increased.

The activity of these procaryotic enzymes, with regard to oxidative DNA repair, has been extensively studied (Chan and Weiss, 1987; Cunningham *et al.*, 1986; Demple, Johnson, and Fung, 1986; Greenberg and Demple, 1989; Kow and Wallace, 1985). Endonuclease III has been shown to cleave at the 3' side of AP sites (Demple *et al.*, 1986). Endonuclease III also appears to posses an N-glycosylase activity for thymine glycol and urea residues; two common products of oxidative damage to DNA. Exonuclease III posses a 3' to 5' nucleolytic activity which may be responsible for removing the sugar fragments generated during oxidative strand breakage ((Demple *et al.*, 1986; Teebor *et al.*, 1988). Exonuclease III is really a poor name for this enzyme which actually expresses 85% of the 5' AP endonucleolytic activity in *E. coli*. Endonuclease IV is also a 5' AP endonuclease.

Eucaryotic investigations are not as far along but several parallels with the results of procaryotic studies have already been published (Doetsch *et al.*, 1986; Doetsch *et al.*, 1987; Helland, Doetsch, and Haseltine, 1986; Johnson and Demple, 1988). Several glycosylases which act on DNA oxidation products have been characterized. A 3'-repair diesterase in yeast is apparently responsible for removing damaged 3' termini left by free radical reactions (Johnson and Demple, 1988). A mammalian endonuclease has been isolated based on its specificity for oxidatively modified DNA (Doetsch *et al.*, 1986; Doetsch *et al.*, 1987). It bears remarkable similarities to the bacterial endonuclease III including: molecular weight (~30 kDa), lack of divalent cation requirement, and substrate specificity. The term "redoxyendonucleases" (as used in Fig. 1) has been proposed for all nucleases which participate in repairing oxidatively damaged DNA (Doetsch *et al.*, 1986).

There is mounting evidence that redoxyendonucleases function in higher eucaryotic cells (Fig. 1). DNA damage which appears in cells as a result of an acute oxidant challenge (including base damage and single-strand breaks) has been shown to diminish as a function of time (Kasai, 1986). These results suggest that a removal of lesions is being carried out by intracellular systems. Oxidatively damaged bases (8-hydroxydeoxyguanosine, thymine glycol, and thymidine glycol) have been measured in animal urine (Richter, 1988b). Again this suggests that there is systematic excision and excretion of oxidized DNA *in vivo*. The vital importance of such DNA excision repair processes was recently highlighted by the selection of DNA repair as the "Molecule of the Year" for 1994 by *Science* magazine (Koshland, 1994).

4. INDUCIBLE DEFENSE AND REPAIR SYSTEMS

Thus far the picture of antioxidant defenses and repair systems I have painted, although extensive, is a static one. In reality, however, both procaryotes and eucaryotes are able to dramatically up-regulate their armory of oxidant protections in response to an oxidative stress. Many researchers have utilized a fairly common cell-culture adaptive response protocol first used in heat-shock studies to study such phenomena. This approach involves first finding an oxidant concentration that is lethal to most of the cells. Next new cultures are exposed to much lower levels of the same oxidant (pretreatment exposures) for various periods of time before being exposed to the normally lethal concentration (the challenge dose). What has now been widely found is that, in cells from *E. coli* to human hepatocytes, pretreatment conditions can be found that will enable cells to survive the subsequent challenge dose. Such adaptive responses to oxidative stress have been shown to involve widespread alterations in gene expression (Adelman *et al.*, 1988; Crawford and Davies, 1994a; Crawford, Edbauer-Nechamen, Lowry, Salmon, Kim, Davies, and Davies, 1994b; Davies and Lin; 1994; Davies, Lowry, and Davies, 1995; Wiese, Pacifici, and Davies, 1995).

In bacteria adaptation to hydrogen peroxide (Christman, Morgan, Jacobson, and Ames, 1985; Christman, Storz, and Ames, 1989; Davies and Lin, 1994; Davies, Wiese, Pacifici, and Davies, 1993; Demple, 1991; Demple and Halbrook, 1983; Farr and Kogoma, 1991; Goerlich, Quillardet, and Hofnung, 1989; Jenkins, Schultz, and Matin, 1988; Kullik and Storz, 1994; Morgan, Christman, Jacobson, Storz, and Ames, 1986; Richter and Loewen, 1981; Storz, Tartaglia, and Ames, 1990a; Storz, Tartaglia, Farr, and Ames, 1990b; Tao, Makino, Yonei, Nakata, and Shinagawa, 1989; Tartaglia, Gimeno, Storz, and Ames, 1992; Tartaglia, Storz, and Ames, 1989; Toledano, Kulik, Trinh, Baird, Schneider, and Storz,1994) has been shown to involve the *oxyR* regulon (Christman *et al.,*1985; Christman *et al.*, 1989; Kullik and Storz, 1994; Morgan *et al.*, 1986; Storz *et al.*, 1990a; Storz *et al.*, 1990b; Tao *et al.*, 1989; Tartaglia *et al.*, 1989; Tartaglia *et al.*, 1992; Toledano *et al.*, 1994). The *oxyR* gene encodes the OxyR protein that can bind to the nine or so target genes whose expression it regulates. In the reduced state the OxyR protein allows a basal level of transcription to occur. When oxidized, during the oxidative stress of H_2O_2 exposure for instance, the OxyR protein dramatically up-regulates transcription of its target genes, including a catalase, an alkyl-hydroperoxide reductase, a glutathione reductase, a non-specific DNA binding protein, and the heat-shock protein DNAK. Each of these enzymes/proteins provides a clear advantage to oxidatively stressed cells. Recent work has revealed the existence of a second H_2O_2 inducible regulon, tentatively called the *oxoR* regulon, that appears to provide inducible protection in *E. coli* (Crawford and Davies, 1994a; Crawford *et al.*, 1994b; Davies and Lin, 1994). When *E. coli* are exposed to superoxide, instead of H_2O_2, the *soxRS* regulon provides inducible protection (Chan and Weiss, 1987; Davies *et al.*, 1993; Demple, 1991; Farr, Natvig, and Kogoma, 1985; Greenberg, Chou, Monach, and Demple, 1991; Greenberg and Demple, 1989; Greenberg, Monach, Chou, Josephy, and Demple, 1990; Kogoma, Farr, Joyce, and Natvig, 1988; Tsaneva and Weiss, 1990; Wu and Weiss, 1991). Genes induced by *soxRS* include a manganese superoxide dismutase, exonuclease IV, glucose 6-phosphate dehydrogenase, fumarase C, and an antisense RNA regulator.

Our understanding of eucaryotic adaptation to oxidative stress is not as advanced as the bacterial work. Limited studies have been performed with yeast, where it is clear that transcriptionally regulated adaptation to H_2O_2 exposure does occur (Collinson and Dawes, 1992; Crawford *et al.*, 1994b; Davies *et al.*, 1995; Davies *et al.*, 1993; Flatter-O'Brien,

Collinson, and Dawes, 1993; Jamieson, 1992; Kuge and Jones, 1994; Spitz, Dewey, and Li, 1987). Limited studies have also been performed in mammalian cell cultures. Mammalian H_2O_2 adaptation is a much slower, less extensive process than that seen for bacteria or yeast, but it does still occur (Gupta and Bhattacharjee, 1988; Laval, 1988; Lu, Maulik, Moraru, Kreutzer, and Das, 1993; Spitz et al., 1987; Wiese et al., 1995). Transcription of several genes has been shown to increase during mammalian responses to oxidative stress (Crawford and Davies, 1997; Crawford, Leahy, Abramova, Lan, Wang, and Davies, 1997; Crawford, Leahy, Wang, Schools, Kochheiser, and Davies, 1996; Crawford, Schools, and Davies, 1996; Crawford, Schools, Salmon, and Davies, 1996; Crawford, Zbinden, Amstad, and Cerutti, 1988; Devary, Gottlieb, Lau, and Karin, 1991; Fornace, Alamo, and Hollander, 1988; Fornace, Nebert, Hollander, Luethy, Papathanasiou, Fargnoli, and Holbrook, 1989; Keyse and Emslie, 1992; Keyse and Tyrrell, 1989; Melendez and Davies, 1996; Rushmore, King, Paulson, and Pickett, 1990; Salo, Donovan, and Davies, 1991; Schull, Heintz, Periasamy, Manohar, Janssen, Marsh, and Mossman, 1991; Stevens and Autor, 1977; Wang, Crawford and Davies, 1996; White, Ghezzi, McMahon, Dinarello, and Repine, 1989). In this laboratory we have shown that HA-1, CHO, V79, C3H 10T1/2, and clone 9 liver cells all adapt to H_2O_2, although to varying degrees (Wiese et al., 1995).

A mammalian genetic library has been constructed that has sequences known to be induced by DNA damage (Fornace et al., 1989). Levels of transcripts which cross hybridize with probes from this library have been shown to increase after H_2O_2 treatment, suggesting oxidative induction of DNA repair enzymes in mammals (Fornace et al., 1988; Fornace et al., 1989). Heme oxygenase (Keyse and Tyrrell, 1989), DT diaphorase or quinone reductase (Rushmore et al., 1990), and a protein–tyrosine phosphatase (Keyse and Emslie, 1992) have all been shown to exhibit peroxide-inducible increases in expression, but the relative importance of any or all of these proteins to actual adaptation remains unclear. Obviously, the contribution of multiple proteins and enzymes to overall oxidative stress adaptation remains an important area for future investigations.

At the moment we are concentrating on H_2O_2 adaptation in the HA-1 cell line (a Chinese Hamster Ovary cell derivative) which relies on de novo protein synthesis for adaptation but does not appear to induce any of the classical antioxidant enzymes. In the HA-1 cell line we find increased expression of the hox-1 (heme oxygenase), gadd45, gadd153, mafG, c-fos, c-myc, c-jun, hsp70, and CL100 tyrosine/threonine phosphatase mRNA's during adaptation to hydrogen peroxide. We also find overexpression of several previously unidentified genes that we have named "adapts" for their potential roles in the adaptive response (Crawford and Davies, 1997a; Crawford et al., 1997b; Crawford et al., 1996a; Crawford et al., 1996b; Crawford et al., 1996c; Melendez and Davies, 1996; Wang et al., 1996). A temporary growth arrest appears important for the successful H_2O_2 adaptation of HA-1 cells, and this growth-arrest includes expression of the gadd153 and gadd45 genes (Giulivi et al., 1994; Kogoma et al., 1988), as well as adapt15, a novel growth-arrest-associated RNA transcript that we discovered and are now investigating (Crawford et al., 1996a; Crawford et al., 1996b). Levels of another novel growth-arrest-associated gene product, adapt33 have also been found to significantly increase during HA-1 cell adaptation to hydrogen peroxide (Wang et al., 1996).

Transient adaptation to hydrogen peroxide also induces the levels of adapt66, which is a mafG nuclear transcription factor/oncogene homologue (Crawford et al., 1996). During hydrogen peroxide adaptation levels of adapt73, which appears highly homologous to the cardiogenic shock-inducible PigHep3, are strongly up-regulated in HA-1 cells (Crawford and Davies, 1997). Finally (for the moment at least) levels of adapt78 are dramatically increased during the HA-1 cell adaptive response (Crawford et al., 1997). Increased levels of *adapt78* are particularly interesting since this gene is a Down Syndrome critical region homologue. The adapt78 protein generated by us using *in vitro* transcription/translation agrees very well with the predicted molecular size of approximately 23,000 Da (Crawford *et al.*, 1997). New preliminary data from this laboratory indicates that *adapt78* expression may be compromised in the *substantia nigra* of human beings who have died with severe Parkinsons Disease [Ermak, G., and Davies, K.J.A., unpublished observations].

5. OXIDATIVE STRESS PRODUCES A WIDE SPECTRUM OF CONCENTRATION-DEPENDENT CELLULAR EFFECTS

Both procaryotic and eucaryotic cells exhibit a wide range of responses to oxidant stress. Very low hydrogen peroxide concentrations actually stimulate the growth of bacteria (Crawford and Davies, 1994a; Crawford *et al.*, 1994b; Davies and Lin, 1994; Davies *et al.*, 1995; Wiese *et al.*, 1995), yeast (Davies *et al.*, 1995), and mammalian cells. Slightly higher peroxide levels cause a temporary growth-arrest and a transient adaptive response in bacteria, yeast, and mammalian cells (Crawford and Davies, 1994a; Crawford *et al.*, 1994b; Davies and Lin, 1994; Davies *et al.*, 1995; Wiese *et al.*, 1995). Higher peroxide levels still, induce a permanent growth-arrest in mammalian cells (Crawford and Davies, 1994a; Crawford *et al.*, 1994b; Wiese *et al.*, 1995) that may be akin to oxidant-induced, age-related declines in function described previously by Ames *et al.* (Ames, Shigenaga, and Hagen, 1993). Even higher hydrogen peroxide concentrations induce a mammalian apoptotic response (Crawford and Davies, 1997d; Crawford, Wang, Schools, Kochheiser, and davies, 1997e), while only the highest peroxide levels cause significant necrosis (Crawford and Davies, 1997d; Crawford *et al.*, 1997e; Wiese *et al.*, 1995). These relationships are depicted in simple form in Figure 3.

Recent studies from this laboratory have suggested that down-regulation and actual degradation of mitochondrial 12S and 16S ribosomal RNA's, and several mitochondrial messenger RNA's, may play an important role in apoptosis (Crawford and Davies, 1997d; Crawford *et al.*, 1997e). Our initial apoptosis studies were performed with hydrogen peroxide, but it is now clear that the classical apoptotic models of staurosporin treatment, or lymphocyte IL-2 withdrawal, also involve down-regulation and degradation of mitochondrial 12S and 16S rRNA's, and several mitochondrial mRNA's, perhaps indicating that oxidative stress is a common factor in apoptosis (Crawford and Davies, 1997d; Crawford

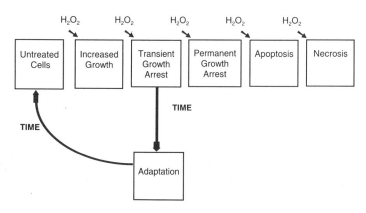

Figure 3. Spectrum of oxidative stress responses.

et al., 1997e). Interestingly, mitochondrial mRNA's encoded by the nuclear genome are unaffected by apoptotic levels of hydrogen peroxide, as are (nuclear encoded) 18S and 28S cytoplasmic ribosomal RNA's (Crawford and Davies, 1997d; Crawford *et al.*, 1997e). In contrast, rRNA's and mRNA's encoded by the mitochondrial genome are significantly down regulated and/or degraded; including 12S and 16S rRNA's ATPase subunit 6 mRNA, cytochrome oxidase subunit I mRNA, NADH dehydrogenase subunit 6 mRNA, and cytochrome oxidase subunit III mRNA (Crawford and Davies, 1997d; Crawford *et al.*, 1997e). Thus, the apoptotic down-regulation/degradation of mitochondrial RNA's appears limited to products of the mitochondrial genome, even though most motochondrial transcripts are encoded in the nucleus. Our recent findings (Crawford and Davies, 1997d; Crawford *et al.*, 1997e), coupled with observations by other groups suggesting rapid derangement of mitochondrial bioenergetic functions during apoptosis (Vayssiere, Petit, and Mignotte, 1994), suggest an important role for mitochondria in the apoptotic pathway.

REFERENCES

Adelman, R., Saul, R. L., and Ames, B. N., 1988, *Proc. Natl. Acad. Sci. USA* **85**:2706–2708.

Ames, B. N., 1989, *Mutat. Res.* **214**:41–46.

Ames, B. N., Shigenaga, M. K., and Hagen, T. M., 1993, *Proc. Natl. Acad. Sci. U.S.A.* **90**:7915–7922.

Brot, N., and Weissbach, H., 1991, *Biofactors* **3**:91–96.

Chan, E., and Weiss, B., 1987, *Proc. Natl. Acad. Sci. USA* **84**:3189–3193.

Crawford, D. R., and Davies, K. J. A., 1997, *Free Radical Biol. Med.* **22**:1295–1300.

Crawford, D. R., and Davies, K. J. A., 1994, Adaptive response and oxidative stress. *Environ. Health Perspect.* **102**:(Suppl. 10), 25–28.

Crawford, D. R., Schools, G. P., Salmon, S. L., and Davies, K. J. A., 1996, *Arch. Biochem. Biophys.* **325**:256–264.

Crawford, D. R., Schools, G. P., and Davies, K. J. A., 1996, *Arch. Biochem. Biophys.* **329**:137–144.

Crawford, D. R., Edbauer-Nechamen, C., Lowry, C. V., Salmon, S. L., Kim, Y. K., Davies, J. M. S., and Davies, K. J. A., 1994, *Methods Enzymol.* **234**:175–217.

Christman, M. F., Morgan, R. W., Jacobson, F. S., and Ames, B. N., 1985, *Cell* **41**:753–762.

Christman, M. F., Storz, G., and Ames, B. N., 1989, *Proc. Natl. Acad. Sci. (USA)* **86**:3484–3488.

Collinson, L. P., and Dawes, I. W., 1992, *J. Gen. Microbiol.* **138**:329–335.

Comporti, M., 1985, **53**:599–623.

Coux, O., Tanaka, K., and Goldberg, A. L., 1996, *Annu. Rev. Biochem.* **65**:801–847.

Crawford, D., Zbinden, I., Amstad, P., and Cerutti, P.,1988, *Oncogene* **3**:27–32.

Crawford, D. R., Leahy, K. P., Wang, Y., Schools, G. P., Kochheiser, J. C., and Davies, K. J. A., 1996, *Free Radical Biol. Med.* **21**:521–525.

Crawford, D. R., and Davies, K. J. A., 1997, *Surgery* **121**:581–587.

Crawford, D. R., Leahy, K. P., Abramova, N., Lan, L., Wang, Y., and Davies, K. J. A., 1997, *Arch. Biochem. Biophys.* **342**:6–12.

Crawford, D. R., Wang, Y., Schools, G. P., Kochheiser, J., and Davies, K. J. A., 1997, *Free Radical Biol. Med.* **22**:551–559.

Cunningham, R. P., Saporito, S. M., Spitzer, S. G., and Weiss, B., 1986, *J. Bacteriol.* **168**:1120–1127.

Davies, K. J. A., 1986, *Free Radical Biol. Med.* **2**:155–173.

Davies, K. J. A., 1993, *Biochem. Soc. Trans.* **21**:346–353.

Davies, K. J. A., and Goldberg, A. L., 1987, *J. Biol. Chem.* **262**:8220–8226.

Davies, K. J. A., and Goldberg, A. L., 1987, *J. Biol. Chem.* **262**:8227–8234.

Davies, K. J. A., 1987, *J. Biol.Chem.* **262**:9895–9901.

Davies, K. J. A., Delsignore, M. E., and Lin, S. W., 1987, *J. Biol. Chem.* **262**:9902–9907.

Davies, K. J. A., and Delsignore, M. E., 1987, *J. Biol. Chem.* **262**:9908–9913.

Davies, J. M. S., Lowry, C. V., and Davies, K. J. A., 1995, *Arch. Biochem. Biophys.* **317**:1–6.

Davies, K. J. A., Lin, S. W., and Pacifici, R. E., 1987, *J. Biol. Chem.* **262**:9914–9920.

Davies, K. J. A., and Lin, S. W., 1988, *Free Radical Biol. Med.* **5**:215–223.

Davies, K. J. A., and Lin, S. W., 1988, *Free Radical Biol. Med.* **5**:225–236.

Davies, K. J. A., 1989, In *Antioxidants in Therapy and Preventive Medicine* (Emerit, I., Packer, L., and Auclair, C., eds.), pp. 503–511, Plenum Press, London.

Davies, K. J. A., 1988, In *Lipofuscin — 1987: State of the Art* (Zs.-Nagy, I., ed.), pp. 109–133, Elsevier, Amsterdam.

Davies, K. J. A., Wiese, A. G., Pacifici, R. E., and Davies, J.M.S., 1993, In *Free Radicals: From Basic Science to Medicine* (Poli, G., Albano, E., and Dianzani, M.U., eds.), pp. 18–30, Birkhäuser-Verlag, Basel/Switzerland.

Davies, K. J. A., and Lin, S. W., 1994, In *Free Radicals in the Environment, Medicine, and Toxicology* (Nohl, H., Esterbauer, H., and Rice-Evans, C., eds.), pp. 563–578, Richelieu Press, London.

Demple, B., and Halbrook, J., 1983, *Nature* **304**:466–468.

Demple, B., and Harrison, L., 1994, *Annu. Rev. Biochem.* **63**:915–948.

Demple, B., 1991, *Ann. Rev. Genet.* **25**:315–337.

Demple, B., Johnson, A., and Fung, P., 1986, *Proc. Natl. Acad. Sci. USA* **83**:7731–7735.

Devary, Y., Gottlieb, R. A., Lau, L. F., and Karin, M., 1991, *Mol. Cell Biol.* **11**:2804–2811.

Doetsch, P. W., Helland, D. E., and Haseltine, W. A., 1986, *Biochemistry* **25**:2212–2220.

Doetsch, P. W., Henner, W. D., Cunningham. R. P., Toney, J. H., and Helland, D. E., 1987, *Mol. Cell. Biol.* **7**:26–32.

Farr, S. B., and Kogoma, T., 1991, *Microbiol. Rev.* **55**:561–585.

Farr, S. B., Natvig, D. O., and Kogoma, T., 1985, *J. Bacteriol.* **164**:1309–1316.

Flatter-O'Brien, J., Collinson, L. P., and Dawes, I. W., 1993, *J. Gen. Microbiol.* **139**:501–507.

Fornace, A. J., Alamo, I., Jr., and Hollander, M. C., 1988, *Proc. Natl. Acad. Sci. (USA)* **85**:8800–8804.

Fornace, A. J., Nebert, D. W., Hollander, M. C., Luethy, J. D., Papathanasiou, M., Fargnoli, J., and Holbrook, N. J., 1989, *Mol. Cell Biol.* **9**:4196–4203.

Fridovich, I., 1978, *Science* **201**:875–880.

Giulivi, C., and Davies, K. J. A., 1993, *J. Biol. Chem.* **268**:8752–8759.

Giulivi, C., Pacifici, R. E., and Davies, K. J. A., 1994, *Arch. Biochem. Biophys.* **311**:329–341.

Goerlich, O, Quillardet, P., and Hofnung, M., 1989, *J. Bacteriol.* **171**:6141–6147.

Grune, T., Reinheckel, T., Joshi, M., and Davies, K. J. A., 1995, *J. Biol. Chem.* **270**:2344–2351.

Greenberg, J. T., and Demple, B., 1989, *J. Bacteriol.* **171**:3933–3939.

Greenberg, J. T., Monach, P. A., Chou, J. H., Josephy, D. P., and Demple, B., 1990, *Proc. Natl. Acad. Sci. (USA)* **87**:6181–6185.

Greenberg, J. T., Chou, J. H., Monach, P. A., and Demple, B., 1991, *J. Bacteriol.* **173**:4433–4439.

Grune, T., Reinheckel, T., Talbot, M., and Davies, K. J. A., 1996, *J. Biol. Chem.* **271**:15504–15509.

Gupta, S. S., and Bhattacharjee, S. B., 1988, *Int. J. Radiat. Biol.* **53**:935–942.

Harman, D., 1956, *J. Gerontol.* **11**:298–300.

Harman, D., 1981, *Proc. Natl. Acad. Sci. USA* **78**:7124–7128.

Helland, D. E., Doetsch. P. W., and Haseltine, W. A., 1986, *Mol. Cell. Biol.* **6**:1983–1990.

Jamieson, D. J., 1992, *J. Bacteriol.* **174**:6678–6681.

Jenkins, D. E., Schultz, J. E., and Matin, A., 1988, *J. Bacteriol.* **170**:3910–3914.

Johnson, A. W., and Demple, B., 1988, *J. Biol. Chem.* **263**:18017–18022.

Kasai, H., Crain, P. F., Kuchino, Y., Nishimura, S., Ootsuyama, A., and Tanooka, H., 1986, *Carcinogenesis* **7**:1849–1851.

Ketterer, B., and Meyer, D. J., 1989, *Mutat. Res.* **214**:33–40.

Keyse, S. M., and Tyrrell, R. M., 1989, *Proc. Natl. Acad. Sci. (USA)* **86**:99–103.

Keyse, S. M., and Emslie, E. A., 1992, *Nature* **359**:644–647.

Kogoma, T., Farr, S. B., Joyce, K. M., and Natvig, D. O., 1988, *Proc. Natl. Acad. Sci. (USA)* **85**:4799–4803.

Koshland, D. E., Jr., 1994, *Science* **266**:1925–1929.

Kow, Y. W., and Wallace, S. S., 1985, **82**:8354–8358.

Kuge, S., and Jones, N., 1994, *EMBO J.* **13**:655–664.

Kullik, I., and Storz, G., 1994, *Redox Report* **1**:23–29.

Laval, F., 1988, *Mutation Res.* **201**:73–79.

Levin, J. D., Johnson, A. W., and Demple, B., 1988, *J. Biol. Chem.* **263**:8066–8071.

Lu, D., Maulik, N., Moraru, I. I., Kreutzer, D. L., and Das, D. K., 1993, *Am. J. Physiol.* **264**:C715–C722.

Lubin, B. H., Shohet, S. B., and Nathan, D. G., 1972, *J. Clin. Invest.* **51**:338–344.

Maiorino, M., Chu, F. F., Ursini, F., Davies, K. J. A., Doroshow, J. H., and Esworthy, R. S., 1991, *J. Biol. Chem.* **266**:7728–7732.

Marcillat, O., Zhang, Y., Lin, S. W., and Davies, K. J. A., 1988, *Biochem. J.* **254**:677–683.

Mead, J. F., 1976, In *Free Radicals in Biology* (Pryor, W. A., ed.), pp. 51–68, Academic Press, New York.

Melendez, J. A., and Davies, K. J. A., 1996, *J. Biol. Chem.* **271**:18898–18903.

Morgan, R. W., Christman, M. F., Jacobson, F. S., Storz, G., and Ames, B. N., 1986, *Proc. Natl. Acad. Sci. (USA)* **83**:8059–8063.

Mossman, B. T., 1991, *J. Biol. Chem.* **266**:24398–24403.

Murakami, K., Jahngen, J. H., Lin, S. H., Davies, K. J. A., and Taylor, A., 1990, *Free Radical Biol. Med.* **8**:217–222.

Pacifici, R. E., Kono, Y., and Davies, K. J. A., 1993, *J. Biol. Chem.* **268**:15405–15411.

Pacifici, R. E., and Davies, K. J. A., 1991, *Gerontology* **37**:166–180.

Pacifici, R. E., Salo, D. C., Lin, S. W., and Davies, K. J. A., 1989, *Free Radical Biol. Med.* **7**:521–536.

Povirk, L. F., and Steighner, R. J., 1989, *Mutat. Res.* **214**:13–22.

Richter, C., 1988, *FEBS Lett.* **241**:1–5.

Richter, H. E., and Loewen, P. C., 1981, *Biochem. Biophys. Res. Commun.* **100**:1039–1046.

Richter, C., Park, J. W., and Ames, B. N., 1988, *Proc. Natl. Acad. Sci. USA* **85**:6465–6467.

Rivett, A. J., 1993, *Biochem. J.* **291**:1–10.

Rushmore, T. H., King, R. G., Paulson, K. E., and Pickett, C. G., 1990, *Proc. Natl. Acad. Sci. (USA)* **87**:3826–3830.

Salo, D. C., Pacifici, R. E., and Davies, K. J. A., 1990, *J. Biol. Chem.* **265**:12751–12757.

Salo, D. C., Donovan, C. M., and Davies, K. J. A., 1991, *Free Radical Biol. Med.* **11**:239–246.

Schraufstätter, I., Hyslop, P. A., Jackson, J. H., and Cochrane, C. G., 1988, *J. Clin. Invest.* **82**:1040–1050.

Sevanian, A. and Hochstein, P., 1985, *Ann. Rev. Nutr.* **5**:365–390.

Sevanian, A., and Kim, E., 1985, *Free Radical Biol. Med.* **1**:263–271.

Sevanian, A., Wratten, M. L., McLeod, L. L., and Kim, E., 1988, *Biochim. Biophys. Acta.* **961**:316–327.

Shull, S., Heintz, N. H., Periasamy, M., Manohar, M., Janssen, Y. M. W., Marsh, J. P., and

Simic, M. G., Bergtold, D. S., and Karam, L. R., 1989, *Mutat. Res.* **214**:3–12.

Spector, A., Kleiman, N. J., Huang, R. C., and Wang, R. R., 1989, *Exptl. Eye Res.* **49**:685–698.

Spitz, D. R., Dewey, W. C., and Li, G. C., 1987, *J. Cell. Physiol.* **131**:364–373.

Stadtman, E. R., 1986, *Trends Biochem. Sci.* **11**:11–12.

Stadtman, E. R., 1993, *Annu. Rev. Biochem.* **62**:797–821.

Starke-Reed, P. E., Oliver, C. N., and Stadtman, E. R., 1989, *Arch. Biochem. Biophys.* **275**:559–567.

Stevens, J. B., and Autor, A. P., 1977, *J. Biol. Chem.* **10**:3509–3514.

Storz, G., Tartaglia, L. A., and Ames, B. N., 1990, *Science* **248**:189–194.

Storz, G., Tartaglia, L. A., Farr, S. B., and Ames, B. N., 1990, *Trends in Genet.* **6**:363–368.

Taylor, A., and Davies, K. J. A., 1987, *Free Radical Biol. Med.* **3**:371–377.

Taylor, A., Daims, M. A., Lee, J., and Surgenor, T., 1982, *Curr. Eye Res.* **2**:47–56.

Tao, K., Makino, K., Yonei, S., Nakata, A., and Shinagawa, H., 1989, *Mol. Gen. Genet.* **218**:371–376.

Tartaglia, L. A., Storz, G., and Ames, B. N., 1989, *J. Mol. Biol.* **210**:709–719.

Tartaglia, L. A., Gimeno, C. J., Storz, G., and Ames, B. N., 1992, *J. Biol. Chem.* **267**:2038–2045.

Teebor, G. W., Boorstein, R. J., and Cadet, J., 1988, *Int. J. Radiat. Biol.* **54**:131–150.

Thomas, J. P., Maiorino, M., Ursini, F., and Girotti, A. W., 1990, *J. Biol. Chem.* **265**:454–461.

Tice, R. R., and Setlow, R. B., 1985, In *Handbook of the Biology of Aging*, 2nd ed., (Finch, C. E., and Schneider, R., eds.) pp. 173–224, Van Nostrand Reinhold Company, New York.

Toledano, M. B., Kulik, I., Trinh, F., Baird, P. T., Schneider, T. D., and Storz, G., 1994, *Cell* **78**:897–909.

Tsaneva, I. R., and Weiss, B., 1990, *J. Bacteriol.* **172**:4197–4205.

Ursini, F., Maiorino, M., Valente, M., Ferri, L., and Gregolin, C., 1982, *Biochim. Biophys. Acta.* **710**:197–211.

Van Kuijk, F. J. G. M., Sevanian, A, Handelman, G. J., and Dratz, E. A., 1987, *Trends Biochem. Sci.* **12**:31–34.

Wallace, S., 1988, *Environ. Mol. Mutagen.* **12**:431–477.

Wang, Y., Crawford, D. R., and Davies, K. J. A., 1996, *Arch. Biochem. Biophys.* **332**:255–260.

Wiese, A. G., Pacifici, R. E., and Davies, K. J. A., 1995, *Arch. Biochem. Biophys.* **318**:1–10.

White, C. W., Ghezzi, P., McMahon, S., Dinarello, C. A., and Repine, J. E., 1989, *J. Appl. Physiol.* **66**:1003–1007.

Wu, J., and Weiss, B., 1991, *J. Bacteriol.* **173**:2864–2871.

Vaca, C. E., Wilhelm, J., and Harms-Ringdahl, M., 1988, *Mutat. Res.* **195**:137–149.

Vayssiere, J. L., Petit, P. X., and Mignotte, B., 1994, *Proc. Natl. Acad. Sci. U.S.A.* **91**:11752–11756.

Zhang, L., Maiorino, M., Roveri, A., and Ursini, F., 1989, *Biochim. Biophys. Acta.* **1006**:140–143.

Zimmerman, W. F., and Keys, S., 1988, *Exp. Eye Res.* **47**:247–260.

24

THE "PUSH–PULL MECHANISM"

Protection against Site-Specific and Transition Metal-Mediated Damage

Mordechai Chevion,[*] Ben-Zhan Zhu, and Eduard Berenshtein

Department of Cellular Biochemistry
Hebrew University
Hadassah Schools of Medicine and Dental Medicine
Jerusalem 91120, Israel

1. INTRODUCTION

Molecular oxygen (O_2) probably appeared on the Earth's surface about 2×10^9 years ago as a result of photosynthetic microorganisms acquiring the ability to split water. Oxygen is now the most abundant element in the biosphere. Its concentration in dry air has risen to 21%. Iron is the second most abundant metal in the Earth's crust whereas copper is more scarce.

A free radical is an atom or a molecule with one or more unpaired electrons. This definition includes the oxygen molecule (a biradical), a hydrogen atom and most of the transition metal ions. To avoid a spin, and possibly, orbital restriction, oxygen accepts electrons one at a time. This considerably slows down the reaction of oxygen with the majority of covalent molecules, which are non-radicals, and can be considered an advantage for life in oxygen. A major disadvantage is, however, that electrons when added singly to oxygen lead to the formation of reactive intermediates, two of which are free radicals. The four electron reduction of oxygen to water gives rise sequentially to the superoxide anion radical ($O_2^{\cdot-}$), hydrogen peroxide (H_2O_2), hydroxyl radical ($\cdot OH$), and water (Halliwell and Gutteridge, 1989).

It has been recognized since the late 19th century that oxygen, which is essential for most living systems, is also inherently toxic (Frank and Massaro, 1980; Cadenas, 1989). In fact, long before this time, one of the two discoverers of oxygen, Joseph Priestly, sug-

* Address for Correspondence: Prof. Mordechai Chevion, The Dr. W. Ganz Chair of Heart Studies, Hebrew University — Hadassah Medical School, P.O.Box 12272, Jerusalem 91120, Israel. Telephone number: 972-2-6758158; 972-2-6758160, Fax number: 972-2-6415848; 972-2-6784010.

Free Radicals, Oxidative Stress, and Antioxidants, edited by Özben.
Plenum Press, New York, 1998.

gested that "as a candle burns much faster in dephlogisticated than in common air, so we might, as may be said, live out too fast, and the animal powers be too soon exhausted in this pure kind of air" (Priestly, 1906).

It was not until 1954 (Gerschman et al., 1954; Gerschman, 1981) that it was proposed that most of the damaging effects of elevated oxygen concentrations in living organisms could be attributed to the formation of free radicals. However, this idea did not capture the interest of many biologists and clinicians until the discovery in 1969 of an enzyme specific for the catalytic removal of $O_2^{\cdot-}$, the superoxide dismutase (SOD) (McCord and Fridovich, 1969). This was a major turning point in the research and understanding of free radicals in biology and medicine. Soon after the discovery of SOD, it was found that $\cdot OH$ appeared to be produced in the superoxide-generating system (Beauchamp and Fridovich, 1970). The mechanism of $\cdot OH$ production was proposed through the following Haber-Weiss reaction:

$$O_2^{\cdot-} + H_2O_2 \longrightarrow \cdot OH + OH^- + O_2$$

Studies of the kinetics of this reaction found it to be very slow at physiological condition (Ferradini et al., 1978). So, the question how is $\cdot OH$ being formed in $O_2^{\cdot-}$-generating system remained to be answered.

The catalytic role of transition metals, particularly iron and copper, was not appreciated until the Pinawa meeting in 1977, when it was presented that adventitious levels of iron in buffer solutions were of the order of 1 μM, and that this level of iron would change the results observed in $O_2^{\cdot-}$-generating system (Czapski and Ilan, 1978; Koppenol et al., 1978). Just as important, it was shown that chelating agents would alter the reactivity of iron in $O_2^{\cdot-}$-generating systems. It was demonstrated that EDTA enhances the reactivity of iron toward $O_2^{\cdot-}$ while diethylenetriaminepentaacetic acid (DTPA) and desferrioxamine (DFO) dramatically slow it down (Buettner et al., 1978; Gutteridge et al., 1979). Based on the above considerations it was suggested that traces of soluble iron or copper can catalyze the transformation of $O_2^{\cdot-}$ to the highly reactive $\cdot OH$, via the metal catalyzed Haber-Weiss reaction (or $O_2^{\cdot-}$-driven Fenton reaction) (Haber and Weiss, 1934; Anbar and Levitzki, 1966; Borg et al., 1978; Halliwell, 1978; Koppenol et al., 1978; McCord and Day, 1978; Borg and Schaich, 1984; Chevion, 1988; Czapski, 1984; Aronovitch et al., 1986; Aust et al., 1985; Goldstein and Czapski, 1986; Halliwell and Gutteridge, 1989; Stadtman, 1990).

$$O_2^{\cdot-} + Fe(III)/Cu(II) \longrightarrow O_2 + Fe(II)/Cu(I) \tag{1}$$

$$Fe(II)/Cu(I) + H_2O_2 \longrightarrow Fe(III)/Cu(II) + OH^- + \cdot OH \tag{2}$$

These metal ions are rather insoluble under physiological conditions, and remain in solution only by becoming complexed to low or high molecular weight cellular components. Consequently, they serve as catalytic centers for free radical production. For example, copper forms stable complexes with many proteins (Shinar et al., 1983), while iron makes stable complexes with nucleotide diphosphates and triphosphates. For iron, partial displacement of organic ligand by hydroxide anions renders $Fe(OH)^+$ suitable for the Fenton reaction. This is on both the kinetic account, where its rate with H_2O_2 is 2×10^6 M^{-1} s^{-1}, and on thermodynamic basis where E° for $Fe(OH)^{2+}/Fe(OH)^+ = 0.304$ V, rather than ~0.77 V, which is usually cited for Fe^{3+}/Fe^{2+}. Once these transition metal ions are completely displaced from such complexes and become aqua-complexes in the bulk solution, they polymerize and precipitate out as unreactive polynuclear structures (Chevion, 1988).

Using the Marcus theory and the experimental data in the literature, it has been known that in most cases the reaction of these metal complexes with H_2O_2 is unlikely to occur via an outer-sphere electron-transfer mechanism. It is suggested that the first step in this process is the formation of a transient complex, which may decompose to an ·OH radical or a higher oxidation state of the metal. Alternatively, it may yield an organic free radical in the presence of organic substrates. Thus, the question whether ·OH radicals are being formed or not, via the Fenton reaction, depends on the relative rates of the decomposition reactions of the metal–peroxide complex, and that of its reaction with organic substrates (Goldstein et al., 1993).

2. THE SITE-SPECIFIC MECHANISM OF METAL-MEDIATED PRODUCTION OF FREE RADICALS

2.1. General

The site-specific mechanism of free radical formation and the subsequently induced damage stem from the assumption that copper or iron complexes are the essential mediators of the damage (Samuni et al., 1980, 1981, 1983, 1984; Gutteridge and Wilkins, 1983; Kohen and Chevion, 1986; Aronovitch et al., 1986; Yamamoto and Kawanishi, 1989). These biologically bound metal ions can undergo redox cycling. Reducing agents such as $O_2^{\cdot-}$, ascorbate, the pyrimidines, thiols, phenolic compounds, hydrazines, as well as other molecules could reduce the metal ion within its complex yielding the cupro or ferro states, and concurrently form $O_2^{\cdot-}$ and/or H_2O_2. Subsequently, these reduced metal ions can react with hydrogen peroxide (the Fenton reaction) yielding the ·OH (Reaction 2). The site-specific metal-mediated mechanism is characterized by the following lines:

i. It explains the funneling of free radical damage to the specific metal-binding sites. Relatively low-reactive reducing agents such as $O_2^{\cdot-}$ or ascorbate, whose life spans are comparatively long, can migrate a relatively long way until they encounter a redox-active metal ion and react with it.

ii. It explains the transformation of rather low-reactive species, such as $O_2^{\cdot-}$ or ascorbate to the highly reactive species such as ·OH, which is known to cause a variety of molecular disruptions (Walling, 1982). Generally, the ·OH is characterized by very high kinetic rate constants for its reaction with a variety of biological molecules or residues. This radical can act by causing breaks in the polymeric backbone of a macromolecule, by abstracting hydrogen, or by addition to a double bond.

iii. It explaithe possible "multi-hit" effects that are often experimentally observed. If we assume that the off-rate of copper or iron is slow enough that its spatial locus is fixed, then the metal can undergo more than a single redox cycle. Alternatively, repeated redox-cycles can take place at the same locus by rapidly exchanging transition metal. In each such a cycle, one ·OH is released and the repeated redox reactions account for the production of multiple ·OH. Because of its high reactivity, it is likely that the ·OH will inflict its damage within a few encounters, i.e., near the site of its formation—the metal binding site. Thus, this mechanism demonstrates how multi-hits by ·OH will take place near the metal binding site.

These features indicate that the site-specific mechanism could explain either amplification or a dampening of the damage caused by a given number of ·OH radicals. If the metal-binding site is important for function or activity, or for spatial orientation or conformation, then the funneling effect will cause marked amplification of the biological damage, because all of the deleterious radicals will be produced and will react at this pertinent site. In contrast, if the metal-binding site is relatively unimportant, the funneling effect will direct the damage to the redundant site, and the recorded biological damage will be noticeably reduced.

The free radical-induced DNA breaks is an illustrative example of the amplification for such a mechanism (Chevion, 1988). DNA damage can take place in a variety of modes, including breaks, depurination–depyrimidation, and chemical modification of the bases or the sugar. When an ·OH radical hits the DNA polymer, it could result in a single-strand break (SSB). The mechanism of such a SSB involves the addition of ·OH to C=C or an abstraction of H-atoms from the sugar moiety. The rate-determining step of this process is the rate of heterolytic cleavage of the phosphate bond. SSBs can be repaired, in most cells, by a variety of repair mechanisms. Thus, a low incidence of SSBs on the DNA is usually efficiently tolerated. In contrast, when the two strands are broken at spatially nearby sites, a double-strand break (DSB) is formed and cannot be properly corrected. This will lead to a loss of genetic information, and eventually, to the cell death. In the presence of copper and ascorbate, the site-specific mechanism of DNA breaks will lead to multi-hit effects at the copper-binding sites. By using an in vitro system of purified DNA, repair processes have been eliminated and precise quantitation of the total number of strand breaks produced during exposure to ascorbate/copper mixture could be conducted. In this mechanism, ascorbate could play a dual role: to produce the necessary hydrogen peroxide in a copper catalyzed reaction, and to reduce DNA-bound copper which would then serve as a center for repeated and site-specific production of ·OH. This has resulted in a high incidence of DSBs, compared with DSBs that were formed by ionizing radiation. It has been shown that more than 100 SSBs are necessary in order to produce one DSB by gamma radiation (Kohen et al., 1986). In contrast, when DNA was exposed to ascorbate and copper and underwent cleavage by the site-specific mechanism, one DSB for every 2.3 SSBs was observed. This high yield of DSBs, representing an increase by a factor of more than 40, can be directly attributed to the multi-hits of ·OH. This increase in probability of DSBs may provide a predictor for the potential hazards to DNA from various xenobiotics.

It should be noted that in some cases, hydroxyl radicals can be produced by organic Fenton reactions, which are metal-independent (Koppenol and Butler, 1985; Nohl and Jordan, 1987).

2.2. The Role of Transition Metal Ions in Converting Low Reactive Molecules to Highly Reactive Species

A number of enzymatic systems and chemical oxidants have been implicated in the activation of carcinogens. Studies have also provided evidence suggesting that transition metals, particularly copper, but also iron and manganese, are capable of directly mediating the metabolic activation of xenobiotics. This can be explained by virtue of the site-specific formation of more reactive species such as ·OH, organic free radicals and other electrophiles (Emerit and Cerutti, 1981; Birnboim, 1982, 1992; Rao, 1991; Rumyantseva et al., 1991; Swauger et al., 1991; Yourtee et al., 1992; Li and Trush, 1993a,b, 1994).

Copper is a redox-active essential element, playing important roles in redox-active centers in a variety of metalloenzymes, such as in cytochrome *c* oxidase, which is critical for oxidative phosphorylation. Other copper-containing enzymes include lysyl oxidase which is involved in the cross-linking of elastin and collagen; tyrosinase, which is needed for melanin pigment formation; dopamine β-hydroxylase, necessary for catecholamine production and therefore nerve and metabolic function; superoxide dismutase, specializing in the disposal of potentially damaging O_2^- produced in normal metabolic reactions; ceruloplasmin, a potential extracellular free radical scavenger as well as a ferroxidase in blood plasma and other extracellular fluids; and tryptophan oxygenase, ascorbate oxidase, polyphenol oxidase and lactase. Copper-containing proteins without known enzymatic function have also been identified, and the association of copper with amino acids, small peptides and perhaps other metabolites have also been reported (Howell and Gawthorne, 1987; Beinert, 1991; Linder, 1991). The physiological roles of copper in mammals and/or humans are associated with erythropoiesis, the development of connective tissues, the development of bones, the development of central nervous system, immune competence and pigmentation. Studies on the interaction of copper with biomolecules have also demonstrated that copper exists in chromosomes (Wacker and Vallee, 1959; Bryan et al., 1981) and is closely associated with DNA bases, particularly G–C sites (Pezzano and Podo, 1980; Agarwal et al., 1989; Geierstanger et al., 1991). DNA-associated copper may be involved in maintaining normal chromosome structure and in gene regulatory processes (Prutz et al., 1990; O'Halloran, 1993). Besides the physiological roles of copper, it is also involved in pathological processes. The involvement of copper in xenobiotic activation and damages has been the interest of recent toxicological studies.

2.3. Involvement of Iron and Copper in Tissue Injury Associated with Ischemia and Reperfusion

Reactive oxygen-derived species (ROS), both non-radical and free radical, have been implicated in tissue injury following ischemia and reperfusion of the heart (McCord, 1985; Garlick et al., 1987; Arroyo et al., 1987; Zweier et al., 1987; Opie, 1989; Vandeplassche et al., 1989; Shlafer et al., 1990), lung (Ayene et al., 1992; Basoglou et al., 1992; Kinnula et al., 1995; Marx and Van Asbeck, 1996), brain (Cao et al., 1988), and kidney (Gamelin and Zager, 1988; Weight et al., 1996), as well as in a variety of other pathologies (Halliwell and Gutteridge, 1984, 1985, 1989, 1990; Aruoma et al., 1991), including apoptotic processes (Oberhammer et al., 1992; Gottlieb et al., 1994). The outcome of the stress is associated with exacerbation of cellular injury, and with immediate and/or programmed cell death. Reperfusion following ischemia occasionally enhances release of intracellular enzymes, excessive influx of Ca^{2+}, breakdown of sarcolemmal phospholipids, and disruption of cell membranes, which result in ultimate cell death (Das, 1993). These sequels of events are known as *reperfusion injury*. The *free radical hypothesis*, the *calcium-overloading hypothesis*, and the *loss of sarcolemmal phospholipid hypothesis* have been proposed, among others, to explain this phenomenon. Many lines of evidence have been published in recent years, supporting the free radical hypothesis.

The free radical hypothesis is based on the original work of McCord (1985), Granger et al. (1981) and others (Halliwell and Gutterige, 1984; Borg et al., 1978; Manson et al., 1983). During ischemia, there is a build-up of adenosine resulting from the of intracellular ATP. It is converted to hypoxanthine and at the same time, during ischemia, a Ca^{2+}-activated protease is thought to catalyze the conversion of xanthine dehydrogenase to xanthine oxidase. This conversion can be caused also under the influence of tumor necrosis factor, interleukins 1 and 3, neutrophil elastase, complement system (C5) and chemotactic peptide N-formyl–Met–Leu–Phe (Bulkley, 1993; Friedl et al., 1989). The ischemic duration required for the conversion of the dehydrogenase to the oxidase varies from organ to organ. It takes about 30 min for kidney (McCord, 1985) and dog heart (Hearse et al., 1986). According to the original proposal reperfusion results in a burst of superoxide anion radicals generated from the oxidation of hypoxanthine to xanthine and uric acid (McCord, 1985). There are several other pathways, which could lead to the production of superoxide radical following ischemia and during the early phase of reperfusion. These include oxidation products of catecholamines, through semiquinones and quinone structures, cytochromes, cytokines and oxidases.

Considering that superoxide radical is relatively low reactive species, it is conceivable that the $O_2^{\cdot-}$ produced during the early phase of reperfusion is converted to the $\cdot OH$ through the catalyses of traces of redox-active labile metal pools.

Circumstantial evidence supporting the causative role of newly mobilized redox active iron in tissue injury has been published (Nayini et al., 1985; Holt et al., 1986; Gower et al., 1989). It has been demonstrated that iron chelation provides protection against tissue injury following ischemia (Aust et al., 1985; Applebaum et al., 1990; Mayers et al., 1985; Ferreira et al., 1990; DeBoer and Clark, 1992), while the addition of iron or copper to the perfusate facilitates injury to reperfused hearts (Bernier et al., 1986; Karwatowska et al., 1992; Powell et al., 1991). Using the isolated rat heart model the involvement of endogenous iron and copper in reperfusion injury was shown (Chevion et al., 1993). Elevated levels of low molecular weight iron (LMWI) have also been reported (Voogd et al., 1992, 1994).

Labile iron pool (LIP) of cells (Breuer et al., 1995, 1996; Cabantchik et al., 1996) constitutes the primary source of metabolic and catalytically reactive iron in the cytosol. LIP is homeostatically regulated in cells, and can be markedly altered following massive mobilization of iron. LIP has been shown to sharply rise following brief exposure to Fe(II) or to oxidative stress. The mobilized iron following ischemia is comprised of two fractions—the major one contains iron in macromolecular structures such as iron containing proteins, and the minor fraction, is LIP. LIP can be measured directly by the extent of quenching of calcein fluorescence (see below). The Lip method, as well as direct measurement of labile copper pool, is awaiting further development. It is important to note that the intracellular levels of *redox-active* iron and copper, rather than the *total* level of these transition metals are important indicators of the susceptibility of the cells and the tissue to the oxidative stress.

2.3.1. Heart. In recent years efforts were invested in trials to identify and quantitative labile pools of iron in tissues subjected to ischemia and reperfusion. Voogd et al. (1992, 1994) have used the desferrioxamine detectable iron pool. These investigators disrupted the tissue under study in the cold and in the presence of DFO, and subsequently quantitated the Fe/DFO by HPLC. The results obtained show that DFO detectable labile iron pool is 2.1 μM in the non-ischemic heart and increases to 55 μM following 45 min of global cardiac ischemia (see Table 1). Both values seem rather high for physiological level and even for injured tissue. It is plausible that during the sample processing, which includes tissue homogenization and cell rupture, even on ice, newly released proteases (and possibly nucleases) digest iron-containing structures and iron is released as a low molecular weight component.

In view of these obvious difficulties, trials to quantitate iron in coronary flow were undertaken (Nohl et al., 1991; Chevion et al., 1993). These studies were based on the assumption that the levels in the coronary flow reflect the intracellular tissue level, and large changes in the intracellular distribution of iron will be manifested in metal leakage from cell into the coronary flow. Both studies have found that the total Fe levels have markedly increased following ischemia (Table 1).

Table 1. Iron levels in isolated rat heart (H) and its coronary flow (CF) and in K562 cells in tissue culture (K)

Description (reported LMWI in μM)	Time of ischemia (min)					System	Other treatment
	0	15–18	30–35	45	60		
Chevion et al., 1993	0.36	0.48	0.90	0.86	0.89	CF	
Voogd et al., 1992, 1994	2.1	13.0	35.0	55.1		H	
Nohl et al., 1991		0.16			0.34	CF	
Ambrosio et al., 1987	6.8		7.2			H	
Breuer et al., 1995, 1996	0.35					K	Iron loading 0.77

Due to the hardships associated with the direct quantitation of LIP, indirect functional measurements have been employed. In these, the coronary flow is used as a source of redox-active metals in the ascorbate-driven conversion of salicylate to dihydroxybenzoate derivatives (DHBA). These indirect measurements also show that the early fraction of reperfusion is rich in redox-active iron and copper (Chevion et al., 1993).

The causative roles of iron and copper in reperfusion injury could be also demonstrated by using myocardial preconditioning. Preconditioning is a protective procedure against extended ischemia, which is attained by a series of short episode of ischemia/reperfusion prior to the extended ischemia.

In a series of unpublished data we found a significant reduction in the mobilization of both iron and copper following extended ischemia in preconditioned hearts, as compared to non-preconditioned groups (Fig. 1).

Hemodynamic recovery was markedly higher in the preconditioned hearts compared to control. The recovery of the "working index" of the preconditioned hearts was markedly higher than non-preconditioned hearts, substantiating the correlation between the degree of metal mobilization and the extent of cardiac damage. The data further support the notion that the mobilized metals play causative role in myocardial reperfusion injury.

What are the cellular sources of iron and copper found in the coronary flow? While these metal ions, in the free state are found only in minute concentrations within the cell, their total levels are at μM range. In addition, we as well as others have shown that the newly mobilized metals are mostly attached to low molecular weight structures. Thus, their source is evidently degraded iron and/or copper-containing proteins. Of special interest was the superoxide dismutase, which is an enzyme associated with protection against free radicals. Alas, we have tried, to identify any SOD activity in the coronary flow, unsuccessfully.

2.3.2. Eye. Free radicals have been considered as possible causative agents in a number of ocular disorders. Some of these disorders are associated with light exposure (such as photic damage to the retina and perhaps age related macular degeneration). Others are associated with ischemic injury, the classic example being retinopathy of prematurity (ROP) and the extreme conditions of retinal artery or vein occlusions. Free radicals may also play a role in conditions where ischemia and reperfusion are more subtle, but none the less critical, such as diabetic retinopathy, sickle cell disease and perhaps glaucoma. In these conditions, repeated events of transient ischemia may well occur. The retina, with its high content of long chain polyunsaturated fatty acids and high rate of oxygen consumption, is especially vulnerable to free radical induced damage.

It is therefore not surprising that in both clinical and experimental systems attempts to curb free radical induced damage have been made. At the experimental level, a number

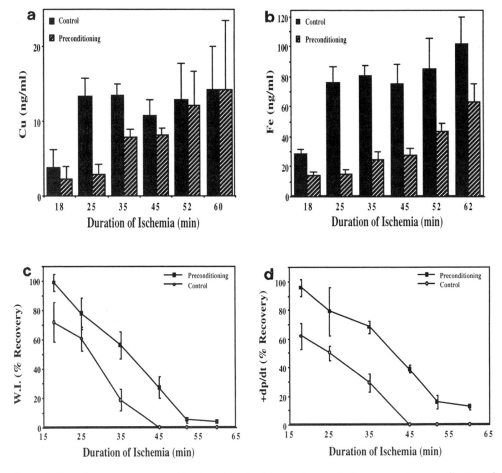

Figure 1. Hemodynamic parameters of hearts and levels of iron and copper in the coronary flow as a function of ischemic duration. Hearts were subjected to ischemia with and without prior preconditioning. **A.** Copper levels in the first coronary flow fraction of reperfusion. **B.** Iron levels in the first coronary flow fraction of reperfusion. **C.** Working Index (W.I.) of the hearts subjected to the ischemia and reperfusion. **D.** +dP/dt of hearts subjected to the ischemia reperfusion.

of studies showed that SOD reduced retinal injury following ischemia and reperfusion in rats and rabbits (Szabo et al., 1991a,b; Nayak et al., 1993). DFO and allopurinol lead to a reduced retinal damage in the isolated cat eye model subjected to ischemia/reperfusion, as measured by electroretinographic (ERG) recordings (Gehbach and Purple, 1994). At the clinical level, vitamin E was given to babies with ROP. Different studies yielded controversial results and the use of vitamin E is not advocated. A large, randomized clinical trial is currently being conducted by the National Eye Institute to evaluate the role of various antioxidants and mineral supplements in the development of macular degeneration. At the present, the use of antioxidants or free radical scavenger in clinical settings of ischemia/reperfusion is not a standard procedure, neither in ocular disorders nor in other organs (such as the heart) where such treatment may be even more relevant.

Using the conversion of salicylate to DHBA, it was possible to show that ·OH is being formed in the early phase of reperfused retina (Ophir et al., 1993). Trials to protect the ischemic retina by different antioxidants were met by varying degrees of success. Treatment by combinations of the chelator DFO and zinc or DFO and gallium, a redox non-active metal, which often neutralize the prooxidant effects of iron and copper, has proven highly effective in minimizing the production of $\cdot$OH and the injury to the retina. This was measured by the ERG, and histologic evaluation, and the maintenance of the endogenous levels of antioxidants and energy metabolites (Ophir et al., 1994; Berenshtein et al., 1996).

3. INTERVENTION AND PREVENTION

Intervention and prevention of biological damage caused by the site-specific mechanism can be accomplished by various approaches.

One classical mode of protection is the use of specific scavengers for $\cdot$OH. Initial predictions for efficient scavenging of homogeneously produced, distributed $\cdot$OH, and could not be met experimentally in biological systems. These findings are in accord with that often the mechanism of damage is the site-specific mechanism. The lack of efficient scavenging could stem from the fact that scavengers usually do not enjoy free access to the metal binding site, because of spatial restrictions and coordination requirements. Thus, the effective concentration of the scavenger at the $\cdot$OH formation site is low, leading to reduced efficiency of scavenging. It should be noted that several $\cdot$OH scavengers such as mannitol, salicylate, and β-mercaptopropionylglycine, have variable degree of transition metal binding ability as well, which is more likely to account for their protective effects than the direct scavenging of $\cdot$OH (Chevion, 1988; Halliwell and Gutteridge, 1989).

Another classical mode of protection is the use of specific enzymes to remove reactive oxygen derived species, H_2O_2 and $O_2^{\cdot-}$. H_2O_2 is an essential component in the site-specific mechanism, and is often produced from $O_2^{\cdot-}$ by spontaneous or enzymatic dismutation. Enzymes such as catalase and glutathione peroxidase can remove H_2O_2 at different concentrations and with different mechanisms, inhibiting the formation of $\cdot$OH, and minimizing molecular and cellular damage (Halliwell, 1990; Halliwell et al., 1992; Rice-Evans and Diplock, 1993). $O_2^{\cdot-}$ can act as a reducing agent for the transition metals. In this regard, $O_2^{\cdot-}$ could be an important reductant in cell even though there are other reducing species at higher concentrations. Because of its size, this small diatomic molecule can penetrate into condensed structures. SOD would generally offer protection in such system by removing superoxide that serves as a reducing agent for transition metals. $O_2^{\cdot-}$ could also be a precursor for H_2O_2. Therefore, in some cases where the rate limiting step for the observed biological damage is the Fenton reaction and H_2O_2 is the limiting substrate, SOD may enhance the biological damage rather than protect against it.

The interaction between $O_2^{\cdot-}$ and nitric oxide could also affect the end result of these free radicals. SOD could influence this reaction, and by this alter the end results.

3.1. "Pull" Mechanism of Protection

Addition of a specific chelator for iron and/or copper, such as desferrioxamine (DFO) or diethylenetriaminepentaacetic acid (DTPA), can remove or 'pull' out labile metal ions from biological binding sites, and thus, protect these sites (Graf et al., 1984; Menasche et al., 1988). The metal may become inaccessible to reducing agents or H_2O_2. DFO was used extensively to treat iron overload patients (Halliwell, 1989; Gabutti and Piga, 1996; Marx and Van Asbeck, 1996) and was shown to protect cells against the poisonous effects of paraquat (Kohen and Chevion, 1985, 1986). DFO was also found to protect against anthracycline cardiotoxicity by iron chelation (Hershko, 1996). Likewise,

1,10-phenanthroline was shown to protect bacterial cells and mammals against hydrogen peroxide-induced toxicity (Mello Filho and Meneghini, 1985; Imlay et al., 1988). Using the retrogradely perfused isolated rat heart, it was demonstrated that the copper-specific chelator neocuproine, provides protection against hydrogen peroxide-induced cardiac damage and against ischemia/reperfusion-induced arrhythmias (Appelbaum et al., 1990). Also, TPEN (*N*,*N*,*N'*,*N'*-tetrakis (2-pyridylmethyl)-ethylenediamine), a heavy metal chelator, provides nearly complete (>90%) protection against ischemia/reperfusion-induced arrhythmias (Appelbaum et al., 1988).

It is known that circulating free iron can be lethal (Eaton, 1996). Humans developed two iron binding proteins to soak up free iron in order to prevent the generation of toxic free radicals. These proteins are transferrin, a high affinity, low-capacity circulating protein (2 atoms of ferric iron per molecule of transferrin). Transferrin receptors on the surface of iron-requiring cell recognize iron-loaded transferrin. Ferritin is a lower-affinity, high-capacity iron-binding protein (maximum of 4500 atoms of iron per molecule of ferritin), for which there are receptors only on the surface of iron-storage cells, such as reticulo-endothethelial cells. Iron is trapped inside the ferritin protein shell as harmless Fe^{3+} (Herbert et al., 1996). Hence the evolution of iron storage and transport proteins provides not only a convenient way of moving iron around the body, but may also be regarded as an antioxidant defense. For example, in normal individuals, per milliliter of serum, there are approximately 300,000 molecules of transferrin per molecule of ferritin, so the concentration of non-transferrin-bound iron ions is negligible (Halliwell and Gutteridge, 1989).

Most plasma copper is attached to the protein ceruloplasmin, which has antioxidant properties. These are partly due to the ability of ceruloplasmin to oxidize Fe^{2+} to Fe^{3+}, which decrease Fe^{2+}-dependent production of $\cdot OH$ from H_2O_2, and Fe^{2+}-dependent lipid peroxidation. The small amount of plasma copper not attached to ceruloplasmin is apparently attached to either histidine, small peptides, or albumin. None of these forms of copper can apparently generate reactive oxidants in solution, and $\cdot OH$ or Cu(III) which are formed by reaction of bound copper ions with H_2O_2, appears to attack the copper-binding site. Indeed, albumin tightly binds copper ions and thus prevents their binding to more important sites. Thus, albumin is acting as an antioxidant in this context (Marx and Chevion, 1986; Halliwell and Gutteridge, 1989).

3.2. "Push" Mechanism of Protection

There is a variable degree of similarity between the coordination chemistry of copper and iron, on the one hand, and, on the other. This particularly relates to the nature of their ligands (Bray and Bettger, 1990). Thus, zinc, a redox inactive metal, could compete and displace ('push' out) Cu^{2+} and Fe^{2+} from various biological sites (Har-El and Chevion, 1989). Once zinc replaces copper, it will divert the site of formation and reaction of free radicals, and would provide an effective protection of this formerly copper-specific site. We have applied this strategy to paraquat-induced cell killing. Using 10-fold excess of Zn over adventitious copper, an effective mediator for paraquat toxicity, the synergistic killing was prevented (Korbashi et al., 1989). Similarly, zinc complexes proved protective against ischemia-induced arrhythmia, using rat heart in the Langendorff configuration (Powell et al., 1990). It appears that in cell damage caused by oxidative stress, endogenous or exogenous copper, and to a lesser degree iron, in the presence of a reducing environment, induces a cyclic Fenton-like reaction which is most deleterious to the cell. The most susceptible sites are the metal binding sites.

3.3. "Pull–Push" Mechanism of Protection

It is proposed that the complex zinc–desferrioxamine (Zn/DFO) ($\log\beta$ = 14) would be the ultimate protector against free radical induced transition metal mediated injury. Zn/DFO could act by both the 'pull' and 'push' mechanisms. Metal free DFO is a randomly oriented linear molecule, consisting of several subunits, which does not penetrate easily into cells. Upon metal binding the DFO within its complex with zinc assumes a well-defined and organized globular structure (Chevion, 1988, 1991). This structure alteration is expected to render the Zn/DFO complex more permeable into cells than the metal-free species. Once within the cell, Zn/DFO would readily exchange with available iron or copper ("pull" mechanism) to yield ferrioxamine or copper–DFO complex, respectively. By this, controlled levels of zinc would be liberated within the cell and could act to further protect by "pushing" out additional redox-active metal ions from their binding sites, and act as a secondary antioxidant. So this combination of "pull" and "push" mechanisms could efficiently prevent the site-specific metal-mediated oxidative damage.

We have investigated the protective effect of another complex that acts via the "push–pull" mechanism. Gallium–DFO complex (Ga/DFO) at low concentrations caused a nearly complete protection against reperfusion injury following global cardiac ischemia of up to 25 min. A nearly complete inhibition of mobilization of copper and iron into the coronaries of the perfused heart was also observed (Fig. 2).

The ability of Ga/DFO (unpublished data) and Zn/DFO (Ophir et al., 1994) to curb free radical formation in the ischemic/reperfused retina was also evaluated. This was accomplished by monitoring the biochemical profile of retinal tissue subjected to ischemia/reperfusion, with or without the drugs, by ERG recordings of retinal function (Fig. 3), and by measuring the levels of conversion of salicylate to its hydroxylated products—2,5-DHBA and 2,3-DHBA. The results obtained were in accord with the proposed high efficiency of the "push–pull" mechanism.

Also we have found the markedly positive effect of the Zn/DFO on the gastric ulceration in the model of social stress (unpublished data: Groisman, Berenshtein and Chevion—Table 2).

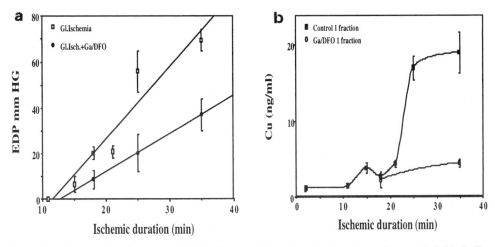

Figure 2. A. The end diastolic pressure (EDP) of hearts subjected to ischemia and reperfusion ± Ga/DFO. **B.** The level of Cu in the first coronary flow fraction of hearts subjected to ischemia and reperfusion ± Ga/DFO.

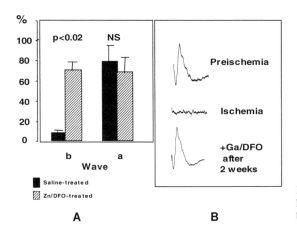

Figure 3. ERG of the retina ± Zn/DFO, 16 hours following reperfusion (A) and ± Ga/DFO, before ischemia, and 2 weeks after ischemia (B).

Table 2. Protection of the "social stress ulcer" by MCJ

Quantity of lesions per 1 stomach (M ± SE)	Type of lesion		
	Ulcers	Erosions	Massive hemorrhages
Control (n = 11)	1.3 ± 0.5	3.6 ± 0.8	4.5 ± 1.3
MCJ — 7.5 mg/kg, i/p (n = 5)	0*	1.2 ± 0.4*	0.4 ± 0.2*
% protection	< 100	< 67	< 91
MCJ — 15 mg/kg, i/p (n = 7)	0.3 ± 0.2	1.6 ± 1.0	4.1 ± 2.0
% protection	< 77	< 57	< 7
MCJ — 25 mg/kg, per os, (n = 7)	0.1 ± 0.1*	0.7 ± 0.3*	0.3 ± 0.2*
% protection	< 89	< 81	< 94

*$p < 0.05$

It should be noted that protecting agents can act in more than a single mode. For example, low concentrations of histidine provide a marked decrease in paraquat-induced cellular killing. This protective effect may be due to a combination of the following modes: i) histidine is an effective ·OH scavenger; ii) an efficient chelator for copper and iron; iii) an efficient donor of hydrogen atoms, and iv) histidine forms a tight complex with H_2O_2 that might be less active to the Fenton reaction. Similarly, while DFO exerts its protection mainly through its chelation of iron, it also acts through scavenging free radicals such as ·OH, $O_2^{\cdot-}$, peroxyl radicals and ferryl radicals (Halliwell, 1989; Darley-Usmar et al., 1989; Green et al., 1993; Denicola et al., 1995; Van Reyk and Dean, 1996). Two new modes of action of DFO have recently been demonstrated: scavenging of semiquinone radical and stimulation of hydrolysis of chlorinated phenols to the corresponding hydroxyl derivatives (Zhu et al., 1997).

REFERENCES

Agarwal, K., Sharma, A. and Talukder, G., 1989, Effects of copper on mammalian cell components, *Chem.–Biol. Interact.* **69**:1–16.

Anbar, M. and Levitzki, B., 1966, Copper-induced radiolytic deactivation of α-amylase and catalase, *Rad. Res.* **27**:32–34.

Appelbaum, Y. J., Kuvin, J., Chevion, M. and Uretzky, G., 1988, TPEN, a heavy metal chelator, protected the isolated perfused rat heart from reperfusion induced arrhythmias, *J. Mol. Cell. Cardiol.* **20** (Supp V):Abstract #32.

Appelbaum, Y. J., Kuvin, J., Borman, J. B., Uretzky, G. and Chevion, M., 1990, The protective role of neocuproine against cardiac damage in isolated perfused rat heart, *Free Rad. Biol. Med.* **8**:133–143.

Aronovitch, J., Samuni, A., Godinger, D. and Czapski, G., 1986, In vivo degradation of bacterial DNA by H_2O_2 and *O*-phenanthroline, In: Rotilio G. ed. *Superoxide and Superoxide Dismutase in Chemistry, Biology and Medicine.* New York, Elsevier. 346–348.

Arroyo, C., Kramer, J., Dickens, B. and Weglicki, W., 1987, Identification of free radicals in myocardial ischemia/reperfusion by spin trapping with nitrone DMPO, *FEBS Lett.* **221**:101–104.

Aruoma, O., Kaur, H. and Halliwell, B., 1991, Oxygen free radicals in human diseases, *J. R. Soc. Health.* **111**:172–177.

Aust, S. D., Morehouse, L. A. and Thomas, C. E., 1985, Role of metals in oxygen radical reaction, *Free Rad. Biol. Med.* **1**:3–25.

Ayene, I. S., Dodia, C. and Fisher, A. B., 1992, Role of oxygen in oxidation of lipid and protein during ischemia/reperfusion in isolated perfused rat lung, *Arch. Biochem. Biophys.* **296**:183–189.

Basoglu, A., Kocak, H., Pac, M., Cerrahogly, M., Bakan, N., Yekeler, I., Yuksek, M. S. and Goksu, S., 1992, Oxygen free radical scavengers and reperfusion injury in dog lung preserved in cold ischemia, *Thorac. Cardiovasc. Surg.* **40**:144–147.

Beauchamp, C. and Fridovich, I., 1970, A mechanism of the production of ethylene from ethanol, *J. Biol. Chem.* **245**:4641–4646.

Beinert, H., 1991, Copper in Biological Systems, A Report from "The 6th Manziana Conference". September 23–27. *J. Inorg. Biochem.* **44**:173–218.

Berenshtein, E., Banin, E., Pe'er, J., Kitrossky, N. and Chevion, M., 1996, Ga/DFO protect retina against reperfusion injury, *VIII Biennial Meeting International Society for Free Radical Research,* Barcelona, Spain. p. 157.

Bernier, M., Hearse, D. J. and Manning, A. S., 1986, Reperfusion-induced arrhythmias and oxygen-derived free radicals. Studies with "anti-free radical" interventions and a free radical-generating system in the isolated perfused rat heart, *Circ. Res.* **58**:331–340.

Birnboim, H. C., 1982, DNA strand breakage in human leukocytes exposed to a tumor promoter, phorbol-myristate acetate, *Science* **215**:1247–1249.

Birnboim, H. C., 1992, Effect of lipophilic chelators on oxyradical-induced DNA strand breaks in human granulocytes: Paradoxical effect of 1,10-phenanthroline, *Arch. Biochem. Biophys.* **294**:17–21.

Borg, D. C., Schaich, K. M., Elmore, J. J. and Bell, J. A., 1978, Cytotoxic reaction of free radical species of oxygen, *Photochem. Photobiol.* **28**:887–907.

Borg, D. C. and Schaich, K. M. 1984, Cytotoxicity from coupled redox cycling of autoxidizing xenobiotics and metals, *Israel J. Chem.* **24**:38–53.

Breuer, W., Epsztein, S., Milligram, P. and Cabantchik, Z. I., 1995, Transport of iron and other transition metals into cells revealed by a fluorescent probe, *Am. J. Physiol.* **268**:C1354–61.

Breuer, W., Epsztein, S. and Cabantchik, Z. I., 1996, Dynamics of the cytosolic chelatable iron pool of K562 cells, *FEBS Lett.* **382**:304–308.

Bryan, S. E., Vizard, D. L., Beary, D. A., La Biche, R. A. and Hardy, K. J., 1981, Partitioning of zinc and copper within subnuclear nucleoprotein particles, *Nucleic Acid Res.* **9**:5811–5823.

Buettner, G. R., Oberley, L. W. and Leuthauser, S. W. H. C., 1978, The effect of iron on the distribution of superoxide and hydroxyl radicals as seen by spin trapping and on the superoxide dismutase assay. *Photochem, Photobiol.* **28**:693–695.

Bulkley, J. B. 1993, Endothelial xanthine oxidase: A radical transducer of inflammatory signals for reticuloendothelial activation, *Br. J. Surg.* **80**:684–686.

Cabantchik, Z. I., Glickstein, H., Milgram, P. and Breuer, W., 1996, A fluorescence assay for assessing chelation of intracellular iron in a membrane model system and in mammalian cells, *Anal. Biochem.* **233**:221–227.

Cadenas, E., 1989, Biochemistry of oxygen toxicity, *Ann. Rev. Biochem.* **58**:79–110.

Cao, W., Carney, J. M., Duchon, A., Floyd, R. A. and Chevion, M., 1988, Oxygen free radical involvement in ischemia and reperfusion injury to brain, *Neurosci. Lett.* **88**:233–238.

Chevion, M., 1988, A site specific mechanism for free radical induced biological damage: The essential role of redox-active transition metals, *Free Rad. Biol. Med.* **5**:27–37.

Chevion, M., 1991, Protection against free radical-induced and transition metal-mediated damage: The use of "pull" and "push" mechanisms, *Free Rad. Res. Comms.* **12–13**:691–696.

Chevion, M., Jiang, Y., Har-El, R., Berenshtein, E., Uretzky, G. and Kitrossky, N., 1993, Copper and iron are mobilized following myocardial ischemia: Possible criteria for tissue injury, *Proc. Natl. Acad. Sci. USA* **90**:1102–1106.

Czapski, G. and Ilan, Y. A., 1978, On the generation of the hydroxyl agent from the superoxide radical: Can the Haber-Weiss reaction be the source of ·OH radicals?, *Photochem. Photobiol.* **28**:651–654.

Czapski, G., 1984, On the use of ·OH scavengers in biological systems, *Israel J. Chem.* **24**:29–32.

Darley-Usmar, V. M., Hersey, A. and Garland, L. G., 1989, A method for comparative assessment of antioxidant as peroxyl radical scavengers, *Biochem. Pharmocol.* **38**:1645–1649.

Das, D. K., 1993, Pathophysiology of reperfusion injury, CRC Press. Boca Raton, FL.

DeBoer, D. A. and Clark, R. E., 1992, Iron chelation in myocardial preservation after ischemia-reperfusion injury: The importance of pretreatment and toxicity, *Ann. Thorac. Surg.* **53**:412–418.

Denicola, A., Souza, J. M., Gatti, R. M., Augusto, O. and Radi R., 1995, Desferrioxamine inhibition of the hydroxyl radical-like reactivity of peroxynitrite: Role of hydroxamic groups, *Free Rad. Biol. Med.* **19**:11–19.

Eaton, J. W., 1996, Iron: The essential poison, *Redox Report* **2**:215.

Emerit, I. and Cerutti, P., 1981, Tumor promoter phorbol-12-myristate-13-acetate induces chromosomal damage via indirect action, *Nature* **293**:144–146.

Ferradini, C., Foos, J. and Houee, C., 1978), The reaction between superoxide anion and hydrogen peroxide, *Photochem. Photobiol.* **28**:697–700.

Ferreira, R., Burgos, M., Milei, J., Llesuy, S., Molteni, L., Hourquebie, H. and Boveris, A., 1990, Effect of supplementing cardioplegic solution with deferoxamine on reperfused human myocardium, *J. Thorac. Cardiovasc. Surg.* **100**:708–714.

Frank, L. and Massaro, D., 1980, Oxygen toxicity, *Am. J. Med.* **69**:117–126.

Friedl, H. P., Till, G. O., Ryan, U. S. and Ward, P. A., 1989, Mediator-induced activation of xanthine oxidase in endothelial cells, *FASEB J.* **3**:2512–2518.

Gabutti, V. and Piga, A., 1996, Results of long-term iron-chelating therapy, *Acta Haematol.* **95**:26–36.

Gamelin, L. M. and Zager, R. A., 1988, Evidence against oxidant injury as a critical mediator of postischemic acute renal failure, *Am. J. Physiol.* **255**:F450–460.

Garlick, P. B., Davies, M. J., Hearse, M. J. and Slater, T. F., 1987, Direct detection of free radicals in the reperfused rat heart using electron spin resonance spectroscopy, *Circ. Res.* **61**:757–760.

Gehbach, P. L. and Purple, R. L., 1994, Enhancement of retinal recovery by conjugated deferoxamine after ischemia-reperfusion, Invest. *Ophthalm. Vis. Sci.* **35**(2):669–676.

Geierstanger, B. H., Kagawa, T. F., Chen, S. L., Quigley, G. T. and Ho, P. S., 1991, Base-specific binding of copper(II) to Z-DNA: The 1.3 A single crystal structure of d(m5CGUAm5CG) in the presence of $CuCl_2$, *J. Biol. Chem.* **266**:20185–20191.

Gerschman, R., Gilbert, D., Nye, S. W., Dwyer, P. and Fenn, W. O., 1954, Oxygen poisoning and X-irradiation: A mechanism in common, *Science* **119**:623–626.

Gerschman, R., 1981, Historical introduction to the "free radical theory" of oxygen toxicity, In: Gilbert, D., ed. *Oxygen and Living Processes, an Interdisplinary Approach.* New York, Springer Verlag. 44–46.

Goldstein, S. and Czapski, G., 1986, The role and mechanism of metal ions and their complexes in enhancing damage in biological systems or in protecting these systems from the toxicity of O_2^-, *Free Rad. Biol. Med.* **2**:3–11.

Goldstein, S., Meyerstein, D. and Czapski, G., 1993, The Fenton reagents, *Free Rad. Biol. Med.* **15**:435–445.

Gottlieb, R. A., Burleson, K. O. and Cloner, R. A., 1994, Reperfusion injury induced apoptosis in rabbit cardiomyocytes, *J. Clin. Invest.* **94**:1621–1628.

Gower, J., Healing, G. and Green, C., 1989, Measurement by HPLC of desferrioxamine-available iron in rabbit kidneys to assess the effect of ischemia on the distribution of iron within the total pool, *Free Rad. Res. Commun.* **5**:291–299.

Graf, E., Mahoney, J. R., Bryant, R. G. and Eaton, J. W., 1984, Iron-catalyzed hydroxyl radical formation. Stringent requirement for free iron coordination site, *J. Biol. Chem.* **259**:3620–3624.

Granger, D. N., Rutili, G. and McCord, J. M., 1981, Superoxide radical in feline intestinal ischemia, *Gastroenterology* **81**:22–29.

Green, E. S. R., Rice-Evans, H., Rice-Evans, P., Davies, M. J., Salah, N and Rice-Evans, C. A., 1993, The efficacy of mono-hydroxamates as free radical scavenging agents compared with di- and tri-hydroxamates, *Biochem. Pharmacol.* **45**:357–366.

Gutteridge, J. M. C., Richmond, R. and Halliwell, B., 1979, Inhibition of the iron-catalyzed formation of hydroxyl radicals from superoxide and of lipid peroxidation by desferrioxamine, *Biochem. J.* **184**:469–472.

Gutteridge, J. M. C. and Wilkins, S., 1983, Copper salt-dependent hydroxyl radical formation damage to proteins acting as antioxidants, Biochim. *Biophys. Acta* **759**:38–41.

Haber, F. and Weiss, J., 1934, The catalytic decomposition of hydrogen peroxide by iron salts, *Proc. R. Soc. London Sec. A.* **147**:332–351.

Halliwell, B., 1978, Superoxide-dependent formation of hydroxyl radicals in the presence of iron chelates, *FEBS Lett.* **92**:321–326.

Halliwell, B. and Gutteridge, J. M. C., 1984, Oxygen toxicity, oxygen radicals, transition metals and disease, *Biochem. J.* **219**:1–14.

Halliwell, B. and Gutteridge, J. M. C., 1985, The importance of free radicals and catalytic metal ions in human disease, Mol. Aspects Med. **8**:89–193.

Halliwell, B. and Gutteridge, J. M. C., 1989, Free Radicals in Biology and Medicine, London, Clarendon Press.

Halliwell, B., 1989, Protection against tissue damage in vivo by desferrioxamine: What is its mechanism of action?, *Free Rad. Biol. Med.* **7**:645–651.

Halliwell, B., 1990, How to characterize a biological antioxidant, *Free Rad. Res. Commun.* **9**:1–32.

Halliwell, B. and Gutteridge, J. M. C., 1990, Role of free radicals and catalytic metal ions in human diseases: An overview, *Methods Enzymol.* **186**:1–85.

Halliwell, B., Gutteridge, J. M. and Cross, C. E., 1992, Free radicals, antioxidants, and human disease: Where are we now?, *J. Lab. Clin. Med.* **119**:598–620.

Har-El, R. and Chevion, M., 1989, Zn(II) protect againfree radical-induced damage: Studies on single and double-strand DNA breakage, *Free Rad. Res. Comms.* **12–13**:509–515.

Hearse, D. J., Manning, A. S., Downey, J. M. and Yellon, D. M., 1986, Xanthine oxidase: A critical mediator of myocardial injury during ischemia and reperfusion? *Acta Physiol. Scand. Suppl.* **548**:65–78.

Herbert, V., Shaw, S. and Jayatilleke, E., 1996, Vitamin C-driven free radical generation from iron, *J. Nutri.* **126**:1213S–1220S.

Hershko, C., Pinson, A. and Link, G., 1996, Prevention of anthracycline cardiotoxicity by iron chelation, *Acta Haematol.* **95**:87–92.

Holt, S., Gunderson, M., Joyce, K., Nayini, N. R., Eyster, G. F., Garitano, A. M., Zonia, C., Krause, G. S., Aust, S. D. and White, B. C., 198, Myocardial tissue iron delocalization and evidence for lipid peroxidation after two hours of ischemia, *Ann. Emerg. Med.* **15**:1155–1159.

Howell, J. M. and Gawthorne, J. M., 1987, *Copper on Animal and Man.* Boca Raton, Florida, CRC Press.

Imlay, J. A., Chin, S. M. and Linn, S., 1988, Toxic DNA damage by hydrogen peroxide through the Fenton reaction in vivo and in vitro, *Science* **240**:640–642.

Karwatowska-Prokopczuk, E., Czarnowska, E. and Beresewicz, A., 1992, Iron availability and free radical induced injury in the isolated ischemic/reperfused rat heart, *Cardiovasc. Res.* **26**:58–66.

Kinnula, V. l., Crapo, J. D. and Raivio, K. O., 1995, Biology of disease. General and disposal of reactive oxygen metabolites in the lung, *Lab Invest.* **73**(1):3–19.

Kohen, R. and Chevion, M., 1985, Paraquat toxicity is enhanced by iron and inhibited by desferrioxamine in laboratory mice, *Biochem. Pharmacol.* **34**:1841–1843.

Kohen, R. and Chevion, M., 1986, Transition metals potentiate paraquat toxicity, *Free Rad. Res. Commun.* **1**:79–88.

Kohen, R., Szyf, M. and Chevion, M., 1986, Quantitation of single and double strand DNA breaks in vitro and in vivo, *Anal. Biochem.* **154**:485–491.

Koppenol, W. H., Butler, J. and Leenwen, J. W. V., 1978, The Haber-Weiss cycle, *Photochem. Photobiol.* **28**:655–660.

Koppenol, W. H. and Butler, J., 1985, Energetics in interconversion reactions of oxyradicals, *Adv. Free Rad. Biol.* **1**:91–131.

Korbashi, P., Katzhendler, J. Saltman, P. and Chevion, M., 1989, Zinc protects *Escherichia coli* against copper-mediated paraquat-induced damage, *J. Biol. Chem.* **264**:8479–8482.

Li, Y. and Trush, M. A., 1993a, Oxidation of hydroquinone by copper: Chemical mechanism and biological effects, *Arch. Biochem. Biophys.* **300**:346–355.

Li, Y. and Trush, M. A.,, 1993b, DNA damage resulting from the oxidation of hydroquinone by copper: Role for a Cu(II)/Cu(I) redox cycle and reactive oxygen generation, *Carcinogenesis* **14**:1303–1311.

Li, Y. and Trush, M. A., 1994, Reactive oxygen-dependent DNA damage resulting from the oxidation of phenolic compounds by a copper-redox cycle mechanism, *Cancer Res.* **54**:1895–1898.

Linder, M. C., 1991, *Biochemistry of Copper.* Plenum Press, New York.

Manson, P. N., Antheneli, R. N., Jim, M., Bulkley, G. B. and Hoopes, J. E., 1983, The role of oxygen-free radicals in ischemic tissue injury in island skin flaps, *Ann. Surg.* **198**:87–90.

Marx, G. and Chevion, M., 1986, Site-specific modification of albumin by free radicals, *Biochem. J.* **236**:397–400.

Marx, J. J. M. and Van Asbeck, B. S., 1996, Use of iron chelators in preventing Hydroxyl radical Damage: Adult respiratory distress syndrome as an experimental model for the pathophysiology and treatment of oxygen-radical-mediated tissue damage, *Acta Haematol.* **95**:49–62.

Mayers, C. L., Weiss, S. J., Krish, M. M. and Shlafer, M., 1985, Involvement of hydrogen peroxide and hydroxyl radical in the oxygen paradox: Reduction of creatine kinase release by catalase, allopurinol or deferoxamine, but not by superoxide dismutase, *J. Mol. Cell. Cardiol.* **17**:675–684.

McCord, J. M. and Fridovich, I., 1969, Superoxide dismutase: An enzyme function for erythrocytes, *J. Biol. Chem.* **244**:6049–6055.

McCord, J. M. and Day, E. D., 1978, Superoxide-dependent production of hydroxyl radical catalyzed by iron–EDTA complex, *FEBS Lett.* **86**:139–143.

McCord, J. M., 1985, Oxygen-derived free radicals in postischemic tissue injury, N. Engl. J. Med. **312**:159–163.

Mello Filho, A. C. D. and Meneghini, R., 1985, Protection of mammalian cells by *O*-phenanthroline from lethal and DNA-damaging effects produced by active oxygen species, *Biochim. Biophys. Acta* **847**:82–89.

Menasche, P., Pasquier, C., Bellucci, S., Lorente, P., Jaillon, P. and Piwnica, A., 1988, Deferoxamine reduced neutrophil mediated free radical production during cardiopulmonary bypass in man, *J. Thorac. Cardiovasc. Surg.* **96**:582–589.

Nayak, M. S., Kita, M. and Marmor, M. F., 1993, Protection of the rabbit retina from ischemic injury by superoxide dismutase and catalase, Invest. Ophthalmol. *Vis. Sci.* **34**:2018–2022.

Nayini, N. R., White, B. C., Aust, S. D., Huang, R. R., Indrieri, R. J., Evans, A. T., Bialek, H., Jacobs, W. A. and Komara, J., 1985, Post resuscitation iron delocalization and malondialdehyde production in the brain following prolonged cardiac arrest, *J. Free Rad. Biol. Med.* **1**:111–116.

Nohl, H. and Jordan, W., 1987, The involvement of biological quinones in the formation of hydroxyl radicals via the Haber-Weiss reaction, *Biorg. Chem.* **15**:374–382.

Nohl, H., Stolze, K., Napetschnig, S. and Ishikawa, T., 1991, Is oxidative stress primarily involved in reperfusion injury of the ischemic heart? *Free Rad. Biol. Med.* **11**:581–588.

O'Halloran, T. V., 1993, Transition metals in control of gene expression, *Science* **261**:715–725.

Oberhammer, F. A., Pavalka, M. and Sharma, S., 1992, Induction of apoptosis in cultured hepatocytes and in regressing liver by transforming growth factor b1, *Proc. Natl. Acad. Sci. USA* **89**:5408–5412.

Ophir, A., Berenshtein, E., Kitrossky, N., Berman, E. R., Photiou, S., Rothman, Z. and Chevion, M., 1993, Hydroxyl radical generation in the cat retina during reperfusion following ischemia, *Exp. Eye Res.* **57**:351–357.

Ophir, A., Berenshtein, E., Kitrossky, N., and Averbukh, E., 1994, Protection of the transiently ischemic cat retina by zinc-desferrioxamine. Invest. Ophthalmol, *Vis. Sci.* **35**:1212–1222.

Opie, L. H., 1989, Reperfusion injury and its pharmacological modification, *Circulation.* **80**:1049–1062.

Pezzano, H. and Podo, F., 1980), Structure of binary complexes of mono- and polynucleotides with metal ions of the first transition group, *Chem. Rev.* **80**:365–399.

Powell, S., Saltmann, P., Uretzky, G. and Chevion, M., 1990, The effect of zinc on reperfusion arrythmias in the isolated perfused rat heart, *Free Rad. Biol. Med.* **8**:33–46.

Powell, S. R., Hall, D. and Shih, A., 1991, Copper loading of hearts increases postischemic reperfusion injury, *Circ. Res.* **69**:881–885.

Priestly, J., 1906, The discovery of oxygen, Chicago, Chicago Press, University of Chicago.

Prutz, W. A., Butler, J. and Land, E. J., 1990, Interaction of Cu(I) with nucleic acids, *Int. J. Radiat. Biol.* **58**:215–234.

Rao, G. S., 1991, Release of 2-thiobarbituric acid reactive products from glutamate-, deoxypuridine or DNA during autoxidation of dopamine in the presence of copper ions, *Pharmacol. Toxicol.* **69**:164–166.

Rice-Evans, C. A. and Diplock A. T., 1993, Current status of antioxidant therapy, *Free Rad. Biol. Med.* **15**:77–96.

Rumyantseva, G. V., Kennedy, C. H. and Mason, R. P., 1991, Trace transition metal-catalyzed reactions in the microsomal metabolism of alkyl hydrazines to carbon-centered free radicals, *J. Biol. Chem.* **266**:21422–21427.

Samuni, A., Kalkstein, A. and Czapski, G., 1980, Does oxygen enhance the radiation-induced inactivation of penicillinase? *Rad. Res.* **82**:65–70.

Samuni, A., Chevion, M. and Czapski, G., 1981, Unusual copper-induced sensitization of the biological damages due to superoxide radicals, *J. Biol. Chem.* **256**:12632–12635.

Samuni, A., Aronovitch, J., Godinger, D., Chevion, M. and Czapski, G., 1983, On the cytotoxicity of vitamin C and metal ions: A site-specific Fenton mechanism, *Eur. J. Biochem.* **137**:119–124.

Samuni, A., Chevion, M. and Czapski, G., 1984, Roles of copper and O_2^- in the radiation induced inactivation of T7 bacteriophage, *Rad. Res.* **99**:562–572.

Shinar, E., Navok, T. and Chevion, M., 1983, The analogous mechanisms of enzymatic inactivation induced by ascorbate and superoxide in the presence of copper, *J. Biol. Chem.* **258**:14778–14783.

Shlafer, M., Brosamer, K., Forder, J. R., Simon, R. H., Ward, P. A. and Grum C. M., 1990, Cerium chloride as a histochemical marker of hydrogen peroxide in reperfused ischemic hearts, *J. Mol. Cell. Cardiol.* **22**:83–97.

Stadtman, E. R., 1990, Metal ion-catalyzed oxidation of proteins: Biochemical mechanism and biological consequences, *Free Rad. Biol. Med.* **9**:315–325.

Swauger, J. E., Dolan, P. M., Zweier, J. L., Kuppusamy, P. and Kensler, T. W., 1991, Role of benzoyoxyl radical in DNA damage mediated by benzoyl peroxide, *Chem. Res. Toxicol.* **4**:233–228.

Szabo, M. E., Dray-Lefaix, M. T., Doly, M. and Braquest, P., 1991a, Free radical-mediated effects in reperfusion injury: A histologic study with SOD and EGB761 in rat retina, *Ophthalmic. Res.* **23**:225–234.

Szabo, M. E., Dray-Lefaix, M. T., Doly, M., Carre, C. and Braquest, P., 1991b, Ischemia and reperfusion-induced histologic changes in the rat retina, Invest. Ophthalmol. *Vis. Sci.* **32**:1471–1478.

Vandeplassche, G., Hermans, C., Thone, F. and Borgers, M., 1989, Mitochondrial hydrogen peroxide generation by NADH-oxidase activity following regional myocardial ischemia in the dog, *J. Mol. Cell. Cardiol.* **21**:383–392.

Van Reyk, D. M. and Dean, R. T., 1996, The iron-selective chelator desferal can reduce chelated copper, *Free Rad. Res.* **24**:55–60.

Voogd, A., Sluiter, W., Eijk, H. G. v. and Koster, J. F., 1992, Low molecular weight iron and the oxygen paradox in isolated rat hearts, *J. Clin. Invest.* **90**:2050–2055.

Voogd, A., Sluiter, W. and Koster, J. F., 1994, The increased susceptibility to hydrogen peroxide of the (post-) ischemic rat heart is associated with the magnitude of the low molecular weight iron pool, *Free Rad. Biol. Med.* **16**:453–458.

Wacker, W. E. C. and Vallee, B. L., 1959, Nucleic acids and metals, *J. Biol. Chem.* **234**:3257–3262.

Walling, C., 1982, The nature of the primary oxidants in oxidation mediated by metal ions, In: King, T. E., Mason, H. S., Morrison, M., eds. *Oxidase and Related Redox Systems*. Oxford, Perganon Press. 85–97.

Weight, S. C., Bell, P. R. F. and Nicholson, M. L., 1996, Renal ischemia-reperfusion injury, *Br. J. Surgery* **83**:162–170.

Yamamoto, K. and Kawanishi, S., 1989, Hydroxyl free radical is not the main active species in site-specific DNA damage induced by copper (II) ion and hydrogen peroxide, *J. Biol. Chem.* **264**:15435–15440.

Yourtee, D. M., Elkins, L. L., Nalvarte, E. L. and Smith, R. E., 1992, Amplification of doxorubicin mutagenicity by cupric ion, *Toxicol. Appl. Pharmacol.* **116**:57–65.

Zhu, B. Z., Har-El, R., Kitrossky, N. and Chevion, M., 1997, New modes of action of desferrioxamine: Scavenging of semiquinone radical and stimulation of hydrolysis of tetrachlorohydroquinone, Free Rad. Biol. Med. in press.

Zweier, J. L., Flaherty, J. T. and Weisfeldt, M. L., 1987. Direct measurement of free radicals generation following reperfusion of ischemic myocardium. *Proc. Natl. Acad. Sci. USA* **84**:1404–1407.

25

PROTECTIVE EFFECTS OF SUPEROXIDE DISMUTASE RELATED TO ITS PREFERENTIAL BINDING TO MONOCYTES

Ingrid Emerit,[1] Jany Vassy,[2] Frédéric Garban,[1] Paulo Filipe,[3] and Joao Freitas[3]

[1]Institut Biomédical des Cordeliers
University of Paris VI
France
[2]Laboratoire d'Analyse d'Images en Pathologie Cellulaire
Hôpital Saint Louis
Paris, France
[3]Department of Dermatology
Hôpital Santa Maria
University of Lisbon
Portugal

Superoxide dismutase (SOD) was used as a drug under the trade name Palosein before the enzymatic nature of this metallo-protein and the importance of superoxide release by inflammatory cells were recognized. The beneficial effects observed in joint disease of horses may have determined the choice of rheumatid arthritis and osteoarthritis as the first human diseases to be treated with this antioxidant enzyme. The first pilot study with bovine Cu–Zn SOD was published by Lund-Olesen and Menander (1974), and the promising results were confirmed in the following by several placebo-controlled double-blind clinical trials in patients with osteoarthritis (Flohé, 1988). However, improvement was not observed for all clinical and laboratory parameters chosen for the study. Intraarticular injection appeared to be more efficient than systemic treatment. Many other diseases were treated with SOD, including ischemia-reperfusion injury, organ transplantation, side effects of radiation and chemotherapy. For a large number of pathologies, however, the reports remained anecdotical. The general acceptance of SOD as a drug was retarded for various reasons, among which the results of pharmacodynamic studies were probably the most important. Because of the rapid clearance of the enzyme through the kidney, the clinical observations of therapeutic effects were doubted. Efforts were made to increase its maintainance in the circulation by binding the enzyme to various macromolecules, in particular polyethylene glycol. There was also considerable discussion with respect to the dosage. The first clinical trials in rheumatic disease were based on the doses found to be

Free Radicals, Oxidative Stress, and Antioxidants, edited by Özben.
Plenum Press, New York, 1998.

efficient in animal models of inflammation and varied between 2–16 mg daily. These models of inflammation had shown a bell-shaped dose response curve, higher doses being less effective than lower doses (Baret *et al.*, 1984; Michelson *et al.*, 1986; Vaille *et al.*, 1990). Similar observations were reported later for SOD application after reperfusion of ischemic organs, where very high doses had been applied by bolus injection (Omar and McCord, 1990). The concept that SOD acts therapeutically according to its documented catalytic function on superoxide anion radicals released into the extracellular space was also doubtful, since there was no correlation between antiinflammatory activity and the level of circulating exogenous SOD (Baret *et al.*, 1984). It was suggested that the anti-inflammatory action of exogenous SOD is due to attachment of the enzyme to the cell membrane (Michelson *et al.*, 1986). This notion was supported by observations from our laboratory, demonstrating prevention of perinuclear halo formation in UVA-exposed fibroblast cultures by pretreatment with exogenous SOD, even after rinsing of the cells and resuspension in fresh, SOD-free medium before irradiation (Emerit *et al.*, 1981). Recent observations of anticlastogenic effects in SOD-pretreated and washed lymphocyte cultures yielded similar results, which stimulated the investigations reported here.

Anticlastogenic (i.e., chromosomal breakage preventing) properties of SOD were reported soon after the discovery of the enzyme. Swedish authors observed that the enzyme, alone or in combination with catalase, decreased the frequency of radiation-induced or spontaneously occuring chromosome breaks in human lymphocyte cultures (Nordenson *et al.*, 1976, 1977). O_2^- was supposed to collaborate with H_2O_2 in the production of OH radicals ultimately responsbile for DNA strand breakage. The role of iron in Fenton-type reactions was proposed also for hydroxyl-radical mediated DNA damage by interaction of peroxides with iron-binding sites on DNA (Mello-Filho and Meneghini, 1984). However, studies done on cellular systems showed that these genotoxic effects were preventable by iron chelators and the hydroxyl radical scavenger dimethysulfoxide, but not by SOD (Ochi and Cerutti, 1987; Blakely *et al.*, 1990).

In contrast to this, our laboratory drew attention to indirect action mechanisms, in which O_2^- played a primary role, since the damage was regularly prevented by SOD alone, while catalase was not or irregularly protective (Emerit, 1994; Emerit *et al.*, 1996). Since inactivated SOD was not protective, the anticlastogenic effect should be related to the catalytic function of the enzyme. Because it did not seem likely that extracellularly produced O_2^- would reach the nucleus without being scavenged by the intracellular SOD abundantly available in the cytosol, we proposed the formation of secondary clastogenic substances as an explanation. Indeed, socalled clastogenic factors (CF) could be isolated from the plasma or tissue culture supernatants of patients with diseases accompanied by oxidative stress, and their damaging effects in cells from healthy persons were regularly prevented by exogenous SOD. Examples of superoxide-mediated clastogenesis and of diseases accompanied by CF formation are described in another chapter of this volume. The biochemical analysis of CF preparations identified various chromosome damaging substances including lipid peroxidation products, inosine triphosphate and cytokines such as tumor necrosis factor alpha. Hydrogen peroxide was not detectable. With the cytochrome c assay, it could be shown that certain clastogenic components have also superoxide stimulating properties. Given that superoxide generation leads to the formation of clastogenic substances generating themselves superoxide, the system is selfsustaining and therefore responsible for long-lasting genotoxic processes. CF formation and action mechanisms of CF action are discussed in detail in a recent review (Emerit, 1994). In this model of an auto-sustaining O_2^- generating process, SOD was supposed to protect by interrupting the vicious circle of CF formation by superoxide and superoxide generation by CF.

During our work on superoxide-mediated clastogenesis, we observed that anticlasto-genic effects of SOD in cells exposed to radiation, certain chemical clastogens or to CF-containing ultrafiltrates persisted despite careful removal of all extracellular SOD. In addition, we noted increases in clastogenic effects as a function of the number of mono-cytes in the culture system. These findings were intriguing and stimulated further investi-gation. In order to examine whether the SOD had bound to the cells during the pretreatment, we used flow cytometry and confocal laser microscopy. Herefore, the en-zyme (native Cu–Zn SOD from bovine erythrocytes, Palleau Production, Chateau Landon France) was labeled with fluorescein isothiocyanate (FITC, Sigma). All cell types present in regular blood and fibroblast cultures were studied. To avoid nonspecific binding, the cells were pretreated with 5% fetal calf serum during 20 min befor exposure to FITC–SOD. After incubation times of variable duration with SOD concentrations varying between 150 and 1500 units/ml, the cells were washed three times in PBS to eliminate all residual free FITC–SOD. Flow cytometry was performed on a FACScan (Becton Dickin-son). FITC fluorescence was detected at 520 nm. Data were collected and anlyzed by Lysis II software (Beckton Dickinson). For confocal laser microscopy, the cells were fixed with 4% formol for 30 min. Nuclei were stained with chromomycin A3 (Sigma) for 30 min in the dark. Confocal fluorescent images were obtained with a MRC-600 confocal scanning laser microscope (Bio-Rad). Excitation at 488 nm and emission at 515 nm were the wavelengths for detection of FITC–SOD, while a combination of two filters (excita-tion at 458 nm, emission at 550 nm) was used for imaging of the chromomycin A3-stained DNA of cell nuclei. For more technical details see Emerit et al. (1996).

As indicated by flow cytometry, monocytes were regularly more fluorescent than lymphocytes and neutrophils. While significant labeling was observed on monocytes with a dose of 150 units/ml, a concentration of 1500 units/ml was necessary to yield significant labeling of lymphocytes and neutrophils for an identical incubation time with FITC–SOD of 1 h (Fig. 1). The fluorescence of monocytes increased with incubation time (Fig. 2). Labeling of lymphocytes was not improved after stimulation with phytohemagglutinin during 1, 4 and 24 h prior to incubation with FITC–SOD. The results obtained with PMN were comparable to those with lymphocytes, while erythrocytes were not labeled at all. Fibroblasts showed significant FITC fluorescence with the lowest dose of 150 units/ml and an incubation time of 2 h.

The data obtained by flow cytometry were confirmed by fluorescence microscopy. In a first approach, monocytes separated from lymphocytes by adherence to plastics were incu-bated with 150 units/ml of FITC–SOD during increasing exposure times. As expected on the basis of the findings with flow cytometry, significant labeling was seen under the fluo-rescence microscope after 15 min incubation. While fluorescence was limited to the cell membrane after this short incubation time, FITC–SOD could be seen in the cytoplasma after 1–2 h of incubation. If the SOD concentration was increased to 1500 units/ml, labeling of the membrane was detectable after 5 min of incubation. At the same SOD concentration and an incubation time of 3 h, practically all monocytes were fluorescent at the membrane and in the cytoplasma. In some cells, there was a circular zone of intense labeling around the nucleus. However, upon focusing through the cell with confocal laser microscopy, no label-ing was seen in the nucleus itself. With the same incubation time of 3 h and the highest con-centraton of 1500 units/ml, which had yielded some degree of labeling in the study using flow cytometry, only rare spots of FITC fluorescence could be seen on PMN. The labelling pattern was quite different for the two cell types. While monocytes showed a punctuate dis-tribution pattern, as it had also been described by Beckman et al. (1988) for FITC–SOD treated endothelial cells, the fluorescence was concentrated in PMN in some rare spots of

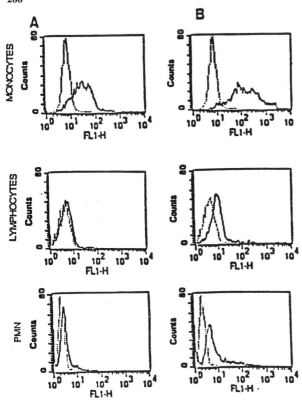

Figure 1. FITC–SOD labelling is dependent on the cell type and the dose. A: 150 units/ml, B: 1500 units/ml.

intense fluorescence situated near the cellular membrane, probably corresponding to phagosomes. While lymphocytes appeared to have the same degree of labeling as PMN in the flow cytometrc study. no FITC could be detected on lymphocytes under the fluorescence microscope. Fibroblasts were not yet studied with this technique.

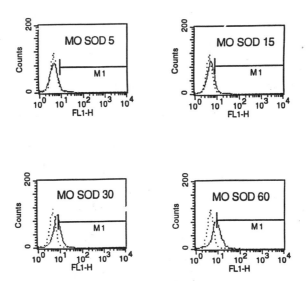

Figure 2. Flow cytometry of monocytes. Time dependent fixation of SOD.

Table 1. Superoxide production by monocytes
and neutrophils after pretreatment with SOD

	Monocytes	Neutrophils
SOD	14 nmol (100%)	58 nmol (100%)
Pretr. 2 hours	4 nmol (–72%)	43 nmol (–27%)
Pretr. 10 min	9 nmol (–38%)	48 nmol (–18%)
Pretr. 5 min	9 nmol (–38%)	54 nmol (–7%)

We wondered whether the different behavior of monocytes and PMN for binding of FITC–SOD was expressed in their capacity to produce O_2^-, when stimulated with TPA after pretreatment with SOD (Table 1). In the first experiments (n = 7), the cells were pretreated with SOD at concentrations between 150 and 1500 units/ml during 2 h, followed by 3 washes. In the cytochrome c assay, untreated monocytes ($0.3–0.5 \times 10^5$ cells) produced 29.03 ± 8.7 nmol O_2^- during 1 h, while SOD-pretreated cells produced only 20.6 and 9.8 nmol respectively for SOD concentrations of 300 and 1500 units/ml. This represented a 30% and 60% reduction of the TPA-stimulated superoxide production. PMN, on the contrary, showed only a reduction from 52.6 nmol to 49.5 nmol for a SOD concentration of 1500 units/ml (difference not statistically significant) (Emerit *et al.*, 1996). There was a good correlation between high SOD binding capacity and reduced superoxide production by these cells. The superoxide production of PMN had been reduced to control values in parallel experiments, where extracellular SOD had been present during the incubation with the stimulant at a concentration of 150 units/ml. Since our experiments with flow cytometry and fluorescence microscopy had indicated that a preincubation of 15 min was sufficient for binding of SOD to the monocytes, we repeated the cytochrome c assay with incubation times shorter than 2 h. Again the superoxide production by pretreated monocytes was reduced. In parallel measurements, this reduction represented 70% for a pretreatment during 2 h, as in our previous experiments, and 38% for pretreatment times of 10 and 5 min. The same experiments done with PMN showed reductions in superoxide production by 18% and 7% only.

One may ask, whether intracellular uptake of SOD is necessary for the observed reduction of superoxide production by monocytes, or whether protective effects are exerted by membrane-bound SOD. Membrane binding of native SOD was rapid, while passage of the enzyme through the membrane with penetration throughout the cytoplasma until the nucleus needed hours. Reduction of superoxide production after pretreatment during 5 min, when no intracellular fluorescence was detectable, suggests that protective effects can be exerted by the membrane-bound enzyme. However, the enzyme may have progressed from the membrane to the cytoplasma during the cytochrome c assay. Further studies with cytochrome reduction times shorter than those used for the present study (1 h) are necessary.

It has been suggested (Kyle *et al.*, 1988; Dini and Rotilio 1989), that cellular uptake of the enzyme occurs via endocytosis, since it was inhibited by various antagonists of endocytosis. In our experiments, cytochalasin B diminished the fluorescence labeling, but did not completely suppress it. Specificity of binding sites for SOD was suggested by the fact that pretreatment of cells with bovine albumin (Dini and Rotilio, 1989) or fetal calf serum (our experiments) did not inhibit binding of SOD. Also FITC-labeled goat anti-mouse immunoglobulins did not bind to human monocytes. The importance of membrane integrity was documented by diminished labeling of fibroblasts, if the cells were detached with trypsin. A cell scraper was therefore used in our study. Trypsinization may be responsible for poor uptake of native SOD reported by others (Michelson and Puget, 1980). Also the fact that we exposed single cell preparations may have contributed to the better fluorescence labeling compared to monolayers.

At least for chromosomal damage mediated via CF formation, the anticlastogenic effects of exogenous SOD appear to be secondary to diminished superoxide production by monocytes and diminished superoxide-mediated production of clastogenic substances. The same is probably true in vivo for CF formation in chronic inflammatory diseases, where activated monocytes play an important role. Small amounts of superoxide elicit monocytes/macrophages to release tumor necrosis factor, which in turn stimulates these cells to produce more superoxide. This cytokine is one of the clastogenic components of CF. In addition, immunologically activated monocytes release free arachidonic acid and various clastogenic metabolites of the AA cascade. It is therefore not astonishing to observe increased CF formation in lymphocyte cultures from polyarthritis patients, when more and more monocytes were added. Addition of polyarthritis monocytes caused also clastogenic effects in normal cells, which were prevented by SOD.

Benefit of SOD treatment in chronic inflammatory diseases can be explained by diminished superoxide production by monocytes. The present data show that binding of exogenous SOD to membranes is very rapid and that beneficial effects such as diminished superoxide production by activated monocytes are already evident after 5 min of contact with the native enzyme. Once fixed on the cells, the enzyme could not be removed even by intensive washing. From these observations, it may be concluded that, under clinical conditions, part of the enzyme is bound to cellular membranes during the first minutes after injection, and that this quantity is sufficient to explain clinical observations of therapeutic effects.

REFERENCES

Baret, A., Jadot, G., Puget, K., 1984, Pharmacocinetic and antiinflammatory properties in the rat of superoxide dismutases from various species. *Biochem. Pharmacol.* **33**:2755–2760.

Beckman, J.S., Minor, R.L., White, C.W., Repine, J.E., Rosen, G.M., Freeman, B.A., 1988, Superoxide dismutase and catalase conjugated to polyethylene glycol increases endothelial enzyme activity and oxidant resistance. *J. Biol. Chem.* **263**:6884–6892.

Blakeley, W.F., Fuciarelli, A.F., Wegher, B.J., Dizdaroglu, M., 1990, Hydrogen peroxide induced base damage in deoxyribonucleic acid. *Radiation Res.* **121**:338–343.

Dini, L., and Rotilio, G., 1989, Electron microscopic evidence for endocytosis of superoxide dismutase by hepatocytes using protein-gold adducts. *Biochem. Biophys. Res. Commun.* **162**:940–944.

Emerit, I., 1994, Reactive oxygen species, chromosome mutation and cancer: Possible role of clastogenic factors in carcinogenesis. *Free Radic. Biol. Med.* **16**:99–109.

Emerit, I., Michelson, A.M., Martin, E., Emerit, J., 1981, Perinuclear halo formation as an indication of phototoxic effects. *Dermatologica* **163**:295–299.

Emerit, I., Garban, F., Vassy, J., Levy, A., Filipe, P., Freitas, J., 1996, Superoxide-mediated clastogenesis and anticlastogenic effects of exogenous superoxide dismutase. *Proc. Natl. Acad. Sci. USA* **93**:12799–12804.

Flohé, L., 1988, Superoxide dismutase for therapeutic use: Clinical experience, dead ends and hopes. *Mol. Cellul. Biochem.* **84**:123–131.

Kyle, M.E., Nakae, D., Sakaida, I., Miccadei, S., Farber, J.L., 1988, Endocytosis of superoxide dismutase is required in order for the enzyme to protect hepatocytes from the cytotoxicity of hydrogen peroxide. *J. Biol. Chem.* **263**:3784–3789.

Lund-Olesen, K., Menander K.B., 1974, Orgoteine, a new antiinflammatory metalloprotein drug: Preliminary evaluation of clinical efficacy and safety in degenerative joint disease. *Curr. Ther. Res.* **16**:706–717.

Mello-Filho, A.C., Meneghini, R., 1984, In vivo formation of single strand breaks in DNA is mediated by the Haber Weiss reaction. *Biochem. Biophys. Acta* **781**:56–63.

Michelson, A.M., Puget, K., 1980, Cell penetration by exogenous superoxide dismutase. *Acta Physiol. Scand. Suppl.* **492**:67–80.

Michelson, A.M., Puget, K, Jadot, G., 1986, Anti-inflammatory activity of superoxide dismutases: Comparison of enzymes from different sources in different models in rats: Mechanism of action. *Free Radic. Res. Commun.* **2**:43–56.

Nordenson, I., 1977., Effect of superoxide dismutase and catalase on spontaneously occuring chromosome breaks in patients with Fanconi's anemia. *Hereditas* **82**:147–149.

Nordenson, I., Beckman, G., Beckman L., 1976, The effect of superoxide dismutase and catalase on radiation-induced chromosome breaks. *Hereditas* **82**:125–128.

Ochi, T., Cerutti, P., 1987, Clastogenic action of hydroperoxy, 5,8,11,13-icosatetraenoic acids in mouse embryo fibroblasts C3H/10 1/2. *Proc. Natl. Acad. Sci. USA* **84**:990–994.

Omar, B.A., McCord, J.M., 1990, The cardioprotective effect of Mn-Superoxide dismutase is lost at high doses in the postischemic isolated rabbit heart. *Free Radic. Biol. Med.* **9**:465–479.

Vaille, A., Jadot, G., Elizagaray, A., 1990, Anti-inflammatory activity of various superoxide dismutases on polyarthritis in the Lewis rat. *Biochem. Pharmacol.* **39**:247–255.

26

UBIQUINOL

An Endogenous Lipid-Soluble Antioxidant in Animal Tissues[*]

Patrik Andrée,[1,2] Gustav Dallner,[1,2] and Lars Ernster[1]

[1]Department of Biochemistry
Arrhenius Laboratories for Natural Sciences
Stockholm University
S-106 91 Stockholm, Sweden
[2]Clinical Research Center
NOVUM
Karolinska Institute
S-141 86 Huddinge, Sweden

1. INTRODUCTION

Ubiquinone (UQ) was first described by Morton and colleagues (Festenstein *et al.*, 1955) as a quinone with a ubiquitous occurrence in various tissues, hence its name (Figure 1). Two years later, Crane *et al.* (1957) identified a quinone that was proposed to be a component of the mitochondrial respiratory chain, functioning as a coenzyme for electron transfer from Complexes I and II to Complex III (Figure 2). As such, it was given the name coenzyme Q. Its structure was determined by Folkers and collegues (Wolf *et al.*, 1958) and found to be identical to ubiquinone.

The biochemical function of ubiquinone as an electron carrier in the mitochondrial respiratory chain was not generally accepted until the late 1960s, when it was demonstrated that depletion of beef heart submitochondrial particles of ubiquinone by pentane extraction caused an inhibition of both the NADH and succinate oxidase activities, and that these could be restored by the reincorporation of the same amount of ubiquinone as was originally present in the particles (Ernster *et al.*, 1969; Norling *et al.*, 1974). A minor fraction of ubiquinone during respiration was found to occur as the semiquinone radical (Bäckström *et al.*, 1970). The finding that there is an appr. 10-fold molar excess of ubiqui-

---

[*] This presentation was based on a chapter by the same authors, to appear in *Reactive Oxygen Species in Biological Systems: Selected Topics*, edited by D.L. Gilbert and C.A Colson, Plenum Publishing Corporation, New York, 1998. (Reproduced with permission of the publishers).

Free Radicals, Oxidative Stress, and Antioxidants, edited by Özben.
Plenum Press, New York, 1998.

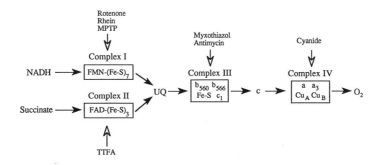

Figure 1. Structure of ubiquinone. The number of isoprene units, n, varies between 6 and 10 in different eucaryotes. It is predominantly 10 in human and 9 in rat.

none in the inner mitochondrial membrane as compared to other respiratory-chain carriers led Kröger and Klingenberg (1970, 1973a,b) to develop the concept of a "pool funtion" of ubiquinone in electron transport. These findings found their explanation through Mitchell's (1975, 1976) proposal of the protonmotive Q cycle (Figure 2) as the mechanism involved in energy conservation at coupling site 2 of the respiratory chain, which is now generally accepted (for review, see Trumpower, 1990).

The earliest observations of an antioxidant function of ubiquinol date back to the 1960s (for reviews, see Beyer and Ernster, 1990; Ernster and Dallner, 1995). In most studies, lipid peroxidation was the parameter to demonstrate an antioxidant effect of ubiquinol, with isolated mitochondria or submitochondrial particles as the test objects. In recent years it was found that the agents used to initiate lipid peroxidation could also damage mitochondrial proteins and DNA, and that ubiquinol may prevent some of these effects. Moreover, it was shown that ubiquinone occurs also in cellular membranes other than mitochondria and in serum low-density lipoprotein (LDL), and that ubiquinol may protect these structures as well from oxidative damage. In general, ubiquinol appears to be the only known lipid-soluble antioxidant that animal cells can synthesize *de novo* and regenerate enzymatically from its oxidized form in the course of its antioxidant function.

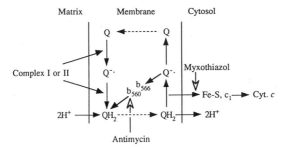

Figure 2. Schematic illustration of the mitochondrial respiratory chain (*top*) and the protonmotive Q cycle (*bottom*). The open arrows indicate the sites of action of some inhibitors.

The present chapter is an overview of the above developments. It also summarizes available information concerning the role of ubiquinol as an antioxidant *in vivo* under various physiological and pathological conditions.

2. PROTECTIVE EFFECT OF UBIQUINOL AGAINST MITOCHONDRIAL LIPID PEROXIDATION, PROTEIN AND DNA OXIDATION

2.1. Lipid Peroxidation: Effects of Ubiquinone and Vitamin E

Lipid peroxidation is one of the earliest recognized and most extensively studied manifestations of oxygen toxicity in biology (for review, see Ernster, 1993). Over the last three decades, peroxidation of polyunsaturated fatty acids induced by reactive oxygen species (ROS)—superoxide radical ($O_2^{\cdot -}$), hydrogen peroxide (H_2O_2), hydroxyl radical (OH$^{\cdot}$) and singlet oxygen ($^1O_2$)—has been studied in great detail as it occurs in various cellular membranes, as well as in serum lipoproteins. In these studies, ROS were generated either non-enzymatically or, in the case of mitochondria and the endoplasmic reticulum, through enzymes present in the membranes of these organelles.

The peroxidation process can be divided into three separate phases: initiation, propagation, and termination.

Initiation takes place through a transition metal-induced (or radiation-induced) abstraction of a hydrogen atom from a methylene group of a fatty acid containing two or more separated double-bonds, giving rise to a carbon-centered alkyl radical (L$^{\cdot}$), with a simultaneous rearrengement of the double-bonds to become conjugated ("diene conjugation"). The L$^{\cdot}$ formed reacts with O_2 at a diffusion-controlled rate (Maillard *et al.*, 1983), giving rise to a peroxyl radical (LOO$^{\cdot}$).

Propagation involves the abstraction of a hydrogen atom from an adjacent unsaturated fatty acid by LOO$^{\cdot}$, resulting in the formation of a lipid hydroperoxide (LOOH) and a new L$^{\cdot}$ radical. LOOH can react with Fe^{2+}, producing the alkoxyl radical (LO$^{\cdot}$). This radical, which is more reactive than LOO$^{\cdot}$, can again reinitiate lipid peroxidation by hydrogen abstraction from an adjacent polyunsaturated fatty acid, with the formation of L$^{\cdot}$ and an alcohol (LOH) as the end product.

LOOH can also undergo degradation into hydrocarbons (ethane, pentane), alcohols, ethers, epoxides, and aldehydes. Among the latter, malondialdehyde (MDA) and 4-hydroxynonenal (4-HNE) are of special importance since they can cross-link phospholipids, proteins and DNA (for review, see Esterbauer *et al.*, 1990).

Termination of the lipid peroxidation process is generally believed to take place by interaction between two radicals, to form a non-radical product.

Beginning in 1975, Takeshige and Minakami (1975, 1979) published a series of studies describing various parameters of lipid peroxidation in beef heart submitochondrial particles induced by NAD(P)H and ADP-Fe^{3+}, a system earlier employed to initiate lipid peroxidation in rat liver microsomes (Hochstein and Ernster, 1963; Hochstein *et al.*, 1964). Takeshige and collaborators (Takayanagi *et al.*, 1980; Takeshige *et al.*, 1980) reported that the lipid peroxidation in submitochondrial particles could be inhibited by the addition of succinate or high concentrations of NADH (but not NADPH). This protective effect was abolished upon removal of the bulk of ubiquinone from the particles by lyophilization and pentane extraction, a procedure described by Ernster *et al.* (1969). Quantitative reincorporation of ubiquinone into the particles by the same procedure restored the ability of NADH or succinate to inhibit lipid peroxidation.

These results were confirmed by Glinn *et al.* (1991, 1997). They found, however, that the protective effect of high concentrations of NADH also occurs in the presence of rotenone and is therefore unlikely to be due to ubiquinol. Rhein, another inhibitor of Complex I, abolished the protection by NADH, and even stimulated lipid peroxidation in the pentane-extracted particles, indicating the occurrence of a lipid-soluble antioxidant component between the rhein- and rotenone-sensitive sites of Complex I (Glinn *et al.*, 1997).

The results by Takeshige and Minakami (1975, 1979) were also confirmed in our laboratory (Forsmark *et al.*, 1991; Ernster *et al.*, 1992) using ascorbate and ADP-Fe^{3+} to induce lipid peroxidation, and NADH or succinate in the presence of antimycin to reduce ubiquinone in the particles. It was found that the pentane extraction method used in these experiments also removed vitamin E (α-tocopherol) from the particles, and thus, that the restoration of the inhibition of lipid peroxidation upon the reincorporation of ubiquinone did not require the presence of vitamin E. These findings eliminated a proposal by Kagan *et al.* (1990) that the antioxidant effect of ubiquinol is dependent on vitamin E.

Incorporation of vitamin E into the extracted particles resulted in an inhibition of lipid peroxidation in the absence of ubiquinone, the extent to which was enhanced by increasing ascorbate concentrations (Ernster *et al.*, 1992). Ubiquinone, when incorporated into the extracted particles together with vitamin E, promoted uptake of the latter, probably by increasing membrane fluidity (Katsikas and Quinn, 1983). After enzymic reduction of ubiquinone, it also amplified the effect of vitamin E as observed at limiting ascorbate concentrations, appearently by regenerating the vitamin from the α-tocopheroxyl radical, in accordance with earlier observations (Kagan *et al.*, 1990; Maguire *et al.*, 1989; Mukai *et al.*, 1990; Maguire *et al.*, 1992; Mukai *et al.*, 1992).

In conclusion, these results demonstrated that ubiquinol can act as an antioxidant both directly, without mediation by vitamin E, and by promoting the antioxidant effect of the latter (Figure 3). Ubiquinol may exert its direct antioxidant effect by preventing initiation, propagation, or both, depending on the prevailing conditions.

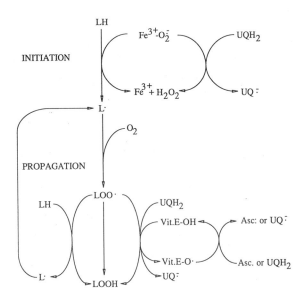

Figure 3. Possible sites of action of ubiquinone, vitamin E and ascorbate on lipid peroxidation initiated by a perferryl radical. From Ernster *et al.*, 1992. LH, polyunsaturated fatty acid; $Fe^{3+}-O_2^{\cdot-}$, perferryl radical; UQH_2, ubiquinol; $UQ^{\cdot-}$, ubisemiquinone; $L^{\cdot}$, carbon-centered radical; $LOO^{\cdot}$, lipid peroxyl radical; Vit.E-OH, vitamin E; Vit.E-O$^{\cdot}$, α-tocopheroxyl radical; Asc, ascorbate; Asc$^{\cdot}$, ascorbyl radical.

2.2. Protein Oxidation and Its Prevention by Ubiquinol

Protein oxidation induced by ROS implies the oxidation of certain amino acid residues of a particular protein. It can take place through different mechanisms, depending on the way of generation of the ROS and the local environment of the protein. Stadtman (1993) has described a "site-specific" initiation based on the involvement of a transition metal, according to the following sequence of reactions: (i) a transition metal, usually iron or copper, is bound to the protein; (ii) the bound metal reacts with H_2O_2 to yield a hydroxyl radical, which oxidizes the amino acid residues at the metal-binding site.

Several authors have described this kind of inactivation of enzymes *in vitro*, and some of the modified amino acid residues have been identified. Davies and associates (Davies, 1987; Davies and Delsignore, 1987; Davies *et al.*, 1987a,b) have studied protein oxidation caused by pulse radiolysis. In this case, no bound transition metal is required. This type of oxidation has been demonstrated with both soluble and membrane-bound proteins (Davies, 1987; Davies and Delsignore, 1987; Davies *et al.*, 1987a,b; Zhang *et al.*, 1990; Grant *et al.*, 1993). In addition, it has been proposed that oxidation of membrane proteins can be mediated by lipid-derived radicals formed during lipid peroxidation (for review, see Wolff *et al.*, 1986).

Oxidative modification of proteins can also occur in a secondary manner. Aldehydes formed as degradation products of lipid peroxides, notably malondialdehyde (MDA) and 4-hydroxynonenal (4-HNE), can bind covalently to amino acid residues of proteins by Schiff's-base or thioether linkage and cause intra- and intermolecular cross-linking (Esterbauer *et al.*, 1991).

Beef heart submitochondrial particles incubated with ascorbate and ADP-Fe^{3+} were found to give rise to protein carbonyl formation parallel to lipid peroxidation (Forsmark-Andrée *et al.*, 1995), the latter measured as formation of thiobarbituric acid reactive substances (TBARS). Both processes could be prevented by the addition of succinate and antimycin, which caused a reduction of endogenous ubiquinol. This effect was abolished by the extraction, and restored by the reincorporation, of ubiquinone (Table 1).

Attempts were made to estimate the extent to which the protein carbonyls originated from direct oxidation of amino acid residues and from secondary modification by way of Schiff's base or thioether formation with MDA or 4-HNE. From results reported by Fors-

Table 1. Formation of TBARS and protein carbonyls (nmol/mg protein) in native, UQ-depleted and UQ-replenished beef heart submitochondrial particles after 30 minutes incubation with ADP-Fe^{3+} and ascorbate, in the absence or presence of succinate and antimycin

SMP	Additions	TBARS (nmol/mg protein)	Carbonyls (nmol/mg protein)
Native			
	None	0.72	1.53
	ADP-Fe^{3+}, asc.	12.15	13.72
	ADP-Fe^{3+}, asc., succ. + antim.	0.63	1.32
UQ-depleted			
	ADP-Fe^{3+}, asc.	13.71	14.53
	ADP-Fe^{3+}, asc., succ. + antim.	12.22	13.76
UQ-replenished			
	ADP-Fe^{3+}, asc.	12.61	13.87
	ADP-Fe^{3+}, asc., succ. + antim.	1.20	2.15

From Forsmark-Andrée *et al.*, 1995.

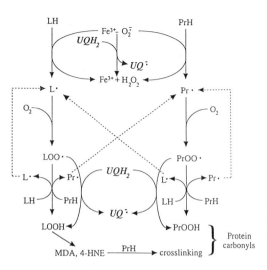

Figure 4. Schematic illustration of the sequence of reactions involved in lipid (LH) and protein (PrH) oxidation as initiated by a perferryl radical, and the conceivable sites of action of ubiquinol (UQH₂) as an antioxidant. From Forsmark-Andrée *et al.*, 1995.

mark-Andrée *et al.* (1995), it appears that the major part of the carbonyl derivatives under the conditions employed were products of direct protein oxidation. However, also the direct oxidation of proteins may be mediated by lipid-derived free radicals (LOO·, LO·), in view of its close kinetic relationship to lipid peroxidation and the protective effect of ubiquinol on both processes. A schematic illustration of the interplay between these two processes is shown in Figure 4.

2.3. Identification of Oxidatively Modified Proteins

SDS-PAGE analysis revealed a complex pattern regarding the effect of ascorbate and ADP-Fe^{3+} on the individual protein components of the particles (Forsmark-Andrée *et al.*, 1995). The most striking change was a shift of a band corresponding to appr. 29 kDa towards a higher molecular weight position. This shift was prevented by succinate and antimycin, *i.e.*, by the reduction of ubiquinone in the particles. The protein corresponding to this band was tentatively identified as the adenine nucleotide translocator (ANT), which is the most abundant protein component of the mitochondrial inner membrane. This protein had been shown to be particularly sensitive to lipid peroxidation (Zwizinski and Schmid, 1992), probably due to its close association with cardiolipin (Beyer and Klingenberg, 1985). This conclusion was subsequently substantiated by immunoblot analysis, by using a monoclonal antibody raised against beef heart ANT (Andrée, 1996).

As revealed by SDS-PAGE analysis, there were also a number of other proteins that were affected in the course of lipid peroxidation and protected by ubiquinol. One component, with a molecular weight of appr. 110 kDa, was subsequently identified by immunoblot analysis as the nicotinamide nucleotide transhydrogenase (NNT) (Forsmark- Andrée *et al.*, 1996).

Mitochondrial proton-translocating NNT has been extensively studied (Hoek and Rydström, 1988; Hatefi and Yamaguchi, 1992) ever since the description of an "energy-dependent" NNT by Danielson and Ernster in 1963 (1963). NNT is the largest protein component and the only known single-subunit proton-translocating enzyme in the mitochondrial inner membrane. Its structure consists of 14 transmembrane α-helices, connecting a 45 kDa N-terminal and a 25 kDa C-terminal domain, both located on the matrix side of the membrane and harbouring the NAD(H) and the NADP(H) binding sites, respectively. The enzyme catalyzes the reaction:

$$NADH_m + NADP^+_m + H^+_c \Leftrightarrow NAD^+_m + NADPH_m + H^+_m$$

where c and m denote the cytosol and the matrix, respectively. It utilizes a proton gradient to shift the equilibrium of the hydride transfer between the two nicotinamide nucleotides, and at the same time to enhance its rate in the forward (left to right) direction. According to a recent proposal by Hatefi and Yamaguchi (Hatefi and Yamaguchi, 1996), the proton gradient acts by changing the binding energy of NADP(H) to the enzyme, in a manner similar to the "binding-change mechanism" proposed by Boyer for ATP synthase (Boyer, 1993).

As recently shown by Fosmark-Andrée *et al.* (1996), treatment of submitochondrial particles with ascorbate and ADP-Fe^{3+} caused an inactivation of NNT which could be prevented by the addition of succinate and antimycin, *i.e.*, by endogenous ubiquinol. In another series of experiments, the effect of peroxynitrite on the NNT reaction was investigated. It was found that treatment of the particles with peroxynitrite caused an inactivation of both the forward and reverse reactions catalyzed by NNT, and that these effects were not prevented by reducing the endogenous ubiquinone (Forsmark-Andrée *et al.*, 1996). Kinetic analysis revealed that treatment with ascorbate and ADP-Fe^{3+} resulted in an increase of the K_m for NADPH, but not for NADH, whereas peroxynitrite treatment gave rise to an increase of both values.

These results suggest that lipid peroxidation is accompanied by an alteration of the binding of NADPH, but not of NADH, to the enzyme, while peroxynitrite affects both substrate-binding sites. It thus appeared that peroxynitrite treatment inhibits the enzyme by affecting both of its extramembraneous domains and thereby its binding affinity for both substrates. In contrast, treatment with ascorbate and ADP-Fe^{3+} may act by perturbing the intramembraneous region of the enzyme through lipid peroxidation, thereby causing a defect in proton translocation and a consequent decrease in NADPH-binding affinity. Support for this conclusion was obtained by analyzing the enzyme by immunoblotting and subsequent trypsin digestion, before and after exposure to different experimental conditions.

2.4. Inactivation of Respiratory Chain and ATP Synthase

Incubation of submitochondrial particles with ascorbate and ADP-Fe^{3+} resulted in an inhibition of the NADH and succinate oxidase activities (Forsmark-Andrée *et al.*, 1997). Investigation of various partial reactions of the respiratory chain revealed that those involving ubiquinone were primarily responsible for the inactivation of the NADH and succinate oxidases. Minor degrees of inhibition were also found with succinate dehydrogenase and cytochrome *c* oxidase, but these could not account for the overall inactivation of the respiratory chain. However, it was found that the content of ubiquinone decreased during lipid peroxidation, and that there was a close correlation between the inactivation of NADH and succinate oxidases and the decrease in ubiquinone content. Reduction of the endogenous ubiquinone, by succinate in the presence of cyanide, prevented both the inactivation of the oxidase activities and the decrease in ubiquinone content (Table 2).

Attempts to reactivate the oxidatively damaged particles by incorporation of ubiquinone, using the pentane procedure (Ernster *et al.*, 1969), were so far unsuccessful. Likewise, little information is available about the breakdown products of ubiquinone, although there are several earlier reports in the literature describing an oxidative breakdown of ubiquinone, and proposing various conceivable pathways (Morimoto *et al.*, 1969; Imada *et al.*, 1970; Morimoto *et al.*, 1970). Preliminary experiments in our laboratory indicate that solanesol and an unidentified polar ring structure may be some of the breakdown products under our experimental conditions.

Table 2. TBARS formation, NADH, and succinate oxidase activities, and ubiquinone content, after 30 minutes of incubation of beef heart submitochondrial particles with or without lipid peroxidation induced by ADP-Fe^{3+} and ascorbate in the absence or presence of succinate and cyanide

Additions	TBARS formation (nmol/mg protein)	NADH oxidase (μatoms O/min/ mg protein)	Succinate oxidase (μatoms O/min/ mg protein)	Ubiquinone content (nmol/mg protein)
None	0.25	1.39	0.73	4.83
ADP-Fe^{3+}, asc.	14.25	0.15	0.21	1.48
ADP-Fe^{3+}, asc., succ. + cyanide	0.52	1.31	0.69	4.37

From Forsmark-Andrée *et al.*, 1997.

In a model system, using ubiquinone incorporated into phospholipid liposomes consisting of either saturated or polyunsaturated phospholipids, it was shown (Forsmark-Andrée *et al.*, 1997) that treatment with ascorbate and ADP-Fe^{3+} resulted in a breakdown of ubiquinone in the latter, but not in the former type of liposomes. These results provide strong evidence that the breakdown of ubiquinone found in the submitochondrial particles (i) is mediated by lipid peroxidation; and (ii) is non-enzymic.

Experiments were also performed in order to assess the effect of lipid peroxidation on the ATP synthase (F_0F_1-ATPase) activity, in both beef heart submitochondrial particles (Andrée, 1996) and whole rat liver mitochondria (Forsmark-Andrée and Ernster, 1994).

In the case of particles, incubation with ascorbate and ADP-Fe^{3+} resulted in an inactivation of the ATPase which was not prevented by reduction of the endogenous ubiquinone. In intact mitochondria, the ATPase activity was also inhibited upon incubation with ascorbate and ADP-Fe^{3+}, but in this case the inhibition was prevented by reduction of the ubiquinone with succinate and antimycin. These findings suggest that in the case of the particles, the radical attack occurs on both the intramembrane proton-translocating F_0 and the protruding catalytic F_1 moieties of the enzyme, the latter probably in a site-specific manner at the ADP-binding sites on the α and/or β subunits, which is inaccessible to ubiquinol. In the case of mitochondria, the oxidative damage is probably restricted to the F_0 moiety of the enzyme, which can be prevented by ubiquinol. In both mitochondria and submitochondrial particles the ATPase activity was oligomycin-sensitive, indicating that the ATP hydrolysis was obligatorily coupled to proton translocation.

2.5. DNA Oxidation

ROS-dependent DNA oxidation is generally believed to proceed in a manner resembling the site-specific oxidation of proteins discussed above. It involves a reaction between H_2O_2 with a transition metal pre-bound to DNA (Halliwell and Auroma, 1991). The damage consists of base oxidation, which can be detected by analysis of the oxidized bases, *e.g.*, 8-hydroxy-deoxyguanosine (8-OH-dG) (Floyd *et al.*, 1986). The measurement of strand-breaks can also be used as an indicator of oxidative DNA damage (Whitaker *et al.*, 1991). In recent years, much attention has been directed to mitochondrial DNA, which is particularly susceptible to oxidative damage (Shigenaga *et al.*, 1994).

Incubation of rat liver mitochondria for 30 minutes in the presence of ascorbate and ADP-Fe^{3+} was found to result in an increased content of 8-OH-dG (Forsmark-Andrée and Ernster, 1994) and a marked induction of DNA strand-breaks (Andrée, 1996) (Figure 5). Both effects were counteracted—the former completely, and the latter partially—by the addition of succinate and antimycin, *i.e.*, by endogenous ubiquinol.

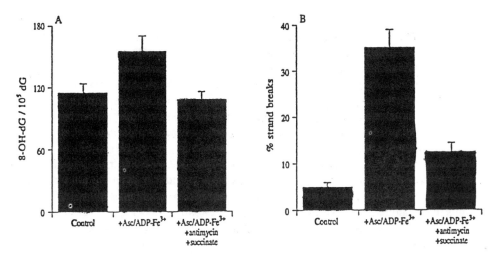

Figure 5. Formation of (A) 8-OH-dG and (B) strand breaks in rat liver mitochondria treated with ascorbate and ADP-Fe^{3+} for 30 min, in the absence or presence of succinate and antimycin. From Forsmark-Andrée and Ernster, 1994, and Andrée, 1996.

2.6. Anti- and Pro-Oxidant Effects of Ubiquinone in Mitochondria

Electron leakage from the respiratory chain giving rise to $O_2^{\cdot-}$ and H_2O_2 has been shown by several investigators, using whole mitochondria, submitochondrial particles or isolated respiratory-chain complexes (for review, see Chance et al., 1979). The site and mechanism of this leakage have recieved much attention, and several components of the respiratory chain have been implicated. Among these, ubiquinone has been proposed to be the major site (Cadenas et al., 1977; Turrens et al., 1985), based on the following lines of evidence: (i) ubisemiquinone is known to occur during respiration (Bäckström et al., 1970; Davies and Hochstein, 1982), and semiquinones in general react rapidly with oxygen, producing $O_2^{\cdot-}$; (ii) NAD(P)H- and/or succinate-supported radical leakage is stimulated 10–15 fold by the presence of antimycin and an uncoupler, indicating that the electron leak occurs on the substrate side of the antimycin-sensitive site of the respiratory chain (Loschen and Flohé, 1971); and (iii) extraction of ubiquinone from mitochondria inhibits, and reincorporation restores, H_2O_2 formation (Boveris and Chance, 1973).

However, the following information argues against the notion that ubisemiquinone is the principal source of electron leakage during mitochondrial respiration: (i) ubisemiquinone formed during mitochondrial respiration is maintained at a relatively high steady-state level, as revealed by EPR measurements (Bäckström et al., 1970; Salerno and Ohnishi, 1980; Salerno et al., 1990), and is bound to specific ubiquinone-binding proteins associated with Complexes I, II and III (Yu and Yu, 1981; King, 1982) which stabilize it against autoxidation; agents that modify these complexes, such as TTFA in the case of Complex II (Salerno and Ohnishi, 1980; Ingledew and Ohnishi, 1977) or antimycin in the case of Complex III (Ohnishi and Trumpower, 1980; Bowyer and Trumpower, 1981), alter this stability of ubisemiquinone (Rich et al., 1990), which may explain the stimulatory effect of TTFA on lipid peroxidation (Glinn et al., 1991; Eto et al., 1992) and of antimycin on H_2O_2 formation (Loschen and Flohé, 1971); (ii) myxothiazol, another inhibitor of Complex III, suppresses H_2O_2 formation (Loschen et al., 1973; von Jagow et al., 1984; Nohl and Jordan, 1986), even though it does not prevent ubiquinone reduction; the difference between the effects of antimycin and myxothiazol on H_2O_2 formation is more likely to be due to their effects on the redox state of cytochrome b_{566}, the latter being reduced in the presence of antimycin but not in the presence of myxothiazol (*cf.* Figure 2); it thus appears likely that it is the reduced cytochrome b_{566}, rather than ubisemiquinone, which is responsible for the electron leak in Complex III; (iii) electron leak during NADH oxidation has been demonstrated to occur in the presence of rotenone (Ramsay and Singer, 1992; Giulivi *et al.*, 1995), rhein (Floridi *et al.*, 1989; Glinn *et al.*, 1991, 1996), and MPTP (Hasegawa *et al.*, 1990; Ramsay and Singer, 1992; Murphy *et al.*, 1995), agents which inhibit Complex I and thus ubiquinone reduction.

In conclusion, the available evidence strongly indicates that electron leakage leading to $O_2^{\cdot-}$ and H_2O_2 formation occurs in Complexes I, II, and III of the respiratory chain, mainly through autoxidation of components other than ubisemiquinone. Like other antioxidants, ubiquinone may obviously also act as a pro-oxidant under special circumstances. However, this does not detract from its importance as a biological antioxidant, especially as this function of ubiquinol is not limited to mitochondria.

3. ANTIOXIDANT FUNCTION OF UBIQUINOL OUTSIDE MITOCHONDRIA

3.1. Intracellular Distribution of Ubiquinone

In view of its role in the respiratory chain, ubiquinone in eukaryotes was initially assumed to be located exclusively in the inner mitochondrial membrane. However, several investigations have revealed that ubiquinone is also present in various other cellular locations (Sastry, 1961; Jayaraman and Ramasarma, 1963; Lang *et al.*, 1986; Kalén *et al.*, 1987), including the Golgi vesicles, the lysosomes, the endoplasmatic reticulum, the peroxisomes, the plasma membrane, and the outer mitochondrial membrane. Table 3 shows the distribution of ubiquinone in various fractions of rat liver homogenate. Notably, this distribution is markedly different from that of α-tocopherol, the ubiquinone/α-tocopherol ratio being highest in the inner and outer mitochondrial membranes and the plasma membrane, and lowest in the microsomes and the Golgi vesicles.

Table 3. Levels of ubiquinones and α-tocopherol
in hepatic subfractions prepared from rats

Fraction	UQ (pmol/mg protein)	α-tocopherol (pmol/mg protein)	UQ/α-tocopherol
Homogenate	1139	165	6.9
Inner mitochondrial membranes	2654	77	34.5
Outer mitochondrial membranes	3800	142	26.8
Rough microsomes	128	102	1.3
Smooth microsomes	178	228	0.8
Golgi vesicles	897	1105	0.8
Lysosomes	1246	421	3.0
Plasma membranes	390	19	20.5
Nuclear fraction	215	95	2.3
Peroxisomes	134	49	2.7
Cytosol	117	47	2.5

From Zhang *et al.*, 1996.

It has also been shown that ubiquinone is discharged to a limited extent across the plasma membrane into the blood, where it is partially bound to LDL (Elmberger *et al.*, 1989). However, in contrast to cholesterol, ubiquinone does not seem to be distributed among different tissues via the circulation.

3.2. Tissue Distribution and Redox State

Analysis of its redox state in various tissues (Åberg *et al.*, 1992) (Figure 6) and in blood (Zhang *et al.*, 1995) has revealed that ubiquinone is present mainly in the reduced form, a fact that may be consistent with its function as an antioxidant. However, the mechanism by which ubiquinone is reduced in locations other than the inner mitochondrial membrane is unclear. Several enzymes have been considered to be involved in ubiquinone reduction, such as DT-diaphorase (Beyer *et al.*, 1996), microsomal NADH-cytochrome b_5 and NADPH-cytochrome P-450 reductases (Cadenas *et al.*, 1992), and NADH dehydrogenases associated with the outer mitochondrial membrane (Sottocasa *et al.*, 1967) and the plasma membrane (Crane *et al.*, 1985). A cytosolic ubiquinone reductase different form DT-diaphorase has recently been described (Takahashi *et al.*, 1995). Reduction of ubiquinone in LDL has been proposed to take place by way of a quinone reductase located in the erythrocyte membrane, utilizing intracellular NADH as the electron donor (Stocker and Suarna, 1993). It has also been proposed that ubiquinone reduction may take place by temporary fusion between different membranes (Takeshige *et al.*, 1980).

In addition to its potential role as an antioxidant, the presence of ubiquinone in membranes other than mitochondria raises the question of possible other functions in these locations. Crane and Morré (1977) have proposed that ubiquinone located in the Golgi apparatus may be involved in vesicle migration associated with secretion. The same group has also shown an involvement of ubiquinone in cell growth via an NADH oxidoreductase-activated signal system (Sun *et al.*, 1992, Crane and Sun, 1993). Furthermore, the neutrophil redox system has been reported to contain ubiquinone (Crawford and Schneider, 1982).

3.3. Ubiquinone Biosynthesis and Its Regulation

The biosynthesis of ubiquinone involves four main processes: (i) synthesis of the benzene ring; (ii) synthesis of the isoprenoid side-chain; (iii) condensation of the side-chain with the ring; and (iv) subsequent modifications of the ring structure, including

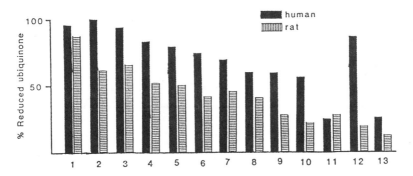

Figure 6. Extent of reduction of ubiquinone from various human and rat tissues. 1. liver; 2. pancreas; 3. intestine; 4. colon; 5. testis; 6. kidney; 7. thyroid; 8. muscle; 9. ventricle; 10. heart; 11. brain; 12. spleen; and 13. lung. From Åberg *et al.*, 1992.

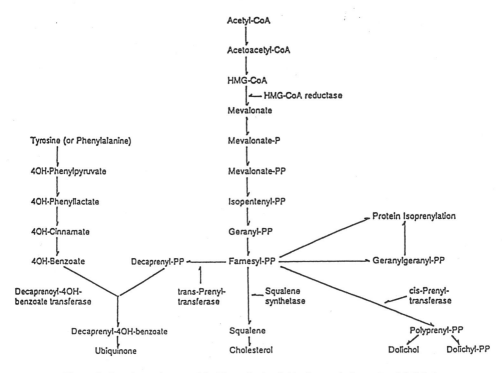

Figure 7. Reaction pathways of the biosynthesis of ubiquinone, cholesterol and dolichol.

hydroxylation, methylation, and decarboxylation steps (for reviews, see Olson and Rudney, 1983; Ericsson and Dallner, 1993; Grünler *et al.*, 1994). The reaction pathways involved in these processes are schematically illustrated in Figure 7.

The benzene ring is synthesized predominantly from tyrosine (in some instances from phenylalanine) and converted through a number of steps into 4-hydroxybenzoate. The polyisoprenoid side-chain is synthesized from acetyl-CoA through a reaction sequence commonly referred to as the mevalonate pathway, leading to the formation of farnesyl-PP. This product constitutes a branching point, serving as a precursor for the side-chain of ubiquinone, decaprenyl-PP, and for two other major products, cholesterol and dolichol. In addition, farnesyl-PP serves as the substrate for protein isoprenylation, either directly or through geranylgeranyl-PP. Decaprenyl-PP undergoes condensation with 4-hydroxybenzoate to form decaprenoyl-4-hydroxybenzoate, which subsequently is converted through several steps to ubiquinone. However, there is still some uncertainity about the order in which these steps take place, and their intracellular localization.

The tissue levels of ubiquinone have been shown to vary according to the metabolic rate. Cold acclimation (Beyer *et al.*, 1962), thyroid hormone treatment (Pedersen *et al.*, 1963; Sterling *et al.*, 1977; Sterling, 1986; Mancini *et al.*, 1989), exercise (Beyer *et al.*, 1984), pancreatectomy (Boveris *et al.*, 1969), peroxisome proliferation (Kalén *et al.*, 1990; Åberg *et al.*, 1994), and myocardial reperfusion (Muscari *et al.*, 1995), all of which affect the overall rate of oxidative metabolism, and thus ROS production, have been shown to alter the tissue ubiquinone content in the same direction as the change in metabolic rate. Significantly, the increase in tissue ubiquinone content following thyroid hormone treatment occurs after the increase in metabolic rate, suggesting that it is an adaptation to, rather than a cause of, the increased oxidative activity (Pedersen *et al.*,

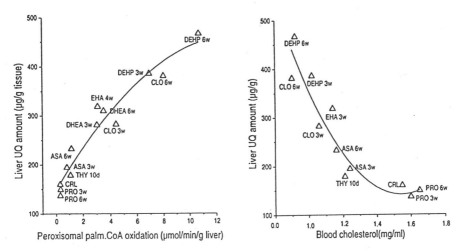

Figure 8. Effects of peroxisome proliferator treatment of rats on ubiquinone content and (A) peroxisomal palmitoyl-CoA oxidation in liver; and (B) blood cholesterol level. From Åberg *et al.*, 1996. CRL, control; PRO, probucol; ASA, acetylsalicylic acid; THY, thyroxine; CLO, clofibrate; DHEA, dehydroepiandrosterone; EHA, 2-ethylhexanoic acid; DEHP, di(2-ethylhexyl)phthalate; w refers to number of weeks of treatment.

1963). In this context, it is of interest to note that no increase in the ubiquinone content of skeletal muscle was found in "Luft's disease" (Luft *et al.*, 1962), a state of severe hypermetabolism of nonthyroid origin, with a defect in mitochondrial respiratory control. In general, there is growing evidence for the existence of a co-ordinate regulation of pro- and antioxidant mechanisms according to the prevailing oxidative stress. In the case of peroxisome proliferator treated rats, there is a positive correlation between the increase in β-oxidation of fatty acids and the ubiquinone content in liver (Figure 8a); and a negative correlation between the plasma cholesterol level and the liver ubiquinone content (Figure 8b) (Åberg *et al.*, 1996).

Ubiquinone levels have been reported to change with increasing age (Beyer *et al.*, 1985; Kalén *et al.*, 1989; Söderberg *et al.*, 1990; Matsura *et al.*, 1991; Edlund *et al.*, 1994), reaching a maximal level at 20 years of age in humans, followed by a decline (Kalén *et al.*, 1989; Söderberg *et al.*, 1990) (Table 4). This decrease in ubiquinone content upon increasing age is consistent with Harman's "free radical theory of aging" (Harman, 1981, 1994), and may be a reflection of the organism's diminished capacity to maintain

Table 4. Age-related changes in the ubiquinone content of human organs (in μg/g wet weight tissue)

Organ	Age group				
	1–3 days	0.7–2 years	19–21 years	39–41 years	77–81 years
Heart	36.7	78.5	110.0	75.0	47.2
Kidney	17.4	53.4	98.0	71.1	64.0
Liver	12.9	45.1	61.2	58.3	50.8
Pancreas	9.2	38.2	21.0	19.3	6.5
Spleen	20.7	30.2	32.8	28.6	13.1
Lung	2.2	6.4	6.0	6.5	3.1
Adrenal	17.5	57.9	16.1	12.2	8.5

From Kalén *et al.*, 1989.

adequate ubiquinol levels in relation to the prevailing need for antioxidant defence. An inverse correlation between longevity and peroxide-producing potential in mammalian tissues has been reported by Cutler (1985).

The regulatory site in ubiquinone biosynthesis has been proposed to be in the mevalonate pathway, *i.e.*, in the biosynthesis of the side-chain (Aiyar and Olson, 1972). Agents that inhibit HMG-CoA reductase have been shown to cause a decrease of the tissue levels of ubiquinone in animals (Willis *et al.*, 1990; Appelkvist *et al.*, 1993) and the plasma levels in humans (Folkers *et al.*, 1990; Elmberger *et al.*, 1991; Watts *et al.*, 1993).

Much interest is currently focused on the effects of various HMG-CoA inhibitors—pravastatin, lovastatin (mevinolin), and related anticholestenemic drugs—on tissue ubiquinone levels *in vivo*. The picture that emerges is that these drugs may reduce ubiquinone levels to various extents depending on the test conditions used (Goldstein and Brown, 1990; Willis *et al.*, 1990; Löw *et al.*, 1992; Appelkvist *et al.*, 1993). An important future aspect of these studies would be to investigate the effects of these drugs under conditions where the organism is exposed to oxidative stress, *e.g.*, by physical training, and thereby to an increased need for ubiquinone biosynthesis, or during aging, when the tissue ubiquinone levels, and thereby the antioxidant capacity, decrease.

An interesting new development in this field is the use of squalestatin 1 as an inhibitor of cholesterol synthesis (Baxter *et al.*, 1992). This fungal product is a potent and specific inhibitor of squalene synthase, *i.e.*, an enzyme that is below the branching point of the mevalonate pathway (see Figure 7). It inhibits cholesterol synthesis selectively, without affecting the synthesis of ubiquinone and dolichol. Moreover, as recently found by Thelin *et al.* (1994), it even enhances ubiquinone synthesis by three- to fourfold (Table 5).

3.4. Biomedical Implications

Tissue ubiquinone levels have been reported to increase in the brain in Alzheimer's disease (Söderberg *et al.*, 1990) and prion disease (Guan *et al.*, 1996), and in preneoplastic nodules (Olsson *et al.*, 1991), and to decrease in hepatocellular cancer (Eggens *et al.*, 1989), cardiomyopathy (Karlsson *et al.*, 1993) and muscle degeneration (Goda *et al.*, 1987; Yamamoto *et al.*, 1987).

Over the last three decades, a large number of studies have been published describing beneficial effects of ubiquinone administration to humans and experimental animals in various diseased states. Among the clinical studies, the most striking effects have been reported on cardiovascular diseases (for review, see Mortensen, 1993) and certain mitochondrial disorders, including mitochondrial encephalomyopathy, lactic acidosis, and strokelike symptoms (MELAS) (Yamamoto *et al.*, 1987), Kearns-Sayre syndrome (KSS)

Table 5. Effect of squalestatin 1 on the relative rates of incorporation of [^{3}H]mevalonate into cholesterol, dolichol and ubiquinone in cultures of Chinese hamster ovary cells

Squalestatin 1 (mM)	% of control		
	Cholesterol	Dolichol	Ubiquinone
0.5	38	118	286
1.0	8	100	387
2.0	1.5	86	345

From Thelin *et al.*, 1994.

(Ogasahara *et al.*, 1986; Bet *et al.*, 1987), and Alper's disease (Fischer *et al.*, 1986). In animal models, an effect of ubiquinone in preventing reperfusion injury of the heart has been demonstrated (Atar *et al.*, 1993; Mortensen, 1993).

Initially, these effects were interpreted in terms of a ubiquinone deficiency and ensuing defect in mitochondrial ATP synthesis in the diseased tissues, which could be repaired by ubiquinone supplementation. An alternative explanation, considered in recent years, is based on the generally recognized antioxidant function of ubiquinol (for review, see Ernster, 1994), and on studies of the uptake of ubiquinone after oral administration (Mohr *et al.*, 1992; Zhang *et al.*, 1995). The latter studies have revealed an increase in ubiquinone level in blood, but not in heart, brain, kidney or skeletal muscle; an increase was found in the liver ubiquinone content, but this was sequestered in the lysosomes. Decreased plasma levels of ubiquinone have been found in patients with mevalonate kinase deficiency, suffering form psychomotor retardation, ataxia, and myopathy (Hübner *et al.*, 1993).

According to reports from two laboratories (Stocker *et al.*, 1991; Yamamoto *et al.*, 1991), ubiquinol-10 protects human LDL from lipid peroxidation more efficiently than does α-tocopherol (Figure 9). Dietary supplementation of humans with ubiquinone-10 has been shown to result in increased levels of ubiquinol-10 within circulating lipoproteins and increased resistance of LDL to the initiation of lipid peroxidation (Mohr *et al.*, 1992; Alleva *et al.*, 1995). Significantly increased LDL-to-ubiquinone ratios are found in patients suffering from ischemic heart disease (Hanaki *et al.*, 1991), which is not altered by treatment with the HMG-CoA reductase inhibitor pravastatin (Hanaki *et al.*, 1993).

On the basis of the available information, it has been proposed that ubiquinone as a dietary supplement may act primarily by elevating the ubiquinone level in blood (Ernster and Dallner, 1995), where it may serve several important functions. Among these are, in addition to an enhanced protection of LDL from oxidation, a prevention of ROS-induced damage caused by neutrophils in inflammatory diseases, and of oxidative injury by endothelial cells resulting from ischemia-reperfusion. These and possibly other protective functions

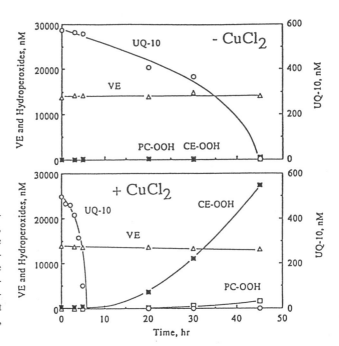

Figure 9. Changes in levels of ubiquinol-10 (UQ-10), α-tocopherol (VE), phosphatidylcholine hydroperoxide (PC-OOH), and cholesteryl ester hydroperoxide (CE-OOH) during the oxidation of human LDL in the absence (*upper panel*) or presence (*lower panel*) of 5 μM cupric chloride at 37°C under air. From Yamamoto *et al.*, 1991.

against ROS-induced damage in the circulation may account for the majority of the reported beneficial effects of ubiquinone administration in experimental and clinical medicine.

4. UBIQUINONE AND REDOX SIGNALLING: FUTURE PERSPECTIVES

The control of gene expression by oxidants, antioxidants, and other factors that influence the intracellular redox state has in recent years become a major field of research in molecular biology (Pahl and Baeuerle, 1994; Burdon, 1995; Sen and Packer, 1996). There is a growing literature describing the mechanisms involved in the adaptation of antioxidant defence to the prevailing oxidative stress, also referred to as "redox signalling." A great deal of information is today available about various transcription factors that are instrumental in this adaptation and are responsible for the regulation of gene expression leading to the synthesis or activation of various antioxidant enzymes. In this context, the antioxidant role of ubiquinone raises the question as to the mechanisms involved in the regulation of its biosynthesis.

As described in the foregoing, there are several lines of evidence indicating that ubiquinone biosynthesis is regulated in response to the prevailing oxidative challenge, and that several diseased states, as well as the physiological process of aging, may be consequences of a defect in this regulation. In view of these lines of evidence, it appears to be of interest to elucidate the mechanisms of this regulation. Does it involve redox signalling and, if so, at what stage?

The findings that mevalonate kinase deficiency leads to a decrease of ubiquinone content in plasma (Hübner *et al.*, 1993), and that inhibition of squalene synthase results in an enhanced ubiquinone biosynthesis (Thelin *et al.*, 1994), suggest that the rate-limiting step is between the former enzyme and the synthesis of farnesyl-PP. Any oxidant-induced stimulation of ubiquinone biosynthesis may thus require an upregulation of some enzyme(s) in this region of the mevalonate pathway. One intriguing possibility is that this regulation is mediated by ubiquinone catabolism. As already discussed, oxidative stress can cause a breakdown of ubiquinone (Forsmark-Andrée *et al.*, 1997). This may have a signalling function for enhanced ubiquinone biosynthesis, similar to the effect of the squalene synthase inhibitor, squalestatin 1 (Thelin *et al.*, 1994). Clearly, we need more information about the biosynthesis and catabolism of ubiquinone before the above questions can be answered.

In conclusion, the exploration of the possible role of ubiquinone in the regulation of antioxidant defence may open new perspectives for future research concerning biological redox signalling.

REFERENCES

Åberg, F., Appelkvist, E.-L., Dallner, G., and Ernster, L. (1992). Distribution and redox state of ubiquinones in rat and human tissues. *Arch. Biochem. Biophys.* **295**:230–234.

Åberg, F., Zhang, Y., Appelkvist, E.-L., and Dallner, G. (1994). Effects of clofibrate, phthalates and probucol on ubiquinone levels. *Chem. Biol. Interact.* **91**:1–14.

Åberg, F., Zhang, Y., Teclebrhan, H., Appelkvist, E.-L., and Dallner, G. (1996). Increases in tissue levels of ubiquinone in association with peroxisome proliferation. Chem. Biol. Interact. **99**:205–218.

Aiyar, A.S. and Olson, R.E. (1972). Enhancement of ubiquinone-9 biosynthesis in rat-liver slices by exogenous mevalonate. *Eur. J. Biochem.* **27**:60–64.

Alleva, R., Tomasetti, M., Battino, M., Curatola, G., Littarru, G.P., and Folkers, K. (1995). The roles of coenzyme Q_{10} and vitamin E on the peroxidation of human low density lipoprotein subfractions. *Proc. Natl. Acad. Sci. USA* **92**:9388–9391.

Andrée, P. (1996). Oxidative stress and Mitochondrial Function. Role of Ubiquinol as Antioxidant. Doctoral Thesis, Karolinska Institute.

Appelkvist, E.-L., Edlund, C., Löw, P., Schedin, S., Kalén, A., and Dallner, G. (1993). Effects of inhibitors of hydroxymethylglutaryl coenzyme A reductase on coenzyme Q and dolichol biosynthesis. *Clin. Investig.* **71**, S97–S102.

Atar, D., Mortensen, S.A., Flachs, H., and Herzog, W.R. (1993). Coenzyme Q10 protects ischemic myocardium in an open-chest swine model. *Clin. Investig.* **71**, S103–S111.

Baxter, A., Fitzgerald, B.J., and Hutson, J.L., McCarthy, A.D., Motteran, J.M., Ross, B.C., Sapra, M., Snowden, M.A., Watson, N.S., Williams, R.J. and Wright, C. (1992). Squalestatin 1, a potent inhibitor of squalene synthase, which lowers serum cholesterol *in vivo. J. Biol. Chem.* **267**:11705–11708.

Bet, L., Bresolin, N., Binda, A., Nador, F., and Ferrante, C. (1987). Cardiac improvement after coenzyme Q10 treatment with Kearns-Sayre syndrome. *Neurology* **37**:202–204.

Beyer, K. and Klingenberg, M. (1985). ADP/ATP carrier protein from beef heart mitochondria has high amounts of tightly bound cardiolipin, as revealed by ^{31}P nuclear magnetic resonance. *Biochemistry* **24**:3821–3826.

Beyer, R.E., Noble, W.M., and Hirschfield, T.J. (1962). Coenzyme Q (ubiquinone) levels of tissues of rats during acclimation to cold. *Can. J. Biochem. Physiol.* **40**:511–518.

Beyer, R.E., Morales-Corral, P.G., Ramp, B.J., Kreitman, K.R., Falzon, M.J., Rhee, S.Y., Kuhn, T.W., Stein, M., Rosenwasser, M.J., and Cartwright, K.J. (1984). Elevation of tissue coenzyme Q (ubiquinone) and cytochrome *c* concentrations by endurance exercise in the rat. *Arch. Biochem. Biophys.* **234**:323–329.

Beyer, R.E., Burnett, B.A., Cartwright, K.J., Edington, D.W., Falzon, M.J., Kreitman, K.R., Kuhn, T.W., Ramp, B.J., Rhee, S.Y., and Rosenwasser, M.J. (1985). Tissue coenzyme Q (ubiquinone) and protein concentrations over the life span of the laboratory rat. *Mech. Ageing Dev.* **32**:267–281.

Beyer, R.E. and Ernster, L. (1990). The antioxidant role of coenzyme Q. In: Highlights in Ubiquinone Research. G. Lenaz, O. Barnabei, A. Rabbi, and M. Battino, eds. (London: Taylor & Francis), pp. 191–213.

Beyer, R.E., Segura Aguilar, J., DiBernardo, S., Cavazzoni, M., Fato, R., Fiorentini, D., Galli, M.C., Setti, M., Landi, L., and Lenaz, G. (1996). The role of DT-diaphorase in the maintenance of the reduced antioxidant form of coenzyme Q in membrane systems. *Proc. Natl. Acad. Sci. USA* **93**:2528–2532.

Boveris, A., Ramos, M.C.P., Stoppani, A.O.M., and Foglia, V.G. (1969). Phosphorylation, oxidation and ubiquinone content in diabetic mitochondria. *Proc. Soc. Exp. Biol. Med.* **132**:170–174.

Boveris, A. and Chance, B. (1973). The mitochondrial generation of hydrogen peroxide. General properties and effect of hyperbaric oxygen. *Biochem. J.* **134**:707–716.

Bowyer, J.R. and Trumpower, B.L. (1981). Rapid reduction of cytochrome c_1 in the presence of antimycin and its implication for the mechanism of electron transfer in the cytochrome bc_1 segment of the mitochondrial respiratory chain. *J. Biol. Chem.* **256**:2245–2251.

Boyer, P.D. (1993). The binding change mechanism for ATP synthase—some probabilities and possibilities. *Biochim. Biophys. Acta* **1140**:215–250.

Burdon, R.H. (1995). Superoxide and hydrogen peroxide in relation to mammalian cell proliferation. *Free Radic. Biol. Med.* **4**:775–794.

Bäckström, D., Norling, B., Ehrenberg, A., and Ernster, L. (1970). Electron spin resonance measurement on ubiquinone-depleted and ubiquinone-replenished submitochondrial particles. *Biochim. Biophys. Acta* **197**:108–111.

Cadenas, E., Boveris, A., Ragan, C.I., and Stoppani, A.O. (1977). Production of superoxide radicals and hydrogen peroxide by NADH-ubiquinone reductase and ubiquinol-cytochrome *c* reductase from beef-heart mitochondria. *Arch. Biochem. Biophys.* **180**:248–257.

Cadenas, E., Hochstein, P., and Ernster, L. (1992). Pro- and antioxidant functions of quinones and quinone reductases in mammalian cells. *Adv. Enzymol.* **65**:97–146.

Chance, B., Sies, H., and Boveris, A. (1979). Hydroperoxide metabolism in mammalian organs. *Physiol. Rev.* **59**:527–605.

Crane, F.L., Hatefi, Y., Lester, R.L., and Widmer, C. (1957). Isolation of a quinone from beef heart mitochondria. *Biochim. Biophys. Acta* **25**:220–221.

Crane, F.L. and Morré, D.J. (1977). Evidence for coenzyme Q function in Golgi membranes. In Biomedical and Clinical Aspects of Coenzyme Q, Vol 1. K. Folkers and Y. Yamamura, eds. (Amsterdam: Elsevier), pp. 3–14.

Crane, F.L., Sun, I.L., Clark, M.G., Grebing, C., and Löw, H. (1985). Transplasma-membrane redox systems in growth and development. *Biochim. Biophys. Acta* **811**:233–264.

Crane, F.L. and Sun, S.E.E. (1993). The essential functions of coenzyme Q. *Clin. Investig.* **71**, S55–S59.

Crawford, D.R. and Schneider, D.L. (1982). Identification of ubiquinone-50 in human neutrophils and its role in microbicidal events. *J. Biol. Chem.* **257**:6662–6668.

Cutler, R.G. (1985). Peroxide-producing potential of tissues: Inverse correlation with longevity of mammalian species. *Proc. Natl. Acad. Sci. USA* **82**:4798–4802.

Danielson, L. and Ernster, L. (1963). Demonstration of a mitochondrial energy-dependent pyridine nucleotide transhydrogenase reaction. *Biochem. Biophys. Res. Commun.* **10**:91–96.

Davies, K.J.A. and Hochstein, P. (1982). Ubisemiquinone radicals in liver: Implications for a mitochondrial Q cycle *in vivo*. *Biochem. Biophys. Res. Commun.* **107**:1292–1299.

Davies, K.J.A. (1987). Protein damage and degradation by oxygen radicals. I. General aspects. *J. Biol. Chem.* **262**:9895–9901.

Davies, K.J.A. and Delsignore, M.E. (1987). Protein damage and degradation by oxygen radicals. III. Modification of secondary and tertiary structure. *J. Biol. Chem.* **262**:9908–9913.

Davies, K.J.A., Delsignore, M.E., and Lin, S.W. (1987a). Protein damage and degradation by oxygen radicals. II. Modification of amino acids. *J. Biol. Chem.* **262**:9902–9907.

Davies, K.J.A., Lin, S.W., and Pacifici, R.E. (1987b). Protein damage and degradation by oxygen radicals. IV. Degradation of denatured protein. *J. Biol. Chem.* **262**:9914–9920.

Edlund, C., Söderberg, M., and Kristensson, K. (1994). Isoprenoids in aging and neurodegeneration. *Neurochem. Int.* **25**:35–38.

Eggens, I., Elmberger, P.G., and Löw, P. (1989). Polyisoprenoid, cholesterol and ubiquinone levels in human hepatocellular carcinomas. *Br. J. Exp. Pathol.* **70**:83–92.

Elmberger, P.G., Kalén, A., Brunk, U.T., and Dallner, G. (1989). Discharge of newly-synthesized dolichol and ubiquinone with lipoproteins to rat liver perfusate and to the bile. *Lipids* **24**:919–930.

Elmberger, P.G., Kalén, A., Lund, E., Reihnér, E., Eriksson, M., Berglund, L., Angelin, B., and Dallner, G. (1991). Effects of pravastatin and cholestyramine on products of the mevalonate pathway in familial hypercholesterolemia. *J. Lipid Res.* **32**:935–940.

Ericsson, J. and Dallner, G. (1993). Distribution, biosynthesis, and function of mevalonate pathway lipids. In: *Subcellular Biochemistry, Volume 21: Endoplasmic Reticulum*. N. Borgese and J.R. Harris, eds. (New York: Plenum Press), pp. 229–272.

Ernster, L., Lee, I.Y., Norling, B., and Persson, B. (1969). Studies with ubiquinone-depleted submitochondrial particles. Essentiality of ubiquinone for the interaction of succinate dehydrogenase, NADH dehydrogenase, and cytochrome *b*. *Eur. J. Biochem.* **9**:299–310.

Ernster, L., Forsmark, P., and Nordenbrand, K. (1992). The mode of action of lipid-soluble antioxidants in biological membranes: Relationship between the effects of ubiquinol and vitamin E as inhibitors of lipid peroxidation in submitochondrial particles. *BioFactors* **3**:241–248.

Ernster, L. (1993). Lipid peroxidation in biological membranes: Mechanisms and implications. In *Active Oxygens, Lipid Peroxides, and Antioxidants*. K. Yagi, ed. (Tokyo: Japan Sci.Soc.Press, and Boca Raton: CRC Press), pp. 1–38.

Ernster, L. (1994). Ubiquinol as a biological antioxidant: A review. In *Oxidative Processes and Antioxidants*. R. Paoletti, Samuelsson, B., Catapano, A.L., Poli, A. and Rinetti, M., eds. (New York: Raven Press), pp. 185–198.

Ernster, L. and Dallner, G. (1995). Biochemical, physiological and medical aspects of ubiquinone function. *Biochim. Biophys. Acta* **1271**:195–204.

Esterbauer, H., Zollner, H., and Schaur, R.J. (1990). Aldehydes formed by lipid peroxidation: Mechanisms of formation, occurrence and determination. In *Membrane Lipid Oxidation*, Vol. I. C.D. Vigo-Pelfrey, ed. (Boca Raton, FL:CRC Press), pp. 239–283.

Esterbauer, H., Schaur, R.F., and Zollner, H. (1991). Chemistry and biochemistry of 4-hydroxynonenal, malonaldehyde and related aldehydes. *Free Radic. Biol. Med.* **11**:81–128.

Eto, Y., Kang, D., Hasegawa, E., Takeshige, K., and Minakami, S. (1992). Succinate-dependent lipid peroxidation and its prevention by reduced ubiquinone in beef heart submitochondrial particles. *Arch. Biochem. Biophys.* **295**:101–106.

Festenstein, G.N., Heaton, F.W., Lowe, J.S., and Morton, R.A. (1955). A constituent of the unsaponifiable portion of animal tissue lipids. *Biochem. J.* **59**:558–566.

Fischer, J.C., Ruitenbeck, W., Gabreels, F.J.M., Janssen, A.J.M., Renier, W.O., Sengers, R.C.A., Stadhouders, A.M., ter Lak, H.J., Trijbels, J.M.F., and Veerkamp, J.H. (1986). A mitochondrial encephalomyopathy: The first case with an established defect at the level of coenzyme Q10. *Eur. J. Pediatr.* **144**:441–447.

Floridi, A., Castiglione, S., and Bianchi, C. (1989). Sites of inhibition of mitochondrial electron transport by rhein. *Biochem. Pharmacol.* **38**:743–751.

Floyd, R.A., Watson, J.J., Wong, P.K., Altmiller, D.H., and Rickard, R.C. (1986). Hydroxyl free radical adduct of deoxyguanosine: Sensitive detection and mechanisms of formation. *Free Rad. Res. Comms.* **1**:163–172.

Folkers, K., Langsjoen, P., Willis, R., Richardson, P., Xia, L.J., Ye, C.Q., and Tamagawa, H. (1990). Lovastatin decreases coenzyme Q levels in humans. *Proc. Natl. Acad. Sci. USA* **87**:8931–8934.

Forsmark, P., Åberg, F., Norling, B., Nordenbrand, K., Dallner, G. and Ernster, L. (1991). Inhibition of lipid peroxidation by ubiquinol in submitochondrial particles in the absence of vitamin E. *FEBS Lett.* **285**:39–45.

Forsmark-Andrée, P. and Ernster, L. (1994). Evidence for a protective effect of endogenous ubiquinol against oxidative damage to mitochondrial protein and DNA during lipid peroxidation. *Mol. Aspects Med.* **15**, S73–S81.

Forsmark-Andrée, P., Dallner, G., and Ernster, L. (1995). Endogenous ubiquinol prevents protein modification accompanying lipid peroxidation in beef heart submitochondrial particles. *Free Radic. Biol. Med.* **19**:749–757.

Forsmark-Andrée, P., Persson, B., Radi, R., Dallner, G., and Ernster, L. (1996). Oxidative modification of mitochondrial nicotinamide nucleotide transhydrogenase in submitochondrial particles. Effect of endogenous ubiquinol. *Arch. Biochem. Biophys.* **336**:113–120.

Forsmark-Andrée, P., Lee, C.P., Dallner, G., and Ernster, L. (1997). Lipid peroxidation and changes in the ubiquinone content ant the respiratory chain enzymes of submitochondrial particles. *Free Radic. Biol. Med.* **22**:391–400.

Giulivi, C., Boveris, A., and Cadenas, E. (1995). Hydroxyl radical generation during mitochondrial electron transfer and the formation of 8-hydroxydeoxyguanosine in mitochondrial DNA. *Arch. Biochem. Biophys.* **316**:909–916.

Glinn, M., Ernster, L., and Lee, C.P. (1991). Initiation of lipid peroxidation in submitochondrial particles: Effects of respiratory inhibitors. *Arch. Biochem. Biophys.* **290**:57–65.

Glinn, M., Lee, C.P., and Ernster, L. (1997). Pro- and anti-oxidant activities of the mitochondrial respiratory chain: Factors influencing NAD(P)H-induced lipid peroxidation. *Biochim. Biophys. Acta* **1318**:246–254.

Goda, S., Hamada, T., Ishimoto, S., Kobayashi, T., Goto, I., and Kuroiwa, Y. (1987). Clinical improvement after administration of coenzyme Q10 in a patient with mitochondrial encephalomyopathy. *J. Neurol.* **234**:62–63.

Goldstein, J.L. and Brown, M.S. (1990). Regulation of the mevalonate pathway. *Nature (London)* **343**:425–430.

Grant, A.J, Jessup, W., and Dean, R.T. (1993). Enhanced enzymatic degradation of radical damaged mitochondrial membrane components. *Free Rad. Res. Comms.* **19**:125–134.

Grünler, J., Ericsson, J., and Dallner, G. (1994). Branch-point reactions in the biosynthesis of cholesterol, dolichol, ubiquinone and prenylated proteins. *Biochim. Biophys. Acta* **1212**:259–277.

Guan, Z., Söderberg, M., Sindelar, P., Prusiner, S.B., Kristensson, K., and Dallner, G. (1996). Lipid composition in scrapie-infected mouse brain: Prion infection increases the levels of dolichyl phosphate and ubiquinone. *J. Neurochem.* **66**:277–285.

Halliwell, B. and Auroma, O.I. (1991). DNA damage by oxygen-derived species. Its mechanism and measurement in mammalian systems. *FEBS Lett.* **281**:9–19.

Hanaki, Y., Sugiyama, S., Ozawa, T., and Ohno, M. (1991). Ratio of low-density lipoprotein cholesterol to ubiquinone as a coronary risk factor. *N. Engl. J. Med.* **325**:814–815.

Hanaki, Y., Sugiyama, S., Ozawa, T., and Ohno, M. (1993). Coenzyme Q10 and coronary artery disease. *Clin. Investig.* **71**, S112–S115.

Harman, D. (1981). The aging process. *Proc. Natl. Acad. Sci. USA* **78**:7124–7128.

Harman, D. (1994). Aging: Prospects for further increases in the functional life span. *Age* **17**:119–146.

Hasegawa, E., Takeshige, K., Oishi, T., Murai, Y., and Minakami, S. (1990). 1-Methyl-4-phenylpyridinium (MPP$^+$) induced NADH-dependent superoxide formation and enhances NADH-dependent lipid peroxidation in bovine heart submitochondrial particles. *Biochem. Biophys. Res. Commun.* **170**:1049–1055.

Hatefi, Y. and Yamaguchi, M. (1992). The energy-transducing nicotinamide nucleotide transhydrogenase. In *Molecular Mechanisms in Bioenergetics*. L. Ernster, ed. (Amsterdam: Elsevier), pp. 265–281.

Hatefi, Y. and Yamaguchi, M. (1996). Nicotinamide nucleotide transhydrogenase: A model for utilization of substrate binding energy for proton translocation. *FASEB J.* **10**:444–452.

Hochstein, P. and Ernster, L. (1963). ADP-activated lipid peroxidation coupled to the TPNH oxidase system of microsomes. *Biochem. Biophys. Res. Commun.* **12**:388–394.

Hochstein, P., Nordenbrand, K., and Ernster, L. (1964). Evidence for the involvement of iron in the ADP-activated peroxidation of lipids in microsomes and mitochondria. *Biochem. Biophys. Res. Commun.* **14**:323–328.

Hoek, J.B. and Rydström, J. (1988). Physiological roles of nicotinamide nucleotide transhydrogenase. *Biochem. J.* **254**:1–10.

Hübner, C., Hoffmann, G.F., Charpentier, C., Gibson, K.M., Finckh, B., Puhl, H., Lehr, H.A., and Kohlschutter, A. (1993). Decreased plasma ubiquinone-10 concentration in patients with mevalonate kinase deficiency. *Pediatr. Res.* **34**:129–133.

Imada, I., Watanabe, M., Matsumoto, N., and Morimoto, H. (1970). Metabolism of ubiquinone-7. *Biochemistry* **9**:2870–2878.

Ingledew, W.J. and Ohnishi, T. (1977). The probable site of action of thenoyltrifluoroacetone on the respiratory chain. *Biochem. J.* **164**:617–620.

Jayaraman, J. and Ramasarma, T. (1963). Intracellular distribution of coenzyme Q in rat liver. *Arch. Biochem. Biophys.* **103**:258–266.

Kagan, V., Serbinova, E., and Packer, L. (1990). Antioxidant effects of ubiquinones in microsomes and mitochondria are mediated by tocopherol recycling. *Biochem. Biophys. Res. Commun.* **169**:851–857.

Kalén, A., Norling, B., Appelkvist, E.-L., and Dallner, G. (1987). Ubiquinone biosynthesis by the microsomal fraction from rat liver. *Biochim. Biophys. Acta* **926**:70–78.

Kalén, A., Appelkvist, E.-L., and Dallner, G. (1989). Age-related changes in the lipid compositions of rat and human tissues. *Lipids* **24**:579–584.

Kalén, A., Appelkvist, E.-L., and Dallner, G. (1990). The effects of inducers of the endoplasmic reticulum, peroxisomes and mitochondria on the amounts and synthesis of ubiquinone in rat liver subcellular membranes. *Chem. Biol. Interact.* **73**:221–234.

Karlsson, J., Liska, J., Gunnes, S., Koul, B., Semb, B., Åström, H., Diamant, B., and Folkers, K. (1993). Heart muscle ubiquinone and plasma antioxidants following cardiac transplantation. *Clin. Investig.* **71**, S76–S83.

Katsikas, H. and Quinn, P.J. (1983). Fluorescence probe studies of the distribution of ubiquinone homologues in bilayers of dipalmitoylglycerophosphocholine. *Eur. J. Biochem.* **131**:607–612.

King, T.E. (1982). Ubiquinone proteins in cardiac mitochondria. In *Functions of Quinones in Energy Conserving Systems*. B.L. Trumpower, ed. (New York: Academic Press), pp. 3–25.

Kröger, A. and Klingenberg, M. (1970). Quinones and nicotinamide nucleotides associated with electron transfer. *Vitam. Horm.* **28**:533–574.

Kröger, A. and Klingenberg, M. (1973a). The kinetics of the redox reactions of ubiquinone related to the electron-transport activity in the respiratory chain. *Eur. J. Biochem.* **34**:358–368.

Kröger, A. and Klingenberg, M. (1973b). Further evidence for the pool function of ubiquinone as derived from the inhibition of the electron transport by antimycin. *Eur. J. Biochem.* **39**:313–323.

Lang, J.K., Gohil, K., and Packer, L. (1986). Simultaneous determination of tocopherols, ubiquinols, and ubiquinones in blood, plasma, tissue homogenates, and subcellular fractions. *Anal. Biochem.* **157**:106–116.

Loschen, G. and Flohé, L. (1971). Respiratory chain linked H_2O_2 production in pigeon heart mitochondria. *FEBS Lett.* **18**:261–264.

Loschen, G., Azzi, A., and Flohé, L. (1973). Mitochondrial H_2O_2 formation: Relationship with energy conservation. *FEBS Lett.* **33**:84–88.

Löw, P., Andersson, M., Edlund, C., and Dallner, G. (1992). Effects of mevinolin treatment on tissue dolichol and ubiquinone levels in the rat. *Biochim. Biophys. Acta* **1128**:253–259.

Luft, R., Ikkos, D., Palmieri, G., Ernster, L., and Afzelius, B. (1962). A case of severe hypermetabolism of nonthyroid origin with a defect in the maintenance of mitochondrial respiratory control: A correlated clinical, biochemical and morphological study. *J. Clin. Invest.* **41**:1776–1804.

Maguire, J.J., Wilson, D.S., and Packer, L. (1989). Mitochondrial electron transport-linked tocopheroxyl radical reduction. *J. Biol. Chem.* **264**:21462–21465.

Maguire, J.J., Kagan, V., Ackrell, B.A., Serbinova, E., and Packer, L. (1992). Succinate-ubiquinone reductase linked recycling of alpha-tocopherol in reconstituted systems and mitochondria: requirement for reduced ubiquinone. *Arch. Biochem. Biophys.* **292**:47–53.

Maillard, B., Ingold, K.U., and Scaniano, J.C. (1983). Rate constants for the reactions of free radicals with oxygen in solution. *J. Am. Chem. Soc.* **105**:5095–5099.

Mancini, A., De Marinis, L., Calabro, F., Sciuto, R., Oradei, A., Lippa, S., Sandric, S., Littarru, G.P., and Barbarino, A. (1989). Evaluation of metabolic status in amiodarone-induced thyroid disorders: Plasma coenzyme Q10 determination. *J. Endocrinol. Invest.* **12**:511–516.

Matsura, T., Yamada, K., and Kawasaki, T. (1991). Changes in the content and intracellular distribution of coenzyme Q homologs in rabbit liver during growth. Biochim. Biophys. Acta **1083**:277–282.

Mitchell, P. (1975). Protonmotive redox mechanism of the cytochrome bc_1 complex in the respiratory chain: Protonmotive ubiquinone cycle. *FEBS Lett.* **56**:1–6.

Mitchell, P. (1976). Possible molecular mechanisms of the protonmotive function of cytochrome systems. *J. Theor. Biol.* **62**:327–367.

Mohr, D., Bowry, V.W., and Stocker, R. (1992). Dietary supplementation with coenzyme Q10 results in increased levels of ubiquinol-10 within circulating lipoproteins and increased resistance of human low-density lipoprotein to the initiation of lipid peroxidation. *Biochim. Biophys. Acta* **1126**:247–254.

Morimoto, H., Imada, I., and Goto, G. (1969). Ubiquinone and related compounds, XV. Photochemical reaction of ubiquinone-7. *Liebigs Ann. Chem.* **729**:184–192.

Morimoto, H., Imada, I., and Goto, G. (1970). Ubiquinone and related compounds, XVI. Photo-oxidation of ubiquinone-7. *Liebigs Ann. Chem.* **735**:65–71.

Mortensen, S.A. (1993). Perspectives on therapy of cardiovascular diseases with coenzyme Q_{10} (ubiquinone). *Clin. Investig.* **71**, S116–S123.

Mukai, K., Kikuchi, S., and Urano, S. (1990). Stopped-flow kinetic study of the regeneration reaction of tocopheroxyl radical by reduced ubiquinone-10 in solution. *Biochim. Biophys. Acta* **1035**:77–82.

Mukai, K., Itoh, S., and Morimoto, H. (1992). Stopped-flow kinetic study of vitamin E regeneration reaction with biological hydroquinones (reduced forms of ubiquinone, vitamin K, and tocopherolquinone) in solution. *J. Biol. Chem.* **267**:22277–22281.

Murphy, M.P., Krueger, M.J., Sablin, S.O., Ramsay, R.R., and Singer, T.P. (1995). Inhibition of Complex I by hydrophobic analogues of N-methyl-4-phenylpyridinium (MPP$^+$) and the use of an ion-selective electrode to measure their accumulation by mitochondria and electron-transport particles. *Biochem. J.* **306**:359–365.

Muscari, C., Biagetti, L., Stefanelli, C., Giordano, E., Guarnieri, C., and Caldarera, C.M. (1995). Adaptive changes in coenzyme Q biosynthesis to myocardial reperfusion in young and aged rats. *J. Mol. Cell Cardiol.* **27**:283–289.

Nohl, H. and Jordan, W. (1986). The mitochondrial site of superoxide production. *Biochem. Biophys. Res. Commun.* **138**:533–539.

Norling, B., Glazek, E., Nelson, B.D., and Ernster, L. (1974). Studies with ubiquinone-depleted submitochondrial particles. Quantitative incorporation of small amounts of ubiquinone and its effects on the NADH and succinate oxidase activities. *Eur. J. Biochem.* **47**:475–482.

Ogasahara, S., Nishikawa, Y., Yorifuji, S., Soga, F., Nakamura, Y., Takahashi, M., Hashimoto, S., Kono, N., and Tarui, S. (1986). Treatment of Kearns-Sayre syndrome with coenzyme Q10. *Neurology* **36**:45–53.

Ohnishi, T. and Trumpower, B.L. (1980). Differential effects of antimycin on ubisemiquinone bound in different environments in isolated succinate-cytochrome *c* reductase complex. *J. Biol. Chem.* **255**:3278–3284.

Olson, R.E. and Rudney, H. (1983). Biosynthesis of ubiquinone. *Vitam. Horm.* **40**:1–43.

Olsson, J.M., Eriksson, L.C., and Dallner, G. (1991). Lipid compositions of intracellular membranes isolated from rat liver nodules in Wistar rats. *Cancer Res.* **51**:3774–3780.

Pahl, H.L. and Baeuerle, P.A. (1994). Oxygen and the control of gene expression. *BioEssays* **16**:497–502.

Pedersen, S., Tata, J.R., and Ernster, L. (1963). Ubiquinone (coenzyme Q) and the regulation of basal metabolic rate by thyroid hormones. *Biochim. Biophys. Acta* **69**:407–409.

Ramsay, R.R. and Singer, T.P. (1992). Relation of superoxide generation and lipid peroxidation to the inhibition of NADH-Q oxidoreductase by rotenone, piericidin A, and MPP$^+$. *Biochem. Biophys. Res. Commun.* **189**:47–52.

Rich, P.R., Jeal, A.E., Madgwick, S.A., and Moody, A.J. (1990). Inhibitor effects on redox-linked protonations of the *b* haems of the mitochondrial *bc*$_1$ complex. *Biochim. Biophys. Acta* **1018**:29–40.

Salerno, J.C. and Ohnishi, T. (1980). Studies on the stabilized ubisemiquinone species in the succinate-cytochrome *c* reductase segment of the intact mitochondrial membrane system. *Biochem. J.* **192**:769–781.

Salerno, J.C., Osgood, M., Liu, Y.J., Taylor, H., and Scholes, C.P. (1990). Electron nuclear double resonance (ENDOR) of the Qc$^{·-}$ ubisemiquinone radical in the mitochondrial electron transport chain. *Biochemistry* **29**:6987–6993.

Sastry, P.S. (1961). Distribution of coenzyme Q in rat liver cell fractions. *Nature* **189**:577–570.

Sen, C.K. and Packer, L. (1996). Antioxidant and redox regulation of gene transcription. *FASEB J.* **10**:709–720.

Shigenaga, M.K., Hagen, T., and Ames, B.N. (1994). Oxidative damage and mitochondrial decay in aging. *Proc. Natl. Acad. Sci. USA* **91**:10771–10778.

Sottocasa, G., Kuylenstierna, B., Ernster, L., and Bergstrand, A. (1967). An electron-transport system associated with the outer mitochondrial membrane of liver mitochondria: A biochemical and morphological study. *J. Cell Biol.* **32**:415–438.

Stadtman, E.R. (1993). Oxidation of free amino acids and amino acid residues in proteins by radiolysis and by metal-catalyzed reactions. *Ann. Rev. Biochem.* **62**:797–821.

Sterling, K., Milch, P.O., Brenner, M.A., and Lazarus, J.H. (1977). Thyroid hormone action: The mitochondrial pathway. *Science* **197**:996–999.

Sterling, K. (1986). Direct thyroid hormone activation of mitochondria: The role of adenine nucleotide translocase. *Endocrinology* **119**:292–295.

Stocker, R., Bowry, V.W., and Frei, B. (1991). Ubiquinol-10 protects human low density lipoprotein more efficiently against lipid peroxidation than does alpha-tocopherol. *Proc. Natl. Acad. Sci. USA* **88**:1646–1650.

Stocker, R. and Suarna, C. (1993). Extracellular reduction of ubiquinone-1 and -10 by human Hep G2 and blood cells. *Biochim. Biophys. Acta* **1158**:15–22.

Sun, I.L., Sun, E.E., and Crane, F.L. (1992). Stimulation of serum-free cell proliferation by coenzyme Q. *Biochem. Biophys. Res. Commun.* **189**:8–13.

Söderberg, M., Edlund, C., Kristensson, K., and Dallner, G. (1990). Lipid compositions of different regions of the human brain during aging. *J. Neurochem.* **54**:415–423.

Takahashi, T., Yamaguchi, T., Shitashige, M., Okamoto, T., and Kishi, T. (1995). Reduction of ubiquinone in membrane lipids by rat liver cytosol and its involvement in the cellular defence system against lipid peroxidation. *Biochem. J.* **309**:883–890.

Takayanagi, R., Takeshige, K., and Minakami, S. (1980). NADH- and NADPH-dependent lipid peroxidation in bovine heart submitochondrial particles. Dependence on the rate of electron flow in the respiratory chain and an antioxidant role of ubiquinol. *Biochem. J.* **192**:853–860.

Takeshige, K. and Minakami, S. (1975). Reduced nicotinamide adenine dinucleotide phosphate-dependent lipid peroxidation by beef heart submitochondrial particles. *J. Biochem.* **77**:1067–1073.

Takeshige, K. and Minakami, S. (1979). NADH- and NADPH-dependent formation of superoxide anions by bovine heart submitochondrial particles and NADH-ubiquinone reductase preparation. *Biochem. J.* **180**:129–135.

Takeshige, K., Takayanagi, K., and Minakami, S. (1980). Reduced coenzyme Q10 as an antioxidant of lipid peroxidation in bovine heart mitochondria. In *Biomedical and Clinical Aspects of Coenzyme Q*, vol. 2. Y. Yamamura, K. Folkers, and T. Ito, eds. (Amsterdam: Elsevier), pp. 15–25.

Thelin, A., Peterson, E., Hutson, J.L., McCarthy, A.D., Ericsson, J., and Dallner, G. (1994). Effect of squalestatin 1 on the biosynthesis of the mevalonate pathway lipids. *Biochim. Biophys. Acta* **1215**:245–249.

Trumpower, B.L. (1990). The protonmotive Q cycle. Energy transduction by coupling of proton translocation to electron transfer by the cytochrome bc_1 complex. *J. Biol. Chem.* **265**:11409–11412.

Turrens, J.F., Alexandre, A., and Lehninger, A.L. (1985). Ubisemiquinone is the electron donor for superoxide formation by complex III of heart mitochondria. *Arch. Biochem. Biophys.* **237**:408–414.

von Jagow, G., Ljungdahl, P.O., Ohnishi, T., and Trumpower, B.L. (1984). An inhibitor of mitochondrial respiration which binds to the cytochrome *b* and displaces quinone from the iron–sulfur protein of the cytochrome bc_1 complex. *J. Biol. Chem.* **259**:6318–6326.

Watts, G.F., Castelluccio, C., Rice-Evans, C.A., Taub, N.A., Baum, H., and Quinn, P.J. (1993). Plasma coenzyme Q (ubiquinone) concentrations in patients treated with simvastatin. *J. Clin. Pathol.* **46**:1055–1057.

Whitaker, S.J., Powell, S.N., and McMillan, T.J. (1991). Molecular assays of radiation-induced DNA damage. *Eur. J. Cancer* **27**:922–928.

Willis, R.A., Folkers, K., Tucker, J.L., Ye, C.Q., Xia, L.J., and Tamagawa, H. (1990). Lovastatin decreases coenzyme Q levels in rats. *Proc. Natl. Acad. Sci. USA* **87**:8928–8930.

Wolf, D.E., Hoffman, C.H., Trenner, N.R., Arison, B.H., Shunk, C.H., Linn, B.O., McPherson, J.F., and Folkers, K. (1958). Coenzyme Q. Structure studies on the coenzyme Q group. *J. Am. Chem. Soc.* **80**:4752–4750.

Wolff, S.P., Garner, A., and Dean, R.T. (1986). Free radicals, lipids, and protein degradation. *Trends Biochem. Sci.* **11**:27–31.

Yamamoto, M., Sato, T., Anno, M., Ujike, H., and Takemoto, M. (1987). Mitochondrial myopathy, encephalopathy, lactic acidosis, and stroke-like episodes with recurrent abdominal symptoms and coenzyme Q10 administration. *J. Neurol. Neurosurg. Psych.* **50**:1475–1481.

Yamamoto, Y., Kawamura, M., Tatsuno, K., Yamashita, S., Niki, E., and Naito, C. (1991). Formation of lipid hydroperoxides in the cupric ion-induced oxidation od plasma and low density lipoprotein. In *Oxidative Dmage and Repair: Chemical, Biological, and Medical Aspects*. K.J.A. Davies, ed. (New York: Pergamon Press), pp. 287–291.

Yu, C.A. and Yu, L. (1981). Ubiquinone-binding proteins. *Biochim. Biophys. Acta* **639**:99–128.

Zhang, Y., Marcillat, O., Giulivi, C., Ernster, L., and Davies, K.J.A. (1990). The oxidative inactivation of mitochondrial electron transport chain components and ATPase. *J. Biol. Chem.* **265**:16330–16336.

Zhang, Y., Aberg, F., Appelkvist, E.-L., Dallner, G., and Ernster, L. (1995). Uptake of dietary coenzyme Q supplement is limited in rats. *J. Nutr.* **125**:446–453.

Zhang, Y., Turunen, M., and Appelkvist, E.-L. (1996). Restricted uptake of dietary coenzyme Q is in contrats to the unrestricted uptake of α-tocopherol into rat organs and cells. *J. Nutr.* **126**:2089–2097.

Zwizinski, C.W. and Schmid, H.H.O. (1992). Peroxidative damage to cardiac mitochondria: Identification and purification of modified adenine nucleotide translocase. *Arch. Biochem. Biophys.* **294**:178–183.

27

LYCOPENE AND β-CAROTENE

Bioavailability and Biological Effects

Helmut Sies and Wilhelm Stahl

Institut für Physiologische Chemie I
Heinrich-Heine-Universität Düsseldorf
P. O. Box 10 10 07
D-40001 Düsseldorf, Germany

INTRODUCTION

Carotenoids are widespread natural colorants with lipophilic properties, and more than 600 different compounds have been identified until know, with β-carotene as the most prominent (Olson and Krinsky, 1995); structures of β-carotene and its acyclic analog lycopene are shown in Fig. 1. Most carotenoids contain an extended system of conjugated double bonds, which is responsible for their color. Carotenoids can be divided in two classes, carotenes which are solely composed of carbon and hydrogen, and oxocarotenoids (xanthophylls) which contain at least one oxygen atom. From a biochemical point of view, carotenoids are grouped as provitamin A and non-provitamin A compounds. The provitamin A carotenoids β-carotene, α-carotene, and β-cryptoxanthin may serve as precursors of retinol and are capable of preventing classical vitamin A deficiency diseases such as xerophthalmia. The biosynthesis of carotenoids is limited to plants and some lower organisms (Young and Britton, 1993), whereas animals are provided with carotenoids from the diet. Important dietary sources of carotenoids for the human are green leafy and orange to red vegetables such as spinach, kale, broccoli, carrots, tomatoes as well as various fruits such as oranges, tangerines, or peaches (Gross, 1987; Gross 1991). More than 35 different carotenoids have been identified in human plasma (Khachik, *et al.,* 1995).

An increased consumption of fruits and vegetables is epidemiologically associated with a decreased risk for several kinds of cancer; elevated β-carotene plasma levels are related to a diminished risk for lung and stomach cancer (Taylor-Mayne, 1996). It has been suggested that carotenoids other than β-carotene might also exhibit protective effects. Several studies have demonstrated that non-provitamin A carotenoids such as lycopene or canthaxanthin show biological activities which might be related to cancer-preventive effects.

Free Radicals, Oxidative Stress, and Antioxidants, edited by Özben.
Plenum Press, New York, 1998.

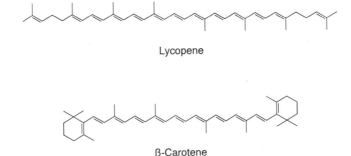

Figure 1. Chemical structure of *all-trans* lycopene and *all-trans* β-carotene.

Biological effects of carotenoids have been discussed according to their functions, actions, and associations.

1. *Functions*: accessory pigments in photosynthesis, protection against photo-sensitization, vitamin A activity.
2. *Actions*: antioxidant activity, enhancement of immune response, inhibition of mutagenesis and transformation, induction of gap junctional communication.
3. *Associations*: decreased risk of some cancers, decreased risk for some cardio-vascular diseases, decreased risk of cataracts and macular degeneration.

With respect to effects of carotenoids on human health, their uptake and distribution in human tissues as well as their metabolism has attracted attention. Some information is available on biokinetic parameters of β-carotene but little is known regarding other caroten-noids (Olson, 1994; Erdman *et al.*, 1993; Wang, 1994).

The efficacy of carotenoid uptake from the diet depends on several factors, such as dietary fat, presence of fiber, or food processing. The particle size of uncooked food also influences carotenoid uptake. The bioavailabilty of carotenoids from pureed or finely chopped vegetables is considerably higher than from whole or sliced raw vegetables. It has also been demonstrated that mild cooking increases the absorption of carotenoids. Heating can cause isomerization of the naturally occurring *all-trans* double bonds to *cis* configuration.

BIOAVAILABILITY OF β-CAROTENE ISOMERS

In the small intestine, ingested carotenoids are incorporated into micelles. Micelles are formed from dietary lipids and bile acids, which facilitate absorption of carotenoids into the intestinal mucosa cell. Thus, the intestinal uptake of these compounds is improved by the coingestion of oil, margarine or butter. The intact carotenoids are incorporated into chylomicrons which are released into the lymphatic system. In blood plasma, carotenoids appear initially in the chylomicron and VLDL fraction, whereas their levels in other lipo-proteins such as LDL and HDL rise at later time points with peak levels at 24–48 h. The major vehicle of hydrocarbon carotenoids is the LDL, while the more polar oxocaro-tenoids are found in LDL and HDL.

Carotenoids tend to isomerize and form a mixture of mono and poly *cis* isomers in addition to the *all-trans* form. Generally the *all-trans* form is predominant in nature, but some organisms such as the halotolerant alga *Dunaliella salina* synthesize considerable amounts of *cis* isomers. The β-carotene pattern of human serum is dominated by the *all-trans* isomer, and only small amounts of *cis*-isomers (about 5% of total β-carotene) are detected (Stahl *et al.,* 1992; Schmitz *et al.,* 1991). However, considerable amounts of different *cis* isomers are found in human tissues (Stahl *et al.,* 1992). With respect to *cis* isomers, interest was mainly focused on the 9-*cis* form since it has been demonstrated that it is an isomer-selective precursor for 9-*cis* retinal. 9-*cis* Retinal can be further oxidized in vivo to yield 9-*cis* retinoic acid which is a selective ligand for the retinoic X receptor (RXR). The RXR-ligand complex acts as a nuclear transcription factor and is involved in the regulation of several genes (Mangelsdorf *et al.,* 1994).

In view of the importance of 9-*cis* retinoic acid in regulatory processes, more information is needed on the uptake, distribution, and metabolism of its precursor 9-*cis* β-carotene. Betatene, an extract of the alga *Dunaliella salina* which contains up to 40% of the 9-*cis* isomer in its β-carotene fraction, was used to study the biokinetics of this compound (Stahl *et al.,* 1993). After ingestion of a single dose of Betatene equivalent to 3.0 µmol of *all-trans* β-carotene/kg body weight and 2.1 µmol of 9-*cis* β-carotene/kg body weight, only the level of the *all-trans* isomer rose in human serum. No significant increase in the 9-*cis* β-carotene level was observed when repeated doses of Betatene were applied, even after ingestion of high amounts of Betatene over a period of 150 days in the treatment erythropoietic protoporphyria (von Laar *et al.,* 1996). Similar results were obtained when analyzing the chylomicron fraction of human serum after Betatene ingestion (Stahl *et al.,* 1995). The data suggest the existence of efficient discriminatory mechanisms between the two isomers operative at the step of absorption from the intestinal lumen.

BIOAVAILABILITY OF LYCOPENE

Lycopene contributes between 21 and 43% of total serum carotenoids. In contrast to other carotenoids, similar lycopene serum levels are detected for men and women. Lower lycopene levels were found with older age and also related to lower non-HDL cholesterol. The consumption of alcohol appears to have no effect on lycopene plasma levels, whereas lower levels of β-carotene are reported.

Regarding absorption of lycopene from dietary sources, surprisingly no increase in lycopene serum levels was observed after the single intake of large amounts of tomato juice (Stahl and Sies, 1992). After ingestion of 180 g or even 700 g of tomato juice corresponding to a single dose of 12 or 80 mg of lycopene, respectively, no change in serum lycopene levels was observed.

In contrast, lycopene plasma levels increased significantly in human serum when processed juice was consumed. Boiling for 1h in the presence of 1% corn oil increased the bioavailability of lycopene from tomato juice significantly. Differences in lycopene bioavailability were also observed when raw tomatoes *vs.* tomato paste were used as a source, with higher lycopene levels in chylomicrons after consumption of tomato paste (Gärtner *et al.,* 1997). Possible factors improving lycopene availabilty from processed tomato products might be the release of the carotenoid by thermally-induced rupture of the cell walls or heat-improved extraction of lycopene into the oily phase of the mixture, with corn oil as vehicle.

Lycopene uptake varies with individuals, and peak serum concentrations were observed at 24–48 h after ingestion of processed tomato juice, comparable to the peak levels after ingestion of β-carotene. Lycopene is eliminated from human plasma with a halflife of about 2–3 days, which is somewhat more rapid than β-carotene.

Hydroxylated derivatives of lycopene have been detected in human serum (Khachik *et al.*, 1996). Tomatoes and tomato-based products are the major source for these compounds. They contain lycopene 5,6-epoxide which might be enzymatically or chemically modified in the organism. The most prominent geometrical isomers are are *all-trans* and 5-*cis* lycopene (Schierle *et al.*, 1997). An overwiew on biological properties of lycopene has recently been published (Stahl and Sies, 1996).

LYCOPENE AND β-CAROTENE IN HUMAN TISSUES

Carotenoids accumulate in human tissues. Interindividual differences in carotenoid levels were reported, but high levels of lycopene and β-carotene were always found in liver, adrenals, and testes, whereas other tissues such as lung and kidney contained less (Stahl *et al.*, 1992; Kaplan *et al.*, 1990; Nierenberg and Nann, 1992; Schmitz *et al.*, 1991). In studies from the US, lycopene levels exceeded that of β-carotene in almost every tissue except the ovary. Higher β-carotene levels than lycopene were found in liver, adrenals, and kidney in a study from Germany, but the lycopene levels in testes exceeded that of β-carotene. The lycopene level in skin was also somewhat higher than that of β-carotene, 0.42 and 0.27 nmol/g, respectively (Nierenberg and Nann, 1992). Higher values have been reported recently with lycopene and β-carotene levels of 1.6 and 1.4 nmol/g skin tissue (Ribaya-Mercado *et al.*, 1995). Although hydrocarbon carotenoids are very lipophilic, their levels in adipose tissue are low in comparison to e.g. liver or testes. Adipose tissue carotenoids contribute considerably to the total content in the organism, since adipose tissue represents about 20% of total body weight.

Absorption and distribution of lycopene into different tissues of rats and monkeys was studied after application of radioactively labelled lycopene. Liver was the major organ for lycopene accumulation (Mathews-Roth *et al.*, 1990).

ANTIOXIDANT ACTIVITY OF CAROTENOIDS

As a normal attribute of aerobic life, the human organism is exposed to a variety of different prooxidants capable to damage biologically relevant molecules, such as DNA, proteins, carbohydrates, and lipids (Sies, 1986; Sies and Stahl, 1995; Halliwell, 1996; Truscott, 1990). Among the various defense strategies, carotenoids are most likely involved in the scavenging of singlet molecular oxygen ($^1O_2$), (Foote and Denny, 1968) and peroxyl radicals (Burton and Ingold, 1984). Further they are effective deactivators of electronically excited sensitizer molecules which are involved in the generation of free radicals and of singlet oxygen.

The interaction of carotenoids with $^1O_2$ depends largely on physical quenching. The energy of singlet molecular oxygen is transferred to the carotenoid molecule to yield ground state oxygen and a triplet excited carotene. Instead of further chemical reactions, the carotenoid returns to ground state dissipating its energy by interactions with the surrounding solvent. Chemical reaction between the excited oxygen and carotenoids is of minor importance, contributing less than 0.05% to the total quenching rate. Since the carotenoids remain intact during physical quenching of $^1O_2$ or excited sensitizers, they can be reutilized in further quenching cycles. Among the various carotenoids, xanthophylls as well as carotenes proved to be efficient quenchers of singlet oxygen, reacting with rate constants that approach diffusion control. The efficacy of carotenoids for physical quenching is related to the number of conjugated double bonds present in the molecule (Foote and Denny, 1968; Stahl *et al.,* 1997). Lutein, zeaxanthin, β-cryptoxanthin, α- and β-carotene, all of which are detected in human serum and tissues, belong to the group of highly active quenchers of $^1O_2$. The most efficient carotenoid is the acyclic carotenoid lycopene (Di Mascio *et al.,* 1989).

Carotenoids are also capable of intercepting peroxyl radicals and thus inhibit lipid peroxidation. In a model system using the formation of radical-initiated hydroperoxides of methyl linoleate, canthaxanthin and astaxanthin were shown to be better and longer lasting antioxidants than β-carotene and zeaxanthin. The antioxidant activity of carotenoids regarding the deactivation of peroxyl radicals likely depends on the formation of radical adducts forming a resonancestabilized carbon-centered radical. A variety of oxidation products have been detected upon β-carotene autoxidation and upon interaction with radicals, including carotenoid epoxides and apocarotenals of different chain length (Kennedy and Liebler, 1991).

The antioxidant activity of carotenoids depends on the oxygen tension present in the system. At low partial pressure of oxygen (15 torr), occurring in most tissues under physiological conditions, β-carotene was found to inhibit oxidation most efficiently.

Carotenoids apparently interact with other lipophilic antioxidants such as vitamin E. In vitro, a synergism regarding the antioxidant activity has been observed in the inhibition of peroxyl radical induced lipid peroxidation, when β-carotene was used together with α-tocopherol (Palozza and Krinsky, 1994).

GAP JUNCTIONAL COMMUNICATION AND CAROTENOIDS

Intercellular signaling is a prerequisite for coordinating biochemical functions in multicellular organisms. Direct signaling is achieved by cell-to-cell channels, called gap junctions (Goodenough *et al.,* 1996). Gap junctions are water-filled pores, connecting the cytosol of two neighboring cells, allowing for exchange of low-molecular mass compounds of < 1000 Da. The channels consist of specific proteins which are coded for by the family of connexin genes. Several physiological functions have been attributed to gap junctional communication (GJC): exchange of nutrients and ions between connected cells, conducting electrical signals, and pathways for signaling compounds.

It has been speculated that GJC might play a role in carcinogenesis (Hotz-Wagenblatt and Shalloway, 1993). Tumor promoters such as phorbol esters are efficient inhibitors of this pathway of communication, while other compounds like carotenoids and retinoids induce GJC. Interestingly, also non-provitamin A carotenoids such as canthaxanthin or lycopene as well as some synthetic carotenoid analogs stimulate intercellular communication in vitro (Zhang *et al.,* 1991, Stahl *et al.,* 1997). The influence of carotenoids on gap junctional communication does not correlate with their antioxidant activity. In cell culture, carotenoids and retinoids reversibly inhibit the progression of carcinogen-initiated fibroblasts to the transformed state (Pung *et al.,* 1988). This inhibitory effect has been found to be related to an increased gap junctional communication induced by these compounds. There is growing evidence that gap junctions play a role in the regulation of morphogenesis, cell differentiation, secretion of hormones, and growth. Recently, we investigated carotenoids of different structure with respect to their capability of stimulating gap junctional communication in murine fibroblasts (Stahl *et al.,* 1997). The series included natural and synthetic carotenoids carrying different substituents at the end of the conjugated double bond system, five- and six-membered ring carotenoids, as well as acyclic polyenes. GJC was measured at day 5 after incubation with the given carotenoid at a concentration of 10 μM. β-carotene, echinenone, canthaxanthin, cryptoxanthin, and 4-hydroxy-β-carotene efficiently induce GJC in murine fibroblasts. A 3- to 5-fold induction was observed in comparison to the solvent control. This is similar to the extent achieved with retinoic acid at 0.1 to 1 μM. The synthetic analog of β-carotene, retro-dehydro-β-carotene, which contains an additional double bond in the molecule, is about as effective as the parent compound itself. The position and nature of the substituent at the six-membered ring appears to have little influence on GJC. Echinenone and 4-OH-β-carotene, carrying a keto- or a hydroxy-group at the 4-position, respectively, exhibit about similar activity. Cryptoxanthin with a OH-group at the 3-position is somewhat less active than the corresponding 4-OH-β-carotene.

Carotenoids containing a five-membered ring system are significantly less active. No stimulatory effects were found for capsorubin, while the stimulation of GJC with violerythrin was only about 1.5-fold. A direct comparison of five- and six-membered ring carotenoids can be taken from the data obtained for dinor-canthaxanthin and canthaxanthin. The six-membered ring carotenoid canthaxanthin is about twice as active than its five-membered ring analog.

No significant influence on GJC was observed for the open-chain polyene compounds C-20, C-30, and C-40 dialdehyde containing 7, 11, and 15 carbon-carbon double bonds, respectively.

The question has been addressed whether these biological activities of carotenoids require the intact compounds or are effects of decomposition products or metabolites of the parent carotenoid. We investigated the biological effects of the non-provitamin A carotenoid canthaxanthin and its decompositon products on GJC (Hanusch *et al.,* 1995). Decomposition fractions of canthaxanthin were isolated by preparative HPLC and shown to be active in the cell communication assay. Two of the decomposition products were identified as *all-trans* and 13-*cis* 4-oxo-retinoic acid. Both isomers of 4-oxo-retinoic acid enhanced GJC in murine fibroblasts (C3H/10T 1/2 cells) accompanied by an increase of the expression of connexin 43 m-RNA. Therefore, it is concluded that the biological activity of canthaxanthin is at least in part due to the formation of active decomposition products such as 4-oxo-retinoic acid. To obtain a significant induction of cell-cell-communication in 10T1/2 cells only 0.1% of the 10 μM incubation mixture with canthaxanthin needed to be converted to 4-oxo-retinoic acid, which is active even at 10 nM.

We suggest that the biological effects of carotenoids are at least in part mediated via retinoids formed during oxidation or generated in metabolic pathways.

ACKNOWLEDGMENTS

Our research was supported by the National Foundation for Cancer Research (Bethesda, MD), the Institut Danone für Ernährung (Rosenheim, Germany) and the Bundesministerium für Bildung, Wissenschaft, Forschung und Technologie (Bonn, Germany).

REFERENCES

Burton, G.W., and Ingold, K.U., 1984, β-Carotene:an unusual type of lipid antioxidant, *Science.* 224: 569–573.
Di Mascio, P., Kaiser, S., and Sies, H., 1989, Lycopene as the most efficient biological carotenoid singlet oxygen quencher, *Arch. Biochem. Biophys.* 274: 532–538.
Erdman, J.W., Bierer, T.L., and Gugger, E.T., 1993, Absorption and transport of carotenoids, *Ann. N. Y. Acad. Sci.* 691: 76–85.

Foote, C.S., and Denny, R.W., 1968, Chemistry of singlet oxygen. VII. Quenching by β-carotene, *J. Am. Chem. Sci.* **90:** 6233–6235.

Gärtner, C., Stahl, W., and Sies, H., 1997, Increased lycopene bioavailability from tomato paste as compared to fresh tomatoes, *Am. J. Clin. Nutr.* in press

Goodenough, D.A., Goliger, J.A., and Paul, D.L., 1996, Connexins, connexons, and intercellular communication, *Annu. Rev. Biochem.* **65:** 475–502.

Gross, J., 1987, *Pigments in fruits*, London, Academic press.

Gross, J., 1991, *Pigments in vegetables*, Van Nordstrand Reinhold, New York.

Halliwell, B., 1996, Antioxidants in human health and disease, *Annu. Rev. Nutr.* **16:** 33–50.

Hanusch, M., Stahl, W., Schulz, W.A., and Sies, H., 1995, Induction of gap junctional communication by 4-oxoretinoic acid generated from its precursor canthaxanthin, *Arch. Biochem. Biophys.* **317:** 423–428.

Hotz-Wagenblatt, A., and Shalloway, D., 1993, Gap junctional communication and neoplastic transformation, *Crit. Rev. Oncogen.* **4:** 541–558.

Kaplan, L.A., Lau, J.M., and Stein, E.A., 1990, Carotenoid composition, concentrations, and relationship in various human organs, *Clin. Physiol. Biochem.* **8:** 1–10.

Kennedy, T.A., and Liebler, D.C., 1991, Peroxyl radical oxidation of β-carotene: formation of β-carotene epoxides, *Chem. Res. Toxicol.* **4:** 290–295.

Khachik, F., Beecher, G.R., and Smith, J.C., 1995, Lutein, lycopene, and their oxidative metabolites in chemoprevention of cancer, *J. Cell. Biochem.* **22:** 236–246.

Khachik, F., Beecher, G.R., Steck, A., and Pfander, H., 1996, Bioavailability, metabolism, and possible mechanism of chemoprevention by lutein and lycopene in humans, in: *International Conference on Food Factors: Chemistry and Cancer Prevention* (Ohigashi,H. ed.) Tokyo, Springer Verlag.

Mangelsdorf, D.J., Umesono, K., and Evans, R.M., 1994, The retinoid receptors, in: *The Retinoids, Biology, Chemistry, and Medicine* (Sporn, M.B., Roberts, A.B., Goodman, D.S. eds.) pp. 319–349, New York, Raven Press.

Mathews-Roth, M.M., Welankiwar, S., Sehgal, P.K., Lausen, N.C.G., Russett, M., and Krinsky, N.I., 1990, Distribution of [^{14}C] canthaxanthin and [^{14}C] lycopene in rats and monkeys, *J. Nutr.* **120:** 1205–1213.

Nierenberg, D.W., and Nann, S.L., 1992, A method for determining concentrations of retinol, tocopherol, and five carotenoids in human plasma and tissue samples, *Am. J. Clin. Nutr.* **56:** 417–426.

Olson, J.A., 1994, Absorption, transport, and metabolism of carotenoids in the human, *Pure Appl. Chem.* **66:** 1011–1016.

Olson, J.A., and Krinsky, N.I., 1995, Introduction: the colorful fascinating world of carotenoids: important biological modulators, *FASEB J.* **9:** 1547–1550.

Palozza, P., and Krinsky, N.I., 1994, Antioxidant properties of carotenoids, in: *Retinoids: from basic science to clinical application* (Livrea,M.A., Vidali,G. eds.) pp. 35–41, Basel, Birkhäuser Verlag.

Pung, A., Rundhaug, J.E., Yoshizawa, C.N., and Bertram, J.S., 1988, β-Carotene and canthaxanthin inhibit chemically- and physically-induced neoplastic transformation in 10T1/2 cells, *Carcinogenesis.* **9:** 1533–1539.

Ribaya-Mercado, J.D., Garmyn, M., Gilchrest, B.A., and Russell, R.M., 1995, Skin lycopene is destroyed preferentially over β-carotene during ultraviolet irradiation in humans, *J. Nutr.* **125:** 1854–1859.

Schierle, J., Bretzel, W., Bühler, I., Faccin, N., Hess, D., Steiner, K., and Schüep, W., 1996, Content and isomeric ratio of lycopene in food and human blood plasma, *Food Chem.* in press

Schmitz, H.H., Poor, C.L., Wellman, R.B., and Erdman, J.W., 1991, Concentrations of selected carotenoids and vitamin A in human liver, kidney and lung tissue, *J. Nutr.* **121:** 1613–1621.

Sies, H., 1986, Biochemistry of oxidative stress, *Angew. Chem. Int. Ed. Engl.* **25:** 1058–1071.

Sies, H., and Stahl, W., 1995, Vitamins E and C, β-carotene, and other carotenoids as antioxidants, *Am. J. Clin. Nutr.* **62:** 1315S–1321S.

Stahl, W., Nicolai, S., Briviba, K., Hanusch, M., Broszeit, G., Peters, M., Martin, H.-D., and Sies, H., 1997, Biological activities of natural and synthetic carotenoids: induction of gap junctional communication and singlet oxygen quenching, *Carcinogenesis.* **18:** 89–92.

Stahl, W., Schwarz, W., and Sies, H., 1993, Human serum concentrations of all-trans-β-carotene and α-carotene nut not 9-cis-β-carotene increase upon ingestion of a natural isomer mixture obtained from *Dunaliella salina* (Betatene), *J. Nutr.* **123:** 847–851.

Stahl, W., Schwarz, W., Sundquist, A.R., and Sies, H., 1992, cis-trans Isomers of lycopene and β-carotene in human serum and tissues, *Arch. Biochem. Biophys.* **294:** 173–177.

Stahl, W., Schwarz, W., von Laar, J., and Sies, H., 1995, all-trans-β-Carotene preferentially accumulates in human chylomicrons and very low density lipoproteins compared with the 9-cis geometrical isomer, *J. Nutr.* **125:** 2128–2133.

Stahl, W., and Sies, H., 1992, Uptake of lycopene and its geometrical isomers is greater from heat-processed than from unprocessed tomato juice in humans, *J. Nutr.* **122:** 2161–2166.

Stahl, W., and Sies, H., 1996, Lycopene: A Biologically Important Carotenoid for the Human?, Arch. Biochem. Biophys. **336:** 1–9

Taylor-Mayne, S., 1996, Beta-carotene, carotenoids, and disease prevention in humans, *FASEB J.* **10:** 690–701.

Truscott, T.G., 1990, The photophysics and photochemistry of the carotenoids, *J. Photochem. Photobiol. B: Biol.* **6:** 359–371.

von Laar, J., Stahl, W., Bolsen, K., Goerz, G., and Sies, H., 1996, β-Carotene serum levels in patients with erythropoietic protoporphyria on treatment with the synthetic all-trans isomer or a natural isomer mixture of β-carotene, *J. Photochem. Photobiol. B:Biol.* **33:** 157–162.

Wang, X.-D., 1994, Review: Absorption and metabolism of β-carotene, *J. Am. Coll. Nutr.* **13:** 314–325.

Young, A., and Britton, G, 1993, *Carotenoids in photosynthesis*, Chappman & Hall, London.

Zhang, L.-X., Cooney, R.V., and Bertram, J.S., 1991, Carotenoids enhance gap junctional communication and inhibit lipid peroxidation in C3H/10T1/2 cells: Relationship to their cancer chemopreventive action, *Carcinogenesis.* **12:** 2109–2114.

28

CAROTENOID PROPERTIES DEFINE PRIMARY BIOLOGICAL ACTIONS AND METABOLISM DEFINES SECONDARY BIOLOGICAL ACTIONS

Norman I. Krinsky

Department of Biochemistry, School of Medicine and
   the Jean Mayer USDA Human Nutrition Research Center on Aging
Tufts University
136 Harrison Avenue
Boston, Massachusetts 02111-1837

1. INTRODUCTION

The biological properties of carotenoids that can be observed and measured in animals, including humans, in plants, in cells grown in culture, in the microbial worlds, or in the classical "test tube" experiments should all be related to their chemical and physical properties (Britton, 1995; Krinsky, 1993). For example, beta-carotene and lycopene, shown in Figure 1, are both $C_{40}H_{56}$ hydrocarbons which share many similar properties. They are very hydrophobic, each have 11 conjugated double bonds, although in the case of lycopene, they are all in a linear polyene chain, whereas with β-carotene, only 9 double bonds are in the linear chain, and 2 additional conjugated double bonds are in the rings. As determined by x-ray crystal analysis of beta-carotene, the ring double bonds are slightly out of plane because of the steric hindrance of the methyl group substituents on and near the ring. In addition, because lycopene is an acyclic compound, it has 2 more non-conjugated double bonds at the termini of the molecule. As will be discussed in section 3, these two carotenoids have different biological actions.

In addition to the hydrocarbon carotenoids, another class contains oxygen, and these carotenoids are known as xanthophylls. They are widely distributed in nature, and are present in our diet and in our serum and tissues.

Free Radicals, Oxidative Stress, and Antioxidants, edited by Özben.
Plenum Press, New York, 1998.

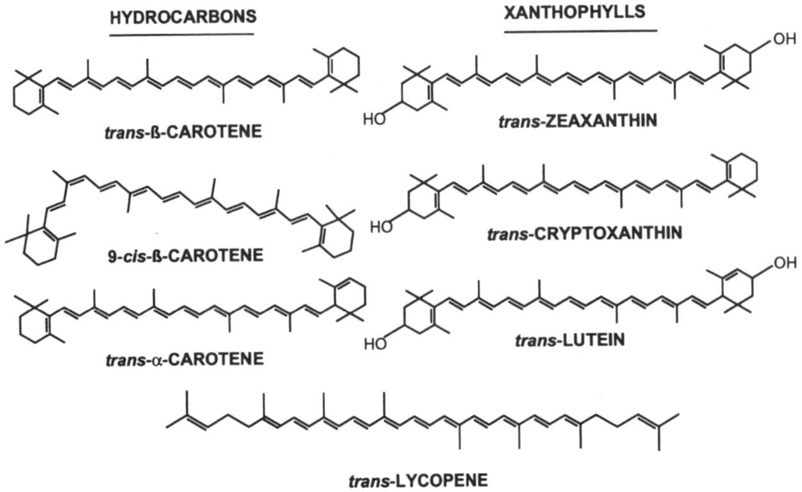

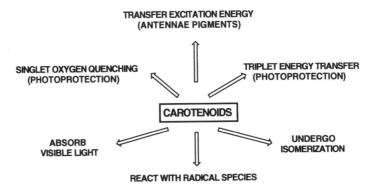

HYDROCARBONS

trans-ß-CAROTENE

9-cis-ß-CAROTENE

trans-α-CAROTENE

XANTHOPHYLLS

trans-ZEAXANTHIN

trans-CRYPTOXANTHIN

trans-LUTEIN

trans-LYCOPENE

Figure 1. The structures of several carotenoids found in human plasma.

2. CAROTENOIDS: CHEMICAL AND PHYSICAL PROPERTIES

Many of the chemical and physical properties of carotenoids have been summarized in Figure 2, in which the bicyclic carotenoid, beta-carotene, has been used as a representative of all carotenoid molecules. All of these properties have been associated with specific biological actions, which are termed Primary Biological Actions, as they can be attributed to the intact carotenoid molecule.

The structural differences between carotenoids are sufficient to allow separation by various chromatographic techniques. High-performance liquid chromatography (HPLC) has be used very extensively for carotenoid separation and isolation.

The major difference between beta-carotene and lycopene resides in the fact that beta-carotene can be metabolized into a series of shortened molecules, culminating in the formation of retinoic acid by an excentric cleavage pathway or to retinal by a central cleavage pathway. As yet we have no evidence of lycopene metabolism under similar circumstances, possibly because we have not looked very carefully. The actions of these metabolites of carotenoids has been termed the Secondary Biological Actions, and will be discussed in Section 4.

TRANSFER EXCITATION ENERGY
(ANTENNAE PIGMENTS)

SINGLET OXYGEN QUENCHING
(PHOTOPROTECTION)

TRIPLET ENERGY TRANSFER
(PHOTOPROTECTION)

CAROTENOIDS

ABSORB
VISIBLE LIGHT

UNDERGO
ISOMERIZATION

REACT WITH RADICAL SPECIES

Figure 2. The chemical and physical properties of carotenoids.

3. PRIMARY BIOLOGICAL ACTIONS

The primary biological actions derive directly from the chemical and physical properties described in Figure 2.

3.1. Light Absorption

Because of their conjugated double bond system, carotenoids absorb both visible and ultraviolet light, and their absorption spectra can be used to characterize them. In some birds and fish pigmentation due to carotenoids can be used for recognition and mating purposes, but this function has not been observed in primates. The only known light-absorbing function by carotenoids in primates may be the accumulation of two dietary xanthophylls, zeaxanthin and lutein (Fig. 1) in the macula region of the fovea centralis in the back of the retina. These yellow carotenoids would absorb blue light before it strikes the cones in the macula, and therefore, can attenuate the light response. In this way, they may be useful in decreasing the chromatic aberration associated with blue light (Landrum *et al.*, 1996; Snodderly, 1995). They could also be acting as antioxidants in this region of the retina, and thus protect the retinal tissue against photooxidative damage.

3.2. Energy Transfer

One of the wonders of nature is the fact that the family of carotenoid pigments that are present in green plants, algae, and photosynthetic bacteria have the capacity either to transfer electronic energy to chlorophyll to extend the useful range of visible light absorption, or to accept electronic energy from an excited donor, and to protect these tissues from light damage.

3.2.1. Singlet Energy Transfer. When carotenoids act as antennae pigments, they absorb visible light in wave length regions that are not accessible to chlorophylls, and form the singlet excited carotenoid (^{1}CAR), shown in Equation 1. The light energy is then transferred to a chlorophyll forming singlet excited chlorophyll (^{1}Chlorophyll) *via* the reactions described in Equation 2.

$$CAR + light \rightarrow {}^1CAR \tag{1}$$

$$^1CAR + Chlorophyll \rightarrow CAR + {}^1Chlorophyll \tag{2}$$

In this process of singlet energy transfer, carotenoids act as accessory pigments, permitting photosynthesis through a broad range of visible wave lengths.

3.2.2. Triplet Energy Transfer. In this case carotenoids act as protective agents, preventing harmful photochemical reaction from destroying tissues, either in plants or in other organisms, including humans. Sensitizers (SEN) are excited by light to form singlet excited species (^{1}SEN) (Equation 3) which can then undergo a conversion to the triplet species (^{3}SEN) through a process known as intersystem crossing (Equation 4). Then, in a subsequent reaction, the triplet sensitizer is de-activated by energy transfer to a carotenoid to form the triplet carotenoid (^{3}CAR) (Equation 5).

$$SEN + light \rightarrow {}^1SEN \tag{3}$$

$$^1\text{SEN–intersystem crossing} \rightarrow {}^3\text{SEN} \tag{4}$$

$$^3\text{SEN} + \text{CAR} \rightarrow \text{SEN} + {}^3\text{CAR} \tag{5}$$

By means of vibrational and rotational interactions with the solvent system the long polyene chain of the ^{3}CAR can dissipate its excess energy with the release of small amounts of heat in each encounter and, ultimately, re-forms the ground state carotenoid (Equation 6).

$$^3\text{CAR} \rightarrow \text{CAR} + \text{Heat} \tag{6}$$

In addition, carotenoids are protective by virtue of their ability to quench singlet excited oxygen ($^1\text{O}_2$), formed photochemically or chemically, by regenerating ground state oxygen ($^3\text{O}_2$) (Equation 7).

$$^1\text{O}_2 + \text{CAR} \rightarrow {}^3\text{O}_2 + {}^3\text{CAR} \tag{7}$$

Again, ^{3}CAR can dissipate its energy into the solvent system. This process regenerates the ground state carotenoid. Finally, there is a recently described process whereby carotenoids can quench excess photic radiation in plants by virtue of the xanthophyll cycle, through a process known as non-photochemical quenching.

3.3. Reactions with Radical Species

It has been known for many years that carotenoids "bleach," i.e., lose their color, when exposed to radicals or to oxidizing species. This phenomenon was clearly demonstrated by Packer et al. (Packer et al., 1981) who used pulse radiolysis to generate the trichloromethyl-peroxyl radical which promptly bleached β-carotene. Krinsky and Deneke (Krinsky and Deneke, 1982) then demonstrated that either β-carotene or canthaxanthin (4,4′-diketo-β,β-carotene) could protect egg yolk liposomes from radical damage induced by the addition of iron to a liposome preparation.

3.3.1. How Can Carotenoids React with Radicals? There are three possible mechanisms for the reaction of carotenoids with radical species (Figure 3). They include 1) addition; 2) electron transfer; or 3) allylic hydrogen abstraction. It was Burton and Ingold who first proposed the addition reaction (Burton and Ingold, 1984). They suggested that a lipid peroxyl radical (ROO·) might add at any place across the polyene chain, resulting in the formation of a resonance-stabilized, carbon-centered radical. Since this radical should be quite stable, it would interfere with the propagating step in lipid peroxidation and would explain the many examples of the antioxidant effect of carotenoids in solution (Palozza and Krinsky, 1992). The proposed reaction is described in Equation 8.

$$\text{ROO·} + \text{CAR} \rightarrow \text{ROO–CAR·} \tag{8}$$

1. Adduct formation: CAR + R· ------> R-CAR·

2. Electron transfer: CAR + R· ------> CAR·$^+$ + R

3. Allylic H abstraction: CAR + R· ------> CAR· + RH

Figure 3. Three possible reactions of radicals with carotenoids.

Electron transfer reactions have also been reported, resulting in the formation of the carotenoid cation radical, CAR·$^+$. While we know very little about the subsequent reactions of this species, it has been observed most frequently by very fast spectroscopic techniques such as laser flash photolysis (Tinkler *et al.*, 1996).

3.3.2. What Are the Products of the Reaction of Radicals with Carotenoids? Grosch *et al.* (Grosch *et al.*, 1976) found a series of volatile carbonyls formed when carotenoids underwent a co-oxidation with soybean lipoxygenase. Several similar compounds were characterized by Handelman *et al.* (Handelman *et al.*, 1991), following the autoxidation of β-carotene, or by Stratton *et al.* (Stratton *et al.*, 1993) when β-carotene was treated with an azo-initiator. The structures of some of these carbonyl derivatives are shown in Figure 4.

3.3.3. Under What Conditions Do Carotenoids Act as Pro-Oxidants? The carbon-centered radical formed by an addition reaction (ROO–CAR·) can react reversibly quite rapidly with oxygen to form a new peroxyl radical, ROO–CAR–OO·, as seen in Equation 9.

$$ROO–CAR· + O_2 \Leftrightarrow ROO–CAR–OO· \tag{9}$$

This new peroxyl radical could then re-initiate lipid peroxidation. Thus the carotenoids would not be acting as antioxidants, but rather as prooxidants, allowing the continuation of the lipid peroxidation chain reaction of lipid peroxidation. This process should be dependent on oxygen tension, as first described by Burton and Ingold (Burton & Ingold, 1984). Several investigators have verified the effects of oxygen tension on the antioxidant/prooxidant balance of β-carotene. In particular, Palozza *et al.* have demonstrated that in rat liver microsomes, to which β-carotene has been added, the carotenoid acts as an antioxidant at 20% oxygen, but as a prooxidant at 100% oxygen (Palozza *et al.*, 1995). Jørgensen and Skibsted also reported that β-carotene was 4 times more effective as an antioxidant at 1% oxygen as at 50% oxygen (Jørgensen and Skibsted, 1993).

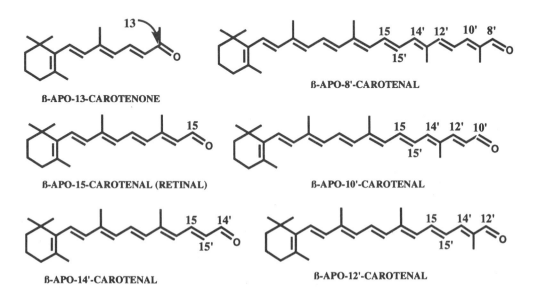

Figure 4. Structures of several apocarotenoids formed *via* radical reactions or autoxidation of β-carotene.

4. SECONDARY BIOLOGICAL ACTIONS

The secondary biological actions of carotenoids include vision, reproduction, growth and differentiation. These actions are related to their metabolites and breakdown products (Figure 5).

4.1. Reproduction

The best known metabolite is the alcohol, vitamin A or retinol, which is involved in reproduction (Appling and Chytil, 1981), and is essential for the growth of human B cells (Buck *et al.*, 1990).

4.2. Vision

The aldehyde retinal is the cofactor for the formation of rhodopsins and iodopsins. This is done by combining retinal with various opsins (Saari, 1994).

4.3. Growth and Differentiation

Growth and differentiation appear to require the use of two retinoic acids, all-*trans*- and 9-*cis*-retinoic acid. Various aspects of differentiation, such as hematopoiesis and pattern formation during embryogenesis, involve the actions of these two retinoic acids in regulating gene expression *via* interaction with specific nuclear retinoic acid receptor proteins (Chambon, 1996). In 1987, the nuclear receptor for all-*trans*-retinoic acid (RAR) was isolated (Giguere *et al.*, 1987; Petkovich *et al.*, 1987), and a few years later, the ligand for another retinoic acid receptor (RXR) was identified as 9-*cis-retinoic* acid (RXR) (Mangelsdorf *et al.*, 1990; Lehmann, 1992 [#4269]).

4.4. Metabolism

There are two pathways that have been proposed for the formation of the retinoic acids. The central cleavage pathway cleaves β-carotene at the central 15,15′-double bond. The reaction through this pathway yields two molecules of retinal (Goodman and Huang, 1965; Olson and Hayaishi, 1965).

The retinal can then be either reversibly reduced to retinol, or irreversibly oxidized to retinoic acid (Crain *et al.*, 1967). In contrast, the excentric cleavage pathway can attack any of the double bonds in the polyene chain (Figure 7), yielding a series of shortened carbonyls, called apo-carotenals. As we now know, it is the formation of both all-*trans*- and

VISION, REPRODUCTION, GROWTH, DIFFERENTIATION

DUE TO METABOLITES AND/OR OXIDATION PRODUCTS SUCH AS:

RETINAL	APO-CAROTENALS
RETINOL	APO-CAROTENOIC ACIDS
RETINOIC ACID	UNKNOWN COMPOUNDS

Figure 5. Secondary actions of carotenoids.

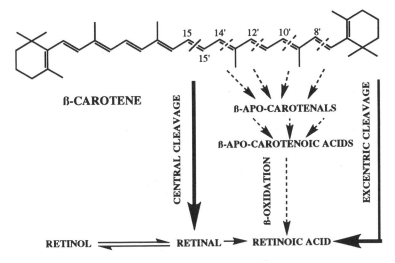

Figure 6. Central and excentric cleavage pathways for carotenoid metabolism.

9-*cis*-retinoic acid that allows the expression of many of the biological properties of retinoids. The formation of retinoids from β-carotene metabolism has been reviewed recently (Wang and Krinsky, 1997). We have been able to demonstrate, both *in vitro* (Wang *et al.*, 1994) and *in vivo* (Hebuterne *et al.*, 1995), the formation of 9-*cis*-retinoic acid from 9-*cis*-β-carotene and this process may serve as a significant pathway for the formation of this important regulator of cellular differentiation. Other workers have demonstrated 9-*cis*-retinal formation from 9-*cis*-β-carotene *in vitro* (Nagao and Olson, 1994).

Although the mechanism is not completely understood, it now appears that the excentric cleavage pathway involves mitochondrial oxidative steps that are very similar to the β-oxidation pathway that oxidize fatty acids (Wang *et al.*, 1996). The β-apo-carotenoic acid intermediates are formed *via* oxidation of the β-apo-carotenal products during excentric cleavage, as shown in Figure 6.

5. AMBIGUOUS BIOLOGICAL REACTIONS

There are many biological actions attributed to carotenoids, but it is not clear whether these actions are due to the intact carotenoid molecule or to a metabolite. I refer to these as "ambiguous actions," until such time as we can specifically identify which molecule is responsible for the observed effect. Among these actions are such important biological phenomena as modification of immunological responses, modification of the incidence of various cancers, and influence coronary heart disease.

- **MODIFY IMMUNOLOGICAL RESPONSES**

- **MODIFY INCIDENCE OF CANCER(S)**

- **MODIFY INCIDENCE OF CARDIOVASCULAR DISEASE**

Figure 7. Ambiguous actions of carotenoids.

5.1. Cancer

A great deal of attention has been placed on the possible relationship between carotenoids and cancer. The interest was stimulated by the article of Peto et al. that raised the question "Can dietary beta-carotene materially reduce human cancer rates?" (Peto *et al.*, 1981). Since then numerous observational epidemiological studies have reported significant decreases in the risk of developing cancer in those individuals who consume diets rich in fruits and vegetables and, therefore, rich in carotenoids. Since it is quite difficult to modify the eating behavior of a controlled population, investigators attempted to answer the Peto question by using supplements of synthetic β-carotene added to the diet in placebo-controlled, double blind studies. The results of the first study to be completed, The Alpha-Tocopherol, Beta-Carotene Cancer Prevention Study (ATBC), were quite disappointing, inasmuch as the authors reported an 18% increase in the incidence of lung cancer among the supplemented group of 29,000 Finnish male smokers (Group, 1994).

In 1996 two other large-scale intervention trials were reported. The Carotene and Retinol Efficacy Trial (CARET), which utilized a cohort of 14,000 smokers and 4,000 asbestos workers who were present or past smokers, terminated their intervention trial early when they observed similar results to the ATBC Study (Omenn *et al.*, 1996). At the same time, the Physician's Health Study (PHS) was completed after 12 years of β-carotene intervention with 22,000 male, US physicians and found neither harm nor benefit from the β-carotene intervention (Hennekens *et al.*, 1996).

How can these results correlate with the original dietary observations? Most probably the fruits and vegetables that were associated with reduced risk for cancer contain many more compounds than β-carotene, and the hope that a single nutrient might modify a process as complex as cancer was not met. That could well explain the results of the PHS study. However, it is more difficult to explain the results of the ATBC study and the results of the aborted CARET study. It is possible that there is some interaction that we are not aware of between high levels of β-carotene intake and the contents of smoke that exacerbates the carcinogenic properties of smoking alone. Only future experiments will be able to determine whether that actually occurs.

5.2. Age-Related Macular Degeneration (AMD)

We have already mentioned the possible role of lutein and zeaxanthin in absorbing blue light in the macular region of the retina. But, in addition to that light-absorbing property of carotenoids, they also appear to play a specific role in preventing AMD, an irreversible form of blindness associated with our elderly populations. Dietary epidemiology clearly indicates that the ingestion of dark green leafy vegetables significantly reduces the risk of developing AMD, and these types of vegetables are the best source of lutein and zeaxanthin in our diet. In addition, studies have indicated that serum lutein and zeaxanthin are also related to a lowered relative risk of neovascular AMD (Group, 1993). Much of this work has been reviewed by Snodderly (Snodderly, 1995).

Recently, we have reported that the dietary intake of total carotenoids was related to the density of the macular pigment in men, whereas only the intake of lutein and zeaxanthin was related to the density of the macular pigment in women (Hammond *et al.*, 1996). We have now been able to demonstrate that it is possible to increase the macular pigment by supplementing the diet with food sources rich in lutein and zeaxanthin, although there appears to be some individuals who are poor absorbers of dietary carotenoids and do not respond to this type of dietary change (Hammond *et al.*, 1997).

6. CONCLUSIONS

Some of the biological actions of carotenoids can be clearly associated with the intact molecule, while others are due to their metabolic or breakdown products. In addition, there is a class of biological actions that have not yet been ascribed to either the intact molecule or to metabolites. These are the areas that will continue to attract the attention of investigators interested in these fascinating and colorful, biological effectors (Olson and Krinsky, 1995).

REFERENCES

Appling, D. R., and Chytil, F., 1981, Evidence for a role of retinoic acid (vitamin A acid) in the maintenance of testosterone production in male rats, *Endocrinol.* **108**:2120–2122.

Britton, G., 1995, Structure and properties of carotenoids in relation to function, *FASEB J.* **9**:1551–1558.

Buck, J., Ritter, G., Dannecker, L., Katta, V., Cohen, S. L., Chait, B. T., and Hämmerling, U., 1990, Retinol is essential for growth of activated human B cells, *J. Exp. Med.* **171**:1613–1624.

Burton, G. W., and Ingold, K. U., 1984, β-Carotene: An Unusual Type of Lipid Antioxidant, *Science* **224**:569–573.

Chambon, P., 1996, A decade of molecular biology of retinoic acid receptors, *FASEB J.* **10**:940–953.

Crain, F. D., Lotspeich, F. J., and Krause, R. F., 1967, Biosynthesis of Retinoic Acid by Intestinal Enzymes of the Rat, *J. Lipid Res.* **8**:249–254.

Giguere, V., Ong, E. S., Segui, P., and Evans, R. M., 1987, Identification of a Receptor for the Morphogen Retinoic Acid, *Nature (London)* **330**:624–629.

Goodman, D. S., and Huang, H. S., 1965, Biosynthesis of Vitamin A with Rat Intestinal Enzymes, *Science* **149**:879–880.

Grosch, W., Laskawy, G., and Weber, F., 1976, Formation of Volatile Carbonyl Compounds and Cooxidation of β-Carotene by Lipoxygenase from Wheat, Potato, Flax, and Beans, *J. Agric. Food Chem.* **24**:456–459.

Group, E. D. C.-C. S., 1993, Antioxidant Status and Neovascular Age-Related Macular Degeneration, *Arch. Ophthalmol.* **111**:104–109.

Group, T. A.-T. B. C. C. P. S., 1994, The effect of vitamin E and beta carotene on the incidence of lung cancer and other cancers in male smokers, *New Engl. J. Med.* **330**:1029–1035.

Hammond, B. R., Jr., Curran-Celentano, J., Judd, S., Fuld, K., Krinsky, N. I., Wooten, B. R., and Snodderly, D. M., 1996, Sex Differences in Macular Pigment Optical Density: Relation to Plasma Carotenoid Concentrations and Dietary Patterns, *Vision Res.* **in press.**

Hammond, B. R., Jr., Johnson, E. J., Russell, R. M., Krinsky, N. I., Yeum, K.-J., Edwards, R. B., and Snodderly, D. M., 1997, Dietary modification of human macular pigment density, *Invest. Ophthalmol. Vis. Sci.* **38**:1795–1801.

Handelman, G. J., van Kuijk, F. J. G. M., Chatterjee, A., and Krinsky, N. I., 1991, Characterization of products formed during the autoxidation of β-carotene, *Free Radic. Biol. Med.* **10**:427–437.

Hebuterne, X., Wang, X.-D., Johnson, E. J., Krinsky, N. I., and Russell, R. M., 1995, Intestinal absorption and metabolism of 9-*cis*-β-carotene *in vivo*: biosynthesis of 9-*cis*-retinoic acid, *J. Lipid Res.* **36**:1264–1273.

Hennekens, C. H., Buring, J. E., Manson, J. E., Stampfer, M., Rosner, B., Cook, N. R., Belanger, C., LaMotte, F., Gaziano, J. M., Ridker, P. M., Willett, W., and Peto, R., 1996, Lack of effect of long-term supplementation with beta carotene on the incidence of malignant neoplasms and cardiovascular disease, *New Engl. J. Med.* **334**:1145–1149.

Jørgensen, K., and Skibsted, L. H., 1993, Carotenoid scavenging of radicals. Effect of carotenoid structure and oxygen partial pressure on antioxidative activity, *Z. Lebensm. Unters. Forsch.* **196**:423–429.

Krinsky, N. I., 1993, Actions of carotenoids in biological systems, *Annu. Rev. Nutr.* **13**:561–587.

Krinsky, N. I., and Deneke, S. M., 1982, The Interaction of Oxygen and Oxy-Radicals with Carotenoids, *JNCI* **69**:205–210.

Landrum, J. T., Bone, R. A., and Kilbum, M. D., 1996, The macular pigment: A possible role in protection from age-related macular degeneration, *Adv. Pharmacol.* **38**:537–556.

Mangelsdorf, D. J., Ong, E. S., Dyck, J. A., and Evans, R. M., 1990, Nuclear Receptor that Identifies a Novel Retinoic Acid Response Pathway, *Nature (London)* **345**:224–229.

Nagao, A., and Olson, J. A., 1994, Enzymatic formation of 9-*cis*, 13-*cis* and all-*trans* retinals from isomers of β-carotene, *FASEB J.* **8**:968–973.

Olson, J. A., and Hayaishi, O., 1965, The Enzymatic Cleavage of β-Carotene Into Vitamin A by Soluble Enzymes of Rat Liver and Intestine., *Proc. Natl. Acad. Sci. USA* **54**:1364–1370.

Olson, J. A., and Krinsky, N. I., 1995, The colorful fascinating world of the carotenoids, *FASEB J.* **9**:1547–1550.

Omenn, G. S., Goodman, G. E., Thornquist, M. D., Balmes, J., Cullen, M. R., Glass, A., Keogh, J. P., Meyskens, F. L., Jr., Valanis, B., Williams, J. H., Jr., Barnhart, S., and Hammar, S., 1996, Effects of a combination of beta carotene and vitamin A on lung cancer and cardiovascular disease, *New Engl. J. Med.* **334**:1150–1155.

Packer, J. E., Mahood, J. S., Mora-Arellano, V. O., Slater, T. F., Willson, R. L., and Wolfenden, B. S., 1981, Free Radicals and Singlet Oxygen Scavengers: Reaction of a Peroxy-radical with β-Carotene Diphenyl Furan and 1,4-Diazobicyclo(2,2,2)-octane, *Biochem. Biophys. Res. Commun.* **98**:901–906.

Palozza, P., Calviello, G., and Bartoli, G. M., 1995, Prooxidant activity of β-carotene under 100% oxygen pressure in rat liver microsomes, *Free Radic. Biol. Med.* **19**:887–892.

Palozza, P., and Krinsky, N. I., 1992, Antioxidant effects of carotenoids *in vitro* and *in vivo*: an overview, *Meth. Enzymol.* **213**:403–420.

Petkovich, M., Brand, N. J., Krust, A., and Chambon, P., 1987, A Human Retinoic Acid Receptor Which Belongs to the Family of Nuclear Receptors, *Nature (London)* **330**:444–450.

Peto, R., Doll, R. J., Buckley, J. D., and Sporn, M. B., 1981, Can dietary β-carotene materially reduce human cancer rates?, *Nature (London)* **290**:201–208.

Saari, J. C., 1994, Retinoids in photosensitive systems, in: *The Retinoids. Biology, Chemistry, and Medicine*, (M. B. Sporn, A. B. Roberts, and D. S. Goodman, eds.), vol. pp. 351–385, Raven, New York.

Snodderly, D. M., 1995, Evidence for protection against age-related macular degeneration by carotenoids and antioxidant vitamins, *Am. J. Clin. Nutr.* **62 (Suppl)**:1448S–1461S.

Stratton, S. P., Schaefer, W. H., and Liebler, D. C., 1993, Isolation and Identification of Singlet Oxygen Oxidation Products of β-Carotene, *Chem. Res. Toxicol.* **6**:542–547.

Tinkler, J. H., Tavender, S. M., Parker, A. W., McGarvey, D. J., Mulroy, L., and Truscott, T. G., 1996, Investigation of carotenoid radical cations and triplet states by laser flash photolysis and time-resolved resonance Raman spectroscopy: Observation of competitive energy and electron transfer, *J. Am. Chem. Soc.* **118**:1756–1761.

Wang, X.-D., and Krinsky, N. I., 1997, The bioconversion of β-carotene into retinoids, in: *Subcellular Biochemistry: Fat Soluble Vitamins*, (P. J. Quinn, eds.), vol. pp. in press, Plenum, London.

Wang, X.-D., Krinsky, N. I., Benotti, P. N., and Russell, R. M., 1994, Biosynthesis of 9-*cis*-retinoic acid from 9-*cis*-β-carotene in human intestinal mucosa in vitro, *Arch. Biochem. Biophys.* **313**:150–155.

Wang, X.-D., Russell, R. M., Liu, C., Stickel, F., Smith, D., and Krinsky, N. I., 1996, β-Oxidation in rabbit liver *in vitro* and in the perfused ferret liver contributes to retinoic acid biosynthesis from β-apo-carotenoic acids, *J. Biol. Chem.* **271**:26490–26498.

29

PREVENTION OF ATHEROSCLEROSIS BY α-TOCOPHEROL IN SMOOTH MUSCLE CELLS BY A MECHANISM INVOLVING SIGNAL TRANSDUCTION MODULATION

Nesrin K. Özer,[1][*] Önder Sirikci,[1] Suzan Taha,[1] N. Kaya Engin,[1]
Daniel Boscoboinik,[2] Sophie Clément,[2] Achim Stocker,[2] and Angelo Azzi[2]

[1]Department of Biochemistry
Faculty of Medicine, Marmara University
81326 Haydarpasa, Istanbul, Turkey
[2]Institut für Biochemie und Molekularbiologie
Universität Bern
Bühlstrasse 28, 3012 Bern, Switzerland

ABSTRACT

α-Tocopherol, the biologically most active form of Vitamin E, decreases in a concentration dependent way, proliferation of smooth muscle cells. At the same concentrations (10–50 μM) it induces inhibition of protein kinase C activity. Proliferation and protein kinase C inhibition by α-tocopherol, the lack of inhibition by β-tocopherol and the prevention by β-tocopherol indicate that the mechanism involved is not related to the radical-scavenging properties of these two molecules, which are essentially equal. All-rac-α-tocopherol is less potent than RRR-α-tocopherol on smooth muscle cell proliferation. probucol (10–50 μM), a potent lipophilic antioxidant, does not inhibit smooth muscle cell proliferation and protein kinase C activity. In rabbit studies, atherosclerosis was induced by a 2% cholesterol-containing, vitamin E poor-diet. Six different groups of rabbits each were received vitamin E, probucol; probucol *plus* vitamin E. After 4 weeks, aortas were analysed for protein kinase C activity. Their media smooth muscle cells exhibited an increase in protein kinase C activity. Vitamin E fully prevented cholesterol induced atherosclerotic lesions and the induction of protein kinase C activity. Probucol was not effective in preventing both cholesterol induced atherosclerotic lesions and the induction of protein kinase C activity. These results

* Correspondence should be addressed to: Prof. Nesrin K. Özer, Department of Biochemistry, Faculty of Medicine, Marmara University, 81326 Haydarpasa, Istanbul, Turkey. Tel.: (90 216) 414 47 33, (90 216) 418 10 47, Fax: (90 216) 418 10 47

Free Radicals, Oxidative Stress, and Antioxidants, edited by Özben.
Plenum Press, New York, 1998.

show that the protective effect of vitamin E against hypercholesterolemic atherosclerosis is not produced by an other antioxidant such as probucol and therefore may not be linked to the antioxidant properties of this vitamin. The effects observed at the level of smooth muscle cells *ex-vivo* and cell culture suggest an involvement of signal transduction events on the onset of atherosclerosis. It is at this level that the protective effect of vitamin E against atherosclerosis is exerted.

INTRODUCTION

Atherosclerosis, the world most common cause of disease and death (Ross, 1993) initiates with the migration and proliferation of media smooth muscle cells to the intima in the arterial wall (Raines and Ross, 1993; Bornfeldt et al., 1994). Antioxidant vitamins, and especially vitamin E, have shown to protect against the progress of atherosclerosis, as documented by epidemiological and intervention studies (Gey, 1990; Rimm et al., 1993; Stampfer et al., 1993). Recently, the Cambridge Heart Antioxidant Study (CHAOS) reported a strong protection by high vitamin E doses against the risk of fatal and non fatal myocardial infarction (Stephens et al., 1996). One of the mechanisms of α-tocopherol action is to protect low density lipoproteins from oxidation by peroxi-radicals (Esterbauer et al., 1992; Carew et al., 1987). Other functions of α-tocopherol are independent of its antioxidant characteristics (Boscoboinik et al., 1991b; Azzi et al., 1993; Boscoboinik et al., 1995; Azzi et al., 1995), such as that of moderating vascular smooth muscle cell proliferation. Smooth muscle cell proliferation is controlled by growth factor receptors, which, *via* a cascade of kinases and phosphatases, produce activation and expression of proteins necessary to the progression and completion of the cell cycle (Muller et al., 1993). Part of the main signal transduction path initiated by growth factors or hormones (reviewed in Azzi et al., 1992), protein kinase C has a complex activation mechanism, superimposed to a permissive phosphorylation (Dutil et al., 1994; Newton, 1995; Borner et al., 1989; Bornancin and Parker, 1996).

While epidemiological data show reduced coronary heart disease risk, cell culture studies carried out in our laboratories, reveal inhibition by α-tocopherol of smooth muscle cell proliferation. The latter event takes place *via* inhibition of protein kinase C activity. α-Tocopherol also inhibits low density lipoprotein stimulated smooth muscle cell proliferation and protein kinase C activity. Since β-tocopherol and other potent antioxidants, are not inhibitory, the effect of α-tocopherol is considered to be due to a non-oxidant mechanism (Özer et al., 1995; Azzi et al., 1993; Özer et al., 1993). Moreover, increased protein kinase C activity and expression in atherosclerotic rabbits was shown (Sirikci et al., 1996)

The following cell culture and animal studies were carried out to investigate the mechanism of α-tocopherol on smooth muscle cell proliferation and effect in the protection against atherosclerosis and the effect of α-tocopherol compared with β-tocopherol and probucol.

MATERIALS AND METHODS

Materials

Growth media and serum for cell culture were obtained from Gibco Laboratories. A7r5 rat aortic smooth muscle cells (SMC) were obtained from the American Type Culture Collection (Rockville, Maryland, USA). Phorbol 12-myristate 13-acetate (PMA) and streptolysin-O (25,000 units) were from Sigma (Buchs, Switzerland). γ^{32}ATP (30 Ci/mmol), methyl-^{3}H thymidine (25 Ci/mmol) were purchased from Amersham International. RRR-α-tocopherol was a generous gift from Henkel Corporation (La Grange, USA), *all-rac*-α-tocopherol and *all-rac*-β-tocopherol from Merck (Darmstadt, Germany). Vitamin E-free rabbit diet and vitamin E (Ephynal) was kindly donated by Dr. U. Moser from Hoffman La Roche (Basel, Switzerland, and probucol was a gift from Marion Merrell Dow Pharmaceuticals Inc., Switzerland. The peptide PLSRTLSVAAKK used as a substrate for the protein kinase C assay was synthesized by Dr. Servis, Epalinges (Switzerland). The kit for measurement of protein kinase C activity was obtained from Upstate Biotechnology Inc., Lake Placid, New York, USA. All other chemicals used were of the purest grade commercially available.

Cell Culture

A7r5 cells are maintained in Dulbecco's modified Eagle's medium (DMEM) supplemented with 10% (v/v) foetal calf serum. Cells in a sub-confluent state are made quiescent by incubation in DMEM containing 0.2% FCS for 48h. Then, cells are washed with phosphate-buffered saline (PBS) and treated as indicated in the figure legends. Viability is determined by the trypan blue dye method.

Determination of Cell Number and [^{3}H]thymidine Incorporation

Quiescent A7r5 cells were restimulated to grow by addition of 10% FCS. Tocopherols and probucol were added to cells at the indicated concentrations. Cell number was determined 48 h later by using an hemocytometer.

To measure DNA synthesis, cells were pulsed for 1 h with [^{3}H]thymidine (1 μCi/well) at the indicated times following restimulation with FCS. After labelling, cells were washed twice with PBS, fixed for 20 min. with ice-cold 5% trichloroacetic acid, and solubilised in 0.1 M NaOH/2% Na$_2$CO$_3$/1% SDS. The radioactivity incorporated into the acid-insoluble material was determined in a liquid scintillation analyser.

Determination of PKC Activity in Permeabilized Cells

Activity of PKC in permeabilized smooth muscle cells was performed according to the procedure of Alexander et al. (Alexander et al., 1990) with minor modifications. Quiescent A7r5 cells were subjected to different treatments as indicated. During the last hour of the preincubation period, cells were treated with 100 nM phorbol 12-myristate 13-acetate. Then, cells were washed twice with PBS, resuspended in intracellular buffer (5.2 mM MgCl$_2$, 94 mM KCl, 12.5 mM HEPES, 12.5 mM EGTA, 8.2 mM CaCl$_2$, pH 7.4) and divided in 220 μl portions (1.5×10^5 cells). Assays were started by adding γ-^{32}P ATP (9 cpm/pmol, final concentration 250 μM), peptide substrate (final concentration 70 μM), and streptolysin-O (0.3 IU). The reaction mixtures were incubated at 37 °C for 10 min. and the reaction was stopped by addition of 100 μl of 25% (w/v) trichloroacetic acid in 2 M acetic acid. After 10 min. on ice, 100 μl aliquots were spotted onto 4.5 cm × 4.5 cm P81 ion exchange chromatographic paper within pre-drawn squares (Whatman International). The paper was then washed twice for 10 min. with 30% (v/v) acetic acid containing 1% phosphoric acid and once with ethanol. The amount of the two washes was 10 ml each per spotted square. The P81 paper was dried, the squares cut and radioactivity was counted in a liquid scintillation analyzer. Basal phosphorylation in the absence of the peptide was subtracted from the experimental data to determine the specific activity.

Animal Experiments

Thirty male New Zealand albino rabbits of 2–4 months of age were assigned randomly to one of the following six groups. All rabbits were fed 100 gr. per day of vitamin E poor diet. Probucol and cholesterol were added to the diet as diethyl ether solution. The control diet was treated with the same amount of pure solvent. All diets were dried of the solvent before use. The concentrations of cholesterol, vitamin E and probucol used were based on previous literature reports (Stein et al., 1989; Ferns et al., 1992; Bocan et al., 1993; Mantha et al., 1993; Konneh et al., 1995).

One group of rabbits was only fed with the diet, without additions or treatments. The second group received injections of 50 mg/kg of vitamin E intramuscularly on alternate rear legs, once daily. The diet of the third group contained 2% cholesterol. The diet of the fourth group 2% cholesterol and the rabbits received injections of 50 mg/kg of vitamin E per day. The fifth group of rabbits received 1% probucol in addition to 2% cholesterol in the daily diet. The sixth group had 2% cholesterol and 1% probucol in the diet and was injected with 50 mg/kg of vitamin E per day.

After 4 weeks plasma samples were taken and thoracic aortas were removed. Samples from thoracic aortas media was taken to measure smooth muscle cell protein kinase C activity by using a non-radioactive kit (Upstate Biotechnology Inc.). Plasma cholesterol level were determined using an automated enzymatic technique. Vitamin E levels were measured by reverse phase high pressure liquid chromatography (Nierenberg and Nann, 1992).

RESULTS

The effect of α-tocopherol on smooth muscle cell proliferation and on protein kinase C is shown in Fig. 1. A parallel inhibition of protein kinase C activity and of proliferation is observed to occur at concentrations of α-tocopherol close to those measured in healthy adults (Gey, 1990; Riemersma et al., 1991). The following experiments were designed to establish if the above effects of α-tocopherol were related to its antioxidant properties, if a direct interaction was responsible for the inhibition, and which of the two events, protein kinase C activity or proliferation inhibition was the cause of the other.

α-Tocopherol at concentrations of 50 μM inhibits rat A7r5 smooth muscle cell proliferation (Fig. 2), while β-tocopherol is ineffective. When α-tocopherol and β-tocopherol are added together, no inhibition of cell growth is seen. Both compounds are transported equally in cells and do not compete each other for the uptake (Azzi et al., 1995). The prevention by β-tocopherol of the proliferation inhibition by α-tocopherol suggests a site-directed event at the basis of α-tocopherol inhibition rather than a general radical scaveng-

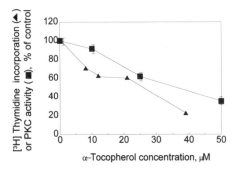

Figure 1. Parallel inhibition of protein kinase C activity and of proliferation in smooth muscle cell. Serum-deprived quiescent cells were stimulated to grow by FCS in the presence or absence of α-tocopherol. After 7 h PKC was measured. For DNA synthesis determination, a 1 h pulse of 1 μCi/well [^{3}H]thymidine was given to the cells 15 h after restimulation with FCS.

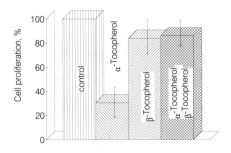

Figure 2. Differential effects of α-tocopherol and β-tocopherol on smooth muscle cell proliferation. Quiescent cells are restimulated to grow with FCS (10%) in the presence of α-tocopherol and/or β-tocopherol (50 μM). After 48 h restimulation cells were counted with hemocytometer. Viability was greater than 95%.

ing reaction. The oxidised product of α-tocopherol, α-tocopherylquinone, is not inhibitory indicating that the effects of α-tocopherol are not related to its antioxidant properties (Azzi et al., 1995). Inhibitory effects of α-tocopherol are also observed in primary human aortic smooth muscle cells (hAOMSC from Clonetics Corp., San Diego, CA and four cell strains provided by Dr. T. Resink, Kantonspital, Basel, Switzerland).

In smooth muscle cells permeabilized with streptolysin-O, to permit the entry of a peptide substrate to measure protein kinase C activity (Fig. 3), α-tocopherol inhibits protein kinase C activity, whereas β-tocopherol is ineffective. When both are present β-tocopherol prevents the inhibitory effect of α-tocopherol. The inhibition by α-tocopherol and the lack of inhibition by β-tocopherol of cell proliferation and protein kinase C activity shows that the mechanism involved is not related to the radical scavenging properties of these two molecules, which are essentially equal (Pryor et al., 1993).

In Fig. 4 the effect of RRR-α-tocopherol and *all-rac*-α-tocopherol on smooth muscle cell proliferation and protein kinase C activity have been studied. The data of Fig. 4 show that 25 μM RRR-α-tocopherol and 50 μM *all-rac*-α-tocopherol gave an equal of inhibition of cell proliferation and protein kinase C activity. Thus it appears that the *all-rac* form of α-tocopherol is 50% as effective as the RRR-α-tocopherol on smooth muscle cells.

Fig. 5 addresses the question whether probucol, a potent lipophilic antioxidant, exerts an effect on smooth muscle cell proliferation and protein kinase C activity. Probucol (10–50 μM) does not inhibit smooth muscle cell proliferation and protein kinase C activity.

The following animal experiments addresses the question if vitamin E can prevent development of atherosclerosis *in vivo* and whether probucol can mimic the effect of α-tocopherol. Cholesterol and vitamin E plasma concentrations of the 6 animal groups are shown in Table 1. 2% cholesterol diet supplementation for 4 weeks resulted in an approximately 20 fold increase of plasma cholesterol. After additional supplementation with vitamin E or probucol, plasma cholesterol increased 12- and 13-fold respectively, relative to control. Plasma vitamin E concentrations were higher in the cholesterol fed rabbits in agreement

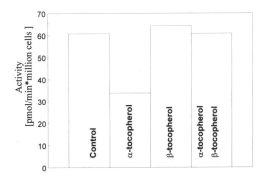

Figure 3. Effect of α-tocopherol and β-tocopherol on smooth muscle cell protein kinase C activity. Quiescent cells are restimulated to grow with FCS (10%) in the presence of α-tocopherol and/or β-tocopherol (50 μM). After 7 h restimulation cells are permeabilized and protein kinase C is measured as described in Materials and Methods. Phorbol myristate acetate (100 nM) is added 60 min before assaying activity. The basal kinase activity is subtracted in all samples and only the PMA-stimulated activity is shown.

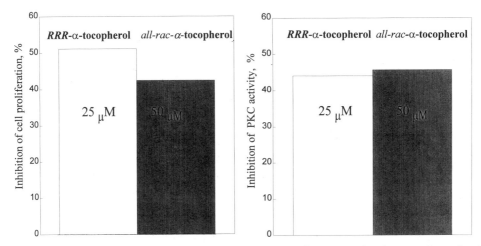

Figure 4. Comparison between the effect of RRR-α-tocopherol and *all-rac-*α-tocopherol on smooth muscle cell proliferation and protein kinase C activity. Quiescent cells are restimulated to grow with FCS (10%) in the presence of indicated amounts of tocopherols. Cell proliferation and protein kinase C activity were measured as indicated under Materials and Methods.

with literature data (Godfried et al., 1989; Wilson et al., 1978; Bitman et al., 1976; Bjorkhem et al., 1991; Morel et al., 1994) but the values corrected for the plasma cholesterol concentrations were of a similar order of magnitude.

Previous results have shown that protein kinase C activity and expression were up-regulated by cholesterol, while vitamin E counteraction, although visible as a tendency, was not statistically significant (Sirikci et al., 1996). In this study (Table 2), in the total absence of vitamin E in the diet, the activity of smooth muscle cell protein kinase C was 8.4 Δ-Absorbance units/min/mg protein. After vitamin E treatment an approximately 50% reduction of protein kinase C activity was seen. With cholesterol supplementation the activity of protein kinase C increased to 10.2 Δ-Absorbance units/min/mg protein. Vitamin E treatment was able in this case to reduce protein kinase C activity values to those meas-

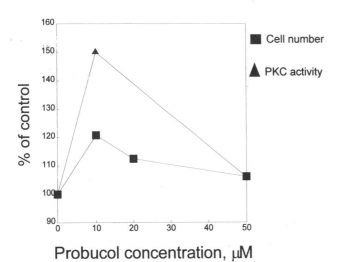

Figure 5. Effect of probucol on smooth muscle cell proliferation and protein kinase C activity. Quiescent cells are restimulated to grow with FCS (10%) in the presence of indicated amounts of probucol. Cell proliferation and protein kinase C activity were measured as indicated under Materials and Methods.

Table 1. Effect of cholesterol probucol and vitamin E treatment
on their plasma levels in rabbits[a]

Group	Cholesterol (mmol/L)	Vitamin E (μmol/L)	Vitamin E/ cholesterol ratio
Control	1.0 ± 0.5	5.2 ± 2.4	5.2
Vitamin E	0.9 ± 0.1	50.7 ± 10.2*	56.3
Cholesterol	21.1 ± 10.7*	14.6 ± 6.5**	0.7
Cholesterol + vitamin E	12.4 ± 6.1*	216.2 ± 70.2*	17.4
Cholesterol + probucol	13.7 ± 6.6*	12.4 ± 4.5**	0.9
Cholesterol + probucol + vitamin E	11.6 ± 3.5*	254.7 ± 64.0*	21.9

[a]The plasma level of cholesterol and vitamin E have been measured in all the five rabbits of the six diet groups. The numbers (mean ± S.D.) represent the plasma values measured after one month diet.
*$p < 0.01$; ** = not statistically significant

ured in the absence of cholesterol (4.5 Δ-Absorbance units/min/mg protein). Smooth muscle cells from probucol *plus* cholesterol treated rabbits showed a protein kinase C activity lower than that in the presence of cholesterol alone, although the data were not statistically significant. Instead, when the rabbits were treated also with vitamin E the protein kinase C value decreased to 5.2 Δ-Absorbance units/min/mg protein and reached significance relative to the cholesterol fed rabbit group.

DISCUSSION

We have observed a specific inhibition of proliferation *in vitro* of smooth muscle cells by α-tocopherol at physiological concentrations. Of particular relevance was the finding of an antiproliferative effect of α-tocopherol on rat and human smooth muscle cells. Proliferation of these cells is fundamental in the onset of accelerated atherosclerosis, after cardiac transplantation and restenosis. Multiple pathways are involved in the proliferative response of smooth muscle cell to an extracellular stimulus, which has rendered difficult to identify precise targets of antiproliferative agents. One of the important ele-

Table 2. Protein kinase C activity from smooth muscle cell homogenates
obtained from differently treated rabbits[a]

Group	Total protein kinase C activity (Δ absorbance/min/mg protein)	
Control	8.4 ± 1.1	
Vitamin E	4.5 ± 2.5	$p < 0.01$°
Cholesterol	10.2 ± 2.4	
Cholesterol + vitamin E	4.5 ± 1.0	$p < 0.02$*
Cholesterol + probucol	6.8 ± 1.8	NS*
Cholesterol + probucol + vitamin E	5.2 ± 1.4	$p < 0.04$*

[a]Aortic medias were minced, homogenised and nuclei sedimented by centrifugation. Supernatants were centrifuged again at $100000 \times g$ to obtain cytosolic fractions. Pellets were used for preparation of membrane fractions. Protein kinase C activity was measured in both fractions. Since they did not show significant membrane/cytosol distribution changes, only the values of total protein kinase C activity are reported. Results are expressed as mean ± S.D. (n = 5).
° referred to control group; * referred to cholesterol group; NS, not statistically significant.

ments in signal transduction cascades is protein kinase C and its inhibition appears to be sufficient to cause inhibition of smooth muscle cell proliferation *in vitro*.

The inhibition of smooth muscle cell proliferation *in vitro* by α-tocopherol at physiological concentrations may explain the finding that *in vivo* smooth muscle cell only proliferate under stress situations (Raines and Ross, 1993; Clowes and Schwartz, 1985). A local or generalised diminution of α-tocopherol concentration, caused by dietary or oxidative factors, can lead to cell growth stimulation and atherosclerosis progress. β-Tocopherol, an antioxidant almost as potent as α-tocopherol, not only does not show any effect at the level of cell proliferation or protein kinase C activity, but rather it prevents the effects of α-tocopherol. Thus, it is legitimate to conclude that the mechanism of action of α-tocopherol, as a regulator of smooth muscle cell proliferation is, is not due to its antioxidant properties. The data discussed above speak for the existence of an α-tocopherol-binding protein, binding α-tocopherol as an agonist and β-tocopherol as an antagonist.

β-Tocopherol, an antioxidant almost as potent as α-tocopherol, show only a minor effect at the level of cell proliferation or protein kinase C activity. Since β-tocopherol prevents the effects of α-tocopherol it seems justifiable to conclude that the mechanism of action of α-tocopherol as a regulator of smooth muscle cell proliferation cannot be associated with its antioxidant properties. Moreover, the above results can be the interpreted in terms of the existence of a common intermediate, a putative α-tocopherol-binding protein, able to bind α- and β-tocopherol with similar affinity, α-tocopherol acting as an agonist and β-tocopherol as an antagonist.

By comparing α-tocopherol with analogous compounds exhibiting similar antioxidant properties such as β-tocopherol or probucol it was concluded that α-tocopherol exerts its action independently of its free-radical scavenger capacity and most probably by interacting with a yet not characterised receptor molecule in smooth muscle cells (Boscoboinik et al., 1991b; Boscoboinik et al., 1994; Boscoboinik et al., 1995).

Several groups have reported that antioxidant vitamins, and especially vitamin E, have an important anti-atherogenic role (Rimm et al., 1993; Stampfer et al., 1993; Gey et al., 1993; Stahelin et al., 1992; Steinberg, 1995). The question posed in this study has been if molecular events in signal transduction can be regulated by cholesterol and vitamin E. Since the importance of protein kinase C in smooth muscle cell proliferation was reported in many studies (Castellot, Jr., et al., 1989; Matsumoto and Sasaki, 1989; Chatelain et al., 1993; Newby et al., 1995) and its proliferation is inhibited by α-tocopherol (Boscoboinik et al., 1995; Boscoboinik et al., 1991a; Boscoboinik et al., 1991b) the activity in the rabbit aorta smooth muscle cells was measured. The present *in vivo* data of cholesterol stimulation and vitamin E inhibition of protein kinase C activity, agree with previous results both obtained *in vitro* (Boscoboinik et al., 1995; Boscoboinik et al., 1991a; Boscoboinik et al., 1991b; Özer et al., 1993; Özer et al., 1995) and *in vivo* (Sirikci et al., 1996). It appears that probucol does not act as a protective agent against atherosclerotic plaque formation in the absence of vitamin E (Özer et al., 1997). It appears also that among the molecular events that may be responsible for the protection by vitamin E against atherosclerotic plaque formation, inhibition of smooth muscle protein kinase C may play a crucial role. In summary, these studies provide a molecular interpretation to the epidemiological information linking a decrease of plasma α-tocopherol with an increased risk of ischemic heart disease.

In conclusion, at physiological concentrations, α-tocopherol effects are indicative of a site-directed recognition mechanism involving a series of events including the binding of α-tocopherol to a "receptor protein", activation (or expression) of a protein phosphatase, dephosphorylation of protein kinase, inhibition of protein kinase C activity and finally by affecting gene transcription inhibition of cell proliferation.

ACKNOWLEDGMENTS

This study was supported by the Marmara University Research Foundation, F. Hoffmann-La Roche A.G., Swiss National Science Foundation and by a UNESCO–MCBN grant.

REFERENCES

Alexander, D.R., Graves, J.D., Lucas, S.C., Cantrell, D.A., and Crumpton, M.J. (1990). A method for measuring protein kinase C activity in permeabilized T lymphocytes by using peptide substrates. Evidence for multiple pathways of kinase activation. Biochem. J. *268*, 303–308.

Azzi, A., Boscoboinik, D., and Hensey, C. (1992). The protein kinase C family. Eur. J. Biochem. *208*, 547–557.

Azzi, A., Boscoboinik, D., Chatelain, E., Özer, N.K., and Stäuble, B. (1993). d-alpha-tocopherol control of cell proliferation. Mol. Aspects. Med. *14*, 265–271.

Azzi, A., Boscoboinik, D., Marilley, D., Özer, N.K., Stäuble, B., and Tasinato, A. (1995). Vitamin E: A sensor and an information transducer of the cell oxidation state. Am. J. Clin. Nutr. *62 Suppl.* 1337S–1346S.

Bitman, J., Weyant, J., Wood, D.L., and Wrenn, T.R. (1976). Vitamin e, cholesterol, and lipids during atherogenesis in rabbits. Lipids *11*, 449–461.

Bjorkhem, I., Henriksson Freyschuss, A., Breuer, O., Diczfalusy, U., Berglund, L., and Henriksson, P. (1991). The antioxidant butylated hydroxytoluene protects against atherosclerosis. Arterioscler. Thromb. *11*, 15–22.

Bocan, T.M., Mueller, S.B., Mazur, M.J., Uhlendorf, P.D., Brown, E.Q., and Kieft, K.A. (1993). The relationship between the degree of dietary-induced hypercholesterolemia in the rabbit and atherosclerotic lesion formation. Atherosclerosis *102*, 9–22.

Bornancin, F. and Parker, P.J. (1996). Phosphorylation of threonine 638 critically controls the dephosphorylation and inactivation of protein kinase Cα. Curr. Biol. *6*, 1114–1123.

Borner, C., Filipuzzi, I., Wartmann, M., Eppenberger, U., and Fabbro, D. (1989). Biosynthesis and posttranslational modifications of protein kinase C in human breast cancer cells. J. Biol. Chem. *264*, 13902–13909.

Bornfeldt, K.E., Raines, E.W., Nakano, T., Graves, L.M., Krebs, E.G., and Ross, R. (1994). Insulin-like growth factor-I and platelet-derived growth factor-BB induce directed migration of human arterial smooth muscle cells via signaling pathways that are distinct from those of proliferation. J. Clin. Invest. *93*, 1266–1274.

Boscoboinik, D., Szewczyk, A., and Azzi, A. (1991a). Alpha-tocopherol (vitamin E) regulates vascular smooth muscle cell proliferation and protein kinase C activity. Arch. Biochem. Biophys. *286*, 264–269.

Boscoboinik, D., Szewczyk, A., Hensey, C., and Azzi, A. (1991b). Inhibition of cell proliferation by alpha-tocopherol. Role of protein kinase C. J. Biol. Chem. *266*, 6188–6194.

Boscoboinik, D., Özer, N.K., Moser, U., and Azzi, A. (1995). Tocopherols and 6-hydroxy-chroman-2-carbonitrile derivatives inhibit vascular smooth muscle cell proliferation by a nonantioxidant mechanism. Arch. Biochem. Biophys. *318*, 241–246.

Boscoboinik, D.O., Chatelain, E., Bartoli, G.M., Stäuble, B., and Azzi, A. (1994). Inhibition of protein kinase C activity and vascular smooth muscle cell growth by d-alpha-tocopherol. Biochim. Biophys. Acta *1224*, 418–426.

Carew, T.E., Schwenke, D.C., and Steinberg, D. (1987). Antiatherogenic effect of probucol unrelated to its hypocholesterolemic effect: evidence that antioxidants in vivo can selectively inhibit low density lipoprotein degradation in macrophage-rich fatty streaks and slow the progression of atherosclerosis in the Watanabe heritable hyperlipidemic rabbit. Proc. Natl. Acad. Sci. U. S. A. *84*, 7725–7729.

Castellot, J.J., Jr., Pukac, L.A., Caleb, B.L., Wright, T.C., Jr., and Karnovsky, M.J. (1989). Heparin selectively inhibits a protein kinase C-dependent mechanism of cell cycle progression in calf aortic smooth muscle cells (published erratum appears in J Cell Biol 1990 Mar; 110(3): 863). J. Cell Biol. *109*, 3147–3155.

Chatelain, E., Boscoboinik, D.O., Bartoli, G.M., Kagan, V.E., Gey, F.K., Packer, L., and Azzi, A. (1993). Inhibition of smooth muscle cell proliferation and protein kinase C activity by tocopherols and tocotrienols. Biochim. Biophys. Acta *1176*, 83–89.

Clowes, A.W. and Schwartz, S.M. (1985). Significance of quiescent smooth muscle migration in the injured rat carotid artery. Circ. Res. *56*, 139–145.

Dutil, E.M., Keranen, L.M., DePaoli-Roach, A.A., and Newton, A.C. (1994). *In vivo* regulation of protein kinase C by trans-phosphorylation followed by autophosphorylation. J. Biol. Chem. *269*, 29359–29362.

Esterbauer, H., Waeg, G., Puhl, H., Dieber Rotheneder, M., and Tatzber, F. (1992). Inhibition of LDL oxidation by antioxidants. EXS. *62*, 145–157.

Ferns, G.A., Forster, L., Stewart Lee, A., Konneh, M., Nourooz Zadeh, J., and Anggard, E.E. (1992). Probucol inhibits neointimal thickening and macrophage accumulation after balloon injury in the cholesterol-fed rabbit. Proc. Natl. Acad. Sci. U. S. A. *89*, 11312–11316.

Gey, K.F. (1990). The antioxidant hypothesis of cardiovascular disease: epidemiology and mechanisms. Biochem. Soc. Trans. *18*, 1041–1045.

Gey, K.F., Moser, U.K., Jordan, P., Stahelin, H.B., Eichholzer, M., and Ludin, E. (1993). Increased risk of cardiovascular disease at suboptimal plasma concentrations of essential antioxidants: an epidemiological update with special attention to carotene and vitamin C. Am. J. Clin. Nutr. *57*, 787S–797S.

Godfried, S.L., Combs, G.F., Jr., Saroka, J.M., and Dillingham, L.A. (1989). Potentiation of atherosclerotic lesions in rabbits by a high dietary level of vitamin E. Br. J. Nutr. *61*, 607–617.

Konneh, M.K., Rutherford, C., Li, S.-R., Änggård, E.E., and Ferns, G.A.A. (1995). Vitamin E inhibits the intimal response to balloon catheter injury in the carotid artery of the cholesterol-fed rat. Atherosclerosis *113*, 29–39.

Mantha, S.V., Prasad, M., Kalra, J., and Prasad, K. (1993). Antioxidant enzymes in hypercholesterolemia and effects of vitamin E in rabbits. Atherosclerosis *101*, 135–144.

Matsumoto, H. and Sasaki, Y. (1989). Staurosporine, a protein kinase C inhibitor interferes with proliferation of arterial smooth muscle cells. Biochem. Biophys. Res. Commun. *158*, 105–109.

Morel, D.W., De la Llera-Moya, M., and Friday, K.E. (1994). Treatment of cholesterol-fed rabbits with dietary vitamins E and C inhibits lipoprotein oxidation but not development of atherosclerosis. J. Nutr. *124*, 2123–2130.

Muller, R., Mumberg, D., and Lucibello, F.C. (1993). Signals and genes in the control of cell-cycle progression. Biochim. Biophys. Acta *1155*, 151–179.

Newby, A.C., Lim, K., Evans, M.A., Brindle, N.P., and Booth, R.F. (1995). Inhibition of rabbit aortic smooth muscle cell proliferation by selective inhibitors of protein kinase C. Br. J. Pharmacol. *114*, 1652–1656.

Newton, A.C. (1995). Protein kinase C: Structure, function, and regulation. J. Biol. Chem. *270*, 28495–28498.

Nierenberg, D.W. and Nann, S.L. (1992). A method for determining concentrations of retinol, tocopherol, and five carotenoids in human plasma and tissue samples. Am. J. Clin. Nutr. *56*, 417–426.

Özer, N.K., Palozza, P., Boscoboinik, D., and Azzi, A. (1993). d-alpha-Tocopherol inhibits low density lipoprotein induced proliferation and protein kinase C activity in vascular smooth muscle cells. FEBS Lett. *322*, 307–310.

Özer, N.K., Boscoboinik, D., and Azzi, A. (1995). New roles of low density lipoproteins and vitamin E in the pathogenesis of atherosclerosis. Biochem. Mol. Biol. Int. *35*, 117–124.

Özer, N.K., Sirikci, Ö., Taha, S., San, T., Moser, U., and Azzi, A. (1997). Effect of vitamin E and probucol on dietary cholesterol-induced atherosclerosis in rabbits. Free Radic. Biol. Med. *In press*.

Pryor, A.W., Cornicelli, J.A., Devall, L.J., Tait, B., Trivedi, B.K., Witiak, D.T., and Wu, M. (1993). A rapid screening test to determine the antioxidant potencies of natural and synthetic antioxidants. J. Org. Chem. *58*, 3521–3532.

Raines, E.W. and Ross, R. (1993). Smooth muscle cells and the pathogenesis of the lesions of atherosclerosis. Br. Heart J. *69*, S30–S37.

Riemersma, R.A., Wood, D.A., Macintyre, C.C., Elton, R.A., Gey, K.F., and Oliver, M. (1991). Risk of angina pectoris and plasma concentrations of vitamins A, C, and E and carotene (see comments). Lancet *337*, 1–5.

Rimm, E.B., Stampfer, M.J., Ascherio, A., Giovannucci, E., Colditz, G.A., and Willett, W.C. (1993). Vitamin E consumption and the risk of coronary heart disease in men (see comments). N. Engl. J. Med. *328*, 1450–1456.

Ross, R. (1993). The pathogenesis of atherosclerosis: a perspective for the 1990s. Nature *362*, 801–809.

Sirikci, Ö., Özer, N.K., and Azzi, A. (1996). Dietary cholesterol-induced changes of protein kinase C and the effect of vitamin E in rabbit aortic smooth muscle cells. Atherosclerosis *126*, 253–263.

Stahelin, H.B., Eichholzer, M., and Gey, K.F. (1992). Nutritional factors correlating with cardiovascular disease: results of the Basel Study. Bibl. Nutr. Dieta. *49*, 24–35.

Stampfer, M.J., Hennekens, C.H., Manson, J.E., Colditz, G.A., Rosner, B., and Willett, W.C. (1993). Vitamin E consumption and the risk of coronary disease in women (see comments). N. Engl. J. Med. *328*, 1444–1449.

Stein, Y., Stein, O., Delplanque, B., Fesmire, J.D., Lee, D.M., and Alaupovic, P. (1989). Lack of effect of probucol on atheroma formation in cholesterol-fed rabbits kept at comparable plasma cholesterol levels. Atherosclerosis *75*, 145–155.

Steinberg, D. (1995). Clinical trials of antioxidants in atherosclerosis: Are we doing the right thing. Lancet *346*, 36–38.

Stephens, N.G., Parsons, A., Schofield, P.M., Kelly, F., Cheeseman, K., Mitchinson, M.J., and Brown, M.J. (1996). Randomised controlled trial of vitamin E in patients with coronary disease: Cambridge Heart Antioxidant Study (CHAOS). Lancet *347*, 781–786.

Wilson, R.B., Middleton, C.C., and Sun, G.Y. (1978). Vitamin E, antioxidants and lipid peroxidation in experimental atherosclerosis of rabbits. J. Nutr. *108*, 1858–1867.

30

POTENTIAL EFFECTS OF DIETARY VITAMIN E IN LABORATORY ANIMAL DIETS ON RESULTS OBTAINED IN MODELS OF DISEASE

Hans-Anton Lehr,[1][*] Peter Vajkoczy,[2] and Michael D. Menger[3]

[1]Institute of Pathology
University of Mainz, Germany
[2]Neurosurgical Clinic
University of Heidelberg at Mannheim, Germany
[3]Institute for Clinical Experimental Surgery
University of Homburg/Saar, Germany

ABSTRACT

This chapter will demonstrate the importance of vitamin E dietary supplements on cell functions, on different biomedical systems, and on diverse pathophysiological conditions, with special emphasis being placed on cardiovascular pathophysiology. These effects are presented as a rationale for the use of vitamin E supplements in diets of farm animals, pets, and also of laboratory animals. The latter point gives reason for considerable problems, which are discussed in the last chapter of this article.

1. VITAMIN E IN CARDIOVASCULAR PATHOPHYSIOLOGY: EFFECTS ON CELLS AND SIGNAL TRANSDUCTION PATHWAYS

1.1. Leukocytes

Vitamin E has diverse effects on blood cells, including leukocytes, platelets, and endothelial cells. In leukocytes, vitamin E exerts pronounced inhibitory (= anti-inflammatory) effects: Vitamin E has been described to reduce the generation of reactive oxygen species, as assessed by lucigenin-amplified chemiluminescence (Herbazcynska et al.,

*Address for correspondence: Prof. Dr. med. Hans-Anton Lehr, Institute of Pathology, Johannes Gutenberg University, Langenbeckstrasse 1, D-55101 Mainz, Germany. Tel: ++-49-6131-173269, Fax ++-49-6131-176604.

Free Radicals, Oxidative Stress, and Antioxidants, edited by Özben.
Plenum Press, New York, 1998.

1994; Devaraj and Jialal, 1996) and of cytokines, such as interleukin (lL-1) (Devaraj and Jialal, 1996), an important stimulus of adhesion molecule expression on endothelial cells (Faruqi et al., 1994) and smooth muscle cells (Wang et al., 1995). It has the ability to reduce leukocyte directed chemotaxis (Luostarinen et al., 1991), neutrophil and monocyte aggregation in response to arachidonic acid and leukotriene B4 (Steiner and Mower, 1982; Villa et al., 1986), and their adhesion to different surfaces, in particular of endothelial cells (Lafuze et al., 1984; Boogaerts et al., 1984; Devaraj et al., 1996). In a recent study, Devaraj and Jialal demonstrated the pronounced inhibitory effect of dietary vitamin E supplementation in human volunteers on monocyte synthesis of superoxide and of IL-I, as well as the formation of thiobarbituric acid reactive substances as a measure of lipid peroxidation. The inhibition of these substances by vitamin E was entirely reversible after a washout period of 6 weeks (Devaraj and Jialal, 1996). Interestingly, vitamin E supplements also significantly reduced the adhesion of monocytes to endothelial cells and this effects was likewise entirely reversible after a six-week washout period (Devaraj and Jialal, 1996). Of particular interest for the purpose of this review is the finding that a dietary vitamin E supplement of only 30 mg vitamin E 1kg diet significantly reduces calcium ionophore stimulated lipoxygenase activity in rat neutrophils, while no additional inhibition could be observed by raising the vitamin E supplement into megadose ranges of up to 3000 mg/kg (Chan et al., 1989). See also Table 1.

1.2. Platelets

Studies on platelets could demonstrate that vitamin E reduces aggregation (Stivastava, 1986; Salonen et al., 1991; Polette and Blache, 1992; de Lorgeril et al., 1994), thromboxane synthesis (Hamelin and Chan, 1983) as well as adhesion to diverse surfaces (Jandak et al., 1988; Steiner et al., 1995). The study by Jandak and coworkers reported that treating human volunteers with dietary vitamin E supplements (400 lU vitamin E/day) significantly reduced the adhesion of human platelets to surfaces coated with collagen type I and type IV as well as with fibronectin. No further inhibition was observed by gradually increasing the vitamin dose up to 1600 lU/day, suggesting that the maximal effect of platelet function was obtained at a daily dose of 400IU per day. See also Table 1.

1.3. Endothelial Cells

Of particular interest are the effects of vitamin E on endothelial cells: in a variety of different models, vitamin E preserves endothelial cell-dependent vasodilatation (Keaney et al., 1993; Andersson et al., 1994; Rubino and Burnstock, 1994), the generation of the vasodilatatory and antiaggregatory prostacyclin (Chan and Leigh, 1981; Umeda et al., 1990), while at the same time reducing the generation in platelets of the pro-aggregatory and vasoconstricting thromboxane (Meydani, 1992). Treatment of endothelial cells in culture with vitamin E reduces the adhesion of leukocytes (Faruqi et al., 1994). A recent study, presented as this year's FASEB-meeting demonstrates a dose-dependent reduction of the expression in human aortic endothelial cells of the adhesion molecules intercellular adhesion molecule 1, vascular cell adhesion molecule 1, and of E-selectin (Wu et al., 1997). This study confirms and extends the previous observation that endothelial cells cultured in the presence of vitamin E show reduced IL-1-induced leukocyte adhesion and reduced levels of E-selectin mRNA and cell surface expression of this adhesion molecule (Faruqi et al., 1994). As a consequence of these and other effects, it could be demonstrated that vitamin E protects endothelial cell injury from diverse noxae, including irradiation

Table 1. Effects of vitamin E on cells and signal transduction pathways

Cell type	Effect	References
Leukocytes		
	Reduced adhesion to surfaces, incl. endothelium	Lafuze et al., 1984
		Boogaerts et al., 1984
		Devaraj and Jialal, 1996
	Reduced aggregation in response to AA or LTB4	Villa et al., 1986
		Steiner and Mower, 1982
	Reduced directed chemotaxis (to fMLP)	Luostarinen et al., 1991
	Reduced generation of ROS (chemiluminescence)	Herbazcynstra et al., 1994
		Devaraj and Jialal, 1996
	Reduced generation of IL-I	Devaraj and Jialal, 1996
	Reduced ionophore-induced LO activity	Chan et al., 1989
	Inhibition of NFkB in T-Iymphocytes	Suzuki and Packer, 1993
Platelets		
	Reduced aggregation to ADP and thrombin	Salonen et al., 1991
		Polette and Blache, 1992
		de Lorgeril et al., 1994
	Reduced adhesion to coated surfaces	Jandak et al., 1988
		Steiner et al., 1995
	Reduced generation of thromboxane	Hamelin et al., 1983
Endothelial cells		
	Reduced adhesion of leukocytes	Faruqi et al., 1994
	Reduced expression of adhesion molecules	Faruqi et al., 1994
		Wu et al., 1997
	Preservation of EC-dependent vasorelaxation	Keaney et al., 1993
		Rubino and Burnstock, 1994
		Andersson et al., 1994
	Increased generation of prostacyclin	Chan and Leigh, 1981
		Umeda et al., 1990
	Reduced generation of thromboxane	Meydani et al., 1992
	Reduced injury induced by	Davidge et al., 1993
	Irradiation	
	Oxidized lipids	Kuzuya et al., 1991
		Balla et al., 1993
		Belcher et al., 1993
	LO-products	Ochi et al., 1992
	Hydrogen peroxide	Block, 1991
		Kaneko et al., 1991
	Linoleic acid	Hennig et al., 1990
		Kaneko et al., 1991
	Adherent leukocytes	Boogaerts et al., 1984
		Lafuze et al., 1984
		Jialal and Grundy, 1993
Signal transduction		
	Reduced DAG generation in endothelial cells	Tran et al., 1994
	Inhibition of PKC	Mahoney and Azzi, 1988
		Faruqi et al., 1994
	Inhibition of PLA2	Caoetal., 1987
		Pentland et al., 1992
	Reduced mobilization of intracellular Calcium	Cox et al., 1980

(Davidge et al., 1993), oxidized lipids (Kuzuya et al., 1991; Balla et al., 1993), lipoxy-genase products (Ochi et al., 1992), hydrogen peroxide (Block, 1991; Kaneko et al., 1991), oxygen (Michiels et al., 1990), linoleic acid (Hennig et al., 1990; Kaneko et al., 1991; Hennig et al., 1996), and, finally from injury by adherent leukocytes (Boogaerts et al., 1984; Lafuze et al., 1984; Jialal and Grundy, 1993). See also Table 1.

1.4. Signal Transduction Pathways

The mechanism by which vitamin E exerts these protective effects on different cell types are only incompletely understood. Several effects of vitamin E on the signal transduction pathways have been demonstrated, including inhibitory effects on the generation of diacylglycerol after thrombin stimulation of endothelial cells (Tran et al., 1994), inhibitory effects on protein kinase C (Mahoney and Azzi, 1988; Sakamoto et al., 1990; Faruqi et al., 1994; A. Azzi et al.; see separate chapter in this volume) and of phospholipase A2 (Cao et al., 1987; Pentland et al., 1992). Vitamin E has been shown to reduce the metabolism of arachidonic acid (Chan et al., 1989; Sakamoto et al., 1990), the mobilization of intercellular calcium (Cox et al., 1980), and more recently the inhibition of NFKB activation (Suzuki and Packer, 1993). See also Table 1.

2. VITAMIN E IN CARDIOVASCULAR PATHOPHYSIOLOGY — EFFECTS ON MODELS OF DISEASE

As a consequence of the above described effects of vitamin E on different cell types, vitamin E has been shown to affect diverse pathophysiological conditions, such as ischemia/reperfusion injury, hypercholesterolemia, hyperglycemia, and diverse pathomechanisms related to organ transplantation.

2.1. Ischemia/Reperfusion Injury

With respect to ischemia/reperfusion injury, it has been shown that vitamin E plasma and tissue levels are reduced following ischemia and reperfusion (Cavarocchi et al., 1986; Yoshikawa et al., 1991; van Jaarsveld et al., 1992). Vitamin E has been shown to reduce the generation of reactive oxygen species and of lipid peroxides after ischemia and reperfusion (Cavarocchi et al., 1986; Massey and Burton, 1989) and to attenuate postischemic creatinine phosphokinase release in isolated perfused rat hearts (Gauduel and Duvelleroy, 1984). This latter finding is of interest since maximal inhibition of creatinine phosphokinase release after myocardial ischemia and reperfusion could be observed in rats at a dietary supplement dose of only 30 mg/kg diet while no further inhibition could be obtained by supplementing significantly higher vitamin E doses of up to 3000 mg vitamin E/kg diet (Gauduel and Duvelleroy, 1984). In a recent article, Rojas and coworkers have shown that NADPHinduced lipid peroxidation of myocardial tissue was significantly reduced in guinea pigs fed a diet with a 150 mg/kg vitamin E supplement — but no further protection was seen in animals fed substantially higher vitamin E doses (Rojas et al., 1997). Also, vitamin E has been shown to prevent the loss of prostacyclin production after ischemia/reperfusion injury in ischemic rat heart (Pyke and Chan, 1990). Another important protective effect of vitamin E lies in its ability to preserve microvascular perfusion in experimental gastric (Kurose et al., 1993) and spinal cord injury (Taoka et al., 1990), after coronary occlusion (Amatuni et al., 1989), and after experimental renal (Defraigne et al.,

1994) and myocardial ischemia-reperfusion injury (Gauduel and Duvelleroy, 1984; Tanabe and Kito, 1989). As a consequence of these different effects of vitamin E in ischemic and reperfused organs, vitamin E has been shown to reduce the ischemia/reperfusion induced tissue damage in different organs, including liver (Marubayashi et al., 1986; Lee and Clemens, 1992), muscle (Ikezawa, 1989), central nervous system (Yamamoto et al., 1983; Fujimoto et al., 1984), kidney (Takenaka et al., 1981; Lee et al., 1993), gastrointestinal tract (Yoshikawa et al., 1991), and the heart (Tanabe and Kito, 1989; Massey and Burton, 1989; Axford-Gatley and Wilson, 1991; van Jaarsveld et al., 1992; Axford-Gatley and Wilson, 1993; Abadie et al., 1993; Haramaki et al., 1993). Most of these effects were accompanied by improved microvascular perfusion of the different postischemic tissues (Gauduel and Duvelleroy, 1984; Engibarian and Agaian, 1989; Amatuni et al., 1989; Taoka et al., 1990). See also Table 2A.

2.2. Hypercholesterolemia

Vitamin E was found to prevent the hypercholesterolemia-induced loss of endothelial-dependent vessel relaxation (Keaney et al., 1993; Andersson et al., 1994; Gilligan et al., 1994), to attenuate atherogenesis associated with hypercholesterolemic diets in laboratory animals (Keaney et al., 1993) and reduce hypercholesterolemia-induced cochlear microvascular damage (Kashiwado et al., 1994). As a potential mechanism for these findings, vitamin E has been shown to prolong the lag time of lipoprotein oxidation in vitro (Esterbauer et al., 1992), in laboratory animals (Keaney et al., 1993), and also in human subjects (Dieber-Rotheneder et al., 1992; Gilligan et al., 1994). See also Table 2A.

2.3. Diabetes

Vitamin E was found to inhibit the increased intercellular generation of reactive oxygen species and advanced glycation end products in hyperglycemic bovine endothelial cells (Giardino et al., 1996). Vitamin E antagonizes the increased production of thromboxane (anti-aggregatory and vasoconstrictive) by platelets of hyperglycemic laboratory animals (Karpen et al., 1982; Zadkova et al., 1993), and the decreased arterial production of prostacyclin (anti-aggregatory, vasodilatory) (Karpen et al., 1982). Vitamin E feeding also has been found to prevent myocardial perivascular fibrosis in diabetic rats (Rösen et al., 1995). In analogy to the findings in hypercholesterolemia, vitamin E was found to counteract the loss of endothelial-dependent vessel relaxation in diabetic rats (Keegan et al., 1995), and to normalize the increased platelet aggregation, blood viscosity, and reduced blood filterability in diabetic human patients (Gerster, 1993; Giugliano et al.,1995). The latter effects were obtained by treating patients with a 300 mg vitamin E supplement per day (Giugliano et al., 1995). Finally, vitamin E was shown to prevent diabetes-induced abnormal retinal blood flow via the diacylglycerol-protein kinase C pathway (Kunisaki et al., 1995). See also Table 2A.

2.4. Organ Transplantation

As far as organ transplantation-related pathomechanisms are concerned, vitamin E levels have been found reduced in plasma and tissue after organ transplantation (Rao et al., 1990; Oda et al., 1992; Goode et al., 1994; Galley et al., 1995). Treating experimental animals with vitamin E was found to inhibit lipid peroxidation and preserve organ function after transplantation of different organs (Demirbas et al., 1993; Ikeda et al., 1994;

Table 2a. Effect of vitamin E in animal models of diverse pathophysiological conditions

Condition	Effect	Reference
Ischemia-reperfusion		
	Reduced ROS production and lipid peroxidation	Cavarocchi et al., 1986
		Massey and Burton, 1989
		Pyke and Chan, 1990
	Reduced release of creatine phosphokinase	Gauduel and Duvelleroy, 1984
	Increased postischemic microvascular blood flow	Gauduel and Duvelleroy, 1984
		Amatuni et al., 1989
		Tanabe and Kito, 1989
		Fngibarian et al., 1989
		Taoka et al., 1990
		Kurose et al.,1993
		Defraigne et al., 1994
	Preservation of prostacyclin generation	Pyke and Chan, 1990
	Reduced tissue injury in	
	Liver	Marubayashi et al., 1986
		Lee and Clemens, 1992
	Muscle	Ikezawa et al., 1989
	CNS	Yamamoto et al., 1993
		Fujimoto et al., 1984
	Kidney	Takenaka et al., 1991
		Lee et al., 1993
	Gl tract	Yoshikawa et al., 1991
	Heart	Massey and Burton, 1989
		Tanabeetal., 1989
		Axford-G. and Wilson, 1991
		van Jaarsveld et al., 1992
		Abadeetal., 1993
		Haramaki et al., 1993
Hyper-cholesterolemia		
	Restoration of EC-dependent vasodilation	Keaney et al., 1993
		Andersson et al., 1994
		Gilligan et al., 1994
	Inhibition of atherogenesis	Keaney et al., 1993
	Reduced microvascular damage	Kashiwado et al., 1994
	Reduced lipoprotein oxidation	Dieber-Rothen et al., 1991
		Fsterbauer et al., 1992
		Keaney et al., 1993
Diabetes/hyperglycemia		
	Restoration of EC-dependent vasorelaxation	Keegan et al., 1995
	Reduced generation of ROS and AGEs	Giardino et al., 1996
	Reduced generation of thromboxane in platelets	Karpen et al., 1982
	Increased generation of prostacyclin in arteries	Karpen et al., 1982
	Reduced myocardial perivascular fibrosis	Rösen et al., 1995
	Reduced platelet aggregation	Gerster, 1993
		Giuliano et al., 1995
	Reduced blood viscosity	Giuliano et al., 1995
	Increased blood filterability	Guiliano et al., 1995
Organ transplantation		
	Reduced lipid peroxidation, improved function	Demirbas et al., 1993
		Ikeda et al., 1993
		Rabi et al., 1993
	Reduced cyclospirin hepatotoxicity	de Lorgeril et al., 1994
	Reduced platelet aggregation	de Lorgeril et al., 1994

reviewed in: Lehr and Messmer, 1996). These experiments were taken into the clinic by Rabi and coworkers who demonstrated significant reduction of post-transplant lipid peroxidation and improved function of transplanted kidney in a group of patients treated prior to declamping with a multi-vitamin formula (Rabi et al., 1993). Finally, treatment of heart transplant recipients with dietary vitamin E significantly reduced the increased platelet aggregation and also cyclosporin hepatotoxicity (de Lorgeril et al., 1994). See also Table 2A.

2.5. Varia

In addition to these effects of vitamin E in ischemia/reperfusion injury, hypercholesterolemia, hyperglycemia and organ transplantation, vitamin E has been found to attenuate endothelial cell damage during experimental multi organ failure (Lu, 1993) and to attenuate the development of adult respiratory distress syndrome in animals as well as in patients (Wolf and Seeger, 1982). Treatment of animals with vitamin E prevented oxygen induced lung toxicity (Jacobson et al., 1990). Likewise, vitamin E significantly reduced experimental paraquat induced lung injury (Suntres et al., 1992), bleomycin-induced pulmonary fibrosis (Kilinc et al., 1993) and adriamycin-induced focal glomerulosclerosis (Washio et al., 1994). In the clinical setting, vitamin E has been found to significantly attenuate retinopathy and intraventricular hemorrhage in oxygen ventilated babies (Shirahata, 1993) and to reduce the clinical manifestations of intermittent claudication in elderly human subjects (Haeger, 1973; Haeger, 1982; Teoh et al., 1992). See also Table 2B.

3. VITAMIN E DIETARY SUPPLEMENTS IN FARM ANIMALS

Taking the above listed effects of vitamin E in *in vitro* models, experimental animals, as well as in patients into consideration, it is not surprising to see that vitamin E has long since found its way into the diets of farm animals.

3.1. Nutritional Diseases

Vitamin E has been found to prevent all kinds of different nutritional diseases, such as muscular dystrophy in chicken (Bieri and Evarts, 1974), as well as degenerative myopathy in cattle (Walsh et al., 1993), and rhabdomyolysis in sheep (Peet and Dickson,

Table 2b. Effect of vitamin E in animal models of diverse pathophysiological conditions

Condition	Effect	Reference
Varia		
	Counteracts experimental multi-organ failure	Lu
	Attenuates experimental and clinical ARDS	Wolf and Seeger
	Prevents oxygen- and paraquat-induced lung injury	Jacobson et al., 1990
		Suntres et al.
	Prevents bleomycin-induced pulmonary fibrosis	Kilinc et al., 1993
	Prevents adriamyclin-induced focal glomeruloscler	Washio et al.
	Prevents O_2-induced retinopathy	Shirahata
	Prevents O_2-induced intraventricular hemorrhage	Shirahata
	Improves clinical intermittent claudication	Haeger
		Haeger
		Teoh

1988). Feeding pigs a vitamin E-supplemented diet prevents the generation of mulberry heart disease, a condition characterized by microangiopathy, endothelial cell damage, leading to myocardial edema and particular hemorrhages, myocarditis, and pulmonary edema (Korpala et al., 1990). Vitamin E supplements prevent encephalomalazia in turkeys (Klein et al., 1994; Fuhrmann and Sallmann, 1995) and exudative diathesis in chicken (Bieri et al., 1974), and, finally, motor neuron disease in horses (Divers et al., 1994). See also Table 3.

3.2. Immune Function

In addition to these disease states, which are a consequence of vitamin E deficiency, it has been recognized that vitamin E supplements improve the immune function of different farm animals, including chicken (Tengerdy et al., 1972; Heinzerling et al., 1974; Heinzerling et al., 1974; Colnago et al., 1984; Blum et al., 1992; Beck et al., 1994), pigs (Ellis and Vorhies, 1976; Peplowski et al., 1980; Nemec et al., 1994), calves (Reddy et al., 1986; St. Laurent et al., 1990), and also of farm fish (Verlhac et al., 1993). For example, Ellis and Vorhies demonstrated a significantly improved antibody production to E. coli bacterim in pigs fed a vitamin E supplemented diet (Ellis and Vorhies, 1976), and Heinzerling and coworkers reported that feeding chicken a vitamin E-supplemented diet (150 and 300 mg/kg diet) significantly reduced E.coli induced mortality (Heinzerling et al., 1974). Likewise, vitamin E supplements have been found to prevent the infection of rainbow trout with Yersinia Ruckeri (Furones et al., 1992), and to effectively prevent mastitis in cattle (Weiss et al., 1990; Hogan et al., 1993; Kolb and Grün, 1993; Politis et al., 1996) and pigs (Mahan, 1994). See also Table 3.

3.3. Animal Performance

Beside the prevention of *nutritional* diseases and the improvement of animal immune function, vitamin E has been found to improve the performance of farm animals, such as increased daily weight gain in cattle (Hutcheson and Cole, 1985; Hicks, 1985; Schäfer et al., 1989; Hill et al., 1990), sheep (Asadian et al., 1996), and also in farm shrimp (He and Lawrence, 1993), and increased milk production in cattle (Wolfram and Flachowsky, 1991). Vitamin E was found to reduce mortality of farm animals during stress and transportation (Lee et al., 1985; Hicks, 1985), and also to improve the quality of the farm animal products: feeding vitamin E-supplemented diets reduces the lipid peroxidation in pork meat and fat (Asghar et al., 1991; Flachowsky et al., 1994; Buckley et al., 1995; Cannon et al., 1996), in milk fat (Goering et al., 1976; Nicholson and St.Laurent, 1991), in butter (Jahreis et al., 1996), in channel catfish fillets (Gatlin et al., 1992; Bai and Gatlin, 1993), rainbow trout (Frigg et al., 1990), and in chicken meat (Marusisch et al., 1975; Bartov and Bornstein, 1977; Lin et al., 1989; Rethwill et al., 1991; Bartov and Frigg, 1992; Frigg et al., 1993). As a consequence, vitamin E prevents the formation of metmyoglobin and thus improves meat color stability, an important measure of meat quality for consumers during display in department stores (Faustmann et al., 1989; Williams et al., 1992; Arnold et al., 1992, Arnold et al., 1993; Kiser et al., 1993; Lanari et al., 1993; Mitsumoto et al., 1993; Wulf et al., 1995; Liu et al., 1996; Smith et al., 1996). Vitamin E was found to reduce the drip loss, a major feature that characterizes what we consider *tenderness of* meat (Ashgar et al., 1991; Augustin et al., 1996), and to improve sensory quality (less rancid off flavor) in milk (St. Laurent, 1990), in chicken meat (Blum et al., 1992, Sheeky et al., 1993; Patterson and Stevenson, 1995), pork meat and sausage (Williams et

Table 3. Effect of vitamin E supplements in farm animals

Nutritional diseases	
Prevents muscular dystrophy in chicken	Bieri et al., 1973
Prevents degenerative myopathy in cattle	Walsh et al., 1993
Prevents mulberry heart disease in pigs	Korpala et al., 1990
Prevents encephalopathy in turkeys	Klein et al., 1994
Prevents exsudative diathesis in chicken	Bieri et al., 1973
Prevents motor neural disease in horses	Divers et al., 1994
Immune function	
Improved immune function	
In chicken	Tengerdy et al., 1972
	Heinzerling et al., 1984
	Heinzerling et al., 1984
In pigs	FIlis and Vorhies, 1976
	Peplowsky et al., 1980
	Nemec et al., 1994
In calves	Reddy et al., 1986
	Pruittetal, 1989
	St. Laurent, 1990
In farm fish	Furones et al., 1992
	Verlhac et al., 1993
Prevents mastitis	
In cattle	Weiss et al.,1990
	Hoganetal, 1993
	Kolb and Grün, 1995
In pigs	Mahan, 1994
Animal performance	
Increased daily weight gain	
In cattle	Hutcheson et al., 1985
	Schafer et al., 1989
	Hill et al., 1993
	Hicks, 1985
In farm shrimp	He and Lawrence, 1993
Increased milk production in cattle	Wolfram & Flachowsky, 1991
Reduced mortality during stress and transportation	Lee et al., 1995
	Hicks, 1995
Quality of products	
Reduced lipid peroxidation in meat and fat	Asghar et al., 1991
	Gatlin et al., 1992
	Bai and Gatlin, 1993
	Buckley et al., 1995
	Cannon et al., 1996
Reduced lipid peroxoidation in milk fat	Goering et al., 1976
	Nicholson et al., 1991
Reduced lipid peroxidation in butter	Jahreis et al., 1996
Improved meat color stability (less metmyoglobin)	Faustmann et al., 1989
	Williams et al., 1992
	Arnold et al., 1992
	Arnold et al., 1993
	Wulf et al., 1995
	Liu et al., 1996
Reduced drip loss (improved tenderness) of meat	Ashgar et al., 1991
	Augustin et al., 1996
Improved sensory quality of meat	Boggio et al., 1985
	St. Laurent, 1990
	Blum et al., 1992
	Williams et al., 1992
	Sheeky et al., 1993
	Patterson et al., 1995
	Smith et al., 1996

al., 1992; Smith et al., 1996), Atlantic farm salmon (Waagboe et al., 1993), and other farm fish (Boggio et al., 1985; Schwarz, 1996). See also Table 3.

While most of the above listed findings have only in the last few years been systematically analyzed and published, it has been known by farmers for many years that supplementation of farm animal diets with 100–250 mg of vitamin E per kg diet are considered optimal for animal health and performance as well as the quality of animal products.

4. VITAMIN E DIETARY SUPPLEMENTS IN PETS

In later years, it was considered that what is good for farm animals must also be good for pets (Fox et al., 1993). The last few years have seen a substantial increase in vitamin E supplements to all kinds of pet diets, and the vitamin E supplements are even used as a major argument in advertisements for these pet diets.

5. VITAMIN E DIETARY SUPPLEMENTS IN LABORATORY ANIMALS

It was therefore not surprising to see that the manufacturers of diets for laboratory animals have likewise supplemented their diets with increasing amounts of vitamin E. Vitamin E was found to increase the fertility and litter size of laboratory animals, to prevent *nutritional* diseases such as testicular degeneration in hamsters (Bieri and Evarts, 1973) and liver necrosis in rats (Bieri and Evarts, 1973). Vitamin E supplements have been found to increase the resistance of rats and mice (Heinzerling et al., 1974; Tanaka et al., 1979) to diverse infections.

The supplementation of animal diets with vitamin E ultimately leads to one key question: what is the normal, non-supplemented vitamin E requirement of laboratory animals. Since some of these animals are almost extinct in the wild, and are fed *artificial* diets in captivity, the basic vitamin requirement is sometimes difficult to assess. For hamsters, it is well known that they feed predominantly on wheat, which they store *(hamster)* in their borrows in large quantities (this fact, and the associated damage inflicted on wheat farmers, has lead to their widespread eradication in most of Europe and North America). Similar dietary preferences have been described for rats and mice. Ground wheat contains approximately 11 mg of vitamin E/kg, while most other components of a balanced rodent diet contain less vitamin E. Systematically analyzing the different food components of the standard laboratory animal diet administered to animals at the National Institute of Health shows that the basic total vitamin E content is roughly 8 mg/kg, which is then supplemented with 22 mg/kg to yield approximately 30 mg vitamin E/kg (National Institutes of Health, 1982). In contrast to these published data for animal diets at the NIH, most diets for laboratory animals contain significantly higher vitamin E supplements, up to 200 mg vitamin E/kg. Yet, vitamin E supplements in lab animal diets may be even higher than what is indicated on the packaging. In a recent study, we have compared vitamin E levels in hamsters fed a standard rodent diet containing a 60 mg/kg vitamin E supplement (Alma, Kempten, Germany). These vitamin E blood levels were compared to those in animals fed a vitamin E depleted diet, and in animals fed the vitamin E depleted diet which was then supplemented with 300, 3000, or 30000 mg vitamin E/kg diet. In this series of diets with different vitamin E levels, the vitamin E plasma levels of the animals fed the standard diet (60 mg vitamin E/kg) had even higher vitamin E levels than animals fed the diet, to which we had added 300 mg of vitamin E per kg (Willy et al., 1995). When asked for an explanation for this observation, a company official indicated, that more vitamin E may be added to the diet to compensate for the loss of antioxidant vitamins during storage. The indication on the package thus only indicates the minimum amount of vitamin E left at the end of the storage period.

In conclusion, laboratory animals today are fed diets which contain substantial vitamin E supplements. While these supplements are added in order to improve the health and fertility of the animals, it should be kept in mind that laboratory animals are bred and raised in order to be used in models of disease and not in order to be resistant to these diseases. With these ideas in mind, we have performed two different experiments trying to answer the question whether the vitamin E supplements in laboratory animal diets may affect results obtained in two different models of disease.

6. CAN VITAMIN E SUPPLEMENTS IN LABORATORY ANIMAL DIETS AFFECT RESULTS OBTAINED IN ANIMAL MODELS OF DISEASE?

6.1. Ischemia/Reperfusion Injury

We have used a dorsal skinfold chamber model in hamsters (Lehr et al., 1991) to study the ischemia/reperfusion injury in a fine striated skin muscle. In this model we have quantified the adhesion of fluorescently labeled leukocytes to the microvascular endothelium, in particular of postcapillary venules, as well as the density of the nutritional capillary network. In analogy to earlier experiments using the same model (Lehr et al., 1991a,b; Menger et al., 1992a,b), we found that ischemia/reperfusion of the muscle resulted in a significant increase in postischemic leukocyte adhesion to the endothelium, as well as a substantial drop in functional capillary perfusion density (Willy et al., 1995). These two parameters of perfusion injury were most pronounced in animals fed a diet containing only the basic amount of 5–10 mg vitamin E/kg (see Figure 1A, B). In contrast, animals fed a standard lab animal diet, containing 60 mg vitamin E/kg diet showed a significant reduction in postischemic leukocyte adhesion, as well as a restoration of postischemic functional capillary density to levels that were no longer significantly different from preischemic baseline values (Willy et al., 1995: see Figure 1A,B). Indeed, the inhibitory effect of 60 mg vitamin E/kg in the standard animal diet was maximal: substantially higher vitamin E supplements (300 mg, 3000 mg, or 30000 mg vitamin E/kg diet) showed no additional inhibition of postischemic leukocyte adhesion and no additional improvement of functional capillary density (Willy et al., 1995; see Figure 1A,B).

6.2. Pancreatic Islet Xenograft Rejection

In a second series of experiments, we have *xeno*transplanted rat pancreatic islets of Langerhans into the observation window of the dorsal skinfold chamber in hamsters and quantified the revascularization and the subsequent rejection of these islets. We found a consistent formation of a microvascular network in the islets during the first two weeks after transplantation (see Figure 2B). In animals fed a normal, non-vitamin E-supplemented diet, the formation of a microvascular network was later followed by the microcirculatory manifestations of xenograft rejection, including massive leukocyte infiltration and adhesion to endothelial cells (see Figure 2A), as well as eventual break-down of the microvascular network and the absorption of the islet (see Figure 2B). In contrast, animals

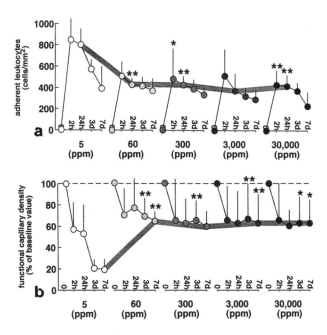

Figure 1. Effect of dietary vitamin E on ischemial reperfusion injury in muscle: leukocyte adhesion to endothelium (A) and functional capillary density (B). A four hour ischemia was induced by means of a silicone stamp to the tissue contained in the observation chamber in hamsters, as described previously (Lehr et al., 1991a). The adhesion of fluorescently stained leukocytes (acridine orange) to the microvascular endothelium of postcapillary venules was visualized and quantified using image analysis before and at defined time intervals after reperfusion (as indicated in the x-axis, figure 1A). Likewise, the density of the fluorescently enhanced capillary network of the muscle tissue (FITC dextran, Mr 150,000) was visualized and quantified before and after ischemia/reperfusion (figure 1B). Experiments were performed in animals fed for two weeks prior to the experiments a control diet containing only the natural vitamin E content, but no vitamin E supplement (5ppm), and in animals, in which this diet was supplemented with 300, 3,000, and 30,000 mg vitamin E per kg diet (mg/kg = ppm). An additional group of animals were treated with a standard rodent diet, containing 60mg vitamin E/kg diet (as indicated on the diet bag). N=7 animals per group, $^*P < 0.05$, $^{**}P < 0.01$, Wilcoxon test. For further details, please see reference (Willy et al., 1995).

fed a diet which was supplemented with 150 mg vitamin E/kg showed a significantly reduced xenograft rejection (reduced leukocyte infiltration and endothelial cell adhesion, see Figure 2A), and a significantly better maintenance of the microvascular network in the xenografts (see Figure 2B). In analogy to the ischemia/reperfusion experiments, addition of significantly higher amounts of vitamin E (8000 mg/kg diet) to the animal diet resulted in no significantly better protection from xenograft rejection (Vajkoczy et al., 1997; see Figure 2A,B)

Had we not been aware of the protection afforded by the vitamin E supplements in the standard laboratory diet and used these animals as controls for the animals fed high doses of vitamin E, we could have written two papers entitled *Vitamin E does not protect from ischemia/reperfusion injury* and *Vitamin E does not affect xenograft rejection*, and they would probably have passed the peer review process and made it into the literature. The results obtained in these two models of disease (ischemia/reperfusion injury and islet xenograft rejection) suggest, that the vitamin E supplements of the *standard* laboratory animal diets may exert a pronounced impact on the manifestation of tissue damage in the described, but also in other models of disease. This pertains perticularly to models of dis-

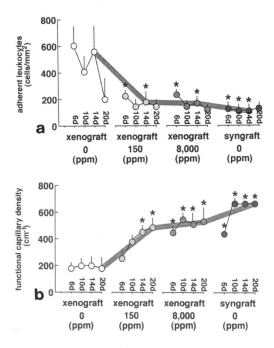

Figure 2. Effect of dietary vitamin E on the rejection of rat pancreatic islets xenografted into hamsters: leukocyte adhesion to endothelium (A) and functional capillary density (B). The adhesion of fluorescently stained leukocytes (acridine orange) to the microvascular endothelium of postcapillary venules was visualized and quantified at defined time intervals after transplantation (as indicated in the x-axis, figure 2A). Likewise, the density of the fluorescently enhanced capillary network of the islet (FITC dextran, Mr 150,000) was visualized and quantified (figure 2B). Experiments were performed in animals fed for two weeks prior to the experiments a control diet containing only the natural vitamin E content, but no vitamin E supplement, and in animals, in which this diet was supplemented with 8,000 mg vitamin E per kg diet (mg/kg = ppm). An additional group of animals were treated with a standard rodent diet, containing 150mg vitamin E/kg diet (as indicated on the diet bag). N=7 animals per group, $^-$P<0.05, **P<0.01, Wilcoxon test. For further details, please see reference (Vajkoczy et al., 1995).

ease with involve — or may involve — the generation and action of reactive oxygen species. It is intriguing to speculate that the increase in vitamin E and other antioxidant supplements in standard laboratory diets, which has occurred over the years, may have accounted for some reports on the absence of protective effects by antioxidants or radical scavengers in animal models of diverse disease pathomechanisms. A similar caveat has been voiced several years ago by Hall and Yonkers, who reported on their inability to reproduce findings in a model of trauma-induced neuronal degeneration, when animals were switched from a non-supplemented diet (9 mg/kg vitamin E) to a diet containing a 70 mg/kg vitamin E supplement (Hall and Yonkers, 1989).

REFERENCES

Abadie, C., Ben Baouali, A., Maupoil, V., and Rochette, L., 1993, An alpha-tocopherol analogue with antioxidant activity improves myocardial function during ischemia reperfusion in isolated working rat hearts, *Free Radic. Biol. Med.* **15**:209–215.

Amatuni, V.G., Matevosian, R. S., Sisakian, S. A., and Arakelian, I. G., 1989, The effect of alphatocopherol and intal on heart capillary bed and lipid peroxide oxidation in experimental necrosis of the rat myocardium, *Cor Vasa* **31**:500–507.

Andersson-T. L., Matz, J., Ferns, G. A., and Anggard, F. F., 1994, Vitamin E reverses cholesterol-induced endothelial dysfunction in the rabbit coronary circulation, *Atherosclerosis* **111**:39–45.

Arnold, R. N., Arp, S. C., Scheller, K. K., Williams, S. N., and Schaefer, D. M., 1993, Tissue equilibration and subcellular distribution of vitamin E relative to myoglobin and lipid oxidation in displayed beef, *J. Anim. Sci.* **71**:105–118.

Arnold, R. N., Scheller, K. K., Arp, S. C., Williams, S. N., Buege, D.R., Schaefer, D. M., 1992, Effect of long or short term feeding of a-tocopherol acetate to Holstein and crossbred beef steers on performance, carcass characteristics and beef color stability, *J. Anim. Sci.* **70**:3055–3065

Asadian, A., Mezes, M., and Mirhadi, S. A., 1996, Effect of vitamins A and E on blood plasma vitamin status and daily body mass gain of different fat-tailed sheep breeds, *Acta.Vet. Hung.* **44**:99–109.

Asghar, A., Gray, J. A., Booren, A. M., Gomaa, E. A., Abouzied, M. M., and Miller, E. F., 1991, Effect of supranutritional dietary vitamin E levels on subcellular deposition of alpha-tocopherol in the muscle and on pork quality, *J. Sci. Food. Agric.* **57**:31–41.

Augustin, C., and Freudenreich, P., 1996, Vitamin E-Einsatz in der Rindermast. *Mitteilungsblatt der BAFF Kulmbach* **132**:149–155.

Axford-Gately, R. A., and Wilson, G. J., 1993, Myocardial infarct size reduction by single high dose or repeated low dose vitamin E supplementataion in rabbits, *Can. J. Cardiol.* **9**:94–98.

Axford-Gatley, R. A., and Wilson, G. J., 1991, Reduction of experimental myocardial infarct size by oral administration of alpha-tocopherol, *Cardiovasc. Res.* **25**:89–92.

Bai, S. C., and Gatli, D. M., 1993, Dietary vitamin E concentration and duration of feeding affect tissue atocopherol concentrations of channel catfish (ictalurus punctatus), *Aquaculture* **113**:129–35.

Balla, J., Belcher, J. D., Balla, G, Jacob, H. S., and Vercellotti, G. M., 1993, Oxidized low-density lipoproteins and endothelium: oral vitamin E supplementation prevents oxidized low-density lipoprotein-mediated vascular injury, *Trans. Assoc. Am. Physicians.* **106**:128–33.

Bartov, I., and Bornstein, S., 1977, Stability of abdominal fat and meat of broilers:the interrelationship between the effects of dietary fat and vitamin E supplements, *Brit. Poultry Sci.* **18**:47–57.

Bartov, I., and Frigg, M., 1992, Effect of high concentrations of dietary vitamin E during various age periods on performance, plasma vitamin E and meat stability of broiler chicks at 7 weeks of age, *Br. Poultry Sci.* **33**:393–402.

Bieri, J.G., and Fvarts, R. P., 1974, Vitamin E activity of gamma-tocopherol in the rat, chick and hamster, *J. Nutr.* **104**:850–857.

Block, F. R.,1991, Hydrogen peroxide alters the physical state and function of the plasma membrane of pulmonary artery endothelial cells, *J. Cell. Physiol.* **146**:362–9.

Blum, J. C., Tourraille, C., Salichon, M. R., Richard, F. H., and Frigg, M., 1992, Effect of dietary vitamin E supplies in broilers. 2. Male and female growth rate, viability, immune response, fat content and meat flavour variations during storage, *Arch. Geflügelkunde* **56**:37–42.

Boggio, S. M., Hardy, R. W., Babbitt, J. K., and Brannon, F. L., 1985, The influence of dietary lipid source and alpha-tocopheryl acetate level on product quality of rainbow trout (Salmo gairdnei), *Aquaculture* **51**:13–24.

Boogaerts, M. A., Van de Broeck, J., Deckmyn, H., Roelant, C., Vermylen, J., and Verwilghen R. L., 1984, Protective effect of vitamin E on immune-triggered, granulocyte mediated endothelial injury, Thromb. *Haemostas.* **51**:89–92.

Buckley, D. J., Morrissey, P. A., and Gray, J. I., 1995, Influence of dietary vitamin E on the oxidative stability and quality of pig meat, *J. Anim. Sci.* **73**:3122–30.

Cannon, J. F., Morgan, J. B., Schmidt, G. R., Tatum, J. D., Sofos, J. N., Smith, G. C., Delmore, R. J., and Williams, S. N., 1996, Growth and fresh meat quality characteristics of pigs supplemented with Vitamin E, *J. Amin. Sci.* **74**:98–105.

Cavarocchi, N. C., Fngland, M. D., O'Brien, J. F., Solis, F., Russo, P, Schaff, H. V., Orszulak, T. A., Pluth, J. R., Kaye, M. P., 1986, Superoxide generation during cardiopulmonary bypass: Is there a role for Vitamin E?, *J. Surg. Res.* **40**:519–527.

Chan, A. C., and Leith, M. K., 1981, Decreased prostacyclin synthesis in vitamin E-deficient rabbit aorta, *Am. J. Clin. Nutr.* **34**:2341–2347.

Chan, A. C., Tran, K., Pyke, D. D., and Powell, W. S., 1989, Effects of dietary vitamin E on the biosynthesis of 5-lipoxygenase products by rat polymorphonuclear leukocytes (PMNL), *Biochem. Biophys. Acta* **1005**: 265–269.

Chen, H.W., Hendrich, S., and Cook, L. R., 1994, Vitamin E deficiency increases serum thromboxane A2, platelet arachidonate and lipid peroxidation in male Sprague-Dawley rats, *Prostagladins. Leukot. Essent. Fatty. Acids.* **51**:11–17.

Colnago, G. L., Jensen, L. S., and Long, P. L., 1984, Effect of selenium and vitamin E on the development of immunity to coccidiosis in chickens, *Poultry Sci.* **63**:1136

Cox, A. C., Rao, G. H., Gerrard, J. M., and White, J. G., 1980, The influence of vitamin E quinone on platelet structure, function, and biochemistry, *Blood* **55**:907–914.

Davidge, S. T., Hubel, C. A., and McLaughlin, M. K., 1993, Cyclooxygenase-dependent vasoconstrictor alters vascular function in the vitamin E-deprived rat, *Circ. Res.* **73**:79–88.

de Lorgeril, M., Boissonnat, P., Salen, P., Monjaud, I., Monnez, C., Guidollet, J., Ferrera, R., Dureau, G., Ninet, J., and Renaud, S., 1994, The beneficial effect of dietary antioxidant supplementation on platelet aggregation and cylosporine treatment in heart transplant recipients, *Trans.* **58**:193–195.

Defraigne, J. 0., Pincemail, J., Detry, 0., Franssen, C., Meurisse, M., and Limet, R., 1994, Preservation of cortical microcirculation after kidney ischemia-reperfusion: value of an iron chelator, *Ann. Vasc. Surg.* **9**:227–228.

Demirbas, A., Bozoklu, S., Ozdemir, A., Bilgin, N., and Haberal, M., 1993, Effect of alpha tocopherol on the prevention of reperfusion injury caused by free oxygen radicals in the canine kidney autotranspantation model, *Transplant. Proc.* **25**:2274

Devaraj, S.,Li, D., and Jialal, I,. 1996, The effects of alpha tocopherol supplementation on monocyte function. Decreased lipid oxidation, interleukin 1 beta secretion, and monocyte adhesion to endothelium, *J. Clin. Invest.* **98**:756–63.

Dieber-Rotheneder, M., Puhl, H., Waeg, G., Striegl, G., and Fsterbauer, H., 1991, Effect of oral supplementation with D-alpha-tocopherol on the vitamin E content of human low density lipoproteins and resistance to oxidation, *J. Lipid Res.* **32**:1325–1332.

Divers, T. J., Mohammed, H. O., Cummings, J. F., Valentine, B. A., De-Lahunta, A., Jackson, C. A., and Summers, B. A., 1994, Equine motor neuron disease: findings in 28 horses and proposal of a pathophysiological mechanism for the disease, *Equine. Vet. J.* **26**:409–15

Ellis, R. P., and Vorhies, M. W., 1976, Effect of supplemental dietary vitamin E on the serologic response of swine to an Escherichia coli bacterin, *J. Am. Vet. Med. Assoc.* **168**:231–232.

Engibarian, A. A., and Agaian, K. A., 1989, Changes of the necrotic area and functional indices of the myocardial microcirculatory bed during the treatment of experimental myocardial infarct with nitroglycerin, alpha-tocopherol and sodium nucleinate, *Patol. Fiziol. Eksp. Ter.* **5**:24–26.

Esterbauer, H., Gebicki, J., Puhl, I., and Jurgens, I., 1992, The role of lipid peroxidation and antioxidants in the oxidative modification of LDKL, *Free Radic. Biol. Med.* **13**:341–390.

Faruqi, R., Motte, C., and DiCorleto, P. 1994. Alpha tocopherol inhibits agonist-induced monocytic cell adhesion and protein kinase C activity in vascular smooth muscle cells. *J. Clin. Invest.* **94**:592–600.

Faustman, C., Cassens, R. G., Schaefer, D. M., Buege, D. R., Williams, S. N., and Scheller, K. K., 1989, Improvement of pigment and lipid stability in Holstein steer beef by dietary supplementation with vitamin E, *J. Food. Sci.* **54**:858–862.

Flachowsky, G., Richter, G. H., Wendemuth, H., Möckel, P., Graf, H., Jahreis, G., and Lubbe, F., 1994, Einfluss von Rapssamen in der Mastrinderernährung auf Fettsäurenmuster, Vitamin-E-Gehalt und oxidative Stabilität des Körperfettes, 7. *Ernährungswiss.* **33**:277–285.

Fox, P. R., Trautwein, F. A., Hayes, K. C., Bond, B, R., Sisson, D. D., and Moise, N. S., 1993, Comparison of taurine, alpha-tocopherol, retinol, selenium, and total triglycerides and cholesterol concentrations in cats with cardiac disease and in healthy cats, *Am. J. Vet. Res.* **54**:563–9.

Frigg, M., Buckley, D. J., Morrissey, P. M., 1993, Influence of alpha tocopherol acetate supplementation on the susceptibility of chicken or pork tissues to lipid oxidation, *Mh. Vet. Med.* **48**:79–83.

Frigg, M., Prabucki, A. L., and Ruhdel, F. U., 1990, Effect of dietary vitamin E levels on oxidative stability of trout fillets, *Aquaculture* **84**:145–158.

Fuhrmann, H., and Sallmann, H. P., 1995, The influence of dietary fatty acids and vitamin E on plasma prostanoids and liver microsomal alkane production in broiler chickens with regard to nutritional encephalomalacia, *J. Nutr. Sci. Vitaminol. Tokyo.* **41**:553–61.

Fujimoto, S., Mizoi, K., Yoshiumoto, T., and Suzuki, J., 1984, The protective effect of vitamin E on cerebral ischemia, *Surg. Neurol.* **22**:449–454.

Furones, M. D., Alderman, D. J., Bucke, D., Fletcher, T. C., Knox, D., and White, A., 1992, Dietary vitamin E and the response of rainbow trout, oncorhynchus mykiss (Walbaum), to infection with yersinia ruckeri, *J. Fish. Biol.* **41**:1037–41.

Galley, H.F., Richardson, N., Howdle, Walker, B.F., and Webster N.R., 1995, Total antioxidant capacity and lipid peroxidation during liver transplantation, *Clin. Sci.* **89**:329–331.

Gatlin, D. M., Ill, Bai, S. C., and Frrickson, M. C:; 1992, Effects of dietary vitamin E and synthetic antioxidants on composition and storage quality of channel catfish, Icta lurus punctatus, *Aquacualture* **106**:323–332.

Gauduel, Y., and Duvelleroy, M. A., 1984, Role of oxygen radicals in cardiac injury due to reoxygenation, *J. Mol. Cell. Cardiol.* **16**:459–470.

Gerster, H., 1993, Prevention of platelet dysfunction by vitamin E in diabetic atherosclerosis, *Z. Ernährungswiss.* **32**:243–261.

Gey, K. F., 1993, Prospects for the prevention of free radical disease, regarding cancer and cardiovascular disease, *Br. Med. Bull.* **49**:679–99.

Giardino, I., Edelstein, D., and Brownlee, M., 1996, BCL-2 expression or antioxidants prevent hyperglycemia-induced formation of intracellular advanced glycation endproducts in bovine endothelial cells, *J. Clin. Invest.* **97**:1422–1428.

Gill, D. R., Smith, R. A., Hicks, R. B., and Ball, R. L., 1986, The effect of vitamin E supplementation on health and performance of newly arrived stocker cattle, *Research Rep. MP-1 18, Stillwater. OK.* 240–243.

Gilligan, D. M., Sack, M. N., Guetta, V., Casino, P. R., Quyyumi, A. A., Rader, D. J., Panza, J. A., and Cannon-R.O. 3rd., 1994, Effect of antioxidant vitamins on low density lipoprotein oxidation and impaired endothelium-dependent vasodilation in patients with hypercholesterolemia, *J. Am. Coll. Cardiol.* **24**:1611–1617.

Giugliano, D., Marfella, R., Verrazzo, G., Acampora, R., Donzella, C., Quatraro, A., Coppola, L., and D'Onofrio, F., 1995, Abnormal rheologic effects of glyceryl trinitrate in patients with non-insulin-dependent diabetes mellitus and reversal by antioxidants, *Ann. Intem. Med.* **123**:338–43.

Goering, H. K., Gordon, C. H., Wrenn, T. R., Bitman, L., King, R. L., and Douglas, F. W., 1976, Effect of feeding protected safflower oil on yield, composition, flavor and oxidative stability of milk, *J. Dairy. Sci.* **59**:416–425.

Goode, H.F., Webster, N.R., Howdle, Leek, J.P., Lodge, J.P., Sadek, S.A., and Walker, B.F., 1994, Reperfusion injury, antioxidants and hemodynamics during orthotopic liver transplantation, *Hepatology* **19**:354–359

Haeger, K., 1973, Walking distance and arterial flow during long term treatment of intermittent claudication with d-alpha tocopherol, *Vasa* **2**:280–287.

Haeger, K.,1982, Long term study of alpha tocopherol in intermittent claudication, *Ann. N. Y. Acad. Sci.* **393**:369–375.

Hall, F. D., and Yonkers, P. A., 1989, Mechanisms of neuronal degeneration secondary to central nervous system trauma or ischemia, *J. Neurotrauma* **6**:227–228.

Halliwell, B., 1993, The role of oxygen radicals in human disease, with particular reference to the vascular system, *Haemostasis* **23**:118–126.

Hamelin, St.- J. S., and Chan, A. C., 1983, A carrier-mediated transport for folate in basolateral membrane vesicles of rat small intestine, *Lipids* **18**:267–269.

Haramaki, N., Packer, L., Assadnazari, H., and Zimmer, G., 1993, Cardiac recovery during post-ischemic reperfusion is improved by combination of vitamin E with dihydrolipoic acid, *Biochem. Biophys. Res. Commun.* **196**:1101–7.

He, H., and Lawrence, A. L., 1993, Vitamin E requirement of penaeus vannamel, *Aquaculture* **118**:245–55.

Heinzerling, R. H., Nockels, C. F., Quarles, C. L., and Tengerdy, R. P., 1974, Protection of chicks against E. coli infection by dietary supplementation with vitamin E, *Proc. Soc. Exp. Biol. Med.* **146**:279–283.

Heizerling, R. H., Tengerdy, R. P., Wick, L. L., and Leuker, C., 1974, Vitamin E protects mice against diplococcus pneumoniae Type I infection, *Infect. Immun.* **10**:1292–1295.

Hennig, B., Boissonneault, G. A., Chow, C. K., Wang, Y., Matulionis, D. H., and Glauert, H. P., 1990, Effect of vitamin E in linoleic acid-mediated induction of peroxisomal enzymes in cultured porcine endothelial cells, *J. Nutr.* **120**:331–7.

Hennig, B., Toborek, M., Joshi-Barve, S., Barger, S. W., Barve, S., Mattson, M. P., and McClain, C. J., 1996, Linoleic acid activates nuclear transcription factor-kappa B (NF-kappa B) and induces NF-kappa B-dependent transcription in cultured endothelial cells, *Am. J. Clin. Nutr.* **63**:322–8.

Herbaczynska-Cedro, K., Wartanowicz, M., Panczenko-Kresowska, B., Cedro, K., Klosiewicz-Wasek, B., and Wasek, W., 1994, Inhibitory effect of vitamins C and E on the oxygen free radical production in human polymorphonuclear leucocytes, *Eur. J. Clin. Invest.* **24**:316–9.

Hicks, R. B., 1985, Effect of nutrition, medical treatments and management practices on health and performance of newly-received stocker cattle, Stillwater. OK. 21–37.

Hill, G. L., and Williams, S. F., 1993, Vitamin E in beef nutrition and meat quality, *Proc. Nutr. Conf. Bloomington* 197–211.

Hogan, J. S., Weiss, W. P., and Smith, K. L., 1993, Role of vitamin E and selenium in host defense against mastitis, *J. Dairy. Sci.* **76**:2795–2803.

Hosoi, Y., Yamamoto, M., Ono, T., and Sakamoto, K., 1993, Prostacyclin production in cultured endothelial cells is highly sensitive to low doses of ionizing radiation, *Int. J. Radiat. Biol.* **63**:631–8.

Hutcheson, D. P., and Cole, N. A., 1985, Vitamin E and selenium for yearling feedlot cattle. *Fed. Proc.* **44**:549

Ikeda, M., Sumimoto K., Urushihara, T., Fukuda, Y., Dohi, K., and Kawasaki, T., 1994, Prevention of ischemic damage in rat pancreatic transplantation by pretreatment with alpha tocopherol, *Transplant. Proc.* **26**:561–562.

Ikezawa, T., 1989, An experimental study of tissue injury associated with reperfusion in ischemic limbs, *J. Jap. Surg. Soc.* **90**:1799–1805.

Jacobson, J. M., Michael, J. R., Jafri, M. H., Jr., and Gurtner, G. H., 1990, Antioxidants and antioxidant enzymes protect against pulmonary oxygen toxicity in the rabbit, *J. Appl. Physiol.* **68:**1252–9.

Jahreis, G., Steinhart, H., Pfalzgraf, A., Flachowsky, G., and Scho~ne, F., 1996, Zur Wirkung von Rapso~Ifutterung an Milchkuhe auf das Fettsaurenspektrum des Butterfettes, *Z. Ernahrungswiss.* **35:**185–190.

Jandak, J., Steiner, M., and Richardson, P. D., 1988, Reduction of platelet adhesiveness by vitamin E supplementation in humans, *Thromb. Res.* **49:**393–404.

Jialal, I., and Grundy, S. M., 1993, Effect of combined supplementation with alpha-tocopherol, ascorbate, and beta-carotene on low-density lipoprotein oxidation, *Circulation* **88:**2780–6.

Kaneko, T., Nakano, S. I., and Matsuo M., 1991, Protective Effect of Vitamin E on Linoleic Acid Hydroperoxide-Induced Injury to Human Endothelial Cells, *Lipids* **26:**345–348.

Karpen, C., Pritchard, K. A., Arnold, J.H, Cornwell, D.G., Panganamala, R.V., 1982, Restoration of prostacyclin/thromboxane A2 balance in the diabetic rat—influence of dietary vitamin E, *Diabetes* **31:**947–951.

Kashiwado, I., Hattori, Y., and Qiao-Y., 1994, Functional and morphological changes in the cochlea of cholesterol fed quinea pigs, *Nippon. Ika. Daigaku. Zasshi.* **61:**321–329.

Keaney, J. F. Jr., Gaziano, J. M., Xu, A., Frei, B., Curran-Celentano, J., Shwaery, G. T., Loscalzo, J., and Vita, J. A., 1993, Dietary antioxidants preserve endothelium-dependent vessel relaxation in cholesterol-fed rabbits, *Proc. Natl. Acad. Sci. USA* **90:**11880–11884.

Keegan, A., Walbank, H., Cotter, M. A., and Cameron, N. F., 1995, Chronic vitamin E treatment prevents defective endothelium-dependent relaxation in diabetic rat aorta, 1995, *Diabetologia.* **38:**1475–1478.

Kilinc, C., Ozcan, O., Karaoz, F., Sunguroglu, K., Kutluay, T., and Karaca L., 1993, Reducing effect of vitamin E on bleomycin-induced lung fibrosis in mice: Biochemical and histological evidence, *Turk. J. Med. Sci.* **17:**275–85.

Kiser, M. L., Williams, S. F., Hill, G. M., and Williams, S. N., 1993, Interrelationship of dietary vitamin E and postmortem storage on visual and oxidative properties of loin steaks and ground beef, *Proc. Recprocal Meat Conf.* 174A.

Klein, D. R., Novilla, M. N., and Watkins, K. L., 1994, Nutritional encephalomalacia in turkeys: diagnostic and growth performance, *Avian. Dis.* **38:**653–659.

Kolb, F., and Grün, F., 1995, Die Bedeutung des Vitamin E und des Selens für das Immunsystem des Rindes, insbesondere für die Eutergesundheit. *Der Praktische Tierarzt* **9:**794–756.

Korpela, H., 1990, Increased myocardial and hepatic iron concentration in pigs with microangiopathy (mulberry heart disease) as a risk factor of oxidative damage, *Ann. Nutr. Metab.* **34:**193–7.

Kunisaki, M., Bursell, S. F., Clermont, A. C., Ishii, H., Ballas, L. M., Jirousek, M. R., Umeda, F., Nawata, H., and King, G. L., 1995, Vitamin E prevents diabetes-induced abnormal retinal blood flow via the diacylglycerol-protein kinase C pathway, *Am. J. Physiol.* **269:**E239–246.

Kurose, I., Fukumura, D., Miura, S., Suematsu, M., Suzuki, M., Sekizuka, F., Nagata, H., Morishita, T., and Tsuchiya, M., 1993, Fluorographic study on the oxidative stress in the process of gastric mucosal injury: Attenuating effect of vitamin E, *J. Gastroenterol. Hepatol.* **8:**254–258.

Kuzuya, M., Naito, M., Funaki, C., Hayashi, T., Asai, K., and Kuzuya, F., 1991, Probucol prevents oxidative injury to endothelial cells, *J. Lipid. Res.* **32:**197–204.

Lafuze, J. F., Weisman, S. J., Al pert, L. A., and Baehner, R. L., 1984, Vitamin E attenuates the effect of FMLP on rabbit circulating granulocytes, *Pediatr. Res.* **18:**536–540.

Lanari, M. C., Cassens, R. G., Schaefer, D. M., and Scheller, K. K., 1993, Dietary vitamin E enhances color and display life of frozen beef from Holstein steers, *J. Food. Sci.* **58:**701–4.

Lee, K. C., Hamel, D. W., and Kunbel, S., 1993, Rodent model of renal ischemia and reperfusion injury: influence of body temperature, seasonal variation, tumor necrosis factor, endogenous and exogenous antioxidants, *Methods. Find. Exp. Clin. Pharmacol.* **15:**153–9.

Lee, R. W., Stuart, R. L., Perryman, K. R., and Ridenour, K. W., 1995, Effect of vitamin supplementation on the performance of stressed beef calves, *J. Anim. Sci.* **61:**425

Lee, S. M., and Clemens, M. G., 1992, Effect of alpha-tocopherol on hepatic mixed function oxidases in hepatic ischemia/reperfusion, *Hepatology* **15:**276–281.

Lehr, H. A., Guhlmann, A., Nolte, D., Keppler, D., and Messmer, K., 1991a, Leukotrienes as mediators in ischemia-reperfusion injury in a microcirculation model in the hamster, *J. Clin. Invest.* **87:**2036–2041.

Lehr, H.A., Hubner, C., Nolte, D., Kohlschutter, A., Messmer, K., 1991b, Dietary fish oil blocks the microcirculatory manifestations of ischemia-reperfusion injury in striated muscle in hamsters. *Proc. Natl. Acad. Sci. USA* **88:**6726–2730.

Lehr, H.A., and Messmer, K., 1996, Rationale for the use of antioxidant vitamins in clinical organ transplantation. *Transplantation* **62:**1197–1199.

Lin, C.F., Gray, J.I., Ashgar, A., Buckley, D.J., Booren, A.M., Flegel, C.J., 1989, Effects of dietary oils and a-tocopherol supplementation on lipid composition and stability of broiler meats, *J. Food. Sci.* **54:**1457–1460.

Liu, Q., Scheller, K. K., Arp, S. C., Schaefer, D. M., and Frigg, M., 1996, Color coordinates for assessment of dietary vitamin E effects on beef color stability, *J. Anim. Sci.* **74**:106–16.

Lu, L. J., 1993, Ultrastructural changes in endothelial cells in rabbits with multiple organ failure (MOF), *Chung. Hua. Ping. Li. Hsueh. Tsa. Chih.* **22**:169–71.

Luostarinen, R., Siegbahn, A., and Saldeen, T., 1991, Effects of dietary supplementation with vitamin E on human neutrophil chemotaxis and generation of LTB4, *Upsala J. Med. Sci.* **96**:103–111.

Mahan, D. C., 1994, Effects of dietary vitamin E on sow reproductive performance over a five-parity period, *J. Anim. Sci.* **72**:2870–9.

Mahoney, C. W., and Azzi, A., 1988, Vitamin E inhibits protein kinase C activity, *Biochem. Biophys. Res. Commun.* **154**:694–697.

Marubayashi, S., Dohi, K., Ochi, K., and Kawasaki, T., 1986, Role of free radicals in ischemic rat liver cell injury: Prevention of damage by alpha-tocopherol administration, *Surgery* **99**:184–192.

Marusisch, W. L., deRitter, E., Ogrinz, E. F., Keating, J., Mitrovic, M., and Bunnel, R. H., 1975, Effect of supplemental vitamin E in control of rancidity in poultry meat, *Poultry. Sci.* 54:831–844.

Massey, K.D., and Burton, K. P., 1989, Alpha-tocopherol attenuates myocardial membrane-related alterations resulting from ischemia and reperfusion, *Am. J. Physiol.* **256**:H1192–H1199.

Menger, M. D., Pelikan, S., Steiner, D., and Messmer, K., 1992a, Microvascular ischemia-reperfusion injury in striated muscle: significance of "reflow paradox", *Am. J. Physiol.* **263**:H1901–H1906.

Menger, M. D., Steiner, D., and Messmer, K., 1992b, Microvascular ischemia-reperfusion injury in striated muscle: significance of "no reflow", *Am. J. Physiol.* **263**:H1892–H1900.

Meydani, M., 1992, Modulation of the platelet thromboxane A2 and aortic prostacyclin synthesis by dietary selenium and vitamin E, *Biol. Trace. Flem. Res.* **33**:79–86.

Michiels, C., Toussaint, 0., and Remade, J.,1990, Comparative study of oxygen toxicity in human fibroblasts and endothelial cells, *J. Cell. Physiol.* **144**:295–302.

Mitsumoto, M., Arnold, R. N., Schaefer, D. M., and Cassens R. G., 1993, Dietary versus postmortem supplementation of vitamin E on pigment and lipid stability in ground beef, *J. Anim. Sc.* **71**:1812–16.

National Institutes of Health, 1982, NIH Rodents, 1980 Catalogue: strains and stocks of laboratory rodents provided by the NIH genetic resource. NIH Publication 83–606. Bethesda, MD.: U.S. Public Health Service.

Nemec, M., Butler, G., Hidiroglou, M., Famworth, E. R., and Nielsen, K., 1994, Effect of supplementing gilts' diets with different levels of vitanin E and different fats on the humoral and cellular immunity of gilts and their progeny, *J. Anim. Sci.* **72**:665–76.

Ochi, H., Morita, I., and Murota, S., 1992, Mechanism for endothelial cell injury induced by 15-hydroperoxyeicosatetraenoic acid, an arachidonate lipoxygenase product, *Biochim. Biophys. Acta.* **1136**:247–52.

Oda, T., Nakai, I., Mituo, M., Yamagishi, H., Oka, T., and Yoshikawa, T.,1992, Role of oxygen radicals and synergistic effect of superoxide dismutase and catalase on ischemia-reperfusion injury of the rat pancreas, *Transplant. Proc.* **24**:797–798

Patterson, R. L., and Stevenson, M. H., 1995, Irradiation-induced off-odour in chicken and its possible control, *Br. Poult. Sci.* **36**:425–41.

Peet, R.L., and Dickson, J., 1988, Rhabdomyolysis in housed, fine-woolled merino sheep, associated with low plasma alpha-tocopherol concentrations, *Aust. Vet. J.* **65**:398–399.

Pentland, A. P., Morrison, A. R., Jacobs, S. C., Hruza, L. L., Hebert, J. S., and Packer, L., 1992, Tocopherol analogs suppress arachidonic acid metabolism via phospholipase inhibition, *J. Biol. Chem.* **267**:15578–15584.

Peplowski, M. A., Mahan, D. C., Murray, F. A., Moxon, A. L., Cantor, A. H., and Ekstrom, K. F., 1980, Effect of dietaray and injectable vitamin E and selenium in weaning swine antigenically challenged with sheep red blood cells, *J. Animal Sci.* **51**:344–351.

Polette, A., and Blache, D., 1992, Effect of vitamin E on acute iron load-potentiated aggregation, secretion, calcium uptake and thromboxane biosynthesis in rat platelets, *Atherosclerosis* **96**:171–9.

Politis, I., Hidiroglou, N., White, J. H., Gilmore, J. A., Williams, S. N., Scherf, H., and Frigg, M., 1996, Effects of vitamin E on mammary and blood leukocyte function, with emphasis on chemotaxis, in periparturient dairy cows, *Am. J. Vet. Res.* **57**:468–471.

Pruitt, S. P., Morrill, J. L., Blecha, F., Reddy, P. G., Higgins, J., and Anderson, N. V., 1989, Effect of supplemental vitamin C and E in milk replacer on lymphocyte and neutrophil function in bull calves, *J. Anim. Sci.* **67**:243A.

Pyke, D. D., and Chan, A. C., 1990, Effects of vitamin E on prostacyclin release and lipid composition of the ischemic rat heart, *Arch. Biochem. Biophys.* **277**:429–433.

Rabi, H., Khoschsorur, G., Colombo, T., Petritsch, P., Rauchenwald, M., Kotringer, P., Tatzber, F., and Esterbauer, H., 1993, A multivitamin infusion prevents lipid peroxidation and improves transplantation performance, *Kidney Int.* **43**:912–917.

Rao, P.N., Walsh, T.R., Makowka, L., Liu, T., Demetris, A.J., Rubin, R.S., Snyder, J.T., Mischinger, H.J., and Stazl, T.F., 1990, Inhibition of free radical generation and improved survival by protection of the hepatic microvascular endothelium by targeted erythrocytes in orthotopic liver transplantation, *Transplantation* 49:1055–1059.

Reddy, P. G., Morrill, J. L., Minocha, H. C., Morrill, M. B., Dayton, A. D., and Frey, R. A., 1986, Effect of supplemental vitamin E on the immune system of calves, *J. Dairy Sci.* 69:164–170.

Reddy, P. G., Morrill, J. L., Minocha, H. C., Stevenson, J. S., 1987, Vitamin E is immunostimulatory in calves, *J. Dairy Sci.* 70:993–999.

Rethwill, C. F., Bruin, T. K., Waibel, P. F., Addis, P. B., 1981, Influence of dietary fat source and vitamin E on market stability of turkeys, *Poultry. Sci.* 60:2466–2474.

Rojas, C., Cadenas, S., Lopez-Torres, M., Perez-Campo, R., and Barja, G. 1997, Increase in heart glutathione redox ratio and total antioxidant capacity and decrease in lipid peroxidation after vitamin E dietary supplementation in guinea pigs. *Free Radic. Biol. Med.* 21:907–918.

Rösen, P., Ballhausen, T., Bloch, W., and Addicks, K., 1995, Endothelial relaxation is disturbed by oxidative stress in the diabetic rat heart: influence of tocopherol as antioxidant, *Diabetologia.* 38:1157–1168.

Rubino, A., and Burnstock, G., 1994, Recovery after dietary vitamin E supplementation of impaired endothelial function in vitamin E-deficient rats, *Br. J. Pharmacol.* 112:515–518.

Sakamoto, W., Fujie, K., Handa, H., Ogihara, T., and Mino, M., 1990, In vivo inhibition of superoxide production and protein kinase C activity in macrophages from vitamin E-treated rats, *Int. J. Nutr. Res.* 60:338–342.

Salonen, J. T., Salonen, R., Seppanen, K., Rinta-Kiikka, S., Kuukka, M., Korpela, H., Alfthan, G., Kantola, M., and Schalch, W., 1991, Effects of antioxidant supplementation on platelet function: A randomized pair-matched, placebo-controlled double-blind trial in men with low antioxidant status, *Am. J. Clin. Nutr.* 53:1222–1229.

Schaeffer, D. M., Scheller, K. K., Arp., S. C., Buege, D. R., and Lane, S. F., 1989, Growth of Holstein steers and beef color as affected by dietary vitamin E supplementation, *J. Anim., Sci.* 68:190A.

Schwarz, F., 1996, Influence of dietary fatty acid composition and vitamin E on fatty acids and alpha-tocopherol in carp (Cyprinus caprio L.) *Arch. Anim. Nutr.* 49:63–71.

Sheehy P. J., Morrissey, P. A., and Flynn, A., 1991, Influence of dietary alpha-tocopherol on tocopherol concentration in chick tissues. *Brit. Poultry Sci.* 32:391–397.

Shirahata, A., 1993, Clinical application of fat soluble vitamins to retinopathy of prematurity and intracranial hemorrhage, *Nippon. Rinsho.* 51:1029–1036.

Smith, G. C., Morgan, J. B., Sofos, J. N., Tatun, J. D., 1996, Supplemental vitamin E in beef cattle diets to improve shelf-life of beef, *Anim. Feed. Sci. Technol.* 59:207–214.

St. Laurent, A., Hidiroglou, M., Snoddon, M., and Nichlson, J. W. G., 1990, Response to dietary vitamin E in the dairy cow and its effect on spontaneous oxidized flavor in milk. *Can. J. Anim. Sci.* 70:561–570.

Steiner, M., and Mower, R., 1982, Mechanism of action of vitamin E on platelet function, *Ann. NY Acad. Sci.* 393:289–299.

Strivastava, K. C., 1986, Vitamin E experts antiaggregatory effects without inhibiting the enzymes of the arachidonic acid cascade in platelets, *Prostagland. Leukotr. Medicine* 21:177–185.

Suntres, Z.F., Herworth, S.R., and Shek, P.N., 1992, Protective effect of liposome-associated alpha-tocopherol against paraquat-induced acute lung toxicity, *Biochem. Pharmacol.* 9:1811–1818.

Suzuki, Y.J., Packer, L. 1993. Inhibition of NFkB activation by vitamin E derivatives. *Biochem. Biophys. Res. Commun.* 193:277–283.

Takenaka, M., Tatsukawa, Y., Dohi, K., Ezaki, H., Matsukawa, K., and Kawasaki, T., 1981, Protective effects of alpha-tocopherol and coenzyme Q10 on warm ischemic dangers of the rat kidney, *Transplantation* 32:137–141.

Tanabe, M., Kito, G. 1989. Effects of CV-361 1, a new free radical scavenger, on ischemic heart failure in conscious dogs. *Jap. J. Pharmacol.* 50:467–476.

Tanaka, T., Fujiwara, H., and Torisu, M., 1979, Vitamin E and immune response: I. Enhancement of helper T cell activity by dietary supplementation of vitamin E in mice, *Immunology* 38:727–734

Taoka, Y., Ikata, T., and Fukuzawa, K., 1990, Influence of dietary vitamin E deficiency on compression injury of rat spinal cord, *J. Nutr. Sci. Vitaminol.* 36:217–226.

Tengerdy, R. P., Heinzerling, R. H., and Nockels, C. F., 1972, Effect of vitamin E on the immune response of hypoxic and normal chicken, *Infect. Immun.* 5:987–989.

Teoh, M. K., Chong, M. K., and Jamaludin, M., 1992, Effects of tocotrienol-rich vitamin E on patients with peripheral vascular disease, *Lipid-Soluble Antoxid.* 606–21.

Toborek, M., Barger, S. W., Mattson, M. P., Barve, S., McClain, C. J., and Hennig, B., 1996, Linoleic acid and TNF-alpha cross-amplify oxidative injury and dysfunction of endothelial cells, *J. Lipid. Res.* 37:123–35.

Tran, K., Proulx, P.R., and Chan, A. C., 1994, Vitamin E suppresses diacylglycerol (DAG) level in thrombin-stimulated endothelial cells through an increase of DAG kinase activity, *Biochim. Biophys. Acta.* 1212:193–202.

Umeda, F., Kunisaki, M., Inoguchi, T., and Nawata, H., 1990, Vitamin E enhances prostacyclin production by cultured aortic endothelial cells, *J. Clin. Biochem. Nutr.* 8:175–183.

Vajkoczy, P., Lehr, H.A., Hubner, C., Arfors, K.E., and Menger, M.D., 1997, Prevention of pancreatic islet xenograft rejection by dietary vitamin E. *Am. J. Pathol.* **150:**1487–1495.

Van Jaarsveld, H., Kuyl, J. M., and Alberts, D. W., 1992, Antioxdant vitamin supplementation of smoke-exposed rats partially protects against myocardial ischemic/reperfusion injury, *Free Radic. Res. Commun.* **17:**263–269.

Verlhac, V., N'Doye, A., Gabaudan, J., Troutaud, D., and Deschaux, P., 1993, vitamin nutrition and fish immunity: influence of antioxidant vitamins (C and E) on immune response of rainbow trout, *Colloq. INRA.* **61:**167–77.

Villa, S., Lorico, A., Morazzoni, G., de Gaetano, G., and Sameraro, N., 1986, Vitamin E and vitamin C inhibit arachidonate-induced aggregation of human peripheral blood leukocytes in vitro, *Agents Actions* **19:**127–131.

Waagboe, R., Sandnes, K., Torrissen, O. J., Sandvin, A., and Lie, O., 1993, Chemical and sensory evaluation of fillets from atlantic salmon (salmo salar) fed three levels of N-3 polyunsaturated fatty acids at two levels of vitamin E, *Food. Chem.* **46:**361–6.

Walsh, D. M., Kennedy, D. G., Goodall, F. A., and Kennedy, S., 1993, Antioxidant enzyme activity in the muscles of calves depleted of vitamin E or selenium or both, *Br. J. Nutr.* **70:**621–30.

Wang, X., Feuerstein, G. Z., Gu, J. L., Lysko, P. G., and Yue, T. L., 1995. IL-1 beta induces expression of adhesion molecules in human vascular smooth muscle cells and enhances adhesion of leukocytes to smooth muscle cells. *Atherosclerosis* **115:**89–98.

Washio, M., Nanishi, F., Okuda, S., Onoyama, K., and Fujishima, M., 1994, Alpha-tocopherol improves focal glomerulosclerosis in rats with adriamycin-induced progressive renal failure, *Nephron.* **68:**347–352.

Weiss, W. P., Hogan, J. S., Smith, K. L., and Hoblet, K. H., 1990, Relationships among selenium, vitamin E and mammary gland health in commercial dairy herds, *J. Dairy. Sci.* **73:**381–390.

Williams, G. M., Frye, T. M., Frigg, M., Schaefer, D. M., Scheller, K. L., and Liu, Q., 1992, Vitamin E, *Meat. Int.* **3:**22.

Willy, C., Thiery, J., Menger, M.D., Messmer, K., Arfors, K.F., and Lehr, H.A. 1995, Impact of vitamin E supplements in standard laboratory animal diet on microvascular manifestations of ischemia/reperfusion injury. *Free Radic. Biol. Med.* **19:**919–926.

Wolf, H. R. D., and Seeger, H. W., 1982, Experimental and clinical results in shock lung treatment with vitamin F, *Ann. NY Acad. Sci.* **393:**392–410.

Wolfram, D., and Flachowsky, G., 1991, Hohe Vitamin-F-Gaben in der Milchkuhfutterung, Proc. 3. Symp. "Vitamine und weitere Zusatzstoffe bei Mensch und Tier", *Stadtroda.* 196–199.

Wulf, D. M., Morgan, J. B., Sanders, S. K., Tatum, J. D., Smith, G, C., and Williams, S., 1995, Effects of dietary supplementation of vitamin E on storage and caselife properties of lamb retail cuts, *J. Anim. Sci.* **73:**399–405.

Yamamoto, M., Shima, T., Uozumi, T., Sogabe, T., Yamada, K., and Kawasaki, T., 1983, A possible role of lipid peroxidation in cellular damages caused by cerebral ischemia and the protective effect of alphatocopherol administration, *Stroke* **14:**977–982.

Yoshikawa, T., Yasuda, M., Ueda, S., Naito, Y., Tanigawa, T., Oyamada, H., and Kondo, M., 1991, Vitamin E in Gastric Mucosal Injury Induced by Ischemia-reperfusion, *Am. J. Clin. Nutr.* **53:**210S–214S.

Zadkova, G. F., Avakian, T. T., and Markov, K. h. M., 1993, Effect of alpha-tocopherol on the development of diabetic angiopathy, platelet aggregation, and status of the prostacyclin-thromboxane system in rats with streptozotocin diabetes, *Probl. Endokrinol. Mosk.* **39:**40–3.

31

EPR SPECTROSCOPY OF PHENOLIC PLANT ANTIOXIDANTS

Wolf Bors, Christa Michel, and Kurt Stettmaier

Institut für Strahlenbiologie
GSF Forschungszentrum für Umwelt und Gesundheit
D-85764 Neuherberg, Germany

1. INTRODUCTION

Plant phenols are mostly secondary metabolites comprising such diverse structures as phenolic acids, e.g., of hydroxycinnamic acid (Pratt, 1993; Foti, Piattelli, Barratta, and Ruberto, 1996) or carnosic acid and its derivates in *Rosmarinus officinalis* (Ho, Ferraro, Chen, Rosen, and Huang, 1994), aromatic lactones such as hydroxycoumarins (Dixon, Moghimi, and Murphy, 1975; Foti et al., 1996; Hoult, Moroney, and Paya, 1994) or iso-quinolines (Hewgill and Pass, 1985a), hydroxy-anthraquinones (Malterud, Farbrot, Huse, and Sund, 1993; Mian, Fratta, Rainaldi, Simi, Mariani, Benetti, and Gervasi, 1991), xan-thones (Ashida, Noguchi, and Suzuki, 1994; Minami, Kinoshita, Fukuyama, Kodama, Yoshizawa, Sugiura, Nakagawa, and Tago, 1994), the large group of flavonoids (Bors, Heller, Michel, and Stettmaier, 1995), and the rather unusual macrocyclic bis-biphenyls (e.g. marchantins) from liverwort (Asakawa, 1990). Substances belonging to each of these structural groups have been shown to act as antioxidants or *vice versa* as cytotoxic agents.

The basic mechanisms for such activities involve the formation of radicals derived from these phenols, called either phenoxyl radicals, semiquinones or, more generally, aroxyl radicals. Functioning as antioxidants, secondary radical formation is an unavoid-able consequence of scavenging of oxygen radicals, most likely peroxyl radicals as the predominant chain carriers during lipid peroxidation (Barclay, 1993; Bors, Erben-Russ, Michel, and Saran, 1990a; Hsieh and Kinsella, 1989):

$$Ph(OH)O^- + ROO^\bullet + H^+ \longrightarrow Ph(OH)O^\bullet + ROOH \qquad [1]$$

Free Radicals, Oxidative Stress, and Antioxidants, edited by Özben.
Plenum Press, New York, 1998.

The cytotoxicity of such compounds, in contrast, might be explained by metal-catalyzed autoxidation of the polyphenol to form an aroxyl radical, which subsequently transfers an electron to oxygen to generate superoxide anions and is oxidized to a quinone itself (Aver'yanov, 1981; Cohen, Heikkila, and MacNamee, 1974),

$$Ph(OH)O^- + Me^{n+} \longrightarrow Ph(OH)O^{\cdot} + Me^{(n-1)+} \tag{2}$$

$$Ph(OH)O^{\cdot} + O_2 \longrightarrow Ph(=O)_2 + O_2^{\cdot-} + H^+ \tag{3}$$

(where $Ph(OH)O^-$, $Ph(OH)O^{\cdot}$, and $Ph(=O)_2$ are phenolate, semiquinone and quinone, respectively). This process has been thoroughly investigated with flavonoids (Hodnick, Milosavljevic, Nelson, and Pardini, 1988b).

While the kinetics of such reactions can be optimally investigated applying the technique of pulse radiolysis with spectrophotometric observation (Bors, Heller, Michel, and Saran, 1990b; Patterson, 1987), it is quite obvious that more detailed information on the structures of these polyphenol radicals can be obtained by the technique of electron paramagnetic (or spin) resonance (EPR/ESR) spectroscopy (Kalyanaraman, 1990; Mason and Chignell, 1982). Furthermore, the formation of oxygen radicals during autoxidation can be verified by EPR spectroscopy combined with the spin trapping technique (Schaich and Borg, 1980). This review is an expansion and update from a recent contribution by our group (Bors, Michel, Stettmaier, and Heller, 1997).

Most studies involving the generation of phenoxyl radicals for the purpose of EPR investigations resort to autoxidation in strongly alkaline aqueous solutions (Cotelle, Bernier, Hénichart, Catteau, Gaydou, and Wallet, 1992; Holton and Murphy, 1980; Jensen and Pedersen, 1983; Kuhnle, Windle, and Waiss, 1969). Oxidation by Ce^{4+} salts has been applied in acidic solutions (Dixon et al., 1975; Hewgill and Pass, 1985b; Russell, Forrester, Chesson, and Burkitt, 1996). Addition of organic solvents, e.g. dimethylsulfoxide or alcohols cause a shift of the coupling constants (Kuhnle et al., 1969), which is even more pronounced if one records the EPR spectra in organic solvents (Klotz, Jülich, Wax, and Stegmann, 1989). To provide a reasonable simile to biological systems, enzymatic catalysis employing either horseradish peroxidase/hydrogen peroxide (HRP/H_2O_2; Ferrari and Laurenti, 1995; Thompson, Norbeck, Olsson, Constantin-Teodosiu, van der Zee, and Moldeus, 1989; Valoti, Sipe, Sgaragli, and Mason, 1989) or tyrosinase (Richard, Cantin-Esnault, and Jeunet, 1995) under *in situ* conditions, are preferred.

EPR spectra of phenoxyl radicals are generally recorded in the X-Band region (about 9.7 GHz). Two types of parameters are normally evaluated: (a) the coupling constants (hyperfine splitting constants), denoting the resonance interactions of the delocalized electron with neighboring hydrogen atoms and (b) the g-factor, giving information on the general type of radical (oxygen- or carbon-centered, metal ion, etc.).

2. EPR SPECTRA OF PHENOXYL RADICALS DERIVED FROM MONOPHENOLS

In this presentation only mono-, di- and polyphenolic compounds are covered, which aside from being natural substances recognized as antioxidants, have been studied by EPR spectroscopy. Thus studies of the synthetic monophenolic antioxidants 2-*tert*-butyl-4-methoxyphenol (BHA) and 2,6-di-*tert*-butyl-4-methylphenol (BHT; Valoti et al.,

1989) or the cytotoxic fungal catechol compound orellaline (Richard et al., 1995) are not discussed further.

2.1. Substituted Phenols

Simple phenoxyl radicals derived from monophenolic compounds are usually too unstable and thus produced at too low steady-state concentrations for EPR spectroscopy. Indeed, only in the case of eugenol (4-allyl-2-methoxyphenol), a component of the essential oils of various plants, could a phenoxyl radical be detected by EPR after generation by HRP/H$_2$O$_2$ in a fast-flow system (Thompson et al., 1989). The radical observed was quite unstable and rapidly formed a quinone methide and polymeric end products.

2.2. Phenolic Alcohols and Acids

A recent study on monomeric precursor radicals during lignin peroxidation applied ammonium Ce^{4+} nitrate in 0.25molar sulfuric acid in a flow system to generate the phenoxyl radicals of p-hydroxycinnamyl alcohol, p-hydroxycinnamic acid and its 3-methoxy (ferulic) and 3,5-dimethoxy (sinapic) derivatives (Russell et al., 1996). Both EPR spectra and coupling constants show, that the carboxyl group does influence the electron spin densities via the vinyl group. As reported below, HRP/H$_2$O$_2$ in slightly alkaline solutions could oxidize only sinapic acid. Yet, as shown in Table 1, the coupling constants in acidic and alkaline solutions are not too dissimilar and most likely represent a protonated, respectively dissociated phenoxyl radical.

2.3. Curcuminoids

The first study of EPR spectra of curcuminoids by Schaich, Fisher, and King (1994) resulted in undefined signals after illumination with bright light or by electrolysis, in effect demonstrating the instability of these radicals. Using EPR/spin trapping, the authors were able to trap a carbon-centered radical rather than the expected phenoxyl radical. More recent studies on a number of natural and synthetic curcuminoids revealed the limitations of the HRP/H$_2$O$_2$ system to generate phenoxyl radicals: only those with two adjacent methoxy groups yielded EPR-observable radical signals, in line with similar results for the model phenolic acids (Fig. 1; Mamero, 1997) and confirming the observations with BHA and BHT (Valoti et al., 1989). Table 1 lists the corresponding coupling constants for a number of phenoxyl radicals, derived from monophenolic natural compounds.

2.4. Monophenolic Hydroxycoumarins, Isoquinolines

The first study on EPR spectra of hydroxycoumarins employed Ce^{4+} oxidation in acidic solution (Dixon et al., 1975). The presence of an oxo function adjacent to the hydroxyl group in the heterocylic ring evidently caused a resonance stabilization of the phenoxyl radicals with mesomeric carbon-centered radicals. Subsequently, Hewgill and Pass (1985a,b) studied bisphenolic derivates of isoquinolines, especially reticuline as a precursor of isoquinoline alkaloids. Oxidation to the phenoxyl radicals was achieved in flow systems either with acidic Ce^{4+} (Hewgill and Pass, 1985b) or alkaline potassium ferricyanide (Hewgill and Pass, 1985a). In the acidic oxidation system, the authors were able to identify phenoxyl radicals derived from 3-hydroxy-4-methoxy-substituted phenols, whereas in line with our own findings for the curcuminoids and model phenolic acids (see above), they did not

Table 1. Coupling constants of phenoxyl radicals derived from plant monophenols

Substance	Substituent				Coupling constant (Gauss)					Reference
	R_1	R_3	R_4	R_5	$-OCH_3$	$H_{2,6}$	$H_{3,5}$	H_α	H_β	
Alkaline pH (8.5–13)										
eugenol	$-CH=CH-CH_3$	$-OCH_3$	$-O^\bullet$	$-H$	1.62 (3)	1.9	4.15	7.97 (2)	—	(1)
p-coumaric acid	$-CH=CH-COO^-$	$-H$	$-O^\bullet$	$-H$	—	1.25 (2)	—	2.3	2.7	(2)
syringic acid	$-COO^-$	$-OCH_3$	$-O^\bullet$	$-OCH_3$	1.26 (6)	0.7 (2)	—	—	—	(2)
sinapic acid	$-CH=CH-COO^-$	$-OCH_3$	$-O^\bullet$	$-OCH_3$	1.25 (6)	1.25 (2)	—	3.5	4.0	(2)
5,5′dimethoxy-curcumin	see Fig. 1	$-OCH_3$	$-O^\bullet$	$-OCH_3$	1.30 (6)	1.30 (2)	—	2.8	6.1	(2)
substance *a*	$-(CH_2)_3-N(C_2H_5)_2$	$-OCH_3$	$-O^\bullet$	$-OCH_3$	1.33 (6)	1.33 (2)	—	6.57 (2)	—	(3)
5′-methoxy-homoorientaline	ethyltetrahydroisoquinoline	$-OCH_3$	$-O^\bullet$	$-OCH_3$	1.33 (6)	1.33 (2)	—	6.8 (2)	—	(3)
Acidic pH (3–5)										
p-coumaric acid	$-CH=CH-COO^-$	$-H$	$-O^\bullet$	$-H$	—	1.68 (2)	5.26, 5.65	3.06	6.36	(4)
ferulic acid	$-CH=CH-COO^-$	$-OCH_3$	$-O^\bullet$	$-H$	1.69 (3)	1.53, 1.70	5.56	2.64	5.68	(4)
sinapic acid	$-CH=CH-COO^-$	$-OCH_3$	$-O^\bullet$	$-OCH_3$	1.33 (6)	0.95, 1.26	—	1.69	4.82	(4)
substance *b*	$-CH_2-COO^-$	$-O^\bullet$	$-OCH_3$	$-H$	1.9 (3)	3.9, 8.8	2.0	—	—	(5)
reticuline	N-methyltetrahydroisoquinoline	$-O^\bullet$	$-OCH_3$	$-H$	2.0 (3)	4.0, 8.65	1.9	—	—	(5)
nor-reticuline	tetrahydroisoquinoline	$-O^\bullet$	$-OCH_3$	$-H$	2.0 (3)	4.0, 8.65	1.9	—	—	(5)

a: 3,5-dimethoxy-4-hydroxyphenyl-N,N-diethylpropylamine, *b*: 3-hydroxy-4-methoxy-phenylacetic acid

References: (1) Thompson et al. J. Biol. Chem. 264, 1016 (1989), (2) Mamero, PhD thesis (1997), (3) Hewgill & Pass Aust. J. Chem. 38, 615 (1985), (4) Russell et al. Arch. Biochem. Biophys. 332, 357 (1996), (5) Hewgill & Pass Aust. J. Chem. 38, 497 (1985).

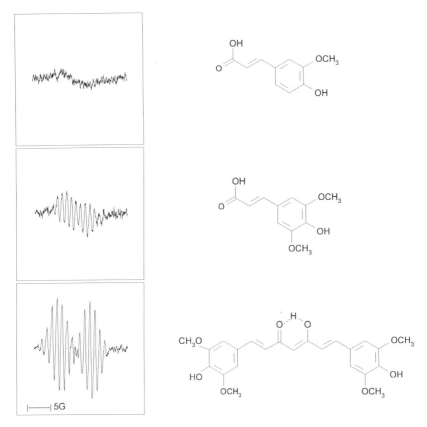

Figure 1. EPR spectra of phenoxyl radicals derived from monophenols: ferulic acid (top), sinapic acid (center) and 5,5′-dimethoxycurcumin (bottom); radicals generated with HRP/H_2O_2 at pH 8.5, HRP 6.7 U/ml, H_2O_2 1 mmolar, ferulic acid 2.58 mmolar, sinapic acid 3.1 mmolar, 5,5′-dimethoxycurcumin 1 mmolar; EPR settings 9.75 kHz, sweep rate 0.15 Gs^{-1}, modulation amplitude 0.4 G.

observe EPR spectra after alkaline oxidation of 3-methoxy-4-hydroxy-substituted isoquinoline derivatives, yet were successful for the sinapic acid analog 3,5-dimethoxy-4-hydroxy derivatives (Hewgill and Pass, 1985a).

3. EPR SPECTRA OF POLYPHENOL AROXYL RADICALS

3.1. Catechols and Hydroquinones

Benzoquinones are quite common plant and fungal constituents (Thomson, 1971). The ease of their univalent reduction or the alkaline oxidation of the respective catechols or hydroquinones was sufficient impetus to initiate EPR studies of the semiquinone intermediates. As already observed for monophenols (see above), the additional presence of methoxy groups in *ortho*-position further stabilized the radicals (Holton and Murphy, 1980). In the case of simple caffeic acid (3,4-dihydroxycinnamic acid) derivatives quite distinct semiquinone spectra were observed and could be identified readily — which could even be applied for the taxonomic screening of *Lycopodium* varieties for their content of these types of substances (Pedersen and Ollgaard, 1982). Caffeic acid esters as components of propolis were similarly identified from their EPR spectra (Rapta, Misik, Stasko, and Vrabel, 1995).

The active principles of *Rosmarinus officinals*, i.e. carnosic acid, carnosol, rosmarol, and derivatives of them turned out to be an interesting group of catechols (Ho et al., 1994). Their potential as food antioxidants has already led to several studies in which these compounds were incorporated into cooking oil (Ho et al., 1994). Interestingly enough, an EPR investigation of the corresponding semiquinone structures (Geoffroy, Lambelet, and Richert, 1994) was the first study, in which the radicals were generated by peroxyl radicals in an autoxidizing methyl oleate solution at elevated temperatures to imitate frying oil. While the radical derived from carnosic acid represented a genuine *o*-semiquinone, temperatures above 110°C converted it into the carnosol radical, which as a keto phenoxyl radical was resonance-stabilized (Geoffroy et al., 1994).

3.2. Hydroxy-Anthraquinones

Anthraquinones comprise the largest group of quinone pigments in plants and other natural sources (Thomson, 1971). The most thorough EPR investigation of the semiquinones of hydroxylated anthraquinones was done by Pedersen and Thomson (1981). Structures were assigned by additivity principles and in some cases by deuterium labelling. In this case the anthrasemiquinone radicals were produced by reduction with sodium dithionite in strongly alkaline solution (pH $\approx$ 12). It turned out, that α-hydroxy groups (i.e., in 1,4,5,8-positions) are not dissociated even at pH 12 due to hydrogen bonds with the central oxo function, whereas dissociation of β-hydroxy groups (in 2,3,6,7-positions) may cause difficulties for the assignment of coupling constants (Pedersen, 1984, 1985).

3.3. Flavonoids

The most comprehensive EPR investigations were performed with this class of substances, both because they are important plant antioxidants, and owing to their structural diversity, they lend themselves quite naturally to structure-activity relationship studies (see Scheme 1). In the first study the radicals were generated by autoxidation in strongly alkaline solutions, adding up to 80% DMSO which further stabilized the radicals (Kuhnle et al., 1969). The coupling constants measured in this investigation generally pointed to the catechol group in the B-ring as the predominant radical-generating site.

Scheme 1. Structural principles of the major flavonoids represented by compounds with 5,7,3',4'-tetra-hydroxy groups lacking or containing the 3-hydroxy group

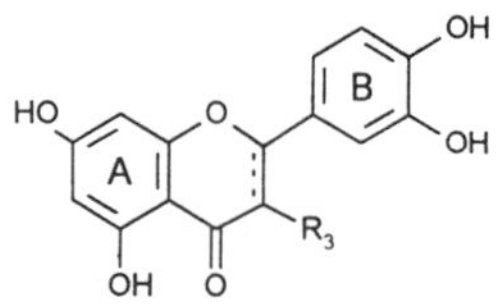

Sub-group	Example	2,3-bond	3-hydroxy	4-oxo
flavan-3-ols	(+)-catechin	saturated	+	−
flavanones	eriodictyol	saturated	−	+
dihydroflavonols	dihydroquercetin	saturated	+	+
flavones	luteolin	unsaturated	−	+
flavonols	quercetin	unsaturated	+	+

Two flavanones, besides ten hydroxy-coumarin derivatives, were the subject of an EPR study, in which the phenols were oxidized by Ce^{4+} in acidic solution (Dixon et al., 1975). Under these conditions aroxyl radicals are produced, which are not necessarily o- or p-semiquinones. In a recent study on synthetic flavonoids with B-ring pyrogallol structure, autoxidation in alkaline solutions was again applied to generate the radicals (Cotelle et al., 1992). The same group recently extended these studies to a rather large number of hydroxy flavones (Cotelle, Bernier, Catteau, Pommery, Wallet, and Gaydou, 1996), observing further hydroxylation in the B-ring at pH > 13 either at 3- or 2'-positions.

To avoid unphysiological pH-conditions, we decided to employ the HRP/H_2O_2 system in slightly alkaline solutions (pH 8.5, which still kept the compounds in solution) for generation of the flavonoid aroxyl radicals *in situ* (Bors et al., 1997; Bors, Heller, Michel, and Stettmaier, 1993). We could basically confirm the coupling constants reported in the literature (Cotelle et al., 1996; Kuhnle et al., 1969; Pedersen, 1985), even though these were obtained in strongly alkaline solutions, thus verifying the B-ring catechol group as the principal radical site (a compilation of the presently known coupling constants of flavonoids can be found in Bors et al., 1997). Yet, our data in the HRP/H_2O_2 system were unequivocal only for the flavanones and dihydroflavonols investigated. In some flavones and flavonols, especially with quercetin whose 2,3-double bond enables strong electron delocalization over the three ring systems (Bors et al., 1990b), the radical observed proved to be unstable. During the recording period or after longer reaction times the spectrum changed considerably, indicating conversion of the primarily formed radical, which made the identification of individual radicals rather difficult. A similar problem evidently occurs also in the strongly alkaline solutions, where the formation of the secondary radicals after further hydroxylation could only be demonstrated after simulating the composite spectrum with a finite amount of the EPR spectrum of the precursor radical (Cotelle et al., 1992; Cotelle et al., 1996).

Applying the technique of 'spin stabilization' to the EPR investigations of flavonoids, i.e. complexing the semiquinone structures with Zn^{2+} or Mg^{2+}, gave quite contrasting results to those with catecholamines. Formation of this complex was described as stabilizing the semiquinone radical structure (Felix and Sealy, 1981; Ferrari and Laurenti, 1995; Kalyanaraman, Premovic, and Sealy, 1987; Stegmann, Bergler, and Scheffler, 1981; Stegmann, Dao-Ba, Stolze, and Scheffler, 1985). The consequence for some flavonoids, however, is a facilitated nucleophilic attack leading to rupture of the C2–C1' bond (Bors et al., 1993). At variance with our results is an earlier report on the stabilization of the semiquinone of myricetin in the presence of Mg^{2+} (Hodnick, Kalyanaraman, Pritsos, and Pardini, 1988a).

A recent electrochemical study combined with EPR spectroscopy of a few flavonoids, which lack hydroxy groups in the B-ring (Rapta et al., 1995), was basically concerned with the redox chemistry and less the antioxidative properties. The predominant anodic two electron-oxidation led to fragmentation of the pyrane ring with carbon-centered radicals trapped by 5,5-dimethyl-1-pyrrolidine N-oxide (DMPO) or 2-methyl-2-nitrosopropane (MNP). Conversely, cathodic *reduction* of chrysin (5,7-dihydroxyflavone) resulted in an anion radical, whose highly resolved EPR spectrum yielded hyperfine splitting constants which demonstrate a strong electron delocalization to the B-ring o- and p-positions from the negatively charged ketyl radical in 4-position. The presence of the 2,3-double bond is therefore sufficient to provide resonance stabilization of the radical, analogous as we have found for the aroxyl radicals generated by *univalent oxidation* (Bors et al., 1990).

An EPR investigation of the alkaline autoxidation of the flavan-3-ols (+)-catechin and (–)-epicatechin. showed further hydroxylation in 2' and 6' positions (Jensen and Pedersen, 1983), and not in 3-position as has been observed for B-ring pyrogallol flavones (Cotelle et al., 1992; Cotelle et al., 1996). The 2,3-*cis* configuration of (–)-epicatechin turned out to be more unstable, with the consequence that no EPR-detectable intermediates but only the final rearrangement products, catechinic acid semiquinones, were observed (Jensen and Pedersen, 1983).

3.4. Tannins

The above-mentioned flavan-3-ols are the monomeric precursors of the flavonoid subgroup procyanidins or proanthocyanIns, which may exist as B-ring pyrogallol derivatives, can be esterified at the 3-hydroxy group with gallic acid or, upon further polymerisation, may form hydrolysable or condensed tannins (Ferreira and Bekker, 1996; Singleton, 1988). Radical autocondensation of condensed tannins was investigated by Masson, Pizzi, and Merlin (1996), using kinetic EPR approaches. The authors were thus less concerned about the identity of the radical species involved. At present, only Guo, Zhao, Li, Shen, and Xin (1996) have attempted to identify the radicals produced from the various epicatechin derivatives. They identified as preferred radical sites the pyrogallol groups rather than the catechol structures and in the case of (–)-epigallocatechin gallate the B-ring pyrogallol to be oxidized before the gallate ester group. Kinetic EPR was likewise employed by Pizzi's group In a very recent study on the antioxidative potential ot these compounds (Noferi, Masson, Merlin, Pizzi, and Deglise, 1997). Alternatively, EPR/spin trapping (see below) experiments also yielded information on the antioxidative potential of procyanidins (Guo et al., 1996).

3.5. Marchantins

Marchantins and related substances have exclusively been detected in liverwort (Asakawa, 1990). As deduced from the structure of marchantin A trimethylether, they probably exist as non-planar macrocyclic ring systems in which four aromatic rings are linked together by both ether and ethyl bridges (Asakawa, 1990; Taira, Takei, Endo, Hashimoto, Sakiya, and Asakawa, 1994). Several of these compounds have shown good antioxidative potential depending on the individual structures, basically due to the presence of one or more catechol groups (Schwartner, Bors, Michel, Franck, Müller-Jakic, Nenninger, Asakawa, and Wagner, 1995). Extending these studies to incorporate the pertinent marchantin radicals, we generated the semiquinones with HRP/H_2O_2 at pH 8.5 (Schwartner, Michel, Stettmaier, Wagner, and Bors, 1996). Yet, only the EPR spectrum of the semiquinone radical of the model bis-catechol compound nor-dihydroguaiaretic acid could be unambiguously assigned with semiquinone formation in both aromatic rings (data not shown). For marchantin B, which contains two catechol moieties in the opposite rings A and C (see Fig. 2), evaluation of the radical structure proved considerably more difficult as several spectra were probably superimposed upon each other.

Marchantin A with only one catechol group gave a rather unspecific EPR spectrum which can be explained, if we assume that ring A is somewhat tilted out of the macrocyclic plane (Taira et al., 1994). Complex formation with Zn^{2+} was also attempted and was successful only for those compounds, which already showed a reasonable spectral resolution in the absence of the complexing cation (Schwartner et al., 1996).

An interesting example of a phenolic antioxidant is provided by Perrottetin D (see Fig. 2). Both the radical-scavenging efficiency and formation of the semiquinone radical are insufficient at room temperature (Schwartner et al., 1996), whereas the antioxidative

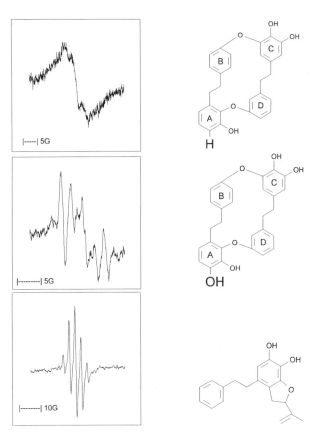

Figure 2. EPR spectra and structures of marchantins and related compounds: top panel — marchantin A, center panel — marchantin B, lower panel — perrettotin D (at 60°C); radicals generated with HRP/H$_2$O$_2$ at pH 8.5, HRP 6.7 U/ml, H$_2$O$_2$ 1 mmolar, all phenols 1 mmolar; EPR settings as in Fig. 1.

potential determined in the arachidonic acid assay at 37° C showed it as one of the most effective compounds (Schwartner et al., 1995). Consequently, when we heated perrottetin D in our EPR experiments, we obtained an excellent signal both with and without Zn^{2+}, which we could show to be due to the fragmentation of the furan ring and formation of a pyrogallol semiquinone with adjacent aliphatic substituents (Schwartner et al., 1996). We assume, that this radical is also the active antioxidant rather than the parent catechol compound (Schwartner et al., 1996).

4. EPR/SPIN TRAPPING RESULTS

As mentioned earlier, EPR spectroscopy of spin adducts obtained from unstable radical intermediates, e.g. O$_2^{\cdot-}$, allows to study autoxidation mechanism as proposed in reaction [3]. This approach was attempted repeatedly with flavonoids and the interpretations of the spin adduct spectra were generally in agreement with this mechanism (Canada, Giannella, Nguyen, and Mason, 1990; Hodnick et al., 1988a). Moreover, the autoxidative and cytotoxic effectiveness correlated well with the electrochemical redox potentials (Hodnick et al., 1988b). In a similar experiment we also found the ·OH-adduct of DMPO, probably resulting from the decay of the DMPO–OOH adduct, after autoxidation of marchantin B (Schwartner et al., 1996). This compound has previously been shown to be cyctotoxic (Asakawa, 1990) and yet to be one of the most effective antioxidative agents in an arachidonic acid autoxidation assay (Schwartner et al., 1995). Thus the mechanism outlined in reactions [2] and [3] (see above) may also be valid for this class of polyphenols derived from primitive plants.

5. *IN SITU* STUDIES

It is obvious that EPR spectroscopy is excellently suited to identify the structures of radicals derived from plant phenols, provided these radicals are sufficiently stable. Concerning *in situ* studies, the technique has been shown to be feasible for simple phenolic acids (Pedersen and Ollgaard, 1982) and hydroxylated naphthoquinones (Pedersen, 1978) even in crude plant extracts. Screening of the plant apoplast or other vacuoles for polyphenol-derived radical species and possibly monitoring their appearance after stress events (Heller, Rosemann, and Sandermann, 1991; Heller, Ernst, Langebartels, and Sandermann, 1995), however, has only rarely been attempted. The few studies undertaken to demonstrate that polyphenol radical species may be detected by the EPR technique in plants themselves (Crawford, Lindsay, Walton, and Wollenweber-Ratzer, 1994; Hendry, Atherton, Seel, and Leprince, 1994; Snijder, Wastie, Glidewell, and Goodman, 1996) were ambiguous, as identification, structural, functional or site-specific correlations remained unknown. Nevertheless, taking the limitations into consideration, EPR spectroscopy may still be developed into a useful method to study plant polyphenol reactions *in situ*.

ACKNOWLEDGMENTS

We appreciate the stimulating and productive discussions with Werner Heller and Manfred Saran.

REFERENCES

Asakawa, Y., 1990, Terpenoids and aromatic compounds with pharmacological activity from bryophytes, *Proc. Phytochem. Soc. Europe* **29**:369–410

Ashida, S., Noguchi, S.F., and Suzuki, T., 1994, Antioxidative components, xanthone derivatives, in *Swertia japonica* Makino, *J. Am. Oil Chem. Soc.* **71**:1095–99

Aver'yanov, A.A., 1981, Generation of superoxide anion radicals and hydrogen peroxide in autoxidation of caffeic acid, *Biochemistry USSR* **46**:210–215

Barclay, L.R.C., 1993, Model biomembranes: quantitative studies of peroxidation, antioxidant action, partitioning, and oxidative stress, *Can. J. Chem.* **71**:1–16

Bors, W., Erben-Russ, M., Michel, C., and Saran, M., 1990a, Radical mechanisms in fatty acid and lipid peroxidation, *NATO ASI Ser.* **A189**:1–16

Bors, W., Heller, W., Michel, C., and Saran, M., 1990b, Flavonoids as antioxidants: determination of radical scavenging efficiencies, *Meth. Enzymol.* **186**:343–354

Bors, W., Heller, W., Michel, C., and Stettmaier, K., 1993, Electron paramagnetic resonance studies of flavonoid compounds. in: *Free Radicals: From Basic Science to Medicine* (G. Poli, M. Albano, and M.U. Dianzani, eds.), pp. 374–387, Birkhäuser, Basel

Bors, W., Heller, W., Michel, C., and Stettmaier, K., 1995, Flavonoids and polyphenols: Chemistry and biology. in: *Handbook on Antioxidants* (E. Cadenas, and L. Packer, eds.), pp. 409–466, Marcel Dekker, New York, NY

Bors, W., Michel, C., Stettmaier, K., and Heller, W., 1997, EPR studies of plant polyphenols. in: *Natural Antioxidants. Chemistry, Health Effects, and Applications* (F. Shahidi, ed.), pp. 346–357, AOCS Press, Champaign, IL

Canada, A.T., Giannella, E., Nguyen, T.D., and Mason, R.P., 1990, The production of reactive oxygen species by dietary flavonols, *Free Radical Biol. Med.* **9**:441–449

Cohen, G., Heikkila, R.E., and MacNamee, D., 1974, The generation of hydrogen peroxide, superoxide radical, and hydroxyl radical by 6-hydroxydopamine, dialuric acid and related cytotoxic agents, *J. Biol. Chem.* **249**:2447–2452

Cotelle, N., Bernier, J.L., Hénichart, J.P., Catteau, J.P., Gaydou, E., and Wallet, J.C., 1992, Scavenger and antioxidant properties of ten synthetic flavones, *Free Radical Biol. Med.* **13**:211–219

Cotelle, N., Bernier, J.L., Catteau, J.P., Pommery, J., Wallet, J.C., and Gaydou, E.M., 1996, Antioxidant properties of hydroxy-flavones, *Free Radical Biol. Med.* **20**:35–43

Crawford, R.M.M., Lindsay, D.A., Walton, J.C., and Wollenweber-Ratzer, B., 1994, Towards the characterization of radicals formed in rhizomes of *Iris germanica*, *Phytochemistry* **37**:979–985

Dixon, W.T., Moghimi, M., and Murphy, D., 1975, ESR study of radicals obtained from the oxidation of naturally occurring hydroxypyrones, *J.Chem. Soc., Perkin II*, **1975**:101–103

Felix, C.C., and Sealy, R., 1981, Photolysis of melanin precursors: Formation of semiquinone radicals and their complexation with diamagnetic metal ions, *Photochem. Photobiol.* **34**:423–425

Ferrari, R.P., and Laurenti, E., 1995, ESR spin-stabilization evidence for alpha-methyldopa and dopamethylester *o*-semiquinones obtained by peroxidasic oxidation: Structural characterization and mechanistic studies, *J. Inorg. Biochem.* **59**:811–825

Ferreira, D., and Bekker, R., 1996, Oligomeric proanthocyanidins: Naturally occurring O-heterocycles, *Nat. Prod. Rep.* **13**:411–433

Foti, M., Piattelli, M., Baratta, M.T., and Ruberto, G., 1996, Flavonoids, coumarins, and cinnamic acids as antioxidants in a micellar system. Structure-activity relationship, *J. Agric. Food Chem.* **44**:497–501

Geoffroy, M., Lambelet, P., and Richert, P., 1994, Radical intermediates and antioxidants: An ESR study of radicals formed on carnosic acid in the presence of oxidized lipids, *Free Radical Res.* **21**:247–258

Guo, Q.N., Zhao, B.L., Li, M.F., Shen, S.R., and Xin, W.J., 1996, Studies on protective mechanisms of four components of green tea polyphenols against lipid peroxidation in synaptosomes, *Biochim. Biophys. Acta* **1304**:210–222

Heller, W., Rosemann, D., and Sandermann, H., 1991, Untersuchungen sekundärer Stoffwechselleistungen von Fichte und Kiefer als möglicher Indikator für Schadeffekte im Höhenprofil des Wank. in: *Proc. 2. Statusseminar der PBWU zum Forschungsschwerpunkt 'Waldschäden'. GSF-Bericht* **26** (M. Reuther, M. Kirchner, and K. Rösel, eds.), pp. 173–181

Heller, W., Ernst, D., Langebartels, C., and Sandermann, H., 1995, Induction of polyphenol biosynthesis in plants during development and environmental stress. in: *Polyphenols '94* (R. Brouillard, M. Jay, A. Scalbert, eds.), pp. 67–78, Ed. INRA, Paris

Hendry, G.A.F., Atherton, N.M., Seel, W., and Leprince, O., 1994, The occurrence of a stable quinone radical accumulating in vivo during natural and induced senescence in a range of plants. *Proc. Roy. Soc. Edinburgh Sect. B* **102**:501–503

Hewgill, F.R., and Pass, M.C., 1985a, The oxidation of (±)-reticuline as studied by ESR spectroscopy. *Aust. J. Chem.* **38**:497–506

Hewgill, F.R., and Pass, M.C., 1985b, The oxidation of some bisphenolic 1-phenethyltetrahydro-isoquinolines studied by ESR spectroscopy, *Aust. J. Chem.* **38**:615–620

Ho, C.T., Ferraro, T., Chen, Q., Rosen, R.T., and Huang, M.T., 1994, Phytochemicals in teas and rosemary and their cancer-preventive properties, *ACS Sympos. Ser.* **547**:2–19

Hodnick, W.F., Kalyanaraman, B., Pritsos, C.A., and Pardini, R.S., 1988a, The production of hydroxyl and semiquinone free radicals during the autoxidation of redox active flavonoids, *Basic Life Sci.* **49**:149–152

Hodnick, W.F., Milosavljevic, E.B., Nelson, J.H., and Pardini, R.S., 1988b, Electrochemistry of flavonoids. Relationships between redox potentials, inhibition of mitochondrial respiration, and production of oxygen radicals by flavonoids, *Biochem. Pharmacol.* **37**:2607–2611

Holton, D.M., and Murphy, D., 1980, The electron spin resonance spectra of semiquinones obtained from some naturally occurring methoxybenzoquinones. *J. Chem. Soc., Perkin II*, **1980**:1757–59

Hoult, J.R.S., Moroney, M.A., and Paya, M., 1994, Actions of flavonoids and coumarins on lipoxygenase and cyclooxygenase, *Meth. Enzymol.* **234**:443–454

Hsieh, R.J., and Kinsella, J.E., 1989, Oxidation of PUFAs: mechanisms, products and inhibition with emphasis on fish, *Adv. Food Nutr. Res.* **33**:233–341

Jensen, O.N., and Pedersen, J.A., 1983, The oxidative transformations of (+)catechin and (−)epicatechin as studied by ESR, *Tetrahedron* **39**:1609–1615

Kalyanaraman, B., 1990, Characterization of *o*-semiquinone radicals in biological systems, *Meth. Enzymol.* **186**:333–342

Kalyanaraman, B., Premovic, P.I., and Sealy, R.C., 1987, Semiquinone anion radicals from addition to amino acids, peptides and proteins to quinones derived from oxidation of catechols and catecholamines. An ESR spin stabilization study, *J. Biol. Chem.* **262**:11080–11087

Klotz, D., Jülich, T., Wax, G., and Stegmann, H.B., 1989, Magnetic Properties of Free Radicals. 17. Semiquinones and Related Species. in: *Landolt-Börnstein Zahlenwerte und Funktionen aus Naturwissenschaften und Technik, Neue Serie, Vol. 17g* (O. Madelung, ed.), Springer, Berlin), pp. 69–394

Kuhnle, J.A., Windle, J.J., and Waiss, A.C., 1969, EPR spectra of flavonoid anion-radicals. *J. Chem. Soc. B*, **1969**:613–616

Malterud, K.E., Farbrot, T.L., Huse, A.E., and Sund, R.B., 1993, Antioxidant and Radical Scavenging Effects of Anthraquinones and Anthrones, *Pharmacology* **47**:77–85

Mamero, B., 1997, Curcumin als Antioxidans: Schwermetallionenvermittelte radikalische Oxidation von Curcuminoiden, *Dissertation, Universität Kiel*

Mason, R.P., and Chignell, C.F., 1982, Free radicals in pharmacology and toxicology. Selected topics, *Pharmacol. Rev.* **33**:189–211

Masson, E., Pizzi, A., and Merlin, A., 1996, Comparative kinetics of the induced radical autocondensation of polyflavonoid tannins. 3. Micellar reactions vs cellulose surface catalysis, *J. Appl. Polymer Sci.* **60**:1655–1664

Mian, M., Fratta, D., Rainaldi, G., Simi, S., Mariani, T., Benetti, D., and Gervasi, P.G., 1991, Superoxide anion production and toxicity in V79 cells of six hydroxy-anthraquinones, *Anticancer Res.* **11**:1071–76

Minami, H., Kinoshita, M., Fukuyama, Y., Kodama, M., Yoshizawa, T., Sugiura, M., Nakagawa, K., and Tago, H., 1994, Antioxidant xanthones from *Garcinia subelliptica*. *Phytochemistry* **36**:501–506

Noferi, M., Masson, E., Merlin, A., Pizzi, A., and Deglise, X., 1997, Antioxidant characteristics of hydrolysable and polyflavonoid tannins: An ESR kinetics study, *J. Appl. Polymer Sci.* **63**:475–482

Patterson, L.K., 1987, Instrumentation for measurement of transient behavior in radiation chemistry. in: *Radiation Chemistry. Principles & Applications*. (Farhataziz, and M.A.J. Rodgers, eds.), pp. 65–96, VCH Verlag, Weinheim

Pedersen, J.A., 1978, Naturally occurring quinols and quinones studied as semiquinones by ESR, *Phytochemistry* **17**:775–778

Pedersen, J.A., 1984, EPR study of hydroxyanthrasemiquinones. β-Hydroxyl proton constants, *J. Magn. Reson.* **60**:136–137

Pedersen, J.A., 1985, CRC *Handbook of EPR spectra from quinones and quinols*, CRC Press, Boca Raton, FL

Pedersen, J.A., and Thomson, R.H., 1981, EPR study of hydroxyanthrasemiquinones. Assignment of hyperfine structure by additivity, *J. Magn. Reson.* **43**:373–386

Pedersen, J.A., and Ollgaard, B., 1982, Phenolic acids in the genus *Lycopodium*. *Biochem. System. Ecol.* **10**:3–9

Pratt, D.E., 1993, Antioxidants indigenous to foods, *Toxicol. Industr. Health* **9**:63–75

Rapta, P., Misik, V., Stasko, A., and Vrabel, I., 1995 Redox intermediates of flavonoids and caffeic acid esters from propolis: An EPR spectroscopy and cyclic voltammetry study, *Free Radical Biol. Med.* **18**:901–908

Richard, J.M., Cantin-Esnault, D., and Jeunet, A., 1995, First electron spin resonance evidence for the production of semiquinone and oxygen free radicals from orellanine, a mushroom nephrotoxin, *Free Radical Biol. Med.* **19**:417–429

Russell, W.R., Forrester, A.R., Chesson, A., and Burkitt, M.J., 1996, Oxidative coupling during lignin polymerization is determined by unpaired electron delocalization within parent phenylpropanoid radicals, *Arch. Biochem. Biophys.* **332**:357–366

Schaich, K.M., and Borg, D.C., 1980, EPR studies in autoxidation. in: *Autoxidation in Food and Biological Systems*. (M.G. Simic, M. Karel, eds.), pp. 45–70, Plenum Press, New York, NY

Schaich, K.M., Fisher, C., and King, R., 1994, Formation and reactivity of free radicals in curcuminoids: an electron paramagnetic resonance study. ACS Sympos. Ser. **547**:204–221

Schwartner, C., Bors, W., Michel, C., Franck, U., Müller-Jakic, B., Nenninger, A., Asakawa, Y., and Wagner, H., 1995, Effect of marchantins and related compounds on 5-lipoxygenase and cyclooxygenase and their antioxidant properties: a structure activity relationship study, *Phytomedicine* **2**:113–117

Schwartner, C., Michel, C., Stettmaier, K., Wagner, H., and Bors, W. (1996) Marchantins and related polyphenols from liverwort: physico-chemical studies of their radical-scavenging properties, *Free Radical Biol. Med.* **20**:237–244

Singleton, V.L., 1988, Wine phenols. *Modern Methods of Plant Analysis* **6**:173–218

Snijder, A.J., Wastie, R.L., Glidewell, S.M., and Goodman, B.A., 1996, Free radicals and other paramagnetic ions in interactions between fungal pathogens and potato tubers. *Biochem. Soc. Trans.* **24**:442–446

Stegmann, H.B., Bergler, H.U., and Scheffler, K., 1981, "Spinstabilisierung" durch Komplexierung: ESR-Untersuchung einiger Catecholamin-Semichinone. *Angew. Chem.* **93**:398–399

Stegmann, H.B., Dao-Ba, H., Stolze, K., and Scheffler, K., 1985, ESR- und ENDOR-Untersuchungen an Catecholaminen und deren Metaboliten als paramagnetische Thallium-Komplexe. *Z. Anal. Chem.* **322**:430–436

Taira, Z., Takei, M., Endo, K., Hashimoto, T., Sakiya, Y., and Asakawa, Y., 1994, Marchantin A trimethyl ether: its molecular structure and tubocurarine-like skeletal muscle relaxation activity, *Chem. Pharm. Bull.* **42**:52–56

Thompson, D., Norbeck, K., Olsson, L.I., Constantin-Teodosiu, D., van der Zee, J., and Moldeus, P., 1989, Peroxidase-catalysed oxidation of eugenol: formation of a cytotoxic metabolite, *J. Biol. Chem.* **264**:1016–1021

Thomson, R.H., 1971, *Naturally Occurring Quinones*. Academic Press, London

Valoti, M., Sipe, H.J., Sgaragli, G., and Mason, R.P., 1989, Free radical intermediates during peroxidase oxidation of 2-*t*-butyl-4-methoxyphenol, 2,6-di-*t*-butyl-4-methylphenol, and related phenol compounds, *Arch. Biochem. Biophys.* **269**:423–432

32

CLASTOGENIC FACTORS AS BIOMARKERS OF OXIDATIVE STRESS

Their Usefulness for Evaluation of the Efficacy of Antioxidant Treatments

Ingrid Emerit

Free Radical Research Group
Institut biomédical des Cordeliers
University Paris VI

1. INTRODUCTION

Clastogenic factors (CF) have been recognized since the early seventieth as an indirect effect of ionizing radiation (Goh and Sumner, 1968; Hollowell and Littlefield, 1968). Because of their persistence in the blood of irradiated persons many years after exposure, they have been considered as risk factors for late effects of radiation, such as cancer and leukemia (Faguet, 1984). Previous work of our laboratory has shown that CF are not specific for irradiated subjects, but found in a variety of other pathological conditions, where they are biomarkers of oxidative stress. Their formation and their clastogenic action are related to increased superoxide production, since both are regularly inhibited by superoxide dismutase (SOD) (Emerit, 1994). They are not single factors, as thought by the first observers, but mixtures of chromosome-damaging pro-oxidant substances. Nevertheless the term "clastogenic factor" has been conserved.

1.1. CF Formation

The strongest argument for the implication of superoxide anion radicals in CF formation came from in vitro models, in which cells were exposed to O_2^- generated by physical or chemical means:

- CF could be produced by irradiation of whole blood or isolated lymphocytes with gamma rays at a rate of 0.46 Gy/min. The lowest dose resulting in CF was 50 cGy. While irradiation of serum did not generate clastogenic materials, irradiation of cells in PBS yielded clastogenic supernatants. If the irradiated cells were

Free Radicals, Oxidative Stress, and Antioxidants, edited by Özben.
Plenum Press, New York, 1998.

washed and resuspended in fresh culture medium, they continued to release CF. The formation as well as the chromosome damaging effects of CF were preventable by SOD (Emerit et al., 1994).

- Not only ionizing radiation, but also near UV light combined with a photosensitizer, such as riboflavin or psoralen, could be used for CF induction. Again the presence of cells was necessary, showing that CF are cellular products. SOD was regularly protective (Alaoui-Youssefi et al., 1994).
- Generation of superoxide by stimulation of a respiratory burst with the tumor promoter TPA (Emerit and Cerutti, 1982) or extracellulary via a xanthine–xanthine oxidase reaction (Emerit et al., 1991) resulted in CF-containing supernatants. Again CF formation occured only in presence of cells and could be regularly inhibited by SOD.

These examples of CF formation in vitro occur also in vivo, as shown below for the various diseases, accompanied by CF.

1.2. Identified Clastogenic Components

The mixture of chromosome damaging agents, called CF,. includes lipid peroxidation products (Emerit et al., 1991), unusual nucleotides of inosine (Auclair et al., 1990) and cytokines, such as tumor necrosis factor alpha (Emerit et al., 1995). The chromosome damaging properties of these endogenous clastogens have been confirmed with the respective commercial standards.

Certain components of CF have not only clastogenic properties, but can also stimulate superoxide production by competent cells. Using the SOD-inhibitable cytochrome c reduction assay (Emerit, 1990) or luminol-enhanced chemiluminescence (Emerit et al., 1995b), we could show that CF stimulate superoxide production by monocytes and neutrophils. Among the identified components, TNF and ITP may be responsible herefore.

Comparison of CF with other biomarkers of oxidative stress, such as TBARS, increased O_2^- production by neutrophils, decreased levels of non-protein plasma thiols, indicated strong correlations between these parameters.

1.3. Clastogenic Action Mechanisms

The consistent anticlastogenic effect of SOD indicates that superoxide radicals are not only involved in CF formation, but produced also on the pathway to chromosomal breakage. The superoxide-stimulating properties of certain CF components will lead to additional clastogenic material in the exposed cultures. Their progressive accumulation explains, why the chromosomal aberrations observed are mainly of the chromatid type despite exposure of the cells in the Go phase of the cell cycle. In agreement herewith, SOD is only anticlastogenic, when added during the first 24h of the cultivation period. According to the model of membrane-mediated chromosome damage, perturbations of the membrane integrity by oxyradicals lead to activation of phospholipase A2 and release of arachidonic acid from membrane phospholipids, which is oxidized enzymatically or non enzymatically to products with clastogenic properties (Emerit, 1991). In addition, oxyradical-mediated increases in intracellular calcium activate lysosomal enzymes, and chromosome damage is induced through the action of nucleases.

The various components of CF may exert clastogenic effects also via superoxide-independent mechanisms. ITP for instance, in addition to its action as a primer of superoxide production, may induce genotoxic effects by inhibition of enzymes involved in DNA replication, since it inhibits DNA topoisomerases by competition with the ATP-binding sites of these enzymes (Kuhns et al., 1988). The aldehyde 4-hydroxynonenal, derived from membrane lipid peroxidation, is a well-known genotoxic agent which inactivates functional SH groups of DNA polymerases and forms mono-adducts with cellular thiols such as glutathione and cysteine (Wavra et al., 1986).

CF formation and CF action are discussed in more detail in a recent review article (Emerit, 1994).

2. METHODS

The endogenous clastogens, called CF, are studied in a cytogenetic test system, as this is usual for exogenous clastogenic agents. They are isolated from the plasma by ultrafiltration through a Millipore or Diaflo ultrafiltration filter. Previously, filters with a cut off at 10000 DA had been used, but when TNF alpha (mol. weight 17000) was recognized as one of the clastogenic components of CF, they were replaced by filters with a cut off at 30000. Experiments with commercial TNF alpha had shown that 75% of the total was retained by a filter with a cut off at 10000 DA.

For evaluation of the clastogenic effects of a sample, regular blood cultures are set up with whole blood from a healthy donor, to which 250 µl of the ultrafiltrate are added (Emerit, 1990b). If this quantity is cytotoxic, the culture is repeated with 100 µl. Since the clastogenic effects are related to oxyradical production, a culture medium poor in free radical scavengers is recommended. TCM 199 or RPMI 1640 are convenient (5 ml per culture tube). The serum used for supplementation (1 ml/culture) should not contain antioxidants from hemolysed erythrocytes. Fetal calf serum is preferable to bovine serum, which may be rich in vitamin E. Lymphocyte proliferation is stimulated by the addition of phytohemagglutinin M or P. After 48 or 72 h of cultivation at 37°C, the mitoses are arrested in metaphase by the addition of colchicine 2 h before harvesting. Microscopic slides are prepared for chromosomal analysis according to standard procedures. The chromosomes of 50 well-spread and complete metaphases (10 on each of 5 coded slides prepared from one culture tube) are examined for the presence of breaks, fragments, exchanges, rings, dicentrics and other morphological abnormalities. Slides with a low mitotic index are not examined, and the culture is repeated. A series of ultrafiltrates is tested the same day on the cultures set up with the blood of the same donor. Two additional control cultures without ultrafiltrate serve for the establishment of the spontaneous chromosomal aberration rate of the donor's lymphocytes (background). The background level of aberrations, determined by the two simultaneous untreated blood cultures, is subtracted from the aberration rate in the ultrafiltrate-treated cultures of the same blood donor. The difference between the two values is called the adjusted clastogenic score (ACS). This way of treating results is necessary in order to allow comparison of results for plasma samples taken in a patient at different time periods.

Several series of experiments, in which background levels were determined for 10 parallel blood cultures of the same donor, ascertained that the variation in background levels of aberrations does not exceed ± 3 aberrations per 50 cells studied, and that the range is the same for two independent observers. Therefore a plasma ultrafiltrate is considered to be clastogenic, if it induces more than 3 aberrations per 50 cells. In agreement herewith, the increase in aberrations induced by ultrafiltrates from a series of 96 healthy blood donors did not exceed 2 additional aberrations for the majority of them. Only with 5% of these normal samples, the increase represented +3, while increases of +4 or higher were not observed (Emerit et al., 1995).

Instead of chromosomal aberrations, other endpoints can be studied such as sister chromatid exchanges, DNA strand breakage or mutations at the HPRT locus (Emerit and Lahoud-Maghani, 1989). They have to be studied on cells, since CF does not induce lesions on isolated DNA. It is also possible to measure the superoxide-stimulating effects instead of the clastogenic effects of of CF preparations. However, we noticed that the superoxide stimulating activity is less stable than the clastogenic activity after freezing. Theoretically one may use the appropriate biochemical assays for their various components, but this would be more time consuming. Also the different components may not always reach detectable levels for the respective assays, while the clastogenic effects are the consequence of synergistic action of all CF components.

3. DISEASES ACCOMPANIED BY CF

3.1. Irradiated Persons

The first reports of CF after radiation exposure were reported for therapeutically and accidentally irradiated persons in Great Britain and the US (Goh and Sumner, 1968, Hollowell and Littlefield (1968). Further reports of the chromosome-breaking effects of plasma came from A-bomb survivors in Japan (Pant and Kamada, 1977). The existence of radiation-induced CF was confirmed in our test system by the study of plasma from adults and children exposed as a consequence of the Chernobyl accident (Emerit et al., 1994 and 1997a). The allowable cumulative dose of 25 cGy was not always respected for Chernobyl accident recovery workers (liquidators), especially those, who worked there during the first months, when monitoring programs and equipment were not adequate. As mentioned above for the in vitro model, radiation doses as low as 50 cGy induce CF. Besides lipid peroxidation products, TNF alpha plays a major role in radiation-induced CF. The response of mononuclear phagocytes, the main source of TNF alpha, was considered in the pathogenesis of radiation-induced inflammatory and noninflammatory reactions (Krivenko et al., 1992). TNF mRNA was increased after exposure of cells to ionizing radiation (Hallahan et al., 1989). Even low doses of radiation primed murine peritoneal macrophages for elevated production of TNF (Iwamoto and McBride, 1994), which then can stimulate other monocytes/macrophages for superoxide production. This feedback mechanisms may explain the longevity of the clastogenic activity in the plasma. CF persisted even more than 30 years in the A-bomb survivors (Pant and Kamada, 1977) and more than 10 years in the liquidators (Emerit et al., 1994) CF formation in irradiated persons may be compared to CF formation in chronic inflammatory diseases, a hypothesis based on the presence of biomarkers of inflammation in irradiated individuals (Neriishi, 1991).

3.2. Chronic Inflammatory Diseases

Connective tissue diseases such as progressive systemic sclerosis, rheumatoid arthritis, systemic lupus erythematosus, periarteritis nodosa, dermatomyositis, chronic inflammatory diseases of the liver (chronic active hepatitis B and C), the digestive tract (ulcerative colitis, Crohn's disease) and the nervous sytem (multiple sclerosis) are accompanied by CF. Clastogenic activity is found not only in the plasma of patients, but also in other body fluids and in the supernatants of cultures set up with blood or fibroblasts from these patients. In systemic lupus erythematosus, CF isolated from patients plasma has photosensitizing properties (Emerit and Michelson, 1981a). Patients' lymphocytes are photosensitive to light in the near UV wave length, and lymphocytes from healthy persons become also photosensitive, when exposed to CF from patients. The exact chemical nature of this photosensitizing component of CF from lupus patients could not be determined. In rheumatoid arthritis, monocytes are immunologically activated. They produce increased amounts of superoxide (Harth et al., 1983) and release increased amounts of free arachidonic acid and of TNF alpha. It is therefore not astonishing that addition of monocytes from rheumatoid arthritis patients to lymphocyte culture from healthy persons results in clastogenic effects. CF could be isolated from the supernatants of these cocultivations as a fonction of the number of monocytes added. This effect was not seen after addition of large numbers of heterologous healthy monocytes (Emerit et al., 1989). Similar observations were made on cocultivations of mononuclear cells from patients with Crohn's ileocolitis with those of a healthy donor (Emerit and Michelson, 1981b). Addition of SOD prevented the clastogenic effects. Inosine triphosphate was detected in the plasma of patients with progressive systemic sclerosis together with an increase in the enzyme adenosine deaminase. Both parameters were correlated with clastogenic activity in patients' plasma (Emerit et al., 1997b). As indicated above, ITP is a primer of superoxide production. Also another enzyme, xanthine oxidase, was found to be increased in this disease (Miesel and Zuber, 1993) and could be another source for superoxide production and CF formation.

3.3. HIV Infection

There is more and more evidence for a role of active oxygen species in the progress of this disease. Antioxidants reduced virus expression, and exposure of latently infected monocytes or CD4+ lymphocytes to oxidative stress was followed by increased reverse transcriptase levels in the culture supernatants (Kalebic et al., 1991). Clastogenic activity was detected with our assay system in plasma samples ultrafiltrated through a filter with a cut off at 10000 DA, retentive for virus particles. These samples were derived from patients with AIDS or from asymptomatic seropositive individuals, indicating that CF formation is an early event in the pathogenetic process (Fuchs and Emerit, 1995). Antiviral medication did not prevent CF formation. There was a good correlation between clastogenic scores and decreases in the levels of plasma thiol and erythrocyte GSH. Antioxidant vitamins, on the other hand, were in the normal range in patients with highly clastogenic plasma. Clastogenic ultrafiltrates upregulated HIV-expression in U1 cells, a chronically HIV-1 infected promonocytic cell line. Exogenous SOD inhibited the clastogenic and the virus-inducing effects of CF (Edeas et al., 1997).

3.4. Psoriasis, PUVA Therapy

Psoriasis is a common skin disease, characterized by hyperproliferation and incomplete differentiation of epidermal keritinocytes. Psoralen plus UVA (PUVA) is one of the treatments proposed for this disease. As mentioned above, exposure of blood cultures from healthy donors to PUVA leads to chromosomal breakage due to CF formation, which can be prevented by SOD (Alaoui-Youssefi et al., 1994). A study of 10 patients submitted to PUVA therapy for severe psoriasis showed that CF formation occurs also in vivvo. The clastogenic activity of plasma ultrafiltrates increased significantly between the first and the last (16th) exposure to PUVA. However, CFwere detectable to a minor degree also in 14 out of 31 patients before exposure. It has been claimed that psoriasis is an autoimmune disorder, and CF formation may occur in these patients as in other autoimmune diseases due to increased superoxide production by phagocytes, formation of lipid peroxidation products and release of cytokines. The level of clastogenic materials can be considerably increased as a consequence of further superoxide production in photosensitized reactions during PUVA. CF may contribute to the well-known risk of photocarcinogenesis following PUVA therapy (Filipe et al., 1997, submitted).

3.5. Familial Mediterranean Fever

Familial Mediterranean Fever, a hereditary disease with autosomal recessive transmission, predominatly affects Armenians and non-Ashkenazi Jews. The disease begins in childhood with paroxysmal attacks of pain and fever accompanied by massive influx of neutrophils into the serosal membranes (Rogers et al., 1989). An increase in the chemotactic lipoxygenase product LTB4, and a decrease in the activity of the inhibitor of chemotaxis C5a in serosal fluids have both been considered as responsible herefore (Gabrielian et al., 1990; Matzner and Brzezinki, 1984). Our laboratory has shown that the chromosomal instability observed in blood cultures of these patients is secondary to circulating clastogenic factors and that SOD as well as lipoxygenase inhibitors reduce the chromosome damaging effects (Emerit et al., 1993). When the clastogenic effects exerted by plasma ultrafiltrates from 20 patients were compared with the spontaneous superoxide production by patients' neutrophils, both parameters were correlated ($r = 0.5235$). CF formation in this disease may be comparable to CF formation in vitro by stimulation of a respiratory burst.

3.6. Ischemia-Reperfusion Injury

It is well established that oxygen-derived free radicals play a crucial role in the pathogenesis of ischemia-reperfusion injury. Upon reoxygenation, superoxide anion radicals are generated by the action of xanthine oxidase on hypoxanthine, which is derived from progressive degradation of adenosine triphosphate during ischemia (McCord et al., 1985). This reaction was used in one of the above described in vitro models for the production of CF. Reperfusion accelerates neutrophil chemotaxis to the myocardium. The invading cells are activated such that they may release a variety of mediators capable of promoting tissue injury. With the SOD-inhibitable cytochrome c assay and with chemiluminescence studies we could show that CF isolated from the plasma after ischemia-reperfusion stimulate the superoxide production by monocytes and neutrophils from control persons. This could be inhibited by allopurinol and by SOD (Emerit et al., 1988, 1995).

3.7. Congenital Breakage Syndromes

Ataxia telangiectasia, Bloom's syndrome and Fanconi anemia (FA) are autosomal recessive disorders, associating chromosome mutation with congenital malformations, immune disturbances and a high cancer incidence. Formation of CF could be demonstrated for all three congenital syndromes (Shaham and Becker, 1981; Emerit and Cerutti, 1981; Emerit et al., 1995). However, the reasons for the prooxidant state is less evident in these syndromes as in the other diseases studied. Fanconi anemia has been most intensively investigated. While low SOD levels were reported for FA erythrocytes and leukocytes, CU-Zn SOD was found normal in FA fibroblasts. However, MnSOD, catalase and glutathione peroxidase activities were consistently higher in FA than in normal fibroblasts. The elevation of these enzyme activities was considered as evidence for an adaptive response to the prooxidant state (Joenje et al., 1981; Gille et al., 1989). Nevertheless, FA fibroblasts appear to be particularly sensitive to hyperoxia (Saito et al., 1993). Freshly drawn FA leukocytes showed an increased luminol-enhanced chemiluminescence response (Korkina et al., 1992). A recent study revealed a significant excess of 80HdG formation in leukocytes DNA from FA homozygotes and heterozygotes (Degan et al., 1995). In agreement herewith, CF were not only present in the plasma of FA homozygotes, but detectable also in plasma of heterozygotes, at the condition to concentrate the ultrafiltrate by a second ultrafiltration through a filter with a cut off at 1000 DA. Spontaneous overproduction of TNF alpha in vitro and in vivo has been described recently as a novel feature of FA (Roselli et al., 1994). TNF was detected in ultrafiltrates from two of our patients.

4. EVALUATION OF ANTIOXIDANTS

Various antioxidants are now being evaluated in longterm clinical trials for disease prevention (Kellof et al., 1992). Cellular and biochemical markers have been introduced as intermediate endpoints for the evaluation of the clinical efficiency of promising drugs. The CF-Test for detection of superoxide-mediated and SOD-inhibitable clastogenesis represents a reliable and sensitive assay for the measurement of circulating pro-oxidants in a patient's plasma. Such cytogenetic studies are currently employed for detection of genotoxic effects in high risk populations. The fact that the plasma can be frozen and studied when convenient, after accumulation of many samples, is advantageous for epidemiologic studies. The assay was used for the first time in an animal model for autoimmune disease, the New Zealand Black (NZB) mouse. This strain harbors a a xenotropic endogenous type C– virus and exhibits increased chromosomal breakage, correlated with the presence of CF in the blood stream. By selective matings, according to chromosome breakage frequency, two NZB sublines could be developed, which differ in CF production, superoxide anion generation by peritoneal macrophages, xenotropic virus expression and the incidence of lymphomas and autoimmune hemolytic anemia (Khan et al., 1990). Intraperitoneal injection of bovine SOD during 3 weeks consistently reduced the chromosomal aberration rates to normal values (Emerit et al., 1981). We had the opportunity to study CF in a case of the classical human autoimmune disease, lupus erythematosus, before and after treatment with SOD. Disappearance of CF from the plasma was correlated with a decrease in DNA antibodies, of the erythrocyte sedimentation rate and with improvement of the clinical condition. Also in 5 cases of rheumatoid arthritis, who received intraarticular injections of SOD, CF were no longer present in the plasma (Camus et al., 1980).

In recent years, CF were used as intermediate endpoints in several pilot studies conducted in Armenian liquidators of the Chernobyl accident with the authorization of the Armenian Ministry of health. Since SOD has to be injected, it appeared not suitable for long-term prophylactic application. We therefore looked for other antioxidants with superoxide scavenging properties. The first study used an extract of Ginkgo biloba leaves (EGb 761), commercialized since many years under the tradename Tanakan (IPSEN, Paris) and standardized for a content of 24% Ginkgo flavone glycosides and 6% Ginkgolides-bilobalides. A CF-Test was performed before the start of the treatment and at different intervals after the 2 months application at the usual dose of 3 × 40 mg/day. Clastogenic activity was reduced to control values, when blood was collected in the first week after arrest of treatment. The benefit of the treatment persisted more than 7 months and even up to 12 months after the end of the treatment. However, 4 of 12 workers, who could be followed up to 12 months, had become CF positive again, demonstrating that the process which produces clastogenic substances was not halted indefinitely (Emerit et al., 1995). In a second study, protective effects of another plant extract with confirmed superoxide-scaveng-

Table 1. Antioxidant treatment of liquidators

	Antioxidant	
	Tanakan*	AOB**
Number of persons	30	20
ACS before tr.	8.1 ± 2.3	10.5 ± 3.0
ACS after tr.		
1st day	2.0 ± 0.6	2.2 ± 1.8
3 months	1.1 ± 2.0	
6 months	2.5 ± 1.9	1.5 ± 1.9
9 months		1.8 ± 1.3
12 months	6.3 ± 4.8	3.0 ± 2.3

*3 × 40 mg daily during 2 months
**3 × 6 g daily during 3 months

ing properties, could be documented. The antioxidant biofactor (AOB) produced by A.O.A. Company Kobe and sold in Japan as a health food, was given to 20 liquidators during 3 months at the usual dose of 3 × 6 g/day. AOB is prepared from various plants including soybean, rice, wheat, green tea, adley, yeast and sesame. It contains various flavonoids such as daizein, genistein and rutin, in addition to small amounts of oligoelements and antioxidant vitamins. Despite arrest of AOB intake, CF did not reach detectable levels, when control blood samples were tested after 6, 9 and 12 months (Table 1). The workers experienced improvement of their general condition and working capacity (Emerit et al., 1997d). Both antioxidants will now be studied in a double blind, placebo-controlled trial. The studies indicated that the use of antioxidants can be discontinuous, what is of considerable interest because of the important costs of longterm prophylactic treatments.

5. CONCLUSION

On the basis of the reported data, CF can be considered as biomarkers of a pro-oxidant state. The CF–Test is useful as an intermediate endpoint in intervention clinical trials. The cytogenetic test system appears preferable to the biochemical analysis of the various clastogenic and pro-oxidant components, since their synergistic action renders the test particularly sensitive. In addition, the finding of chromosomal aberrations demonstrates that the antioxidant defences and the DNA repair system are overwhelmed. The study of the clastogenic effects of a patient's plasma on cells of healthy donors, instead of studying chromosome damage in the cells from the exposed person, has the advantage that the plasma can be frozen and studied when convenient. The detection of CF indicates that the chromosomal damage observed in a patient's cells is not transient, and that the person will be exposed to clastogenic effects as long as the vicious circle of CF and superoxide production is not interrupted.

REFERENCES

Alaoui-Youssefi, A., Aroutiounian, R., Emerit I., 1994, Chromosome damage in PUVA-treated human lymphocytes is related to active oxygen species and clastogenic factors, Mutat. Res. **309**: 185–191.
Auclair, C., Gouyette, A., Levy, A., Emerit, I., 1990, Clastogenic inosine nucleotides as components of the chromosome breakage factor in scleroderma patients, Arch. Biochem. Biophys. **278**: 238–244.

Camus, J.P., Emerit, I., Michelson, A.M., Prier, A., Koeger, A.C., Merlet, C., 1980, Superoxide dismutase et polyarthrite rhumatoide, Revue du Rhumatism 47: 489–492.

Degan, P., Bonassi, S., De Caterina, M., Korkina, L., Pinto, L., Scopascasa, F., Zatterale, A., Calzone, R., Pagano, G., 1995, In vivo accumulation of 8-hydroxy-2′-deoxyguanosine in DNA correlates with release of reactive oxygen species in Fanconi's anemia families; Carcinogenesis 16: 735–742.

Edeas, M.A., Emerit, I., Khalfoun, Y., Lazizi, Y., Cernjavski, L., Levy, A., Lindenbaum, A. 1997, Clastogenic factors in plasma of HIV-infected patients activate HIV-1 replication in vitro. Inhibition by superoxide dismutase, 23: in press

Emerit, I., 1990 a, Superoxide production by clastogenic factors, in *Free Radicals, Lipoproteins and Membrane Lipids* (A. Crastes de Paulet, L. Douste-Blazy and R. Paoletti, eds.), pp. 99–104, Plenum Press, New York.

Emerit, I., 1990 b, Clastogenic factors: Detection and assay, in: *Methods in Enzymology,* (L. Packer and A.N. Glazer, eds;), pp. 555–564, Academic Press, New York.

Emerit, I., 1991, Membrane-mediated chromosome damage anf formation of clastogenic factors, in:*Membrane lipid oxidation* (C. Vigot-Pelfrey, ed.), pp. 33–43, CRC Press, Boca Raton.

Emerit, I., 1994, Reactive oxygen species, chromosome mutation and cancer: possible role of clastogenic factors in carcinogenesis, Free Radic. Biol. Med. 16: 99–109.

Emerit, I. and Michelson, A.M. 1981a, Mechanisms of photosensitivity in systemic lupus erythematosus patients, Proc. Natl. Acad. Sci., USA, 8: 2537–2540.

Emerit, I. and Michelson, A.M. 1981b, Chromosomal breakage in Crohn's disease, in: *Recent Advances in Crohn's Disease* (Pena, A.S., Weterman, I.T., Booth, CC., Strober, W., eds.), pp. 225–229, Martinus Nijhoff Publishers, The Hague.

Emerit, I. and Cerutti, P. 1981, Clastogenic activity from Bloom's syndrome fibroblast cultures, Proc. Natl. Acad. Sci. USA 78: 1868–1872.

Emerit, I. and Cerruti, P., 1982, Tumor promoter phorbol-myristate acetate induces a clastogenic factor in human lymphocytes, Proc. Natl. Acad. Sci. USA, 79: 7509–7513.

Emerit, I. and Lahoud-Maghani, 1989, Mutagenic effects of TPA-induced clastogenic factor in Chinese hamster cells, Mutat. Res. 214: 97–104.

Emerit, I., Levy, A., Michelson, A.M. 1981, Effect of superoxide dismutase on the chromosomal instability of New Zealand black mice, Cytogenet. Cell Genet. 30: 65–69.

Emerit, I., Fabiani, J.N., Ponzio, O., Murday, A., Lunel, F., Carpentier, A. 1988, Clastogenic factor in ischemia-reperfusion injury during open-heart surgery: Protective effect of Allopurinol, Ann. Thorac. Surg. 46: 619–624.

Emerit, I., Levy, A. and Camus, J.P. 1989, Monocyte-derived clastogenic factor in rheumatoid arthritis, Free Radic. Biol. Med. 6: 245–250.

Emerit, I., Khan, S.H. and Esterbauer, H. 1991, Hydroxynonenal, a component of clastogenic factors? Free Radic. Biol. Med. 10: 371–377.

Emerit, I., Aroutiounian, R., Sarkisian, T., Torossian, E. and Panossian, A.G. 1993, Oxyradical-related chromosome damage in patients with Familial Mediterranean Fever, Free Radic. Biol. Med. 15:265–271.

Emerit, I., Levy, A., Cernjavski, L., Aroutiounian, R., Panassian, A., Pogossian, A., Mejlumian, H., Sarkisian, T., Gulkandanian, M., Quastel, M., Goldsmith, J., Riklis, E., Kordysh, R., Poliak, S. Merklin, 1994, Transferable clastogenic activity in plasma from persons exposed as salvage personnel of the Chernobyl reactor. J. Cancer Res. Clin. Oncol. 120: 558–561.

Emerit, I., Levy, A., Pagano, G., Pinto, L., Calzone, R., Zatterale, A. 1995a, Transferable clastogenic activity in plasma from patients with Fanconi anemia, Hum. Genet. 96: 14–20.

Emerit, I., Fabiani, J.N., Levy, A., Ponzio, O., Conti, M., Brasme, B., Bienvenu, P., Hatmi, M.1995b, Plasma from patients exposed to ischemia reperfusion contains clastogenic factors and stimulates the chemiluminescence response of normal leukocytes, Free Radic. Biol. Med. 19: 405–415.

Emerit, I., Oganesian, N., Sarkisian, T., Arutiounian, R., Pogossian, A., Asrian, K., Levy, A., Cernjavski, L. 1995c, Clastogenic factors in the plasma of Chernobyl accident recovery workers: anticlastogenic effect of Ginkgo biloba extract, Radiation Res. 144: 198–205.

Emerit, I., Garban, F., Vassy, J., Levy, A., Filipe, P., Freitas, J. 1996, Superoxide-mediated clastogenesis and anticlastogenic effects of exogenous superoxide dismutase, Proc. Natl. Acad. Sci. USA 93: 12799–12804.

Emerit, I., Quastel, M., Goldsmith, J., Merkin, L., Levy, A., Cernjavski, L., Alaoui-Youssefi, A., Pogossian, A., Riklis, E. 1997a, Clastogenic factors in the plasma of children exposed at Chernobyl, Mutat. Res. 373: 47–54.

Emerit, I., Filipe, P., Meunier, P., Auclair, C., Freitas, J., Deroussent, A., Gouyette, A., Fernandez, A. 1997b, Clastogenic factors in plasma of scleroderma patients: aa biomarker of oxidative stress, Dermatology, 194: 140–146.

Emerit, I., Oganesian, N., Aroutiounian, R., Pogossian, A., Sarkisian, T., Cernjavski, L., Levy, A., Feingold, J. 1997d, Oxidative stress-related clastogenic factors in plasma from Chernobyl liquidators: protective effects of antioxidant plant phenols, vitamins and oligoelements, Mutat. Res. in press.

Faguet, G.G., Reichard, S.M., Welter, D.A., 1984, Radiation-induced clastogenic plasma factors, Cancer Genet. Cytogenet. **12**: 73–83.

Filipe, P., Emerit, I., Alaoui-Youssefi, A., Levy, A., Cernjavski, L., Freitas, J., Cirne de Castro, J.L. 1997, Oxyradical-mediated clastogenic plasma factors in psoriasis. Increase in clastogenic activity after PUVA, submitted.

Fuchs, J., Emerit, I., 1995, Clastogenic factors in plasma of HIV-infected patients, Free Radic. Biol. Med. **19**: 843–848.

Gabrielian, E., Grigorian, S., Davidian, D., Mchitarian, G., Panossian, A., 1990, Leukotrienes B4 and C4 in blood plasma of patients with Familial Mediterranean Fever, Bull. Exp. Biol. Med. **110**: 296–297.

Gille, J.J.P., Wortelboer, H.M., Joenje, H., 1989, Antioxidant status of Fanconi fibroblasts, in: *Hyperoxia-induced oxidative stress in mammalian cell cultures*, (J.J.P. Gille ed.), pp. 79–88, Free University Press, Amsterdam.

Goh, K.O., Sumner, K., 1968, Breaks in normal human chromosomes are they induced by a transferable substance in the plasma of irradiated persons exposed to total-body irradiation?, Radiation Res. **6**: 51–60.

Harth, M., Keown, P.A., Orange, J.F., 1983, Monocyte-dependent excited oxygen radical generation in rheumatoid arthritis, J. Rheumatol. **10**: 701–707.

Hallahan D.E., Spriggs, D.R., Beckett, M.A., Kufe, D., Weichselbaum, R.R., 1989, Increased tumor necrosis factor alpha mRNA after cellular exposure to ionizing radiation, Proc. Natl. Acad. Sci. USA **886**: 10104–10107.

Hollowell, H.G., Littlefield, L.G., 1968, Chromosome damage induced by plasma from irradiated patients. An indirect effect of X-ray, Proc. Soc. Exp. Biol. Med. **129**: 240–244.

Iwamoto, K.S., McBride, W.H., 1994, Production of 13-hydroxy-octadecadienoic acid and tumor necrosis factor alpha by murine peritoneal macrophages in response to irradiation. Radiation Res. **139**: 103–108.

Joenje, H., Arwert, F., Erikson, A.W., DeKoning, H., Ostra, A.B., 1981, Oxygen dependence of chromosomal aberrations in Fanconi's anemia, Nature, **290**: 142–143.

Kalebic, T., Kinter, A., Poli, G., Anderson, M.E., Meister, A., Fauci, A.S., 1991, Suppression of human immunodeficiency virus expression in chronically infected monocytic cells by glutathione, glutathione ester, and N-acetylcysteine, Proc. Natl. Acad. Sci. USA **88**: 986–990.

Kellof, G.J., Boone, C.W., Malone, W.F., Steele, V.E., 1992, Chemoprevention clinical trials, Mutat. Res. **267**: 291–295.

Khan, S.H., Emerit, I., Feingold, J., 1990, Superoxide and hydrogen peroxide production by macrophages of NZB mice, Free Radic. Biol. Med. **8**: 339–345.

Korkina, L.G., Samochatova, E.V., Mashan, A.A., Suslova, T.B., Cheremissina, Z.P., Afanas"ev, I.B., 1992, J. Leukocyte Biol. **52**: 357–363.

Krivenko, S., Dryk, S., Komarovskaya, M., Karkanitsa, L., 1992, Ionizing radiation increases TNF/cachectin production in human peripheral mononuclear cells in vitro, Int. J. Hematol. **55**: 127–130.

Kuhns, D.B., Wright, D.G., Nath, J., Kaplan, S.S., Basford, R.E. ATP induces transient elevations of Ca^{2+} in human neutrophils and primes cells for enhanced O_2^- production, Lab. Invest. **58**: 448–53.

Matzner, Y., Brzezinki, R., 1984, C5a inhibitor deficiency in peritoneal fluid from patients with Familial Mediterranean Fever, New Engl. J. Med. **311**: 287–290.

McCord, J.M., 1985, Oxygen-derived free radicals in postischemic tissue injury, N. Engl. J. Med. **312**: 159–162.

Miesel, R., Zuber, M. 1993, Elevated levels of xanthine oxidase in serum of patients with inflammatory and autoimmune rheumatic diseases, Inflammation **17**: 551–561.

Neriishi, K., Possible involvement of a free radical mechanism in late effects of A-bomb radiation. Proc. 5th, Int. Congr. on Oxygen Radicals: Active Oxygen, Lipid Peroxides and Antioxidants, Kyoto, 1991, Abstr.

Pant, G.S., Kamada, N., 1977, Chromosome aberrations in normal human leukocytes induced by the plasma of exposed inividuals. Hiroshima J. Med. Sci. **26**: 149–154.

Rogers, D.B;, Shohat, M., Petersen, G.M., Bickal, J., 1989, Familial Mediterranean Fever in Armenians: Autosomal recessive inheritance with high gene frequency. Amer. J. Med. Genet. **34**: 168–172.

Roselli, F, Sanceau, J., Gluckman, E., Wietzerbin, J., Moustacchi, E., 1994, Abnormal lymphokine production: A novel feature of the genetic disease Fanconi anemia. II. In vitro and in vivo spontaneous overproduction of tumor necrosis factor alpha. Blood, **83**: 1216–1225.

Saito, H., Hammond, A.T., Moses, R.E., 1993, Hypersensitivity to oxygen is a uniform and secondary defect in Fanconi anemia cells. Mutat. Res. **294**: 255–262.

Shaham, M., Becker, Y., 1986, The ataxia telangiectasia clastogenic factor is a low molecular weight peptide. Hum. Genet. **75**: 197–208.

Wavra, E., Zollner, H., Schaur, R.J., Tillian, H.M., Schauenstein, E., 1986, The inhibitory effect of 4-hydroxynonenal on DNA polymerases alpha and beta from rat liver and rapidly dividing Yoshida ascites hepatoma. Cell Biochem. Function **4**: 31–36.

33

CHEMILUMINESCENCE MEASUREMENTS FOR THE DETECTION OF FREE RADICAL SPECIES

A. Süha Yalçın, Goncagül Haklar, Belgin Küçükkaya, Meral Yüksel, and Gönül Dalaman

Department of Biochemistry
Faculty of Medicine
Marmara University
İstanbul, Turkey

1. INTRODUCTION

Reactive oxygen species (ROS) have been implicated in the pathogenesis of a variety of human diseases (Halliwell and Gutteridge, 1989). Increased appreciation of occurence of ROS, their injury potential and pathogenic role in several disease states demands quantitative methods which are diagnostic of the process and meet basic analytical criteria regarding accuracy, reliability, sensitivity and specificity (Halliwell and Grootveld, 1987). However, because of their reactive nature and short half lives, it is hard to quantitate ROS. Instead analyses of secondary or end products produced by their attack on lipids, proteins or other cellular components are preferred. These indirect methods usually give misleading results due to their poor specificity and sensitivity.

2. SPONTANEOUS CHEMILUMINESCENCE

Chemiluminescence (CL) is the production of light generated from chemical sources (Van Dyke and Castranova, 1987; Wulff, 1983). It is a universal property of organic substances able to undergo an oxidative reaction sufficiently exothermic to produce an emitting state (Figure 1). Generally the light produced is visible and has a wavelength between 400–600 nm. However, this light can be UV or infrared which would have a shorter or longer wavelength than the visible, respectively. CL may be utilized as a direct noninvasive method for measuring ROS. In the 1960s several Soviet investigators reported studies of "dark chemiluminescence" from living tissues (Baremboin et al., 1969). Later, Boveris et al. (1981) have summarized the knowledge of ultraweak CL caused by lipid peroxides, and have concluded that the sources of emission were the relaxation of singlet oxygen

Free Radicals, Oxidative Stress, and Antioxidants, edited by Özben.
Plenum Press, New York, 1998.

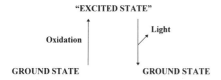

Figure 1. Schematic diagram showing light production from certain molecules in the excited state. The initial change from ground state to an excited state could be accompanied by an oxidation.

($^1O_2$) to triplet ground state ($^3O_2$) and the relaxation of excited state carbonyl groups to ground state. On the other hand, Howes and Steele (1971) reported that CL was generated by rat liver microsomes incubated with NADPH and oxygen. They, too, proposed that $^1O_2$ was involved in the CL. It was also reported that CL was generated by polymorphonuclear leukocytes stimulated by bacteria (Allen, 1982).

3. ENHANCED CHEMILUMINESCENCE

Because of potential variability and low intensity of native CL, investigators introduced the use of enhancer compounds (Allen, 1982; Allen and Loose, 1976; Minkenberg and Ferber, 1978). These compounds were selected primarily because of their high quantum efficiency and photon yield after oxidation. Luminol and lucigenin are two enhancers which are widely used in CL studies (Wulff, 1983). They function as by-stander substrates for oxygenation when added to an *in vitro* biological system and form high levels of excited-state products and CL. Luminol and lucigenin react with oxidants to form 3-aminophthalate and N-methylacridone, respectively. The excited electrons in these compounds revert to their ground state with the emission of energy as light (CL), which can be detected by photomultipliers. The sensitivities of luminol and lucigenin are different. Luminol detects H_2O_2, ˙OH, hypochlorite, peroxynitrite and lipid peroxy radicals, whereas lucigenin is particularly sensitive to the superoxide radical.

4. INSTRUMENTATION

For low-level light detection there is, at present, no alternative to photomultiplier tubes (Van Dyke and Castranova, 1987; Wulf, 1983). An important specification of photomultiplier tube is quantum efficiency. Photons hitting the photocathodes release single photoelectrons. Quantum efficiency is the ratio of photoelectrons released to the number of photons incident on the cathode area. The cathode type mostly used is bialkali, which is common in liquid scintillation counters (LSC). Bialkali photocathodes have excellent quantum efficiency around 400 to 450 nm even at room temperature. This is very adequate for luminol and lucigenin emission. LSCs are designed to count events of nuclear decay by bundles of several hundred photons per disintegration. The sample is measured by two photomultipliers to minimize background caused by thermal noise of the detector and by single photons, and the background is filtered off by a so-called coincidence circuit, counting only events which send photons at the same time to both detectors. However, LSCs set in out-of-coincidence mode are suitable for cellular CL measurements. This maneuver allows the instrument to measure single photon events. LSCs have a mechanical drawback in measuring the fast reactions due to delay (drop of the vial to the detector) of the discrete sampling system. Another disadvantage of LSCs is the poor temperature control of the system. Luminometers which are specially designed equipments for CL measurements are a better choice for assay because they are more efficient at counting photons and they possess a high degree of temperature regulation. In addition they can be linked to computer control for data handling.

5. DATA EXPRESSION

There are several different ways of expressing data (Van Dyke and Castranova, 1987). A typical curve with the maximum point (cpm), time to the maximum point (min) and the area under the curve (AUC) is shown in Figure 2.

AUC can be calculated by integrating the area under the curve by the trapezoidal rule.

$$\text{Area} = \Delta t[(y_1/2) + y_2 + y_3 + \ldots y_{n-1} + (y_n/2)]$$

where, Δt is the time interval, n is the total number of data points, and $y_1, y_2, y_3, \ldots, y_{n-1}, y_n$ represent the data points in the order in which they were recorded.

6. APPLICATIONS

In our initial studies, spontaneous CL was measured in human erythrocytes. Erythrocyte suspensions were incubated with cumene hydroperoxide (CumOOH) and the time course of changes in lipid peroxidation and CL intensity was determined (Yalçın et al., 1990). It was shown that CL measurements can be used as an indicator of CumOOH-induced lipid peroxidation. Chain-breaking antioxidants and –SH compounds were able to suppress CL formation (Yalçın et al., 1992). Recently, we have applied CL for the quantitation of ROS production in different samples from human as well as animal tissues.

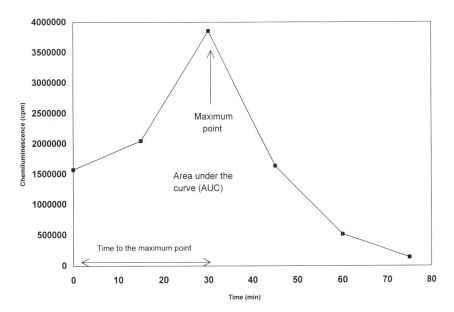

Figure 2. A typical curve showing different ways of expressing chemiluminescence data (maximum point, time to the maximum point and the area under the curve).

6.1. ROS Production by Spermatozoa of Infertile Patients

ROS can be produced by human spermatozoa. It has been suggested that peroxidative damage of the human spermatozoa membrane is important in the pathogenesis of human male infertility. We have used luminol and lucigenin enhanced CL for measuring ROS production by spermatozoa of idiopathically infertile patients and healthy donors (Alkan et al., 1997). ROS formation was detected in 89% of idiopathically infertile patients. Origin of ROS was investigated by Percoll separation of the semen samples. It was shown that the sources of ROS were largely immotile and abnormal spermatozoa.

6.2. Involvement of ROS in Helicobacter Pylori Gastritis

Helicobacter pylori (Hp) plays an important role in the pathogenesis of primary or unexplained gastritis in children as well as in adults. The exact mechanism by which Hp produces gastric mucosal injury has not been identified. One possibility is that Hp and its soluble proteins activate neutrophils which release inflammatory cytokines and ROS. We have measured ROS production by luminol and lucigenin enhanced CL in antral biopsy samples taken from children with complaint of unexplained abdominal pain (Ertem et al., 1996). Commercial urease test confirmed the presence of Hp. CL was present only in Hp (+) patients. Samples were evaluated for the severity of gastritis and the degree of neutrophilic infiltration. It was found that only severe neutrophilic infiltration was accompanied by a significant increase in ROS production.

We have also studied antral biopsy samples from adult patients having diagnostic upper gastrointestinal endoscopy. We have observed significant increases in both luminol and lucigenin enhanced CL where phagocyte mediated ROS formation had the greatest contribution. Effectiveness of therapies in retarding ROS formation was also studied. One group of patients received Omeprazole (H^+/K^+-ATPase inhibitor) while the other received Amoxicillin (an antibiotic) in addition to Omeprazole. After one month of treatment a second biopsy sample was obtained. ROS production was depressed in the second group which indicated that dual drug treatment was more effective (Haklar et al., 1996b).

6.3. ROS in Diet-Induced Experimental Atherosclerosis

ROS have been implicated in the pathogenesis of atherosclerosis. We have investigated the effects of a high cholesterol diet on ROS production and evaluated the effects of vitamin E and probucol. Rabbits were given a vitamin E deficient diet supplemented with 2% cholesterol. Treatment groups received vitamin E (50 mg/kg) and probucol (1%) in addition to 2% cholesterol. After 4 weeks of treatment, ROS formation was measured by luminol and lucigenin enhanced CL in aortic rings. Cholesterol feeding increased both luminol and lucigenin enhanced CL. Vitamin E and probucol were effective as scavengers but the effect of vitamin E was more pronounced (Haklar et al., 1996a).

6.4. ROS in Acute Pancreatitis

Acute pancreatitis remains as an important surgical problem with considerable morbidity and mortality. Its treatment is difficult and pathogenesis poorly understood. There have been reports on the role of free radicals in the pathogenesis of acute pancreatitis. However, studies on direct detection of ROS in an acute pancreatitis model is still lacking. Several models have been developed to study acute pancreatitis. We have monitored ROS and NO production in two different models: cerulein induced and duct ligation pancreatitis (Haklar et al., 1997b). Acute pancreatitis induction led to an increase in both luminol and lucigenin enhanced CL and NO production. We have also tested the effects of antioxidants in therapy and observed that combined vitamin E and vitamin C therapy after induction of pancreatitis decreased ROS formation.

6.5. ROS in Mesenteric Ischemia-Reperfusion

ROS are involved in various gastrointestinal pathologies including gastrointestinal injuries after ischemia-reperfusion (I/R) and hemorrhagic shock. We have detected ROS production by CL and investigated the roles of NO donor and a competitive antagonist of NO synthase in a rat mesenteric I/R model. Both luminol and lucigenin enhanced CL were significantly increased in I/R. In treatment groups, L-arginine and L-NAME significantly reduced oxidative injury (Haklar et al., 1997a).

6.6. CL as a Marker of Peritoneal Infections

Continuous ambulatory peritoneal dialysis (CAPD) is a widely accepted treatment for end-stage renal failure. However, the high incidence peritonitis has been a major problem. Polymorphonuclear leukocytes (PMN) play an important role in peritonitis. During phagocytosis PMNs form superoxide anions. In addition, stimulated PMNs emit light. We have recently studied CL formation in peritoneal fluids from CAPD patients. Peritoneal fluids were obtained from both non-infected patients and those presenting with acute peritonitis. Luminol and lucigenin enhanced CL of peritoneal fluids were higher in CAPD patients with infection compared to those without infection. Luminol amplified CL was considerably inhibited by sodium azide which inhibits neutrophil myeloperoxidase. In addition, patients with high values were followed at regular intervals and CL measurements were considerably suppressed under antibiotic treatment (Dalaman et al., 1997). These results indicate that CL measurements can be used as an early marker for peritonitis in CAPD patients.

REFERENCES

Allen, R.C., 1982, Biochemiexcitation, in: *Chemiluminescence and the Study of Biological Generation of Excited States*, (W. Adam, and G. Cilento, eds.) pp. 309–326, Academic Press, New York.

Allen, R.C, and Loose, L.D., 1976, Phagocytic activation of a luminol-dependent chemiluminescence in rabbit alveolar and peritoneal macrophages, *Biochem. Biophys. Res. Commun.* **69**: 245–253.

Alkan, İ., Şimşek, F., Haklar, G., Kervancıoğlu, E., Yalçın, A.S., and Akdaş, A., 1997, Reactive oxygen species production by spermatozoa of idiopathic infertile patients: relationship with seminal plasma antioxidants, *Am. J. Urol.* **157**: 140–143.

Baremboin, G.M, Domannskii, A.N, and Turovenov, K.K. (eds.), 1969, *Luminescence of Biopolymers and Cells*, Plenum Press, New York.

Boveris, A., Cadenas, E., and Chance, B., 1981, Ultraweak chemiluminescence: a sensitive assay for oxidative radical reactions, *Fed. Proc. Fed. Am. Soc. Exp. Biol.* **40**: 195–200.

Dalaman, G., Sipahiu, A., Haklar, G., Özener, Ç., Akoğlu, E., and Yalçın, A.S., 1996, Chemiluminescence measurements as an early diagnostic marker of peritoneal infections in continuous ambulatory peritoneal dialysis patients, VIII Biennial Meeting International Society for Free Radical Research, Barcelona, Spain (October 1–5, 1996), Abstract Book, p. 46.

Ertem, D., Yüksel, M., Haklar, G., Çelikel, Ç., Özgüven, E., Yalçın, A.S., and Pehlivanoğlu, E., 1996, Involvement of reactive oxygen metabolites in the pathogenesis of Helicobacter pylori gastritis in children, VIII Biennial Meeting International Society for Free Radical Research, Barcelona, Spain (October 1–5, 1996), Abstract Book, p. 313–314.

Haklar, G., Şirikçi, Ö., Özer, N.K., and Yalçın, A.S., 1996a, Chemiluminescent detection of reactive oxygen species in diet induced atherosclerosis: effect of vitamin E and probucol, Second International Conference on Clinical Chemiluminescence, Berlin, Germany (April 27–30, 1996), Abstract Book, P-7A.

Haklar, G., Yüksel, M., Küçük, M., Aksoy, N., Yenice, N., and Yalçın, A.S., 1996b, Measurement of reactive oxygen species in Helicobacter pylori gastritis: comparison of different treatment protocols, Second International Conference on Clinical Chemiluminescence, Berlin, Germany (April 27–30, 1996), Abstract Book, P-4D.

Haklar, G., Ulukaya-Durakbaşa, Ç., Yüksel, M., Dağlı, T., and Yalçın, A.S., 1997a, Chemiluminescent detection of oxygen free radicals and nitric oxide in mesenteric ischemia/reperfusion: modulation by nitric oxide donors and antagonists. SFRR Europe Summer Meeting, Abano Terme, Italy (June 26–28, 1997), Abstract Book, pp. 69–70.

Haklar, G., Yüksel, M., Soybir, G., Fincan, K., Adaş, G.T., Küpelioğlu, R., and Yalçın, A.S., 1997b, Oxygen free radicals and nitric oxide in the pathogenesis of acute pancreatitis: chemiluminescent detection of radicals and protective effect of scavengers, SFRR 1997 Summer Meeting, Abano Terme, Italy (June 26–28, 1997), Abstract Book, pp. 217–219.

Halliwell, B., and Grootveld, M., 1987, The measurement of free radical reactions in humans, *FEBS Lett.* **213**: 9–14.

Halliwell, B., and Gutteridge, J.M.C., 1989, *Free Radicals in Biology and Medicine*, 2nd ed., Oxford University Press, London.

Howes, R.M., and Steele, R.H., 1971, Microsomal chemiluminescence induced by NADPH and its relation to lipid peroxidation, *Res. Comm. Chem. Pathol. Pharmacol.* **2**: 619–625.

Minkenberg, I., and Ferber, E., 1978, Lucigenin-dependent chemiluminescence as a new assay for NAD(P)H oxidase activity in particulate fractions of human polymorphonuclear leucocytes, *J. Immun. Methods* **23**: 315–325.

Van Dyke, K., and Castranova, V. (eds.), 1987, *Cellular Chemiluminescence,* CRC Press, London.

Wulff, K., 1983, Luminometry, in: *Methods of Enzymatic Analysis*, 3rd ed. (H.U. Bergmeyer, J. Bergmeyer, M. Grassl, eds.), pp. 340–368, Verlag Chemie, Weinheim.

Yalçın, A.S., Sabuncu, N., and Emerk, K., 1990, Chemiluminescence measurements as an indicator of cumene hydroperoxide-induced lipid peroxidation in human erythrocytes, Hematol. Rev., **4**: 147–150.

Yalçın, A.S., Sabuncu, N., and Emerk, K., 1992, Cumene hydroperoxide-induced chemiluminescence in human erythrocytes: effect of antioxidants and sulfhydryl compounds, *Int. J. Biochem.* **24**: 499–502.

INDEX